NINTH EDITION

HOW
HUMANS
EVOLVED

NINTH EDITION

HOW
HUMANS
EVOLVED

Robert Boyd and Joan B. Silk

Arizona State University

W. W. NORTON & COMPANY

Independent Publishers Since 1923

W. W. Norton & Company has been independent since its founding in 1923, when William Warder Norton and Mary D. Herter Norton first published lectures delivered at the People's Institute, the adult education division of New York City's Cooper Union. The firm soon expanded its program beyond the Institute, publishing books by celebrated academics from America and abroad. By midcentury, the two major pillars of Norton's publishing program—trade books and college texts—were firmly established. In the 1950s, the Norton family transferred control of the company to its employees, and today—with a staff of four hundred and a comparable number of trade, college, and professional titles published each year—W. W. Norton & Company stands as the largest and oldest publishing house owned wholly by its employees.

Editor: Jake Schindel
Project Editor: Caitlin Moran
Editorial Assistant: Mia Davis
Managing Editor, College: Marian Johnson
Managing Editor, College Digital Media: Kim Yi
Production Manager: Benjamin Reynolds
Media Editor: Chris Rapp
Associate Media Editor: Ariel Eaton
Editorial Media Assistant: Lindsey Heale
Marketing Manager, Anthropology: Kandace Starbird
Design Director: Lissi Sigillo
Designer: Lissi Sigillo
Photo Editor: Ted Szczepanski
Permissions Manager: Bethany Salminen
Permissions Clearing: Josh Garvin
Composition: MPS North America LLC
Illustrations: Imagineeringart.com and Dragonfly Media Group
Manufacturing: Transcontinental

Permission to use copyrighted material is included in the credits section of this book.

Library of Congress Cataloging-in-Publication Data

Names: Boyd, Robert, 1948– author. | 1 Silk, Joan B., author. | 2 W. W. Norton
    & Company.
Title: How humans evolved / Robert Boyd and Joan B. Silk, Arizona State
    University.
Description: Ninth Edition. | New York: W. W. Norton & Company, 2020. |
    Includes in
Identifiers: LCCN 2020029530 | **ISBN 9780393427967** (Paperback) | ISBN
    9780393533088 (ePub)
Subjects:
Classification: LCC GN281 .B66 2020 | DDC 599.93/8—dc23
LC record available at https://lccn.loc.gov/2020029530

ABOUT THE AUTHORS

ROBERT BOYD

has written widely on evolutionary theory, focusing especially on the evolution of cooperation and the role of culture in human evolution. His book *Culture and the Evolutionary Process* received the J. I. Staley Prize. He is the coauthor of *Not by Genes Alone*, *A Different Kind of Animal*, and several edited volumes. He has also published numerous articles in scientific journals and edited volumes. He is currently a professor in the School of Human Evolution and Social Change at Arizona State University.

JOAN B. SILK

has conducted extensive research on the social lives of monkeys and apes, including extended fieldwork on chimpanzees at Gombe Stream Reserve in Tanzania and on baboons in Kenya and Botswana. She is also interested in the application of evolutionary thinking to human behavior. She is the coeditor of *The Evolution of Primate Societies* and has published numerous articles in scientific journals and edited volumes. She is currently a professor in the School of Human Evolution and Social Change at Arizona State University.

CONTENTS

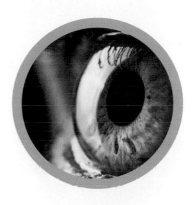

PART THREE: THE HISTORY OF THE HUMAN LINEAGE

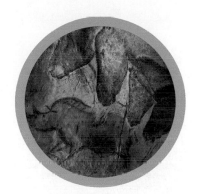

PREFACE

How Humans Evolved focuses on the processes that have shaped human evolution. This approach reflects our training and research interests. As anthropologists, we are interested in the evolutionary history of our own species, *Homo sapiens*, and the diversity of contemporary human societies. As evolutionary biologists, we study how evolution works to shape the natural world. In this book, we integrate these two perspectives. We use current theoretical and empirical work in evolutionary theory, population genetics, and behavioral ecology to interpret human evolutionary history. We describe the changes that have occurred as the human lineage has evolved, and we consider why these changes may have happened. We try to give life to the creatures that left the bones and made the artifacts that paleontologists and archaeologists painstakingly excavate by our focus on the processes that generate change, create adaptations, and shape bodies and behavior. We also pay serious attention to the role of evolution in shaping contemporary human behavior. There is considerable controversy over evolutionary approaches to human behavior within the social sciences, but we think it is essential to confront these issues openly and clearly. Positive responses to the first eight editions of *How Humans Evolved* tell us that many of our colleagues endorse this approach.

One of the problems in writing a textbook about human evolution is that there is considerable debate on many topics. Evolutionary biologists disagree about how new species are formed and how they should be classified; primatologists argue about whether large primate brains are adaptations to social or ecological challenges and whether reciprocity plays an important role in primate societies; paleontologists disagree about the taxonomic relationships among early hominin species and the emergence of modern humans; and those who study modern humans disagree about the meaning and significance of race, the role of culture in shaping human behavior and psychology, the adaptive significance of many aspects of modern human behavior, and a number of other things. Sometimes multiple interpretations of the same data can be defended; in other cases, the facts seem contradictory. Textbook writers can confront this kind of uncertainty in two different ways. They can weigh the evidence, present the ideas that best fit the available evidence, and ignore the alternatives. Or they can present opposing ideas, evaluate the logic underlying each idea, and explain how existing data support each of the positions. We chose the second alternative, at the risk of complicating the text and frustrating readers looking for simple answers. We made this choice because we believe that this approach is essential for understanding how science works. Students need to see how theories are developed, how data are accumulated, and how theory and data interact to shape our ideas about how the world works. We hope that students remember this long after they have forgotten many of the facts that they will learn in this book.

New in the Ninth Edition

The study of human evolution is a dynamic field. No sooner do we complete one edition of this book than researchers make new discoveries that fundamentally change our view of human evolution. New developments in human evolutionary studies require regular updates of the textbook. This is especially true in the genetics of human evolution. Rapid progress in the technology of gene sequencing and in the recovery of DNA from ancient fossils has produced an avalanche of new data.

Accordingly, we invited Prof. Kevin Langergraber, an expert in molecular genetics and our colleague at Arizona State University, to help us revise these areas of the textbook. His contributions are especially evident in the updated discussions of ancient DNA in Chapter 13 and the completely overhauled discussion of genetic variation among groups and the genetics of race in Chapter 14. We have also added new material in the following chapters:

- Chapter 5: Updated coverage of gorilla and primate ranging and territoriality.
- Chapter 6: A major update of the discussion of male and female reproductive strategies, sexual selection, and mating systems.
- Chapter 7: Updated coverage of cooperative breeding in comparative perspective.
- Chapter 9: A new box on how fossils are formed.
- Chapter 10: Now includes new material on *Australopithecus anamensis* in Ethiopia.
- Chapter 11: An updated discussion of evidence for fire.
- Chapter 12: Expanded coverage of evidence of *Homo heidelbergensis* hunting. We also added new discussions of *H. luzonensis*; complex tool making by Neanderthals, including use of adhesives created by anaerobic distillation of birch tar; evidence for Neanderthal use of clothing; and the function of Neanderthal/human genetic differences.
- Chapter 13: Discussion of new evidence for modern humans in Apidima, Greece, more than 200 ka. We also revised the treatment of whole-genome evidence about human migrations out of Africa and updated the discussion of introgression between Neanderthals and modern humans including beneficial genes. We also expanded the discussion of the Gravettian–Solutrean–Magdalenian sequence in Europe.
- Chapter 14: Extensively revised mainly thanks to Kevin Langergraber's efforts, including an updated discussion of the evolution of lactase persistence and expanded discussions of heritability and twin studies, methods of genomewide association studies (GWAS), and genetic variation among groups. There's also a new discussion of consumer genetic tests.
- Chapter 15: An updated discussion of learning and evolution, including a new section on prepared learning about plants; new discussion of human kinship systems; and an updated discussion of human mate choice that is based on recent cross-cultural data.

Ancillary Materials for Teaching and Learning

Visit wwnorton.com/instructors to download resources.

 INQUIZITIVE

InQuizitive

Authored by Eric Schniter (California State University, Fullerton), Caitlyn Placek (Ball State University), and Jacqueline Eng (Western Michigan University)

InQuizitive is a formative, adaptive learning tool that improves student understanding of important learning objectives by personalizing quiz questions for each student. Engaging, game-like elements built into InQuizitive motivate students as they learn.

InQuizitive includes a variety of question types that test student knowledge in different ways and enrich the user experience. Performance-specific feedback creates teaching moments that help students understand their mistakes and get back on the right track. Animations, videos, and other resources built into InQuizitive allow students to review core concepts as they answer questions.

InQuizitive is easy to use. Instructors can assign InQuizitive out of the box or use intuitive tools to customize the learning objectives they want students to work on. Students can access InQuizitive on computers, tablets, and smartphones, making it easy to study on the go. InQuizitive can also be assigned through the campus learning management system (LMS), with student results reporting to the LMS gradebook.

For this Ninth Edition, InQuizitive includes more than 60 new higher-level critical thinking questions distributed across all chapters to improve students' comprehension of concepts. In addition, InQuizitive now incorporates even more real-world videos, animations, and visuals to help guide student learning of the topics they find most complex.

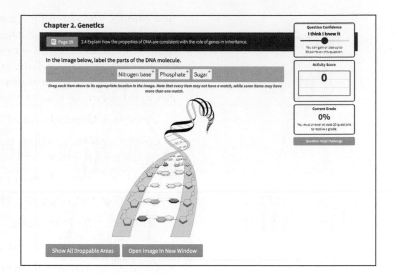

Guided Learning Explorations

Authored by Kristin Rauch (California State University, Sacramento) and Meradeth Snow (University of Montana)

New Guided Learning Explorations (GLEs) are scaffolded activities on eight compelling topics that get students thinking critically about and applying the course concepts as an anthropologist would. In Level 1 (Warm-up), they work through a refresher of the core science concepts for the given topic. In Level 2 (Applying the Concepts), they build on those concepts through more rigorous questions that engage more critical thinking. In Level 3 (Anthropology in Practice), they apply their concept knowledge to real-world scenarios as an anthropologist would. The GLEs can be assigned as stand-alone activities or in tandem with InQuizitive activities, which provide a concept-mastery foundation for the application-oriented GLEs. Students can access the GLEs on computers, tablets, and smartphones, and they can also be assigned through the campus learning management system (LMS).

The topics covered in the GLEs include

- continuous traits
- speciation and ancestral/derived traits
- mating systems and sexual dimorphism
- life history trade-offs and brain size
- coalescence
- increase in foraging complexity
- race and modern human variation
- evolution of culture and cooperation

Student Access Codes

Student access codes to InQuizitive and the Guided Learning Explorations are automatically included with all new texts in any format. Access to these resources can also be inexpensively purchased as a stand-alone option at **digital.wwnorton.com/howhumans9**. Contact your W. W. Norton sales representative to learn more.

Biological Anthropology Animations and Videos

Animations of key concepts from the text, as well as curated real-world videos, are available to instructors and students in several ways, including within the ebook

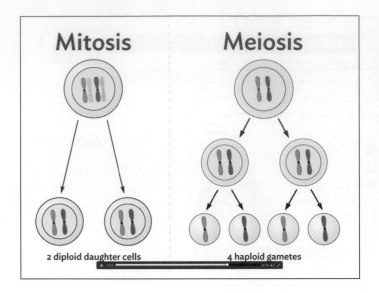

and at **digital.wwnorton.com/howhumans9**. They can also be accessed via the campus learning management system (LMS). These are brief, easy to use, and great for explaining and helping students better visualize and understand concepts, either in class or as a self-study tool. Select videos and animations are also incorporated into InQuizitive and the Guided Learning Explorations to allow students to engage with them in an interactive manner.

Resources for Your Learning Management System (LMS)

Norton provides resources for online or hybrid courses in a variety of formats, including Blackboard, Canvas, Brightspace/D2L, and Moodle. With just a simple download from our instructor's website, instructors can bring high-quality Norton digital media into a new or existing online course. Contact your local Norton representative to learn about our easy integration options for a single sign-on and gradebook-reporting solution.

Lecture PowerPoint Slides

The PowerPoint slides for this edition retain their richly illustrated format and extensive lecture notes. Designed to cover the core material in a highly visual way using photography and art from the text, these PowerPoint slides bring the concepts in each chapter to life with additional photography and design. These slides also include a lecture script in the notes field. Download these resources from wwnorton.com/instructors.

Update PowerPoint Service

Authored by Jennifer Spence (Texas State University)

To help cover what is new in the discipline, at the start of each term (Fall, Winter, and Spring) Norton provides a new set of supplemental lectures, notes, and assessment material covering current and breaking research. This material is available for download at wwnorton.com/instructors.

JPEGs of Art

To help you improve your class, JPEGs of all the art and photographs are available for download from the Norton instructor website.

Instructor's Manual

Authored by Susan Kirkpatrick Smith (Kennesaw State University)

The Instructor's Manual provides an overview of each chapter's key concepts with additional explanation for topics that students may find more challenging, as well as answers to the end-of-chapter Study Questions found in the text. For the Ninth Edition, the Instructor's Manual also features class activities and assignments to help students better understand and apply anthropological concepts. The Instructor's Manual is available for download at the Norton instructor website.

Test Bank

Authored by Tracy Betsinger (State University of New York, Oneonta)

The Test Bank offers teachers approximately 80–90 multiple-choice and essay questions (organized by difficulty level and topic) for each chapter. Every question is keyed according to Bloom's knowledge types and the corresponding learning objective from the textbook. The Test Bank is available in Norton Testmaker.

Norton Testmaker brings Norton's high-quality testing materials online. Create assessments for your course from anywhere with an Internet connection, without downloading files or installing specialized software. Search and filter test bank questions by chapter, type, difficulty, learning objectives, and other criteria. You can also edit questions or create your own. Then, easily export your tests to Microsoft Word or Common Cartridge files for import into your LMS.

Ebook: Same Great Book, a Fraction of the Price

An affordable and convenient alternative, Norton ebooks retain the content and design of the print book and allow students to highlight and take notes with ease, print chapters as needed, read online or offline, and search the text. Instructors can even take notes in their ebooks that can be shared with their students.

Acknowledgments

Over the past 25 years, many of our colleagues have provided new information, helpful comments, and critical perspectives that have enriched this book. We are grateful for all those who have responded to our requests for photographs, clarifications, references, and opinions. For the Ninth Edition, we are very grateful for the contributions of Kevin Langergraber. For the Eighth Edition, we are grateful to our ASU colleagues, particularly William Kimbel, Kaye Reed, and Gary Schwartz, for their generous help with revisions of Part Three. We also thank Melissa Wilson Sayers and Luca Pagani for their assistance with material in Chapter 13. For the Seventh Edition, we thank Curtis Marean for reviewing Chapter 13 and Kim Hill for reading Chapter 16. For the Sixth Edition, we thank Christopher Kirk for reviewing Chapter 5, Leanne Nash for reviewing Chapters 4 and 8, Roberto Delgano for reviewing Chapters 6 and 7, and Carol Ward and Jeremy DeSilva for help with Chapter 9. For the Fourth Edition, Laura MacLatchy provided help with the Miocene apes in Chapter 10, Dan Fessler and David Schmitt gave us access to material for Chapter 16, and Kermyt Anderson dug up original data for figures in Chapter 17. Steven Reznik reviewed our discussion of the rapid evolution of placentas in the minnows he studies and kindly provided an image. Leslie Aiello helped with our discussion of hominin developmental rates. For help with the Third Edition, we thank Carola Borries, Colin Chapman, Richard Klein, Cheryl Knott, Sally McBrearty, Ryne Palombit, Steve Pinker, Karin Stronswold, and Bernard Wood. For help with the Second Edition, we also thank Tom Plummer, Daniel Povinelli, Beverly Strassman, and Patricia Wright. We remain grateful for the help we received for the First Edition from Leslie Aiello, Monique Borgerhoff Mulder, Scott Carroll, Dorothy Cheney, Glenn Conroy, Martin Daly, Robin Dunbar, Lynn Fairbanks, Sandy Harcourt, Kristin Hawkes, Richard Klein, Phyllis Lee, Nancy Levine, Jeff Long, Joseph Manson, Henry McHenry, John Mitani, Jocelyn Peccei, Susan Perry, Steve Pinker, Tom Plummer, Tab Rasmussen, Mark Ridley, Alan Rogers, Robert Seyfarth, Frank Sulloway, Don Symons, Alan Walker, Tim White, and Margo Wilson.

A number of people have provided reviews of all or parts of the text. We thank the following: Stephanie Anestis, Thad Bartlett, AnnMarie Beasley, Rene Bobe, Barry Bogin, Doug Broadfield, Bryce Carlson, Joyce Chan, Margaret Clarke, Raffaella Commitante, Steve Corbett, Julie Cormack, Douglas Crews, Roberto Delgado, Arthur Durband, Charles Edwards, Daniel Fessler, Donald Gaff, Renee Garcia, Susan Gibson, Peter Gray, Mark Griffin, Corinna Guenther, Sharon Gursky, Kim Hill, Kevin Hunt, Andrew Irvine, Trine Johansen, Andrea Jones, Barbara King, Alice Kingsnorth, Richard Klein, Jeremy Koster, Kristin Krueger, Darrell La Lone, Clark Larsen, Lynette Leidy, Joseph Lorenz, Laura MacLatchy, Lara McCormick, Emily McDowell, Elizabeth Miller, Shannon Mills, John Mitani, Gilliane Monnier, Peer Moore-Jansen, M. J. Mosher, Martin Muller, Marilyn Norconk, Ann Palkovich, Amanda Wolcott

Paskey, James Paterson, Michael Pietrusewsky, John Polk, Barbara Quimby, Corey Ragsdale, Kristin Rauch, Denné Reed, Ulrich Reichard, Michael Robertson, Stacey Rucas, Michael Schillaci, Eric Schnicter, Liza Shapiro, Bet Shook, Eric Smith, Craig Stanford, Horst Steklis, Joan Stevenson, Mark Stoneking, Rebecca Storey, Rebecca Stumpf, Roger Sullivan, Yanina Valdos, Timothy Weaver, Elizabeth Weiss, Jill Wenrick, Patricia Wright, and Alexandra Zachieja. Although we are certain that we have not satisfied all those who read and commented on parts of the book over the years, we have found all of the comments to be helpful as we revised the text.

Richard Klein provided us with many exceptional drawings of fossils that appear in Part Three—an act of generosity that we continue to appreciate. We also give special thanks to Neville Agnew and the Getty Conservation Institute for granting us permission to use images of the Laetoli conservation project for the cover of the Second Edition.

We also acknowledge the thousands of students and dozens of teaching assistants at ASU and UCLA who have used various versions of this material over the years. Student evaluations of the original lecture notes, the first draft of the text, and the first seven editions were helpful as we revised and rewrote various parts of the book. The teaching assistants helped us identify many parts of the text that needed to be clarified, corrected, or reconsidered.

We thank all the people at Norton who helped us produce this book, particularly our current editor, Jake Schindel, and his predecessors, Leo Wiegman, Pete Lesser, Aaron Javsicas, and Eric Svendsen. We are also grateful to the editorial assistant, Mia Davis, and the project editor, Caitlin Moran, as well as the digital media team: Chris Rapp, Ariel Eaton, Lindsey Heale, Jessica Awad, and Rachel Mayer. We would also like to thank all of the other people who saw the book through the production, permissions, and marketing processes, including Josh Garvin, Ben Reynolds, Bethany Salminen, Lissi Sigillo, Kandace Starbird, and Ted Szczepanski.

PROLOGUE

WHY STUDY HUMAN EVOLUTION?

Origin of man now proved—Metaphysics must flourish—He who understand
baboon would do more toward metaphysics than Locke.

—*Charles Darwin, M Notebook, August 1838*

In 1838, Charles Darwin discovered the principle of evolution by natural selection and revolutionized our understanding of the living world. Darwin was 28 years old, and it was just 2 years since he had returned from a 5-year voyage around the world as a naturalist on HMS *Beagle* (**Figure P.1**). Darwin's observations and experiences during the journey had convinced him that biological species change through time and that new species arise by the transformation of existing ones, and he was avidly searching for an explanation of how these processes worked.

In late September of the same year, Darwin read Thomas Malthus's *An Essay on the Principle of Population*, in which Malthus (**Figure P.2**) argued that human populations invariably grow until they are limited by starvation, poverty, and disease. Darwin realized that Malthus's logic also applied to the natural world, and this intuition inspired the conception of his theory of evolution by natural selection. In the intervening century and a half, Darwin's theory has been augmented by discoveries in genetics and amplified by studies of the evolution of many types of organisms. It is now the foundation of our understanding of life on Earth.

FIGURE P.1

When this portrait of Charles Darwin was painted, he was about 30 years old. He had just returned from his voyage on HMS *Beagle* and was still busy organizing his notes, drawings, and vast collections of plants and animals.

FIGURE P.2

Thomas Malthus was the author of *An Essay on the Principle of Population*, a book Charles Darwin read in 1838 that profoundly influenced the development of his theory of evolution by natural selection.

FIGURE P.3

Sir Isaac Newton discovered the laws of celestial mechanics, a body of theory that resolved age-old mysteries about the movements of the planets.

This book is about human evolution, and we will spend a lot of time explaining how natural selection and other evolutionary processes have shaped the human species. Before we begin, it is important to consider why you should care about this topic. Many of you will be working through this book as a requirement for an undergraduate class in biological anthropology and will read the book in order to earn a good grade. As instructors of a class like this ourselves, we approve of this motive. However, there is a much better reason to care about the processes that have shaped human evolution: Understanding how humans evolved is the key to understanding why people look and behave the way they do.

The profound implications of evolution for our understanding of humankind were apparent to Darwin from the beginning. We know this today because he kept notebooks in which he recorded his private thoughts about various topics. The quotation that begins this prologue is from the *M Notebook*, begun in July 1838, in which Darwin jotted down his ideas about humans, psychology, and the philosophy of science. In the nineteenth century, metaphysics involved the study of the human mind. Thus Darwin was saying that because he believed humans evolved from a creature something like a baboon, it followed that an understanding of the mind of a baboon would contribute more to an understanding of the human mind than would all of the works of the great English philosopher John Locke.

Darwin's reasoning was simple. Every species on this planet has arisen through the same evolutionary processes. These processes determine why organisms are the way they are by shaping their morphology, physiology, and behavior. The traits that characterize the human species are the result of the same evolutionary processes that created all other species. If we understand these processes and the conditions under which the human species evolved, then we will have the basis for a scientific understanding of human nature. Trying to comprehend the human mind without an understanding of human evolution is, as Darwin wrote in another notebook that October, "like puzzling at astronomy without mechanics." By this, Darwin meant that his theory of evolution could play the same role in biology and psychology that Isaac Newton's laws of motion had played in astronomy. For thousands of years, stargazers, priests, philosophers, and mathematicians had struggled to understand the motions of the planets without success. Then, in the late 1600s, Newton discovered the laws of mechanics and showed how all of the intricacies in the dance of the planets could be explained by the action of a few simple processes (**Figure P.3**).

In the same way, understanding the processes of evolution enables us to account for the stunning sophistication of organic design and the diversity of life and to understand why people are the way they are. As a consequence, understanding how natural selection and other evolutionary processes shaped the human species is relevant to all of the academic disciplines that are concerned with human beings. This vast intellectual domain includes medicine, psychology, the social sciences, and even the humanities. Beyond academia, understanding our own evolutionary history can help us answer many questions that confront us in everyday life. Some of these questions are relatively trivial: Why do we sweat when hot or nervous? Why do we crave salt, sugar, and fat, even though large amounts of these substances cause disease (**Figure P.4**)? Why are we better marathon runners than we are mountain climbers? Other questions are more profound: Why do only women nurse their babies? Why do we grow old and eventually die? Why do people look so different around the world? As you will see, evolutionary theory provides answers or insights about all of these questions. Aging, which eventually leads to death, is an evolved characteristic of humans and most other creatures. Understanding how natural selection shapes the life histories of organisms tells us why we are mortal, why our life span is about 70 years, and why other species live shorter lives. In an age of horrific ethnic conflicts and growing respect for multicultural diversity, we are constantly reminded of the variation within the human species. Evolutionary analyses tell us that genetic differences between human groups are relatively minor and that our notions of race and ethnicity are culturally constructed categories, not biological realities.

All of these questions deal with the evolution of the human body. However, understanding evolution is also an important part of our understanding of human behavior and the human mind. The claim that understanding evolution will help us understand contemporary human behavior is much more controversial than the claim that it will help us understand how human bodies work. But it should not be. The human brain is an evolved organ of great complexity, just like the endocrine system, the nervous system, and all of the other components of the human body that regulate our behavior. Understanding evolution helps us understand our mind and behavior because evolutionary processes forged the brain that controls human behavior, just as they forged the brain of the chimpanzee and the brain of the salamander.

FIGURE P.4

A strong appetite for sugar, fat, and salt may have been adaptive for our ancestors, who had little access to these foods. We have inherited these appetites and now have easy access to sugar, fat, and salt. As a consequence, many of us suffer from obesity, high blood pressure, diabetes, and heart disease.

One of the great debates in Western thought centers on the essence of human nature. One view is that people are basically honest, generous, and cooperative creatures who are corrupted by an immoral economic and social order. The opposing view is that we are fundamentally amoral, egocentric beings whose antisocial impulses are held in check by social pressures. This question turns up everywhere. Some people believe that children are little barbarians who are civilized only through sustained parental effort; others think that children are gentle beings who are socialized into competitiveness and violence by exposure to negative influences such as toy guns and violent television programs (**Figure P.5**). The same dichotomy underpins much political and economic thought. Economists believe that people are rational and selfish, but other social scientists, particularly anthropologists and sociologists, question and sometimes reject this assumption. We can raise an endless list of interesting questions about human nature: Does the fact that, in most societies, women rear children and men make war mean that men and women differ in their innate predispositions? Why do men typically find younger women attractive? Why do some people neglect and abuse their children, while others adopt and lovingly raise children who are not their own?

Understanding human evolution does not reveal the answers to all of these questions or even provide a complete answer to any one of them. As we will see, however, it can provide useful insights about all of them. An evolutionary approach does not imply that behavior is genetically determined or that learning and culture are unimportant. In fact, we will argue that learning and culture play crucial roles in human behavior. Behavioral differences among peoples living in different times and places result mainly from flexible adjustments to different social and environmental conditions. Understanding evolution is useful precisely because it helps us understand why humans respond in different ways to different conditions.

FIGURE P.5

One of the great debates in Western thought focuses on the essential elements of human nature. Are people basically moral beings corrupted by society or fundamentally amoral creatures socialized by cultural conventions, social strictures, and religious beliefs?

Overview of the Book

Humans are the product of organic evolution. By this we mean there is an unbroken chain of descent that connects every living human being to a bipedal, apelike creature that walked through the tall grasses of the African savanna 3 million years ago (Ma); to a monkeylike animal that clambered through the canopy of great tropical forests covering much of the world 35 Ma; and, finally, to a small, egg-laying, insect-eating mammal that scurried about at night during the age of the dinosaurs, 100 Ma. To understand what we are now, you have to understand how this transformation took place. We tell this story in four parts.

Part One: How Evolution Works

More than a century of hard work has given us a good understanding of how evolution works. The transformation of apes into humans involved the assembly of many new, complex adaptations. For example, in order for early humans to walk upright on two legs, there had to be coordinated changes in many parts of their bodies, including their feet, legs, pelvis, backbone, and inner ear. Understanding how natural selection gives rise to such complex structures and why the genetic system plays a crucial role in this process is essential for understanding how new species arise. Understanding these processes also allows us to reconstruct the history of life from the characteristics of contemporary organisms.

Part Two: Primate Ecology and Behavior

In the second part of the book, we consider how evolution has shaped the behavior of nonhuman primates—an exercise that helps us understand human evolution in two ways. First, humans are members of the primate order: We are more similar to other primates, particularly the great apes, than we are to wolves, raccoons, or other mammals. Studying how primate morphology and behavior are affected by ecological conditions helps us determine what our ancestors might have been like and how they may have been transformed by natural selection. Second, we study primates because they are an extremely diverse order and are particularly variable in their social behavior. Some are solitary, others live in pair-bonded groups, and some live in large groups that contain many adult females and males. Data derived from studies of these species help us understand how social behavior is molded by natural selection. We can then use these insights to interpret the hominin fossil record and the behavior of contemporary people (**Figure P.6**).

Part Three: The History of the Human Lineage

General theoretical principles are not sufficient to understand the history of any lineage, including our own. The transformation of a shrewlike creature into the human species involved many small steps, and each step was affected by specific environmental and biological circumstances. To understand human evolution, we have to reconstruct the actual history of the human lineage and the environmental context in which these events occurred. Much of this history is chronicled in the fossil record. These bits of mineralized bone, painstakingly collected and reassembled by paleontologists, document the sequence of organisms that link early mammals to modern humans. Complementary work by geologists, biologists, and archaeologists allows us to reconstruct the environments in which the human lineage evolved (**Figure P.7**).

Part Four: Evolution and Modern Humans

Finally, we turn our attention to modern humans and ask why we are the way we are. Why is the human species so variable? How do we acquire our behavior? How has evolution shaped human psychology and behavior? How do we choose our mates? Why do people commit infanticide? Why have humans succeeded in inhabiting every corner of Earth when other species have more limited ranges? We will explain how an

FIGURE P.6

We will draw on information about the behavior of living primates, such as this chimpanzee, to understand how behavior is molded by evolutionary processes, to interpret the hominin fossil record, and to draw insights about the behavior of contemporary humans.

FIGURE P.7

Fossils painstakingly excavated from many sites in Africa, Europe, and Asia provide us with a record of our history as a species. Two million years ago in Africa, there were a number of apelike species that walked bipedally but still had ape-size brains and apelike developmental patterns. These are the fossilized remains of *Homo habilis*, a species that some think is ancestral to modern humans.

understanding of evolutionary theory and a knowledge of human evolutionary history provide a basis for addressing questions like these.

The history of the human lineage is a great story, but it is not a simple one. The relevant knowledge is drawn from many disciplines in the natural sciences, such as physics, chemistry, biology, and geology, and from the social sciences, mainly anthropology, psychology, and economics. Learning this material is an ambitious task, but it offers a very satisfying reward. The better you understand the processes that have shaped human evolution and the historical events that took place in the human lineage, the better you will understand how we came to be and why we are the way we are.

PART
ONE

HOW EVOLUTION
WORKS

1

ADAPTATION BY NATURAL SELECTION

CHAPTER OBJECTIVES

By the end of this chapter you should be able to:

A   Describe why our modern understanding of the diversity of life is based on the ideas of Charles Darwin.

B   Explain how competition, variation, and heritability lead to evolution by natural selection.

C   Explain why natural selection sometimes causes species to become better adapted to their environments.

D   Explain why natural selection can produce change or cause species to remain the same over time.

E   Describe how natural selection can produce very complex adaptations such as the human eye.

F   Assess why natural selection usually works at the level of the individual, not at the level of the group or species.

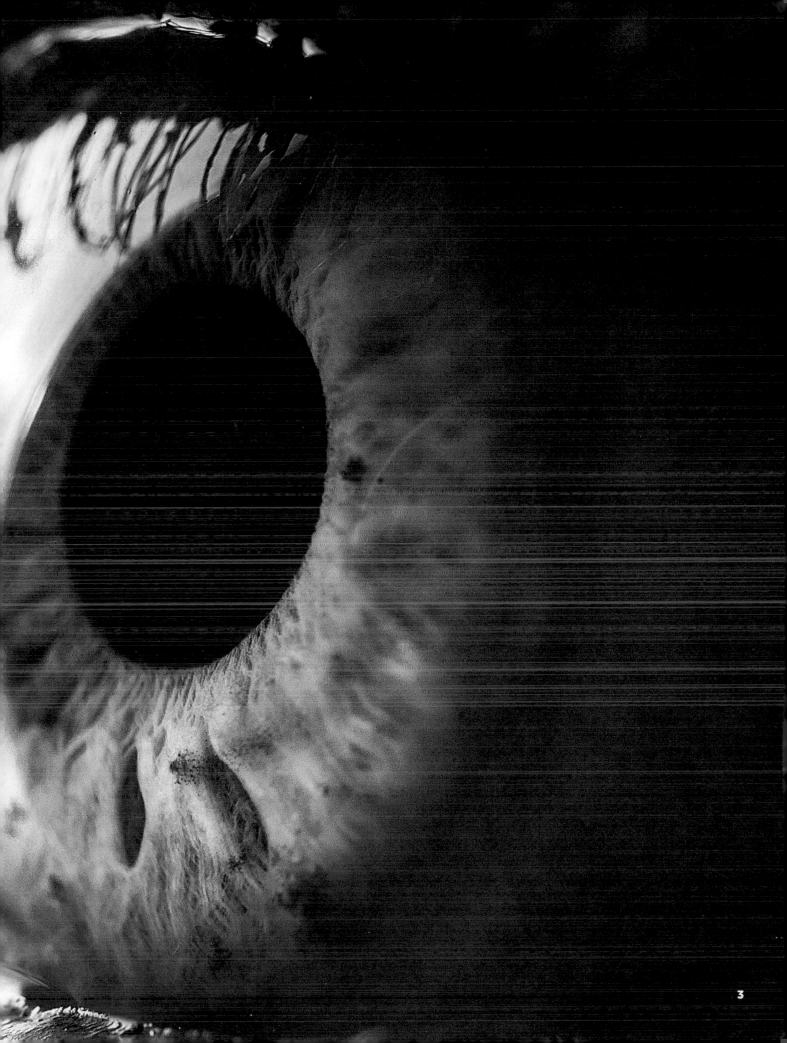

Explaining Adaptation before Darwin

Animals and plants are adapted to their conditions in subtle and marvelous ways. Even the casual observer can see that organisms are well suited to their circumstances. For example, fish are clearly designed for life underwater, and certain flowers are designed to be pollinated by particular species of insects. More careful study reveals that organisms are more than just suited to their environments: They are complex machines, made up of many exquisitely constructed components, or **adaptations**, that interact to help the organism survive and reproduce.

The human eye provides a good example of an adaptation. Eyes are amazingly useful: They allow us to move confidently through the environment, to locate critical resources such as food and mates, and to avoid dangers such as predators and cliffs. Eyes are extremely complex structures made up of many interdependent parts (**Figure 1.1**). Light enters the eye through a transparent opening, then passes through a diaphragm called the iris, which regulates the amount of light entering the eye and allows the eye to function in a wide range of lighting conditions. The light then passes through a lens that projects a focused image on the retina on the back surface of the eye. Several kinds of light-sensitive cells then convert the image into nerve impulses that encode information about spatial patterns of color and intensity. These cells are more sensitive to light than the best photographic film. The detailed construction of each of these parts of the eye makes sense in terms of the eye's function: seeing. If we probed into any of these parts, we would see that they, too, are made of complicated, interacting components whose structure is understandable in terms of their function.

Differences between human eyes and the eyes of other animals make sense in terms of the types of problems each creature faces. Consider, for example, the eyes of fish and humans (**Figure 1.2**). The lens in the eyes of humans and other terrestrial mammals is much like a camera lens; it is shaped like a squashed football located near the front of the eye and has the same index of refraction (a measure of light-bending capacity) throughout. In contrast, the lens in fish eyes is a sphere located at the center of the curvature of the retina, and the index of refraction of the lens increases smoothly from the surface of the lens to the center. It turns out that this kind of lens, called a spherical gradient lens, provides a sharp image over a full 180° visual field, a very short focal length, and high light-gathering power—all desirable properties. Terrestrial creatures like us cannot use this design because light is bent when it passes from the air through the cornea (the transparent cover of the pupil), and this fact constrains the design of the remaining lens elements. In contrast, light is not bent when it passes from water through the cornea of aquatic animals, and the design of their eyes takes advantage of this fact.

FIGURE 1.1

A cross section of the human eye.

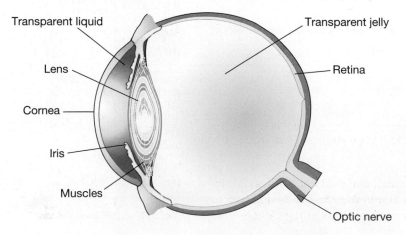

Transparent liquid

Lens

Cornea

Iris

Muscles

Transparent jelly

Retina

Optic nerve

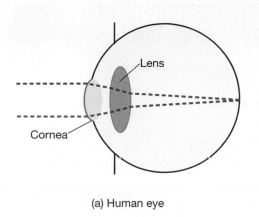

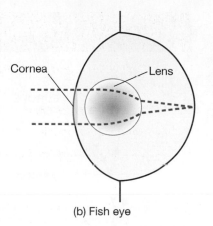

(a) Human eye (b) Fish eye

FIGURE 1.2

(a) Like those of other terrestrial mammals, human eyes have more than one light-bending element. A ray of light entering the eye (*dashed lines*) is bent first as it moves from the air to the cornea and then again as it enters and leaves the lens. (b) In contrast, fish eyes have a single lens that bends the light throughout its volume. As a result, fish eyes have a short focal length and high light-gathering power.

Before Darwin, there was no scientific explanation for the fact that organisms are well adapted to their circumstances.

As many nineteenth-century thinkers were keenly aware, complex adaptations such as the eye demand a different kind of explanation than that for other natural objects. This is not simply because adaptations are complex, since many other complicated objects exist in nature. Adaptations require a special kind of explanation because they are complex in a particular, highly improbable way. For example, the Grand Canyon, with its maze of delicate towers intricately painted in shades of pink and gold, is byzantine in its complexity (**Figure 1.3**). Given a different geologic history, however, the Grand Canyon might be quite different—different towers in different hues—yet we would still recognize it as a canyon. The particular arrangement of painted towers of the Grand Canyon is improbable, but the existence of a spectacular canyon with a complex array of colorful cliffs in the dry sandstone country of the American Southwest is not unexpected at all; in fact, wind and water have produced many such canyons in this region. In contrast, any substantial changes in the structure of the eye would prevent the eye from functioning, and then we would no longer recognize it as an eye. If the cornea were opaque or the lens on the wrong side of the retina, then the eye would not transmit visual images to the brain. It is highly improbable that natural processes would randomly bring together bits of matter having the detailed structure of the eye because only an infinitesimal fraction of all arrangements of matter would be recognizable as a functioning eye.

In Darwin's day, most people were not troubled by this problem because they believed that adaptations were the result of divine creation. In fact, the theologian William Paley used a discussion of the human eye to argue for the existence of God in his book *Natural Theology,* published in 1802. Paley argued that the eye is clearly *designed* for seeing, and where there is design in the natural world, there certainly must be a heavenly designer.

Although most scientists of the day were satisfied with this reasoning, a few, including Charles Darwin, sought other explanations.

Darwin's Theory of Adaptation

FIGURE 1.3

Although an impressive geologic feature, the Grand Canyon is much less remarkable in its complexity than the eye.

Charles Darwin was expected to become a doctor or clergyman, but instead he revolutionized science.

Charles Darwin was born into a well-to-do, intellectual, and politically liberal family in England. Like many prosperous men of his time, Darwin's father wanted his son to become a doctor. But after failing at the prestigious medical school at the University of

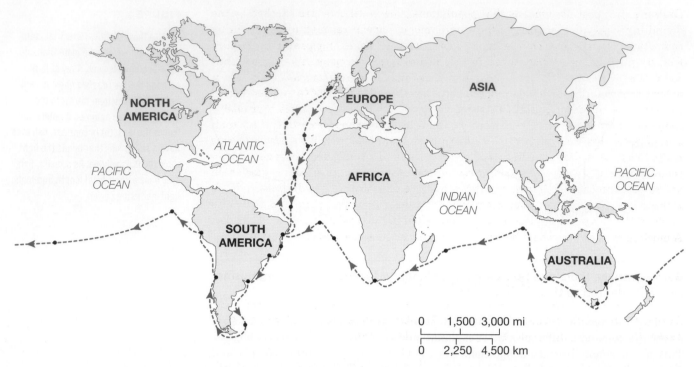

FIGURE 1.4

Darwin circumnavigated the globe during his 5-year voyage on HMS *Beagle*.

Edinburgh, Charles went on to Cambridge University, resigned to becoming a country parson. He was, for the most part, an undistinguished student—much more interested in tramping through the fields around Cambridge in search of beetles than in studying Greek and mathematics. After graduation, one of Darwin's botany professors, John Stevens Henslow, provided him with a chance to pursue his passion for natural history as a naturalist on HMS *Beagle*.

The *Beagle* was a Royal Navy vessel whose charter was to spend 2 to 3 years mapping the coast of South America and then to return to London, perhaps by circling the globe (**Figure 1.4**). Darwin's father forbade him to go, preferring that Charles get serious about his career in the church, but Darwin's uncle (and future father-in-law) Josiah Wedgwood II intervened. The voyage was the turning point in Darwin's life. His work during the voyage established his reputation as a skilled naturalist. His observations of living and fossil animals ultimately convinced him that plants and animals sometimes change slowly through time and that such evolutionary change is the key to understanding how new species come into existence. This view was rejected by most scientists of the time and was considered heretical by the general public.

Darwin's Postulates

Darwin's theory of adaptation follows from three postulates: (1) the struggle for existence, (2) the variation in fitness, and (3) the inheritance of variation.

In 1838, shortly after the *Beagle* returned to London, Darwin formulated a simple mechanistic explanation for *how* species change through time. His theory follows from three postulates:

1. The ability of a population to expand is infinite, but the ability of any environment to support populations is always finite.

2. Organisms within populations vary, and this variation affects the ability of individuals to survive and reproduce.

3. This variation is transmitted from parents to offspring.

Darwin's first postulate means that populations grow until they are checked by the dwindling supply of resources in the environment. Darwin referred to the resulting competition for resources as "the struggle for existence." For example, animals require food to grow and reproduce. When food is plentiful, animal populations grow until their numbers exceed the local food supply. Because resources are always finite, it follows that not all individuals in a population will be able to survive and reproduce. According to the second postulate, some individuals will possess traits that enable them to survive and reproduce more successfully (producing more offspring) than others in the same environment. The third postulate holds that if the advantageous traits are inherited by offspring, then these traits will become more common in succeeding generations. Thus traits that confer advantages in survival and reproduction are retained in the population, and traits that are disadvantageous disappear. When Darwin coined the term **natural selection** for this process, he was making a deliberate analogy to the artificial selection practiced by animal and plant breeders of his day. A much more apt term would be "evolution by variation and selective retention."

An Example of Adaptation by Natural Selection

Contemporary observations of Darwin's finches provide a particularly good example of how natural selection produces adaptations.

In his autobiography, first published in 1887, Darwin claimed that the curious pattern of adaptations he observed among the several species of finches that live on the Galápagos Islands off the coast of Ecuador—now referred to as "Darwin's finches"—was crucial in the development of his ideas about evolution (**Figure 1.5**). Some evidence suggests that Darwin was actually confused about the Galápagos finches during his visit, and that they played little role in his discovery of natural selection. Nonetheless, Darwin's finches hold a special place in the minds of most biologists.

Biologists Peter and Rosemary Grant of Princeton University conducted a landmark study of the ecology and evolution of one particular species of Darwin's finches on one of the Galápagos Islands. The study is remarkable because the Grants were able to directly document how Darwin's three postulates led to evolutionary change. The island, Daphne Major, is home to the medium ground finch (*Geospiza fortis*), a small bird that subsists mainly by eating seeds (**Figure 1.6**). The Grants and their

[a]

[b]

FIGURE 1.6

The medium ground finch, *Geospiza fortis,* uses its beak to crack open seeds.

colleagues caught, measured, weighed, and banded nearly every finch on the island each year of their study—some 1,500 birds in all. They also kept track of critical features of the birds' environment, such as the distribution of seeds of various sizes, and they observed the birds' behavior.

A few years into the Grants' study, a severe drought struck Daphne Major (**Figure 1.7**). During the drought, plants produced far fewer seeds, and the finches soon depleted the stock of small, soft, easily processed seeds, leaving only large, hard seeds that were difficult to process (**Figure 1.8**). The bands on the birds' legs enabled the Grants to track the fate of individual birds during the drought, and the regular measurements that they had made of the birds allowed them to compare the traits of birds that survived the drought with the traits of those that perished. The Grants also kept detailed records of the environmental conditions, which allowed them to determine how the drought affected the birds' habitat. It was this vast body of data that enabled the Grants to document the action of natural selection among the finches of Daphne Major.

The Grants' data show how the processes identified in Darwin's postulates lead to adaptation.

The events on Daphne Major embodied all three of Darwin's postulates. First, the supply of food on the island was not sufficient to feed the entire population, and many finches did not survive the drought. From the beginning of the drought in 1976 until the rains came nearly 2 years later, the population of medium ground finches on Daphne Major declined from 1,200 birds to only 180.

Second, beak depth (the top-to-bottom dimension of the beak) varied among the birds on the island, and this variation affected the birds' survival. Before the drought began, the Grants and their colleagues had observed that birds with deeper beaks were able to process large, hard seeds more easily than could birds with shallower beaks. Deep-beaked birds usually focused their efforts on large seeds, whereas shallow-beaked birds usually focused their efforts on small seeds. The open bars in the histogram in **Figure 1.9a** show what the distribution of beak sizes in the population was like before the drought. The height of each open bar represents the number of birds with beaks in a given range of depths—for example, 8.8 to 9.0 mm or 9.0 to 9.2 mm. During the drought, the relative abundance of small seeds decreased, forcing shallow-beaked birds to shift to larger and harder seeds. Shallow-beaked birds were then at a distinct disadvantage because it was harder for them to crack the seeds. The distribution of individuals within the population changed during the drought because finches with deeper beaks were more likely to survive than were finches with shallow beaks (**Figure 1.9b**). The shaded portion of the histogram in Figure 1.9a shows what the distribution of beak depths would have been like among the survivors. Because many birds died,

FIGURE 1.7

Daphne Major (a) after a year of good rains and (b) after a year of very little rain.

[a]

[b]

there were fewer remaining in each category. However, mortality was not random. The proportion of shallow-beaked birds that died greatly exceeded the proportion of deep-beaked birds that died. As a result, the shaded portion of the histogram shows a shift to the right, which means that the average beak depth in the population increased. Thus, the average beak depth among the survivors of the drought was greater than the average beak depth in the same population before the drought.

Third, parents and offspring had similar beak depths. The Grants discovered this by capturing and banding nestlings and recording the identity of the nestlings' parents. When the nestlings became adults, the Grants recaptured and measured them. The Grants found that, on average, parents with deep beaks produced offspring with deep beaks (**Figure 1.10**). Because parents were drawn from the pool of individuals that survived the drought, their beaks were, on average, deeper than those of the original residents of the island, and because offspring resemble their parents, the average beak depth of the survivors' offspring was greater than the average beak depth before the drought. This means that, through natural selection, the average **morphology** (an organism's size, shape, and composition) of the bird population changed so that birds became better adapted to their environment. This process, operating over approximately 2 years, led to a 4% increase in the mean beak depth in this population (**Figure 1.11**).

Selection preserves the status quo when the most common type is the best adapted.

So far, we have seen how natural selection led to adaptation as the population of finches on Daphne Major evolved in response to changes in their environment. Will this process continue forever? If it did, eventually all the finches would have deep enough beaks to efficiently process the largest seeds available. However, large beaks have disadvantages as well as benefits. The Grants showed, for instance, that birds with large beaks are less likely to survive the juvenile period than are birds with small beaks, probably

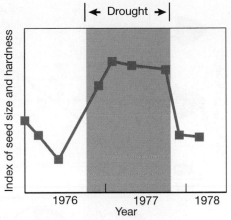

FIGURE 1.8

During the 2-year drought, the size and hardness of seeds available on Daphne Major increased because birds consumed all of the desirable small, soft seeds, leaving mainly larger and harder seeds. Each point on this plot represents an index of seed size and hardness at a given time.

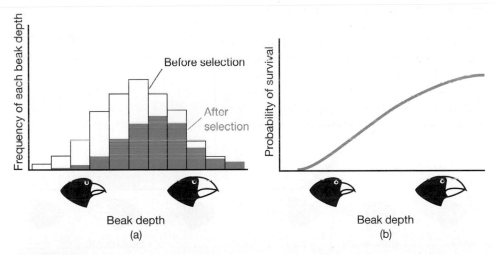

FIGURE 1.9

How directional selection increased mean beak depth among medium ground finches on Daphne Major. (a) The heights of the bars represent the numbers of birds whose beak depths fall within each of the intervals plotted on the *x* axis, with beak depth increasing to the right. The open bars show the distribution of beak depths before the drought began. The shaded bars show the distribution of beak depths after a year of drought. Notice that the number of birds in each category has decreased. Because birds with deep beaks were less likely to die than were birds with shallow beaks, the peak of the distribution shifted to the right, indicating that the mean beak depth had increased. (b) The probability of survival for birds of different beak depths is plotted. Birds with shallow beaks are less likely to survive than are birds with deep beaks.

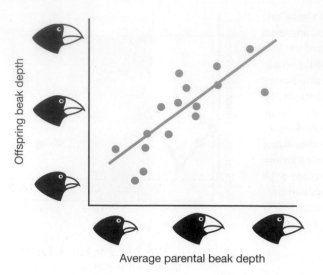

FIGURE 1.10

Parents with deeper-than-average beaks tend to have offspring with deeper-than-average beaks. Each point represents one offspring. Offspring beak depth is plotted on the *y* axis (deeper beaks farther up the axis), and the average of the two parents' beak depths is plotted on the *x* axis (deeper beaks farther to the right).

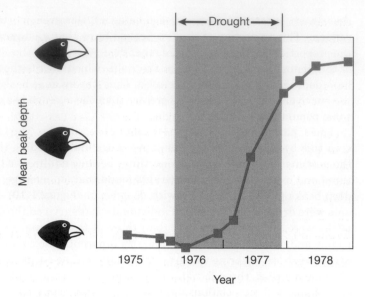

FIGURE 1.11

The average beak depth in the population of medium ground finches on Daphne Major increased during the drought of 1976–1977. Each point plots an index of average beak depth of the population in a particular year. Deeper beaks are plotted higher on the *y* axis.

because large-beaked birds require more food (**Figure 1.12**). Evolutionary theory predicts that, over time, selection will increase the average beak depth in the population until the costs of larger-than-average beak size exceed the benefits. At this point, finches with the average beak size in the population will be the most likely to survive and reproduce, and finches with deeper or shallower beaks than the new average will be at a disadvantage. When this is true, beak size does not change, and we say that an **equilibrium** exists in the population in regard to beak size. The process that produces this equilibrium state is called **stabilizing selection**. Notice that even though the average

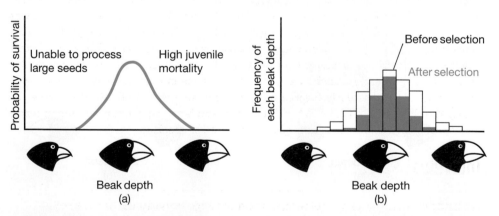

FIGURE 1.12

When birds with the most common beak depth are most likely to survive and reproduce, natural selection keeps the mean beak depth constant. (a) Birds with deep or shallow beaks are less likely to survive than are birds with average beaks. Birds with shallow beaks cannot process large, hard seeds, and birds with deep beaks are less likely to survive to adulthood. (b) The open bars represent the distribution of beak depths before selection, and the shaded bars represent the distribution after selection. As in Figure 1.9, notice that there are fewer birds in the population after selection. Because birds with average beaks are most likely to survive, however, the peak of the distribution of beak depths does not shift, and mean beak depth remains unchanged.

characteristics of the beak in the population will not change in this situation, selection is still going on. Selection is required to change a population, and selection is also required to keep a population the same.

It might seem that beak depth would also remain unchanged if this trait had no effect on survival (or put another way, if there were no selection favoring one type of beak over another). Then all types of birds would be equally likely to survive from one generation to the next, and beak depth would remain constant. This logic would be valid if selection were the only process affecting beak size. However, real populations are also affected by other processes that cause **traits**, or **characters**, to change in unpredictable ways. We will discuss these processes further in Chapter 3. The point to remember here is that populations do not remain static over the long run unless selection is operating.

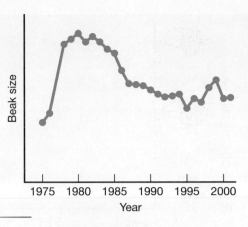

FIGURE 1.13

An index of mean beak size on Daphne Major for 1975–2001.

Evolution does not always lead to change in the same direction.

Natural selection has no foresight; it simply causes organisms to change so that they are better adapted to their current environment. Often environments fluctuate over time, and when they do, selection may track these fluctuations. We see this kind of pattern in the finches of the Galápagos Islands in the years since the original study. During this time there have been dry periods (1976–1978), but there have also been wet periods (1983–1985), when small, soft, easily processed seeds were exceedingly abundant. During wet years, selection favors smaller beaks, reversing the changes in beak size and shape wrought by natural selection during the drought years. As **Figure 1.13** shows, beak size wobbled up and down during the Grants' long study of the medium ground finch on Daphne Major.

Species are populations of varied individuals that may or may not change over time.

As the Grants' work on Daphne Major makes clear, a species is not a fixed type or entity. Species change in their general characteristics from generation to generation according to the postulates Darwin described. Before Darwin, however, people thought of species as unchanging categories, much the same way that we think of geometric figures: A finch could no more change its properties than a triangle could. If a triangle acquired another side, it would not be a modified triangle, but rather would become a rectangle. In much the same way, to biologists before Darwin, a changed finch was not a finch at all. Ernst Mayr, a distinguished evolutionary biologist, called this pre-Darwinian view of immutable species "essentialism." According to Darwin's theory, a **species** is a dynamic *population* of individuals. The characteristics of a particular species will be static, or unchanged, over a long time only if the most common type of individual is consistently favored by stabilizing selection. Both **stasis** (staying the same) and change result from natural selection, and both require explanation in terms of natural selection. Stasis is not the natural state of species.

Individual Selection

Adaptation results from the competition among individuals, not between entire populations or species.

Selection produces adaptations that benefit *individuals*. Such adaptation may or may not benefit the population or species. In the case of simple morphological characters such as beak depth, selection probably does allow the population of finches to compete more effectively with other populations of seed predators. However, this need not be the case. Selection often leads to changes in behavior or morphology that increase the reproductive success of individuals but decrease the average reproductive success of the group, population, and species.

FIGURE 1.14

A female macaque holds her infant.

The fact that almost all organisms produce many more offspring than are necessary to maintain the species provides an example of the conflict between individual and group interests. Suppose that a female monkey, on average, produces 10 offspring during her lifetime (**Figure 1.14**). In a stable population, only two of these offspring will survive and reproduce on average. From the point of view of the species, the other eight are a waste of resources. They compete with other members of their species for food, water, and sleeping sites. The demands of a growing population can lead to serious overexploitation of the environment, and the species as a whole might be more likely to survive if all females produced fewer offspring. This does not happen, however, because natural selection among individuals favors females who produce many offspring.

To see why selection on individuals will lead to this result, let's consider a simple hypothetical case. Suppose the females of a particular species of monkey are maximizing individual reproductive success when they produce 10 offspring. Females that produce more than or less than 10 offspring will tend to leave fewer descendants in the next generation. Further suppose that the likelihood of the species becoming extinct would be lowest if females produced only two offspring apiece. Now suppose that there are two kinds of females. Most of the population is composed of low-fecundity females that produce just two offspring each, but a few high-fecundity females produce 10 offspring each. (**Fecundity** is the term demographers use for the ability to produce offspring.) High-fecundity females have high-fecundity daughters, and low-fecundity females have low-fecundity daughters. The proportion of high-fecundity females will increase in the next generation because such females produce more offspring than do low-fecundity females. Over time, the proportion of high-fecundity females in the population will increase rapidly. As fecundity increases, the population will grow rapidly and may deplete available resources. The depletion of resources, in turn, will increase the chance that the species becomes extinct. However, this outcome is irrelevant to the evolution of fecundity before the extinction because natural selection results from competition among individuals, not competition among species.

The idea that natural selection operates at the level of the individual is a key element in understanding adaptation. In discussing the evolution of social behavior in Chapter 7, we will encounter several additional examples of situations in which selection increases individual success but decreases the competitive ability of the population.

The Evolution of Complex Adaptations

The example of the evolution of beak depth in the medium ground finch illustrates how natural selection can cause adaptive change to occur rapidly in a population. Deeper beaks enabled the birds to survive better, and deeper beaks soon came to predominate in the population. Beak depth is a fairly simple character, lacking the intricate complexity of an eye. As we will see, however, the accumulation of small variations by natural selection can also give rise to complex adaptations.

Why Small Variations Are Important

There are two categories of variation: continuous and discontinuous.

It was known in Darwin's day that most variation is continuous. An example of **continuous variation** is the distribution of heights in people. Humans grade smoothly from one extreme to the other (short to tall), with all the intermediate types (in this case, heights) represented. However, Darwin's contemporaries also knew about **discontinuous variation**, in which several distinct types exist with no intermediates. In humans, height is also subject to discontinuous variation. For example, a genetic condition called achondroplasia causes affected individuals to be much shorter than other people, have proportionately shorter arms and legs, and bear a variety of other distinctive features (Peter Dinklage, who played Tyrion Lannister in *Game of Thrones*, has this condition). Discontinuous variants are usually quite rare in nature. Nonetheless, many of Darwin's contemporaries who were convinced of the reality of evolution believed that new species arise as discontinuous variants.

Discontinuous variation is not important for the evolution of complex adaptations because complex adaptations are extremely unlikely to arise in a single jump.

Unlike most of his contemporaries, Darwin thought that discontinuous variation did not play an important role in evolution. A hypothetical example, described by the Oxford University biologist Richard Dawkins in his book *The Blind Watchmaker*, illustrates Darwin's reasoning. Dawkins recalls an old story in which an imaginary collection of monkeys sits at typewriters happily typing away. Lacking the ability to read or write, the monkeys strike keys at random. Given enough time, the story goes, the monkeys will reproduce all the great works of Shakespeare. But Dawkins points out that this is not likely to happen in the lifetime of the universe, let alone the lifetime of one of the monkey typists. To illustrate why it would take so long, Dawkins presents these illiterate monkeys with a much simpler problem: reproducing a single line from *Hamlet*, "Methinks it is like a weasel" (III.ii). To make the problem even simpler for the monkeys, Dawkins ignores the difference between uppercase and lowercase letters and omits all punctuation except spaces. There are 28 characters (including spaces) in the phrase. Because there are 26 characters in the alphabet and Dawkins is keeping track of spaces, each time a monkey types a character, there is only a 1-in-27 chance that it will type the right character. There is also only a 1-in-27 chance that the second character will be correct. Again, there is a 1-in-27 chance that the third character will be right, and so on, up to the twenty-eighth character. Thus the chance that a monkey will type the correct sequence at random is 1/27 multiplied by itself 28 times, or

$$\underbrace{\frac{1}{27} \times \frac{1}{27} \times \frac{1}{27} \times \cdots \times \frac{1}{27}}_{28 \text{ times}} \approx 10^{-40}$$

This is a *very* small number. To get a feeling for how small a chance there is of the monkeys typing the sentence correctly, suppose that a very fast computer could generate 100 billion (10^{11}) characters per second and run for the lifetime of Earth—about 4 billion years, or 10^{17} seconds. Even at that pace and given that much time, the chance of the computer randomly typing the line "Methinks it is like a weasel" even once during the whole of Earth's history would be about 1 in 1 trillion! Typing the whole play is obviously astronomically less likely, and although *Hamlet* is a very complicated thing, it is much less complicated than a human eye. There's no chance that a structure like the human eye would arise by chance in a single trial. If it did, it would be, as the astrophysicist Sir Fred Hoyle is reported to have said, like a hurricane blowing through a junkyard and chancing to assemble a Boeing 747.

Complex adaptations can arise through the accumulation of small random variations by natural selection.

Darwin argued that continuous variation is essential for the evolution of complex adaptations. Once again, Richard Dawkins provides an example that makes Darwin's reasoning clear. Again imagine a room full of monkeys and typewriters, but now the rules of the game are different. The monkeys type the first 28 characters at random, and then during the next round they attempt to copy the same initial string of letters and spaces. Most of the sentences are just copies of the previous string, but because monkeys sometimes make mistakes, some strings have small variations, usually in only a single letter. During each trial, the monkey trainer selects the string that most resembles Shakespeare's phrase "Methinks it is like a weasel" as the string to be copied by all the monkeys in the next trial. This process is repeated until the monkeys come up with the correct string. Calculating the exact number of trials required to generate the correct sequence of characters is quite difficult, but it is easy to simulate the process on a computer. Here's what happened when Dawkins performed the simulation. The initial random string was

```
            WDLMNLT DTJBKWIRZREZLMQCO P
```

After one trial Dawkins got

```
            WDLMNLT DTJBSWIRZREZLMQCO P
```

After 10 trials:

```
            MDLDMNLS ITJISWHRZREZ MECS P
```

After 20 trials:

```
            MELDINLS IT ISWPRKE Z WECSEL
```

After 30 trials:

```
            METHINGS IT ISWLIKE B WECSEL
```

After 40 trials:

```
            METHINKS IT IS LIKE I WEASEL
```

The exact phrase was reached after 43 trials. Dawkins reports that it took his 1985-vintage Macintosh only 11 seconds to complete this task.

Selection can give rise to great complexity starting with small random variations because it is a *cumulative* process. As the typing monkeys show us, it is spectacularly unlikely that a single random combination of keystrokes will produce the correct sentence. However, there is a much greater chance that some of the many *small* random changes will be advantageous. The combination of reproduction and selection allows the typing monkeys to accumulate these small changes until the desired sentence is reached.

Why Intermediate Steps Are Favored by Selection

The evolution of complex adaptations requires all of the intermediate steps to be favored by selection.

There is a potent objection to the example of the typing monkeys. Natural selection, acting over time, can lead to complex adaptations, but it can do so only if each small change along the way is itself adaptive. Although it is easy to assume that this is true in a hypothetical example of character strings, many people have argued that it is unlikely for every one of the changes necessary to assemble a complex organ such as the eye to be adaptive. An eye is useful, it is claimed, only after all parts of the complexity have been assembled; until then, it is worse than no eye at all. After all, what good is 5% of an eye?

Darwin's answer, based on the many adaptations for seeing or sensing light that exist in the natural world, was that 5% of an eye *is* often better than no eye at all. It is quite possible to imagine that a very large number of small changes—each favored by selection—led cumulatively to the wonderful complexity of the eye. Living mollusks, which display a broad range of light-sensitive organs, provide examples of many of the likely stages in this process:

1. Many invertebrates have a simple light-sensitive spot. Photoreceptors of this kind have evolved many times from ordinary epidermal (surface) cells—usually cells with microscopic hairlike projections (cilia) whose biochemical machinery is light sensitive. Those individuals whose cells are more sensitive to light are favored when information about changes in light intensity is useful. For example, a drop in light intensity may often indicate that a predator is in the vicinity.

2. The second step is for the light-sensitive spot to form a depression (**Figure 1.15a**). When the cells form a depression, the light does not hit all of the cells at the same time, which provides information about the direction from which the light is coming. The surface of organisms is variable, and those individuals whose photoreceptors are in depressions will be favored by selection in environments in which such information is useful. For example, mobile organisms may need better information about what is happening in front of them than do immobile ones.

3. Through a series of small steps, the depression could deepen (**Figure 1.15b**), and each step could be favored by selection because better directional information would be available with the deepening depression.

4. If the depression got deep enough (**Figure 1.15c**), it could form images on the light-sensitive tissue, much the way pinhole cameras form images on photographic film. In settings in which detailed images are useful, selection could then favor the elaboration of the neural machinery necessary to interpret the image.

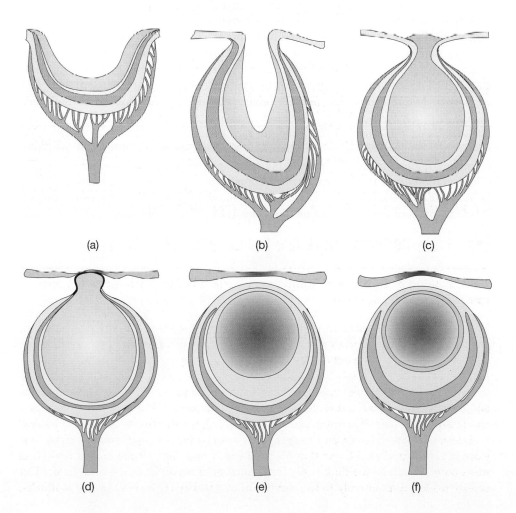

(a)   (b)   (c)

(d)   (e)   (f)

FIGURE 1.15

Living gastropod mollusks illustrate all of the intermediate steps between a simple eye cup and a camera-type eye. (a) The eye pit of a limpet, *Patella* sp.; (b) the eye cup of Beyrich's slit shell, *Pleurotomaria beyrichii*; (c) the pinhole eye of a California abalone, *Haliotis* sp.; (d) the closed eye of a turban shell, *Turbo creniferus*; (e) the lens eye of the spiny dye-murex, *Murex brandaris*; (f) the lens eye of the Atlantic dog whelk, *Nucella lapillus*. (Lens is shaded in e and f.)

5. The next step is the formation of a transparent cover (**Figure 1.15d**). This change might be favored because it protects the interior of the eye from parasites and mechanical damage.

6. A lens could evolve through gradual modification of either the transparent cover or the internal structures within the eye (**Figure 1.15e** and **f**).

Notice that evolution produces adaptations like a tinkerer, not an engineer. New organisms are created by small modifications of existing organisms, not by starting with a clean slate. Clearly many beneficial adaptations will not arise because they are blocked at some step along the way when a particular variation is not favored by selection. Darwin's theory explains how complex adaptations can arise through natural processes, but it does not predict that every possible adaptation, or even most, will occur. This is not the best of all possible worlds; it is just one of many possible worlds.

Sometimes distantly related species have independently evolved the same complex adaptation, absent in their common ancestor, suggesting that the evolution of complex adaptations by natural selection is not a matter of mere chance.

The fact that natural selection constructs complex adaptations like a tinkerer might lead you to think that the assembly of complex adaptations is a chancy business. If even a single step were not favored by selection, the adaptation could not arise. Such reasoning suggests that complex adaptations are mere coincidence. Although chance does play a very important role in evolution, the power of cumulative natural selection should not be underestimated. The best evidence that selection is a powerful process for generating complex adaptations comes from a phenomenon called **convergence**, the evolution of similar adaptations in unrelated groups of animals.

The similarity between the marsupial faunas of Australia and South America and the placental faunas of the rest of the world provides a good example of convergence. In most of the world, the mammalian fauna is dominated by **placental mammals**, which nourish their young in the uterus during long pregnancies. Both Australia and South America, however, became separated from an ancestral supercontinent, known as Pangaea, long before placental mammals evolved. In Australia and South America, **marsupials** (nonplacental mammals, like kangaroos, that rear their young in external pouches) came to dominate the mammalian fauna, filling all available mammalian niches. Some of these marsupial mammals were quite similar to the placental mammals on the other continents. For example, there was a marsupial wolf in Tasmania that looked very much like placental wolves of Eurasia, even sharing subtle features of their feet and teeth (**Figure 1.16**). These marsupial wolves became

FIGURE 1.16

The marsupial wolf that lived in Tasmania until early in the twentieth century (drawn from a photograph of one of the last living animals). Similarities with placental wolves of North America and Eurasia illustrate the power of natural selection to create complex adaptations. Their last common ancestor was probably a small, insectivorous, shrewlike creature.

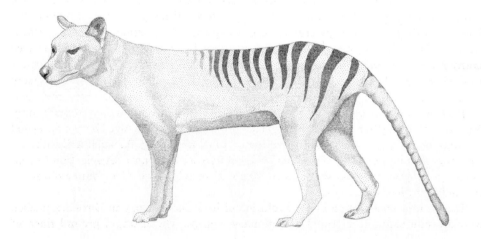

[a]

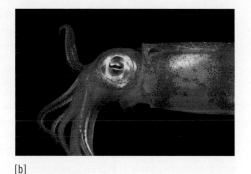
[b]

FIGURE 1.17

Complex eyes with lenses have evolved independently in several kinds of aquatic animals, including the (a) moon wrasse and (b) squid.

extinct in the 1930s. Similarly, in South America a marsupial saber-toothed cat independently evolved many of the same adaptations as the placental saber-toothed cat that stalked North America 10,000 years ago. These similarities are more impressive when you consider that the last common ancestor of marsupial and placental mammals was a small, nocturnal, insectivorous creature, something like a shrew, that lived about 120 million years ago (Ma). Thus, selection transformed a shrew step by small step, each step favored by selection, into a saber-toothed cat—and it did it twice. This cannot be coincidence.

The evolution of eyes provides another good example of convergence. Remember that the spherical gradient lens is a good lens design for aquatic organisms because it has good light-gathering ability and provides a sharp image over the full 180° visual field. Complex eyes with lenses have evolved independently eight times in distantly related aquatic organisms: once in fish, once in cephalopod mollusks such as squid, several times among gastropod mollusks such as the Atlantic dog whelk, once in annelid worms, and once in crustaceans (**Figure 1.17**). These are very diverse creatures whose last common ancestor was a simple creature that did not have a complex eye. Nonetheless, in every case they have evolved very similar spherical gradient lenses. Moreover, no other lens design is found in aquatic animals. Despite the seeming chanciness of assembling complex adaptations, natural selection has achieved the same design in every case.

Rates of Evolutionary Change

Natural selection can cause evolutionary change that is much more rapid than we commonly observe in the fossil record.

In Darwin's day, the idea that natural selection could change a primate into a human, much less that it might do so in just a few million years, was unthinkable. Even though people are generally more accepting of evolution today, many still think of evolution by natural selection as a glacially slow process that requires many millions of years to accomplish noticeable change. Such people often doubt that there has been enough time for selection to accomplish the evolutionary changes observed in the fossil record. And yet, as we will see in later chapters, most scientists now believe that humans evolved from an apelike creature in only 5 million to 10 million years. In fact, some of the rates of selective change observed in contemporary populations are far faster than necessary for such a transition. The puzzle is not whether there has been enough time for natural selection to produce the adaptations that we observe. The real puzzle is why the change observed in the fossil record was so slow.

The Grants' observation of the evolution of beak morphology in Darwin's finches provides one example of rapid evolutionary change. The medium ground finch of

FIGURE 1.18

We can trace the relationship among various species of Darwin's finches by analyzing their protein polymorphisms. Species that are closely linked in the phylogenetic tree are more similar to one another genetically than to other species because they share a more recent common ancestor. The tree does not include 3 of the 14 species of Darwin's finches.

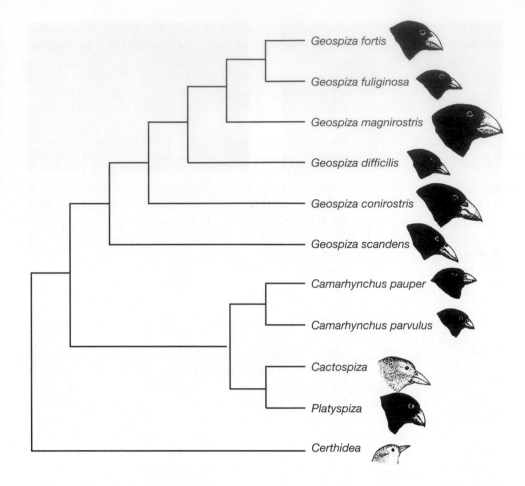

Geospiza fortis

Geospiza fuliginosa

Geospiza magnirostris

Geospiza difficilis

Geospiza conirostris

Geospiza scandens

Camarhynchus pauper

Camarhynchus parvulus

Cactospiza

Platyspiza

Certhidea

FIGURE 1.19

The large ground finch (*Geospiza magnirostris*) has a beak that is nearly 20% deeper than the beak of its close relative, the medium ground finch (*Geospiza fortis*). At the rate of evolution observed during the drought of 1976–1977, Peter Grant calculated that selection could transform the medium ground finch into the large ground finch in less than 46 years.

Daphne Major is one of 14 species of finches that live in the Galápagos. Evidence suggests that all 14 are descended from a single South American species that migrated to the newly emerged islands about half a million years ago (**Figure 1.18**). This doesn't seem like a very long time. Is it possible that natural selection created 14 species in only half a million years?

To start to answer the question, let's calculate how long it would take for the medium ground finch (*Geospiza fortis*) to come to resemble its closest relative, the large ground finch (*Geospiza magnirostris*), in beak size and weight (**Figure 1.19**). The large ground finch is 75% heavier than the medium ground finch, and its beak is about 20% deeper. Remember that beak size increased about 4% in 2 years during the 1977 drought. The Grants' data indicate that body size also increased by a similar amount. At this rate, Peter Grant calculated that it would take between 30 and 46 years for selection to increase the beak size and body weight of the medium ground finch to match those of the large ground finch. But these changes occurred in response to an extraordinary environmental crisis. The data suggest that selection doesn't generally push consistently in just one direction. Instead, in the Galápagos, evolutionary change seems to go in fits and starts, moving traits one way and then another. So let's suppose that a net change in beak size like the one that occurred during 1977 occurs only once every century. Then it would take about 2,000 years to transform the medium ground finch into the large ground finch—still a very rapid process.

Similar rates of evolutionary change are observed elsewhere when species invade new habitats. For example, about 100,000 years ago a population of elk (called "red deer" in Great Britain) colonized the island of Jersey, off the French coast, and then became isolated, presumably by rising sea levels. By the time the island was reconnected with the mainland approximately 6,000 years later, the red deer had shrunk to the size of a large dog. University of Michigan paleontologist Philip Gingerich compiled data on the rate of evolutionary change in 104 cases in which species invaded new habitats. These rates ranged from a low of zero (that is, no change) to a high of 22% per year, with an average of 0.1% per year.

The changes the Grants observed in the medium ground finch are relatively simple: The birds and their beaks just got bigger. More complex changes usually take longer to evolve, but several kinds of evidence suggest that selection can produce big changes in remarkably short periods.

One line of evidence comes from artificial selection. Humans have performed selection on domesticated plants and animals for thousands of years, and while for most of this period this selection was not deliberate (that is, not influenced by active human intervention), more recent deliberate selection by humans has led to rapid rates of evolution in certain species. There are many familiar examples. All domesticated dogs, for instance, are believed to be descendants of wolves. Scientists are not sure when dogs were domesticated, but 15,000 years ago is a good guess, which means that over a few thousand generations, selection changed wolves into Pekingese, beagles, greyhounds, and Saint Bernards. In reality, though, most of these breeds were created fairly recently as the products of directed breeding. Darwin's favorite example of artificial selection was the domestication of pigeons. In the nineteenth century, pigeon breeding was a popular hobby, especially among working people who competed to produce showy birds (**Figure 1.20**). Pigeon fanciers created a menagerie of wildly different forms, all descended from the rather plain-looking rock pigeon. Darwin pointed out that these breeds were often so different that biologists would surely have classified them as members of different species if they had been discovered in nature. Yet they had been produced by artificial selection within a few hundred years.

[a]

[b]

[c]

[d]

FIGURE 1.20

In Darwin's day, pigeon fanciers created many new breeds of pigeons, including (a) pouters, (b) fantails, and (c) carriers, all from (d) the common rock pigeon.

FIGURE 1.21

Fish in the genus *Poeciliopsis* include small minnows such as *Poeciliopsis occidentalis,* shown here.

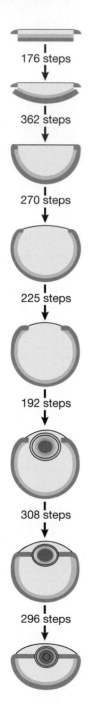

176 steps

362 steps

270 steps

225 steps

192 steps

308 steps

296 steps

Rapid evolution of a complex feature has also been documented in a recent study of a group of very closely related species of fish from the genus *Poeciliopsis* (**Figure 1.21**). These small minnows can be found in tropical lowland streams, high-altitude lakes, and desert springs and streams in Mexico and Central America. All the species in this genus bear live young, but the sequence of events between fertilization, the initial union of the egg and sperm cells, and birth varies. In most species, females endow the eggs with nutrients before fertilization. As the young develop, they consume this endowment. In a few species, however, females continue to provide nutrients to their unborn offspring throughout development by using tissues that are analogous to mammalian placentas. When these offspring are born, they can be more than 100 times the mass of the egg at fertilization. Biologist David Reznick of the University of California, Riverside, and his colleagues have shown that these placental tissues evolved independently in three groups of species within the genus *Poeciliopsis*. Genetic data indicate that one of these species groups diverged from ancestors lacking placental tissue only 0.75 Ma, and the other two diverged less than 2.4 Ma. These time estimates actually represent the time since these species shared a common ancestor, and they set an upper bound on the amount of time required for the placenta to evolve. The generation time for these fish ranges from 6 months to a year, so this complex adaptation evolved in fewer than a million generations.

A third line of evidence comes from theoretical studies of the evolution of complex characters. Dan-Eric Nilsson and Susanne Pelger of Lund University in Sweden built a mathematical model of the evolution of the eye in an aquatic organism. They started with a population of organisms, each with a simple eyespot, a flat patch of light-sensitive tissue sandwiched between a transparent protective layer and a layer of dark pigment. They then considered how every possible small (1%) deformation of the shape of the eyespot affects the resolving power of the eye. They determined which 1% change had the greatest positive effect on the eye's resolving power and then repeated the process again and again, deforming the new structure by 1% in every possible way at each step. The results are shown in **Figure 1.22**. After 538 changes of 1% each, a simple concave eye cup evolves; after 1,033 changes of 1%, crude pinhole eyes emerge; after 1,225 changes of 1%, an eye with an elliptical lens is created; and after 1,829 changes, the process finally comes to a halt because no small changes increase resolving power. The result is an eye with a spherical gradient lens just like those in fish and other aquatic organisms. As Nilsson and Pelger pointed out, 1,829 changes of 1% add up to a substantial amount of change. For instance, 1,829 changes of 1% would lengthen a 10-cm (4-in.) human finger to 8,000 km (5,500 miles)—about the distance from Los Angeles to New York and back. Nonetheless, making very conservative assumptions about the strength of selection, Nilsson and Pelger calculated that this would take only about 364,000 generations. For organisms with short generations, the complete structure of the eye can evolve from a simple eyespot in less than a million years, a brief moment in evolutionary time.

By comparison, most changes observed in the fossil record are much slower. Human brain size has roughly doubled in the past 2 million years—a rate of change of 0.00005% per year. This is 10,000 times slower than the rate of change that the Grants observed in the Galápagos. Moreover, such slow rates of change typify what can be observed from the fossil record. As we will see, however, the fossil record is

FIGURE 1.22

A computer simulation of the evolution of the eye generates this sequence of forms. Between each pair of forms is the number of 1% changes (or steps) necessary to transform the upper form into the lower one. The eye begins as a flat patch of light-sensitive tissue (*red*) that lies between a transparent layer (*light blue*) and a layer of dark pigmented tissue (*black*). After 176 steps, each of which increases resolving power, a shallow eye cup is formed. After 362 additional steps, the eye cup deepens. Eventually, a spherical gradient lens evolves, leading to a camera-type eye. The entire process involves about 1,800 changes of 1%.

incomplete. It is quite likely that some evolutionary changes in the past were rapid, but the sparseness of the fossil record prevents us from detecting them.

Darwin's Difficulties Explaining Variation

Darwin's *On the Origin of Species*, published in 1859, was a best seller during his day, but his proposal that new species and other major evolutionary changes arise by the accumulation of small variations through natural selection was not widely embraced. Most educated people accepted the idea that new species arise through the transformation of existing species, and many scientists accepted the idea that natural selection is the most important cause of organic change (although by the turn of the twentieth century even this consensus had broken down, particularly in the United States). But only a minority endorsed Darwin's view that major changes occur through the accumulation of small variations.

Darwin couldn't convince his contemporaries that evolution occurred through the accumulation of small variations because he couldn't explain how variation is maintained.

Darwin's critics raised a telling objection to his theory: The actions of blending inheritance (described in the next paragraph) and selection would both inevitably deplete variation in populations and make it impossible for natural selection to continue. These were potent objections that Darwin was unable to resolve in his lifetime because he and his contemporaries did not yet understand the mechanics of inheritance.

Everyone could readily observe that many of the characteristics of offspring are an average of the characteristics of their parents. Most people, including Darwin, believed this phenomenon to be caused by the action of **blending inheritance**, a model of inheritance that assumes the mother and father each contribute a hereditary substance that mixes, or "blends," to determine the characteristics of the offspring. Shortly after publication of *On the Origin of Species*, a Scottish engineer named Fleeming Jenkin published a paper in which he clearly showed that, with blending inheritance, there could be little or no variation available for selection to act on. The following example shows why Jenkin's argument was so compelling. Suppose a population of one species of Darwin's finches displays two forms: tall and short. Further suppose that a biologist controls mating so that every mating is between a tall individual and a short individual. Then, with blending inheritance, all of the offspring will be the same intermediate height, and their offspring will be the same height as they are. All of the variation for height in the population will disappear in a single generation. With random mating, the same result will occur, though it will take longer. If inheritance were purely a matter of blending parental traits, then Jenkin would have been right about its effect on variation. However, as we will see in Chapter 3, genetics can account for the fact that offspring are intermediate between their parents without assuming any kind of blending.

Another problem arose because selection works by removing variants from populations. For example, if finches with small beaks are more likely to die than are finches with large beaks, over many generations only birds with large beaks will be left. There will be no variation for beak size, and Darwin's second postulate holds that without variation there can be no evolution by natural selection. For example, suppose the environment changes so that individuals with small beaks are less likely to die than are those with large beaks. The average beak size in the population will not decrease because there are no small-beaked individuals. Natural selection destroys the variation required to create adaptations.

Even worse, as Jenkin also pointed out, there was no explanation of how a population might evolve beyond its original range of variation. The cumulative evolution of complex adaptations requires populations to move far outside their original range of

[a]

[b]

[c]

FIGURE 1.23

(a) The wolf is the ancestor of all domestic dogs, including (b) the mini-dachshund and (c) the Saint Bernard. These transformations were accomplished in several thousand generations of artificial selection.

variation. Selection can cull away some traits from a population, but how can it lead to new types not present in the original population? This apparent contradiction was a serious impediment to explaining the logic of evolution. How could elephants, moles, bats, and whales all descend from an ancient shrewlike insectivore unless there were a mechanism for creating new variants not present at the beginning? For that matter, how could all the different breeds of dogs have descended from their one common ancestor, the wolf (**Figure 1.23**)?

Remember that Darwin and his contemporaries knew there were two kinds of variation: continuous and discontinuous. Because Darwin believed that complex adaptations could arise only through the accumulation of small variations, he thought discontinuous variants were unimportant. However, many biologists thought that the discontinuous variants, called "sports" by nineteenth-century animal breeders, were the key to evolution because they solved the problem of the blending effect. For example, suppose that a population of green birds has entered a new environment in which red birds are better adapted. How can evolution shift the population from green to red? Some of Darwin's critics believed that any new variant that emerged in the population of green birds that was only slightly red would have only a small advantage and the color change would be rapidly swamped by blending. In contrast, an all-red bird would have a large enough selective advantage to overcome the effects of blending and could increase its frequency in the population.

Darwin's letters show that these criticisms worried him greatly. Although he tried a variety of counterarguments, he never found one that was satisfactory. The solution to these problems required an understanding of genetics, which was not available for another half-century. As we will see, it was not until well into the twentieth century that geneticists came to understand how variation is maintained and Darwin's theory of evolution became generally accepted.

CHAPTER REVIEW

Key Terms

adaptations (p. 4)	traits (p. 11)	continuous variation (p. 12)	placental mammals (p. 16)
natural selection (p. 7)	characters (p. 11)	discontinuous variation (p. 13)	marsupials (p. 16)
morphology (p. 9)	species (p. 11)		blending inheritance (p. 21)
equilibrium (p. 10)	stasis (p. 11)	convergence (p. 16)	
stabilizing selection (p. 10)	fecundity (p. 12)		

Study Questions

1. It is sometimes observed that offspring do not resemble their parents for a particular character, even though the character varies in the population. Suppose this were the case for beak depth in the medium ground finch.
   (a) What would the plot of offspring beak depth against parental beak depth look like?
   (b) Plot the mean depth in the population among (i) adults before a drought, (ii) the adults that survived a year of drought, and (iii) the offspring of the survivors.

2. Many species of animals engage in cannibalism. This practice certainly reduces the ability of the species to survive. Is it possible that cannibalism could arise by natural selection? If so, with what adaptive advantage?

3. Some insects mimic dung. Ever since Darwin, biologists have explained this behavior as a form of camouflage: Selection favors individuals who most resemble dung because they are less likely to be eaten. The late Harvard paleontologist Stephen Jay Gould objected to this explanation. He argued that although selection could perfect such mimicry once it evolved, it could not cause the resemblance to arise in the first place. "Can there be any edge," Gould asked, "to looking 5% like a turd?" (Dawkins 1996, p. 81). Can you think of a reason why looking 5% like a turd would be better than not looking at all like a turd?

4. In the late 1800s an American biologist named Hermon Bumpus collected many sparrows that had been killed in a severe ice storm. He found that birds whose wings were about average in length were rare among the dead birds. What kind of selection is this? What effect would this episode of selection have on the mean wing length in the population?

5. Critics of Darwin's theory argued that it couldn't explain how variation is maintained in populations. Explain the basis for their criticism.

6. Some insect larvae look a bit like snakes, with symmetric spots that look something like eyes and a pointed end that looks something like a snake's head. The biologist who described these snakelike features hypothesized that they deter predation by birds, which are wary of attacks by snakes. But in order for this to work, birds must be fairly gullible. They must think that the larvae really are snakes, not larvae that are trying to look like snakes. Explain why natural selection might favor gullibility over skepticism in the larvae's predators.

7. Many people find it implausible that a complex organ such as the human eye can be the product of a random, undirected process such as evolution by natural selection. Explain how the metaphor Dawkins offers of the typing monkeys helps to explain how complexity can arise.

8. Explain how competition, variation, and inheritance are central to Darwin's theory of evolution by natural selection.

9. Most people think of evolution as a very slow process that requires millions of years to produce noticeable change. Explain why this view is not necessarily correct.

10. If you see no change in the mean value of a trait from one generation to another, is it reasonable to conclude that selection is not operating on that trait?

Further Reading

Bergstrom, C. and L.A. Dugatkin. 2016. *Evolution.* 2nd ed. New York: Norton.

Browne, J. 1995. *Charles Darwin: A Biography,* vol. I: *Voyaging.* New York: Knopf.

Dawkins, R. 2015. *The Blind Watchmaker: Why the Evidence of Evolution Reveals a Universe without Design.* New York: Norton.

Dennett, D. C. 1995. *Darwin's Dangerous Idea: Evolution and the Meanings of Life.* New York: Simon & Schuster.

Ridley, M. 1996. *Evolution.* 2nd ed. Cambridge, Mass.: Blackwell Science.

Weiner, J. 1994. *The Beak of the Finch: A Story of Evolution in Our Time.* New York: Knopf.

 Review this chapter with personalized, interactive questions through INQUIZITIVE.

2

GENETICS

CHAPTER OBJECTIVES

By the end of this chapter you should be able to:

A Describe how experiments by Gregor Mendel revealed the logic of inheritance.

B Explain how Mendel's principles follow from the machinery of cell replication.

C Explain why genes affecting different traits are sometimes linked.

D Explain how the properties of DNA are consistent with the role of genes in inheritance.

E Describe how genes control the structure of proteins and influence the properties of organisms.

F Explain how gene regulation allows the same genes to control the development and function of many different parts of the body.

Mendelian Genetics

Although none of the main participants in the nineteenth-century debate about evolution knew it, the key experiments necessary to understand how genetic inheritance really worked had already been performed by an obscure monk, Gregor Mendel (**Figure 2.1**), living in what is now the Czech Republic. The son of peasant farmers, Mendel was recognized by his teachers as an extremely bright student, and he enrolled in the University of Vienna to study the natural sciences. While he was there, Mendel received a first-class education from some of the scientific luminaries of Europe. Unfortunately, Mendel had an extremely nervous disposition: Every time he was faced with an examination, he became physically ill, taking months to recover. As a result, he was forced to leave the university, and then he joined a monastery in the city of Brno, more or less because he needed a job. Once there, Mendel continued to study inheritance, an interest he had developed in Vienna.

By conducting careful experiments with plants, Mendel discovered how inheritance works. Between 1856 and 1863, using the common garden pea plant (**Figure 2.2**), Mendel isolated several traits with only two forms, or **variants**. For example, one of the traits he studied was pea color. This trait had two variants: yellow and green. He studied pea texture as well, a trait that also had two variants: wrinkled and smooth. Mendel cultivated populations of plants in which these traits bred true, meaning that the traits did not change from one generation to the next. For example, **crosses** (matings) between plants that bore green peas always produced offspring with green peas, and crosses between plants that bore yellow peas always produced offspring with yellow peas. Mendel performed many crosses between these kinds of true-breeding peas, or seeds.

Before going further, we need to establish a way to keep track of the results of the matings. Geneticists refer to the original founding population as the **F_0 generation**, the offspring of the original founders as the **F_1 generation**, and so on. In this case, the original true-breeding plants constitute the F_0 generation, and the plants created by crossing true-breeding parents constitute the F_1 generation. The offspring of the F_1 generation will be the **F_2 generation**.

In one set of Mendel's experiments with garden peas, a series of crosses between green and yellow variants yielded offspring that all bore yellow peas, matching only one of the parent plants (**Figure 2.3**). Mendel's next step was to perform crosses among the offspring of these crosses. When members of the F_1 generation (all of which bore yellow peas) were crossed, some of the offspring produced yellow peas and some produced green peas. Unlike most of the people who had experimented with plant crosses before, Mendel performed many of these kinds of crosses and kept careful count of the numbers of each kind of individual that resulted. These data showed that in the F_2 generation, there were three individuals with yellow peas for every one with green peas.

Mendel was able to formulate two principles that accounted for his experimental results.

Mendel derived two insightful conclusions from his experimental results:

1. The observed characteristics of organisms are determined jointly by two particles, one inherited from the mother and one from the father. The American geneticist T. H. Morgan later named these particles **genes**.

FIGURE 2.1

Gregor Mendel, about 1884, 15 years after abandoning his botanical experiments.

FIGURE 2.2

Mendel's genetic experiments were conducted on the common edible garden pea.

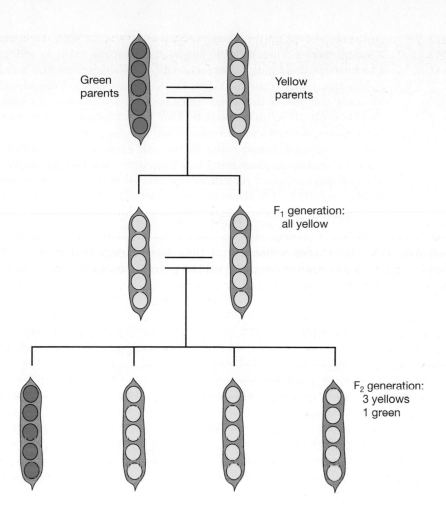

FIGURE 2.3

In one of Mendel's experiments, crossing true-breeding lines of green and yellow peas led to all yellow offspring. Crossing the F_1 individuals led to an F_2 generation with a 3:1 ratio of yellow to green individuals.

Green parents

Yellow parents

F_1 generation: all yellow

F_2 generation: 3 yellows 1 green

2. Each of these two particles, or genes, is equally likely to be transmitted when **gametes** (eggs and sperm) are formed. Modern scientists call this **independent assortment**.

These two principles account for the pattern of results in Mendel's breeding experiments, and as we shall see, they are the key to understanding how variation is preserved.

Cell Division and the Role of Chromosomes in Inheritance

Nobody paid any attention to Mendel's results for more than 30 years.

Mendel thought that his findings were important, so he published them in 1866 and sent a copy of the paper to Karl Wilhelm von Nägeli, a prominent botanist. Nägeli was studying inheritance and should have understood the importance of Mendel's experiments. Instead, Nägeli dismissed Mendel's work, perhaps because it contradicted his own results or because Mendel was an obscure monk. Soon after this, Mendel was elected abbot of his monastery and was forced to give up his experiments. His ideas did not resurface until the turn of the twentieth century, when several botanists independently replicated Mendel's experiments and rediscovered the mechanics of inheritance.

In 1896, the Dutch botanist Hugo de Vries unknowingly repeated Mendel's experiments with poppies. Instead of publishing his results immediately, however, he

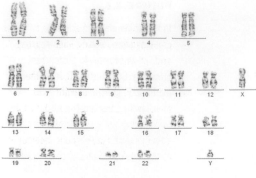

FIGURE 2.4

When a human cell divides, 23 pairs of chromosomes appear in its nucleus, including a pair of sex chromosomes (X and Y for a male, as shown here). Different chromosomes can be distinguished by their shape and by the banding patterns created by dyes that stain the chromosomes.

FIGURE 2.5

In all plants and animals, every cell contains a body called the nucleus (the solid red circle in the center of this magnified image). The nucleus contains the chromosomes.

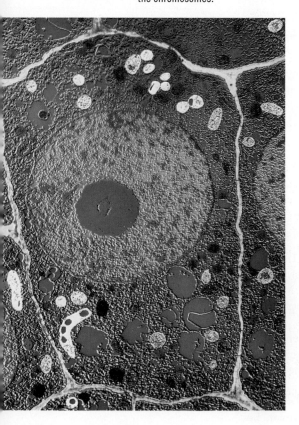

cautiously waited until he had replicated his results with more than 30 plant species. Then in 1900, just as de Vries was ready to send off a manuscript describing his experiments, a colleague sent him a copy of Mendel's paper. Poor de Vries; his hot new results were already 30 years old! About the same time, two other European botanists, Carl Correns and Erich Tschermak, also duplicated Mendel's breeding experiments, derived similar conclusions, and discovered that they too had been scooped. Correns and Tschermak graciously acknowledged Mendel's primacy in discovering how inheritance worked, but de Vries was less magnanimous. He did not cite Mendel in his treatise on plant genetics and refused to sign a petition advocating the construction of a memorial in Brno commemorating Mendel's achievements.

When Mendel's results were rediscovered, they were widely accepted because scientists now understood the role of chromosomes in the formation of gametes.

By the time Mendel's experiments were rediscovered in 1900, it was well known that virtually all living organisms are built out of cells. Moreover, careful embryological work had shown that all the cells in complex organisms arise from a single cell through the process of cell division. Between the time of Mendel's initial discovery of the nature of inheritance and its rediscovery at the turn of the twentieth century, a crucial feature of cellular anatomy was discovered: the **chromosome**. Chromosomes are small linear bodies contained in every cell and replicated during cell division (**Figure 2.4**). Moreover, scientists had also learned that chromosomes are replicated in a special kind of cell division that creates gametes. As we will see in later sections, this research provides a simple material explanation for Mendel's results. Our current model of cell division, which was developed in small steps by several scientists, is summarized in the sections that follow.

Mitosis and Meiosis

Ordinary cell division, called mitosis, creates two copies of the chromosomes present in the nucleus.

When plants and animals grow, their cells divide. Every cell contains within it a body called the **nucleus** (plural, *nuclei*; **Figure 2.5**); when cells divide, their nuclei also divide. This process of ordinary cell division is called **mitosis**. As mitosis begins, a cloud of material begins to form in the nucleus, and gradually this cloud condenses into several linear chromosomes. The chromosomes can be distinguished under the microscope by their shape and by how they stain. (Stains are dyes added to cells in the laboratory to allow researchers to distinguish parts of a cell.) Different organisms have different numbers of chromosomes, but in **diploid** organisms, chromosomes come in homologous pairs (pairs whose members have similar shapes and staining patterns). All primates, including humans, are diploid, but other organisms have a variety of arrangements. Diploid organisms also vary in the number of chromosome pairs their cells have. The fruit fly *Drosophila* has 4 pairs of chromosomes, humans have 23 pairs, and some organisms have many more.

Two features of mitosis suggest that the chromosomes play an important role in determining the properties of organisms. First, the original set of chromosomes is duplicated so that each new daughter cell has an exact copy of the chromosomes present in its parent. This means that, as an organism grows and develops through a sequence of mitotic divisions,

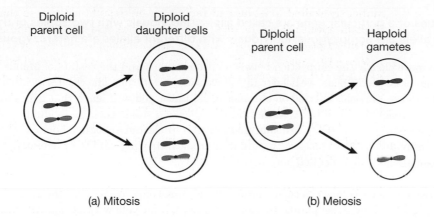

FIGURE 2.6

Diploid cells have *n* pairs of homologous chromosomes; *n* varies widely among species, but here *n* = 1. Members of homologous pairs may differ in the alleles they carry at specific sites along the chromosome. (a) Mitosis duplicates the chromosomes. (b) Meiosis creates gametes that carry only one member of each homologous pair of chromosomes.

every cell will have the same chromosomes that were present when the egg and sperm united. Second, the material that makes up the chromosome is present even when cells are not dividing. Cells spend little of their time dividing; most of the time they are in a "resting" period, doing what they are supposed to do as liver cells, muscle cells, bone cells, and so on. During the resting period, chromosomes are not visible. However, the material that makes up the chromosomes is always present in the cell.

In meiosis, the special cell division process that produces gametes, only half of the chromosomes are transmitted from the parent cell to the gamete.

The sequence of events that occurs during mitosis is quite different from the sequence of events during **meiosis**, the special form of cell division leading to the production of gametes. The key feature of meiosis is that each gamete contains only *one* copy of each chromosome, whereas cells that undergo mitosis contain a *pair* of **homologous chromosomes**. Cells that contain only one copy of each chromosome are said to be **haploid (Figure 2.6)**. When a new individual is conceived, a haploid sperm from the father unites with a haploid egg from the mother to produce a diploid **zygote**. The zygote is a single cell that then divides mitotically over and over to produce the millions and millions of cells that make up an individual's body.

Chromosomes and Mendel's Experimental Results

Mendel's two principles can be deduced from the assumption that genes are carried on chromosomes.

In 1902, less than 2 years after the rediscovery of Mendel's findings, Walter Sutton, a young graduate student at Columbia University, made the connection between chromosomes and the properties of inheritance revealed by Mendel's principles. Recall that the first of Mendel's two principles states that an organism's observed characteristics are determined by particles acquired from each of the parents. This concept fits with the idea that genes reside on chromosomes because individuals inherit one copy of each chromosome from each parent. The idea that observed characteristics are determined by genes from both parents is consistent with the observation that mitosis transmits a copy of both chromosomes to every daughter cell, so every cell contains copies of both the maternal and the paternal chromosomes. Mendel's second principle states that genes segregate independently. The observation that meiosis involves the creation of gametes with only one of the two possible chromosomes from each homologous pair is consistent with two notions: (1) that one gene is inherited from each parent and (2) that each of these genes is equally likely to be transmitted to gametes. Not everyone agreed with Sutton, but over the next 15 years, Morgan and his colleagues at Columbia performed many experiments that proved Sutton right.

Varieties of a particular gene are called alleles. Individuals with two copies of the same allele are homozygous; individuals with different alleles are heterozygous.

To see more clearly the connection between chromosomes and the results of Mendel's experiments with the peas, we need to introduce some new terms. The word *gene* is used to refer to the particles carried on chromosomes. Later you will learn that genes are made of a molecule called DNA. **Alleles** are varieties of a single gene. Individuals with two copies of the same allele are **homozygous** for that allele and are called homozygotes. When individuals carry copies of two different alleles, they are said to be **heterozygous** for those alleles and are called heterozygotes.

Consider the case in which all the yellow individuals in the parental generation carry two genes for yellow pea color, one on each chromosome. We will use the symbol *A* for this allele. Thus these plants are homozygous (*AA*) for yellow pea color. All of the individual plants with green peas are homozygous for a different allele, which we denote *a*, so green-pea plants are *aa*. As we will see, this is the only pattern that is consistent with Mendel's model. What happens if we cross two yellow-pea plants with each other, or two green-pea plants with each other? Because they are homozygous, all of the gametes produced by the yellow parents will carry the *A* allele. This means that all of the offspring produced by the crossing of two *AA* parents will also be homozygous for that *A* allele and therefore will also produce yellow peas. Similarly, all of the gametes produced by parents that are homozygous for the *a* allele will carry the *a* allele; when the gametes of two *aa* parents unite, they will produce only *aa* individuals with green peas. Thus we can explain why each type breeds true.

A cross between a homozygous dominant parent and a homozygous recessive parent produces all heterozygotes in the F₁ generation.

Next let's consider the offspring of a mating between a true-breeding green parent and a true-breeding yellow parent (**Figure 2.7**). The green parent produces only *a* gametes, and the yellow parent produces only *A* gametes. Thus every one of their offspring inherits an *a* gamete from one parent and an *A* gamete from the other parent. According to

FIGURE 2.7

In Mendel's experiments, crosses of two true-breeding lines of the garden pea produced offspring that all had yellow peas. All of the gametes produced by homozygous *AA* parents carry the *A* allele. Similarly, all of the gametes produced by homozygous *aa* parents carry the *a* allele. All of the zygotes from an *AA* × *aa* mating get an *A* from one parent and an *a* from the other parent. Thus all F₁ offspring are *Aa*, and because *A* is dominant, they all produce yellow peas.

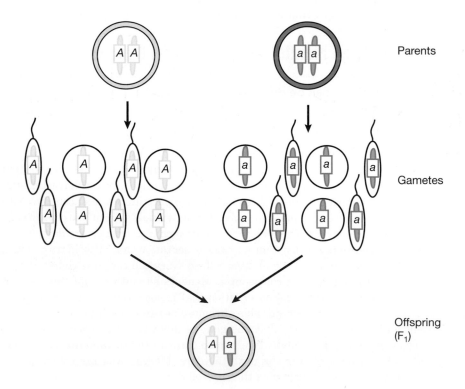

Parents

Gametes

Offspring (F₁)

Mendel's model, all the individuals in the F$_1$ generation will be *Aa*. (We could write this as *aA*, but by convention the *A* usually comes first.) Given that Mendel discovered that all offspring of such crosses have yellow peas, it must be that heterozygotes bear yellow peas. To describe these effects, geneticists use the following four terms:

1. **Genotype** refers to the particular combination of genes or alleles that an individual carries.

2. **Phenotype** refers to the observable characteristics of the organism, such as the color of the peas in Mendel's experiments.

3. The *A* allele is **dominant** because individuals with only one copy of the dominant allele have the same phenotype, yellow peas, that individuals with two copies of that allele have.

4. The *a* allele is **recessive** because it has *no* effect on phenotype in heterozygotes.

As you can see in **Table 2.1**, *AA* and *Aa* individuals have the same phenotype but different genotypes; knowing an individual's observable characteristics, or phenotype, does not necessarily tell you its genetic composition, or genotype.

A cross between heterozygous parents produces a predictable mixture of all three genotypes.

Now consider the second stage of Mendel's experiment: crossing members of the F$_1$ generation with each other to create an F$_2$ generation (**Figure 2.8**). We have seen that every individual in the F$_1$ generation is heterozygous, *Aa*. This means that half of their gametes contain a chromosome with an *A* allele, and the other half of their gametes

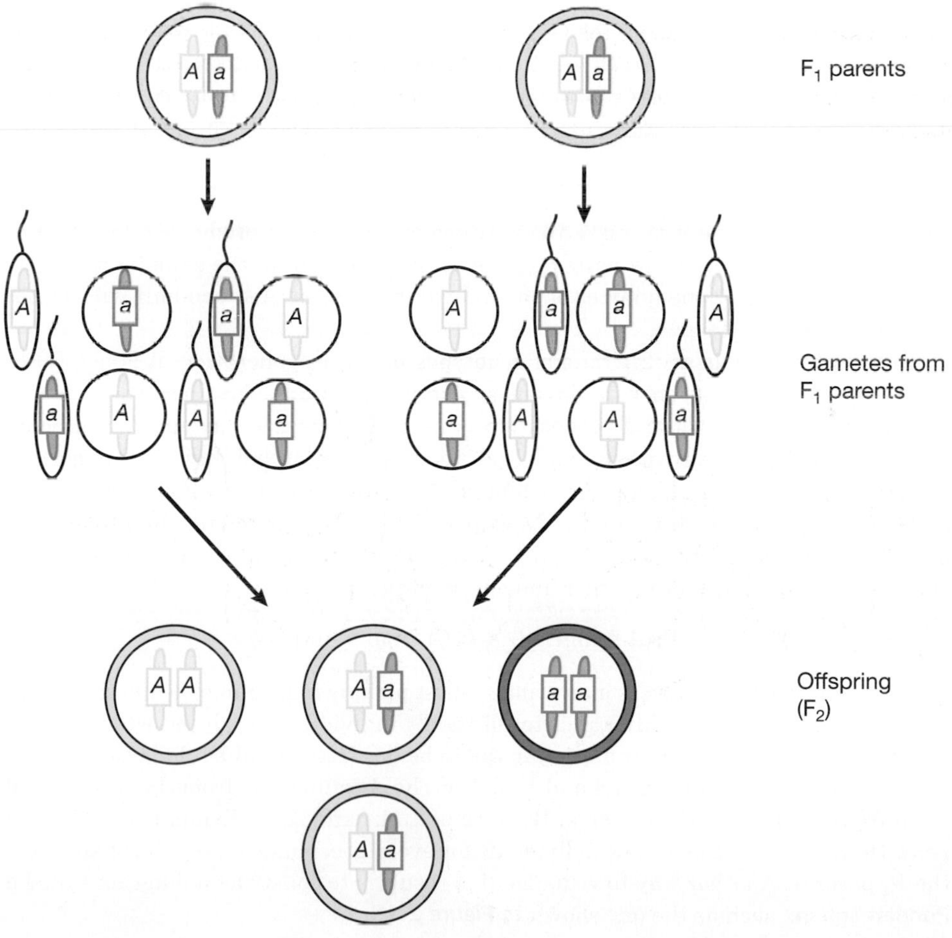

TABLE 2.1	
Genotype	**Phenotype**
AA	Yellow
Aa	Yellow
aa	Green

The relationship between genotype and phenotype in Mendel's experiment on pea color.

FIGURE 2.8

Crosses of the F$_1$ heterozygotes yielded a 3:1 ratio of phenotypes in Mendel's experiments. All the parents are heterozygous, or *Aa*. This means that half of the gametes produced by each parent will carry the *A* allele, and the other half of the gametes produced by each parent will carry the *a* allele. Thus one-quarter of the zygotes will be *AA*, half will be *Aa*, and the remaining quarter will be *aa*. Because *A* is dominant, three-quarters of the offspring ($\frac{1}{4}$ *AA* + $\frac{1}{2}$ *Aa*) will produce yellow peas.

F$_1$ parents

Gametes from F$_1$ parents

Offspring (F$_2$)

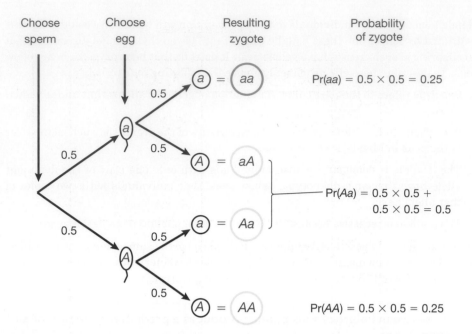

$Pr(aa) = 0.5 \times 0.5 = 0.25$

$Pr(Aa) = 0.5 \times 0.5 + 0.5 \times 0.5 = 0.5$

$Pr(AA) = 0.5 \times 0.5 = 0.25$

FIGURE 2.9

This event tree shows why there is a 1:2:1 genotypic ratio among offspring in the F_2 generation. Imagine forming a zygote by first choosing a sperm and then an egg. The numbers along each branch give the probability that a choice at the previous node ends on that branch. The circles at the end of each branch represent the resulting zygote, and the color of the ring denotes the phenotype: yellow or green peas. The first node represents the choice of a sperm. There is a 50% chance of selecting a sperm that carries the *A* allele and a 50% chance of selecting a sperm that carries the *a* allele. Now choose an egg. Once again, there is a 50% chance of selecting an egg that carries the *A* allele and a 50% chance of selecting an egg that carries the *a* allele. Thus the probability of getting both an *A* sperm and an *A* egg is $0.5 \times 0.5 = 0.25$. Similarly, the chance of getting an *a* egg and an *a* sperm is $0.5 \times 0.5 = 0.25$. The same kind of calculation shows that there is a 25% chance of getting an *a* sperm and an *A* egg and a 25% chance of getting an *A* sperm and an *a* egg. Thus the probability of getting an *Aa* zygote is 50%.

---

contain a chromosome with an a allele. (Remember, meiosis produces haploid gametes.) On average, if we draw pairs of gametes at random from the gametes produced by the F_1 generation, one-quarter of the individuals will be AA, one-half will be Aa, and one-quarter will be aa.

To see why there is a 1:2:1 ratio of genotypes in the F_2 generation, it is helpful to construct an event tree, such as the one shown in **Figure 2.9**. First pick the paternal gamete. Every male in the F_1 generation is heterozygous: He has one chromosome with an A allele and one with an a allele. Thus, there is a probability of 1/2 of getting a sperm that carries an A, and a probability of 1/2 of getting a sperm that carries an a. Suppose you select an A by chance. Now pick the maternal gamete. Once again you have a 1/2 chance of getting an A and a 1/2 chance of getting an a. The probability of getting two As, one from the father and one from the mother, is

$$Pr(AA) = Pr(A \text{ from Dad}) \times Pr(A \text{ from Mom}) = \tfrac{1}{2} \times \tfrac{1}{2} = \tfrac{1}{4}$$

If you repeated this process many times—first picking male gametes and then picking female gametes—roughly a quarter of the F_2 individuals produced would be AA. Similar reasoning shows that a quarter would be aa. Half would be Aa because there are two ways of combining the A and a alleles: The A could come from the mother and the a from the father, or vice versa. Because both AA and Aa individuals have yellow peas, there will be three yellow individuals for every green one among the offspring of the F_1 parents. Another way to visualize this result is to construct a diagram called a **Punnett square**, such as the one shown in **Figure 2.10**.

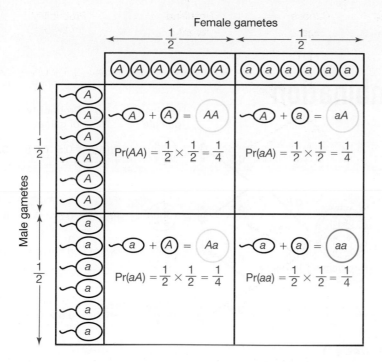

Female gametes

Male gametes

$\Pr(AA) = \frac{1}{2} \times \frac{1}{2} = \frac{1}{4}$

$\Pr(aA) = \frac{1}{2} \times \frac{1}{2} = \frac{1}{4}$

$\Pr(aA) = \frac{1}{2} \times \frac{1}{2} = \frac{1}{4}$

$\Pr(aa) = \frac{1}{2} \times \frac{1}{2} = \frac{1}{4}$

FIGURE 2.10

This diagram, called a Punnett square, provides another way to see why there is a 1:2:1 genotypic ratio among offspring in the F_2 generation. The horizontal axis is divided in half to reflect the equal proportions of eggs carrying A or a alleles, and the vertical axis is divided in half to reflect the equal proportions of sperm carrying A or a alleles. The areas of the squares formed by the intersection of the vertical and horizontal dividing lines give the proportion of zygotes resulting from each of the four possible fertilization events: AA = 0.25 (top left square), Aa = 0.25 (bottom left square), aA = 0.25 (top right square), and aa = 0.25 (bottom right square). It doesn't matter whether the allele comes from the mother or the father; to Aa and aA zygotes are the same, thus zygotes will have genotypes in the ratio 1:2:1.

Linkage and Recombination

Mendel also performed experiments involving two traits, and he believed the experiments showed that separate traits segregate independently.

Mendel also performed experiments that involved two characters, or traits. For example, he crossed individuals that bred true for two traits: pea (or seed) color and seed texture. He crossed plants with smooth, yellow seeds and plants with wrinkled, green seeds. All of the F_1 individuals were smooth and yellow, but the F_2 individuals occurred in the following ratio:

9 smooth yellow:3 smooth green:3 wrinkled yellow:1 wrinkled green

This experiment is important because it demonstrates that sexual reproduction shuffles genes that affect different traits, thereby producing new combinations of traits—a phenomenon called **recombination**. This process is extremely important for maintaining variation in natural populations. We will return to this issue in Chapter 3.

To understand Mendel's experiment, recall that each of the parental traits (seed color and seed texture) breeds true, which means that each parent is homozygous for the alleles controlling the respective traits. For seed color, the yellow line is *AA*, and the green line is *aa*. For seed texture, the smooth line is *BB*, and the wrinkled line is *bb*. Thus in the parental generation (F_0) there are only two genotypes: *AABB* (smooth yellow) and *aabb* (wrinkled green). By the F_2 generation, sexual reproduction has generated all 16 possible genotypes and two new phenotypes: smooth green and wrinkled yellow. The details are given in **A Closer Look 2.1**.

The 9:3:3:1 ratio of phenotypes tells us that the genes determining seed color and the genes determining seed texture each segregate independently, as Mendel's second principle predicts. Thus knowing that a gamete has the *A* allele tells us nothing about whether it will have *B* or *b*; it is equally likely to carry either of these alleles. Mendel's experiments convinced him that all traits segregate independently. Today, however, we know that independent segregation occurs mainly when the traits measured are controlled by genes that reside on different chromosomes.

2.1 **More on Recombination**

The first step to a deeper understanding of recombination is to see the connections among Mendel's two-trait experiment, independent segregation, and chromosomes. Mendel crossed a smooth-yellow (*AABB*) parent and a wrinkled-green (*aabb*) parent to produce members of an F_1 generation. Smooth-yellow parents produce only *AB* gametes, and wrinkled-green parents produce only *ab* gametes; all members of the F_1 generations are, therefore, *AaBb*. If we assume that the genes for seed color and seed texture enter gametes independently, then each of the four possible types—*AB*, *Ab*, *aB*, and *ab*—will be represented by one-quarter of the gametes produced by members of the F_1 generation (**Figure 2.11**).

With this information, we can construct a Punnett square predicting the proportions of each genotype in the F_2 generation (**Figure 2.12**). Once again, we divide the vertical and horizontal axes in proportion to the frequency of each type of gamete, in this case dividing each axis into four equal parts. The areas of the rectangles formed by the intersection of the vertical and horizontal dividing lines give the proportions of zygotes that result from each of the 16 possible fertilization events. Because the area of each cell in this matrix is the same, we can determine the phenotypic ratios of zygotes that result from each of the 16 possible fertilization events in this example by simply counting the squares that contain each phenotype. There are nine smooth-yellow squares, three wrinkled-yellow squares, three smooth-green squares, and one wrinkled-green square. Thus the 9:3:3:1 ratio of phenotypes that Mendel observed

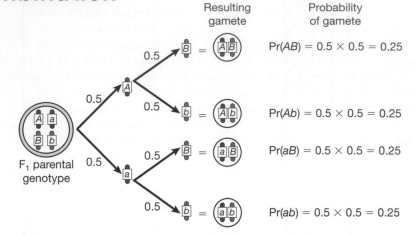

FIGURE 2.11

This event tree shows why an F_1 parent is equally likely to produce all four possible gametes when genes for two traits are carried on different chromosomes. The first node represents the choice of the chromosome carrying the gene for seed color (*A* or *a*), and the second node represents the choice of the chromosome that carries the gene for seed texture (*B* or *b*). The number along each branch gives the probability that the choice at the previous node ends on that branch. The circles at the end of each branch represent the resulting gametes.

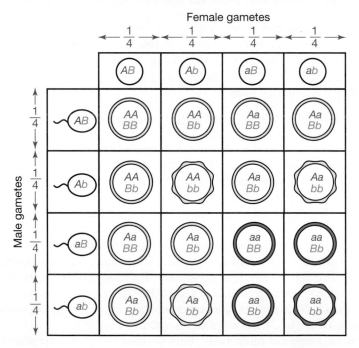

FIGURE 2.12

This Punnett square shows why there is a 9:3:3:1 phenotypic ratio among offspring of the F_2 generation when the genes for the two traits, seed color and seed texture, are carried on different chromosomes. The rings in each square show the color (green or yellow) and seed texture (wrinkled or smooth) of the phenotype associated with each genotype.

was consistent with the assumption that genes on different chromosomes segregate independently. If the genes controlling these traits did not segregate independently, the ratio of phenotypes would be different, as we will see next.

During meiosis, chromosomes frequently become damaged, break, and recombine. This process, called crossing-over, creates chromosomes with combinations of genes not present in the parent (see text). In the preceding example, remember that all members of the F_1 generation were $AaBb$. Now suppose that the locus controlling seed color and the locus controlling seed texture are carried on the same chromosome. Further assume that when the chromosomes are duplicated during meiosis, a fraction r of the time there is crossing-over (**Figure 2.13**) and a fraction $1 - r$ of the time there is no crossing-over. Next, chromosomes segregate independently into gametes. The types of chromosomes present in the parental generation, AB and ab, each occur in a fraction $(1 - r)/2$ of the gametes; the novel, recombinant types Ab and aB each occur in a fraction $r/2$ of the gametes (**Figure 2.14**).

Now we can use a Punnett square to calculate the frequency of each of the 16 possible genotypes in the F_2 generation (**Figure 2.15**). As before, we divide the vertical and horizontal axes in proportion to the relative frequency of each type of gamete, and the areas of the rectangles formed by the intersection of these grid lines give the frequency of each genotype. If recombination rates are low, most members of the F_2 generation will be one of the three genotypes (that is, $AABB$, $AaBb$, or $aabb$), just as if there were only two alleles, AB and ab. If recombination rates are high, however, more of the novel recombinant genotypes will be produced.

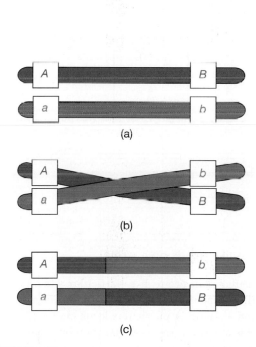

FIGURE 2.13

Crossing-over during meiosis sometimes leads to recombination and produces novel combinations of traits. Suppose that the A allele leads to yellow seeds and the a allele to green seeds, whereas the B allele leads to smooth seeds and the b allele to wrinkled seeds. (a) Here an individual carries one AB chromosome and one ab chromosome. (b) During meiosis, the chromosomes are damaged and crossing-over occurs. (c) Now the A allele is paired with b, and the a allele is paired with B.

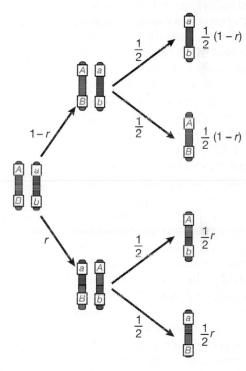

FIGURE 2.14

This event tree shows how to calculate the fraction of each type of gamete that will be produced when genes are carried on the same chromosome. The first node represents whether crossing-over takes place. There is a probability r that crossing-over takes place and produces novel trait combinations and a probability $1 - r$ that there is no crossing-over. At the second node, chromosomes are randomly assigned to gametes. The value given above each branch represents the probability of reaching that branch from the previous node. The likelihood of forming each type of genotype is the product of the probabilities along each pathway.

(continued)

(continued)

FIGURE 2.15

This Punnett square shows how to calculate the frequency of each phenotype among offspring in the F_2 generation if the genes for seed color and seed texture are carried on the same chromosome. The horizontal axis is divided according to the proportion of each type of egg: AB, Ab, aB, and ab. The vertical axis is divided according to the proportion of each type of sperm: AB, Ab, aB, and ab. When genes are carried on the same chromosome, the frequency of each type of gamete is calculated as shown in Figure 2.14. The areas of the rectangles formed by the intersection of the vertical and horizontal dividing lines give the proportions of zygotes that result from each of the 16 possible fertilization events.

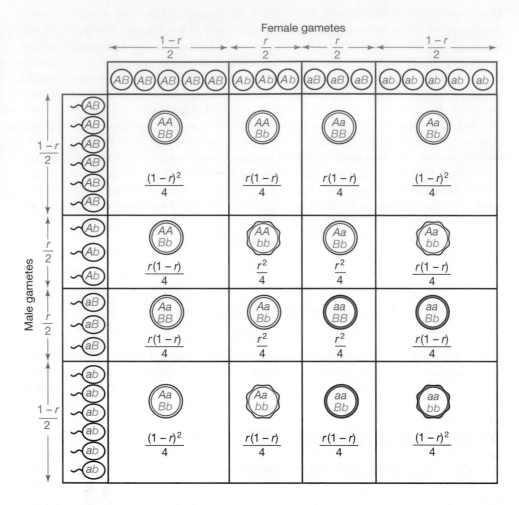

Genes are arranged on chromosomes like beads on a string.

It turns out that the genes for a particular trait occur at a particular site on a particular chromosome. Such a site is called a **locus** (plural, *loci*). Loci are arranged on the chromosomes in a line like beads on a string. A particular locus may hold any of several alleles. The gene for seed color is always in the same position on a particular chromosome, whether it codes for green or for yellow seeds. Keep in mind that the locus for seed color may be located on one chromosome and the locus for seed texture located on another chromosome. All of the genes carried on all of the chromosomes are referred to as the **genome**.

Traits may not segregate independently if they are affected by genes on the same chromosome.

Mendel's conclusion that traits segregate independently is true only if the loci that affect the traits are on different chromosomes, as links between loci on the same chromosome alter the patterns of segregation. When loci for different traits occur close to each other on the same chromosome, they are **linked**; loci on different chromosomes are **unlinked**. You might think that genes at two loci on the same chromosome would always segregate together as if they were a single gene. If this were true, then a gamete receiving a particular chromosome from one of its parents would get all of the same genes that occurred on that chromosome in its parent. There would be no recombination. However, chromosomes frequently tangle and break as they are replicated during meiosis. Thus chromosomes are not always preserved intact, and genes on one chromosome are sometimes shifted from one member of a homologous pair to the other (see Figure 2.13). We call this process **crossing-over**. Linkage reduces, but does not eliminate, the rate of recombination. The rate at which recombination generates novel combinations of genes at two loci on the same chromosome depends on the likelihood that a crossing-over event will occur. If two loci are located close together on the chromosome, crossing-over will be rare and the rate of recombination will be low. If the two loci are located far apart, then crossing over will be common and the rate of recombination will approach the rate for genes on different chromosomes. This process is discussed more fully in A Closer Look 2.1.

Molecular Genetics

Genes are segments of a long molecule called DNA, which is contained in chromosomes.

In the first half of the twentieth century, biologists learned a lot about the cellular events that take place during meiosis and mitosis and began to understand the chemistry of reproduction. For instance, by 1950 it was known that chromosomes contain two structurally complex molecules: protein and **deoxyribonucleic acid**, or **DNA**. It had also been determined that the particle of heredity postulated by Mendel was DNA, not protein, though exactly how DNA might contain and convey the information essential to life was still a mystery. Then in 1953, two young biologists at Cambridge University, Francis Crick and James Watson, made a discovery that revolutionized biology. Using data produced by Rosalind Franklin and Maurice Wilkins of King's College London, they deduced the structure of DNA. Watson and Crick's elucidation of the structure of DNA was the wellspring of a great flood of research that continues to provide a deep and powerful understanding of how life works at the molecular level. We now know how DNA stores information and how this information controls the chemistry of life. This knowledge explains why heredity leads to the patterns Mendel described in pea plants and why there are sometimes new variations.

Understanding the chemical nature of the gene is critical to the study of human evolution: (1) Molecular genetics links biology to chemistry and physics, and (2) molecular methods help us reconstruct the evolutionary history of the human lineage.

Modern molecular genetics, the product of Watson and Crick's discovery, is a field of great intellectual excitement. Every year yields new discoveries about how living things work at the molecular level. This knowledge is deeply important because it links biology to chemistry and physics. One of the grandest goals of science is to provide a single consistent explanatory framework for the way the world works. We want to place evolution in this grand scheme of scientific explanation. It is important to be able to explain not only how new species of plants and animals arise but also how a wide range of phenomena evolves—from the origin of stars and galaxies to the rise of complex societies. Modern molecular biology is profoundly important because it connects physical and geochemical evolution to Darwinian processes.

Molecular genetics also provides data that help biologists and anthropologists reconstruct evolutionary history. As we will see in Chapter 4, comparing the DNA sequences of different species allows us to reconstruct their evolutionary histories. For example, this kind of analysis tells us that humans share a more recent common ancestor with chimpanzees than members of either species share with gorillas. The same data tell us that the last common ancestor of chimpanzees and humans lived between 5 million and 7 million years ago. Patterns of variation in DNA sequences within species are also informative. In Chapter 13, we will see that patterns of genetic variation within the human species allow anthropologists to figure out when the first modern humans left Africa and where they went. In Chapter 14, we will see that the patterns of genetic variation also provide important clues about how natural selection has shaped adaptation within the human species.

In this section, we provide a brief and highly selective introduction to molecular genetics. Our aim is to give enough background to allow students to understand the molecular evidence about human evolution. Students who want a richer understanding of the science of molecular genetics and its important implications for human societies should consult the Further Reading section at the end of this chapter.

Genes Are DNA

DNA is unusually well suited to be the chemical basis of inheritance.

The discovery of the structure of DNA was fundamental to genetics because the structure itself implied how inheritance must work. Each chromosome contains a single DNA molecule roughly 2 m (about 6 ft.) long that is folded up to fit in the nucleus. DNA molecules consist of two long strands twisted around each other in a double helix. Each strand has a "backbone" of alternating sequences of sugar and phosphate molecules. Attached to each sugar is one of four molecules, collectively called **bases**: **adenine**, **guanine**, **cytosine**, or **thymine**. The two strands of DNA are held together by very weak chemical bonds, called *hydrogen bonds*, which connect some of the bases on different strands. Thymine bonds only with adenine, and guanine bonds only with cytosine (**Figure 2.16**).

The repeating four-base structure of DNA allows the molecule to assume a vast number of distinct forms. Each DNA configuration is exactly like a message written in an alphabet with letters that stand for each of the four bases (T for thymine, A for adenine, G for guanine, and C for cytosine). Thus

TCGGTAGTAGTTACGG

is one message and

ATCCGGATGCAATCCA

is another message. Because the DNA in a single chromosome is millions of bases long, there is room for a nearly infinite variety of messages.

These messages would be of no consequence if they were not preserved over time and transmitted faithfully. DNA is uniquely suited to this task. The staggering numbers of DNA molecules that could exist in nature are equally stable chemically. Although DNA is not the only complex molecule with many alternative forms, other molecules have forms that are less stable than others. Such molecules would be unsuitable for carrying information because the messages would become garbled as the molecules changed toward a more stable form. DNA is unusual because all its nearly infinite numbers of forms are equally stable.

In addition to preserving a message faithfully, hereditary material must be replicable. Without the ability to make copies of itself, the genetic message could not be spread to offspring, and natural selection would be impossible. DNA is replicated within cells by a highly efficient cellular machinery. It first unzips the two strands and then, with the help of other specialized molecular machinery, it adds complementary bases to each of the strands until two identical sugar and phosphate backbones are

FIGURE 2.16

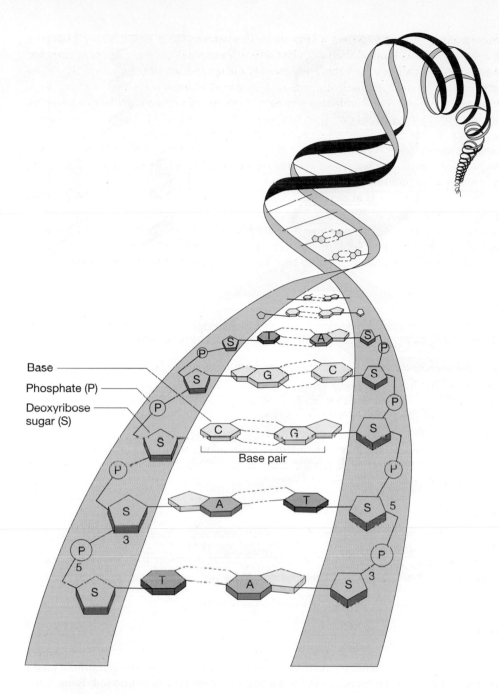

The chemical structure of DNA consists of two long backbones made of alternating sugar and phosphate molecules. One of four bases—adenine (A), guanine (G), cytosine (C), or thymine (T)—is attached to each of the sugars. The two strands are connected to each other by hydrogen bonds (*dotted lines*) between certain pairs of bases. Thymine bonds only to adenine, and guanine bonds only to cytosine.

Base

Phosphate (P)

Deoxyribose sugar (S)

Base pair

built (**Figure 2.17**). There are also mechanisms that "proofread" the copies and correct errors that crop up. These proofreading mechanisms are astoundingly accurate; they miss only one error in every *billion* bases replicated.

The message encoded in DNA affects phenotypes in several ways.

For DNA to play an important role in evolution, different DNA messages must lead to different phenotypes. Here the story gets more complicated because DNA affects phenotypes in several ways. The three most important ways that DNA affects phenotypes are the following:

1. DNA in **protein-coding genes** (or protein-coding sequences) specifies the structure of proteins. Proteins are large molecules made up of a chain of amino acids. They play many important roles in the machinery of life. In particular, many proteins are **enzymes** that regulate much of the biochemical machinery of organisms.

FIGURE 2.17

When DNA is replicated, the two strands of the parent DNA are separated and two daughter DNA strands are formed.

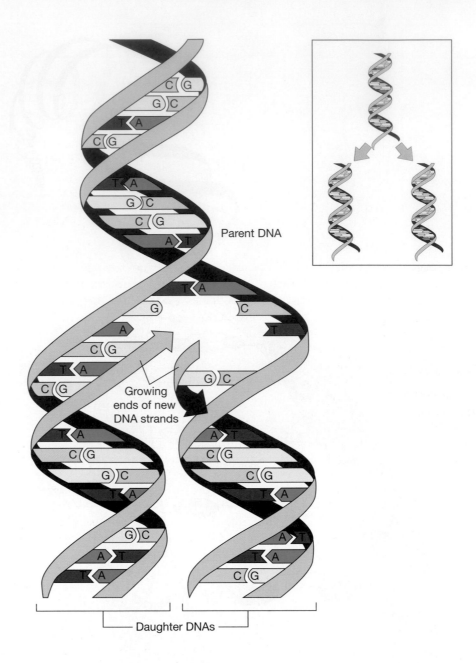

Parent DNA

Growing ends of new DNA strands

Daughter DNAs

2. DNA in **regulatory genes** (or regulatory sequences) determines the conditions under which the message encoded in a protein-coding gene will be expressed. Regulatory genes play a crucial role in shaping the differentiation of cells during development.

3. DNA specifies the structure of several kinds of ribonucleic acid (RNA) molecules that perform important cellular functions. These include RNA molecules that play a crucial role in the machinery of protein synthesis and other RNA molecules that help regulate the expression of other genes.

Some Genes Code for Proteins

Enzymes, an important type of protein, influence an organism's biochemistry.

The cells and organs of living things are made up of a great many chemical compounds, and it is this combination of compounds that gives each organism its characteristic form and structure. All organisms use the same raw materials, but they achieve different results. How does this happen?

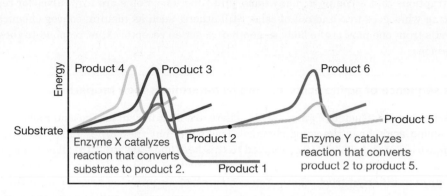

FIGURE 2.18

Enzymes control the chemical composition of cells by catalyzing some chemical reactions but not others. In this hypothetical example, the molecules that provide the initial raw material (called a substrate) of the pathway could undergo four different reactions yielding different molecules, labeled products 1 through 4. However, because enzyme X is present, the reaction that yields product 2 proceeds much more rapidly than the other reactions (note that this reaction has the lowest activation energy), and all of the substrate is converted to that product. Product 2 could then undergo two different reactions yielding products 5 and 6. When enzyme Y is present, it lowers the activation energy, thus causing the reaction yielding product 5 to proceed much more rapidly, and only product 5 is produced. In this way, enzymes link products and reactants into pathways that satisfy particular chemical functions, such as the extraction of energy from glucose.

The answer is that enzymes present in the cell determine what the raw materials are transformed into when cells are built (**Figure 2.18**). The best way to understand how enzymes determine the characteristics of organisms is to think of an organism's biochemical machinery as a branching tree. Enzymes act as switches to determine what will happen at each node and thus what chemicals will be present in the cell. For example, glucose serves as a food source for many cells, meaning that it provides energy and a source of raw materials for the construction of cellular structures. Glucose might initially undergo any one of a great many slow-moving reactions. The presence of particular catalytic enzymes will determine which reactions occur rapidly enough to alter the chemistry of the cell. For example, some enzymes lead to the metabolism of glucose and the release of its stored energy. At the end of the first branch, there is another node representing all of the reactions that could involve the product(s) of the first branch. Again, one or more enzymes will determine what happens next.

This picture has been greatly simplified. Real organisms take in many kinds of compounds, and each compound is involved in a complicated tangle of branches that biochemists call pathways. Real **biochemical pathways** are complex. One set of enzymes causes glucose to be shunted to a pathway that yields energy. A different set of enzymes causes the glucose to be shunted to a pathway that binds glucose molecules together to form glycogen, a starch that functions to store energy. The presence of a different set of enzymes would lead to the synthesis of cellulose, a complex molecule that provides the structural material in plants. Enzymes play roles in virtually all cellular processes—from the replication of DNA and division of cells to the contraction and movement of muscles.

Proteins play several other important roles in the machinery of life.

Some proteins have crucial structural functions in living things. For example, your hair is made up mainly of a protein called keratin, and your ligaments and tendons are strengthened by another protein called collagen. Other proteins act as tiny mechanical contraptions that accomplish many important functions. Some are tiny valves for regulating what goes into and out of cells. Still others, such as insulin, convey chemical signals from one part of the body to another or act as receptors that respond to these messages.

The sequence of amino acids in proteins determines their properties.

Proteins are constructed of amino acids. There are 20 different amino acid molecules. All **amino acids** have the same chemical backbone, but they differ in the chemical composition of the side chain connected to this backbone (**Figure 2.19**). The sequence of amino acid side chains, called the **primary structure** of the protein, is what makes one protein different from others. You can think of a protein as a very long railroad train in which there are 20 kinds of cars, each representing a different amino acid. The primary structure is a list of the types of cars in the order that they occur.

To do their work, proteins are folded in complex ways. The three-dimensional shape of the folded protein, called the **tertiary structure**, is crucial to its catalytic function.

FIGURE 2.19

All amino acids share the same chemical backbone, here colored *blue*. They differ according to the chemical structure of the side group attached to the backbone, here colored *tan*.

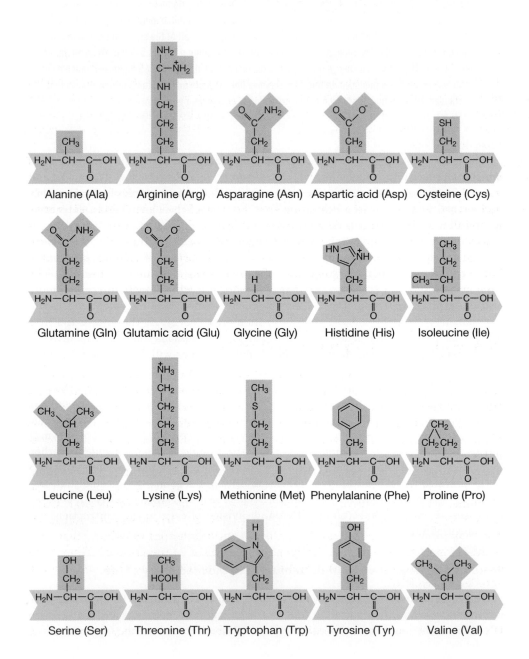

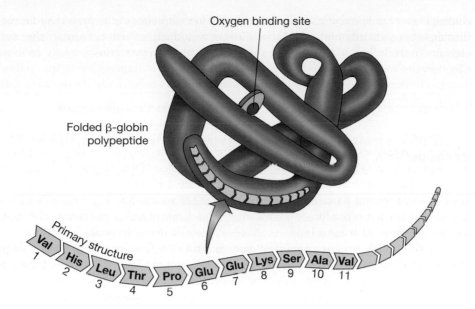

Oxygen binding site

Folded β-globin polypeptide

Primary structure

Val 1　His 2　Leu 3　Thr 4　Pro 5　Glu 6　Glu 7　Lys 8　Ser 9　Ala 10　Val 11

FIGURE 2.20

The primary and tertiary structures of hemoglobin, a protein that transports oxygen on red blood cells, are illustrated. The primary structure is the sequence of amino acids making up the protein. The tertiary structure is the way the protein folds into three dimensions.

The way the protein folds depends on the sequence of amino acid molecules that make up its primary sequence. This means that the function of an enzyme depends on the sequence of the amino acids that make up the enzyme. (Proteins also have secondary structure and sometimes quaternary structure; to keep things simple, we will ignore those levels here.)

These ideas are illustrated in **Figure 2.20**, which shows the folded shape of part of a **hemoglobin** molecule, a protein that transports oxygen from the lungs to the tissues via red blood cells. As you can see, the protein folds into a roughly spherical glob, and oxygen is bound to the protein near the center of the glob. **Sickle-cell anemia**, a condition common among people in West Africa and among African Americans, is caused by a single change in the primary sequence of amino acids in the hemoglobin molecule. Glutamic acid is the sixth amino acid in normal hemoglobin molecules, but in people afflicted with sickle-cell anemia, valine is substituted for glutamic acid. This single substitution changes the way that the molecule folds and reduces its ability to bind oxygen.

DNA specifies the primary structure of protein.

Now we return to our original question: How does the information contained in DNA—its sequence of bases—determine the structure of proteins? Remember that DNA encodes messages in a four-letter alphabet. Researchers determined that these letters are combined into three-letter "words" called **codons**, each of which specifies a particular amino acid. Because there are four bases, there are 64 possible three-letter combinations for codons (4 possibilities for the first letter of a codon times 4 possibilities for the second letter of a codon times 4 possibilities for the third letter of a codon, or $4 \times 4 \times 4 = 64$). Of these codons, 61 are used to code for the 20 amino acids that make up proteins. For example, the DNA codons CGT, CGG, CGA, and CGC all code for alanine; CTA and CTG code for asparagine; and so on. The remaining three codons are "punctuation marks" that mean either "start, this is the beginning of the protein" or "stop, this is the end of the protein." Thus if you can identify the base pairs, it is a simple matter to determine what proteins are encoded on the DNA.

You might be wondering why several codons code for the same amino acid. This redundancy serves an important function. Because several processes can damage DNA and cause one base to be substituted for another, redundancy decreases the chance that a random change will alter the primary sequence of the protein produced. Proteins represent complex adaptations, so we would expect most changes to be deleterious (harmful). But because the code is redundant, many substitutions have no effect on the message of a particular stretch of DNA. The importance of this redundancy is underscored by the fact that the most common amino acids are the ones with the most codon variants.

Before DNA is translated into proteins, its message is first transcribed into messenger RNA.

DNA can be thought of as a set of instructions for building proteins, but the real work of synthesizing proteins is performed by other molecules. The first step in the translation of DNA into protein occurs when a facsimile of one of the strands of DNA, which will serve as a messenger or chemical intermediary, is made, usually in the cell's nucleus. This copy is **ribonucleic acid**, or **RNA**. RNA is similar to DNA, except that it has a slightly different chemical backbone, and the base **uracil** (denoted U) is substituted for thymine. RNA comes in several forms, many of which aid in protein synthesis. The form of RNA used in this first step is **messenger RNA (mRNA)**.

The ribosome then synthesizes a particular protein by reading the mRNA copy of the gene.

Meanwhile, amino acid molecules are bound to a different kind of RNA called **transfer RNA (tRNA)**. Each tRNA molecule has a triplet of bases, called an **anticodon**, at a particular site (**Figure 2.21**). Each type of tRNA is bound to the amino acid whose mRNA codon binds to the anticodon on the tRNA. For example, one of the mRNA codons for the amino acid alanine is the base sequence GCA, which binds only to the anticodon CGU. Thus the tRNA with the anticodon CGU binds only to the amino acid alanine.

The next step in the process involves the **ribosomes**. Ribosomes are small cellular **organelles**. Composed of protein and RNA, organelles are cellular components that perform a particular function, analogous to the way organs such as the liver perform a function for the body as a whole. The mRNA first binds to a ribosome at a binding site and then moves through the binding site one codon at a time. As each codon of mRNA enters the binding site, a tRNA with a complementary anticodon is drawn from the complex soup of chemicals inside the cell and bound to the mRNA. The amino acid bound to the other end of the tRNA is then detached from the tRNA and added to one end of the growing protein chain. The process repeats for each mRNA codon, continuing until the end of the mRNA molecule passes through the ribosome. Voilà! A new protein is ready for action.

In eukaryotes, the DNA that codes for proteins is interrupted by noncoding sequences called introns.

So far, our description of protein synthesis applies to almost all organisms. However, most of this information was learned through the study of *Escherichia coli,* a bacterium that lives in the human gut. Like other bacteria, *E. coli* belongs to a group of organisms called the **prokaryotes** because it does not have a chromosome or a cell nucleus. In prokaryotes, the DNA sequence that codes for a particular protein is uninterrupted. A stretch of DNA is copied to RNA and then translated into a protein. For many years, biologists thought that the same would be true of **eukaryotes** (organisms such as plants, birds, and humans that have chromosomes and a cell nucleus).

Beginning in the 1970s, new recombinant DNA technology allowed molecular geneticists to study eukaryotes. These studies revealed that in eukaryotes, the segment of DNA that codes for a protein is almost always interrupted by at least one—and

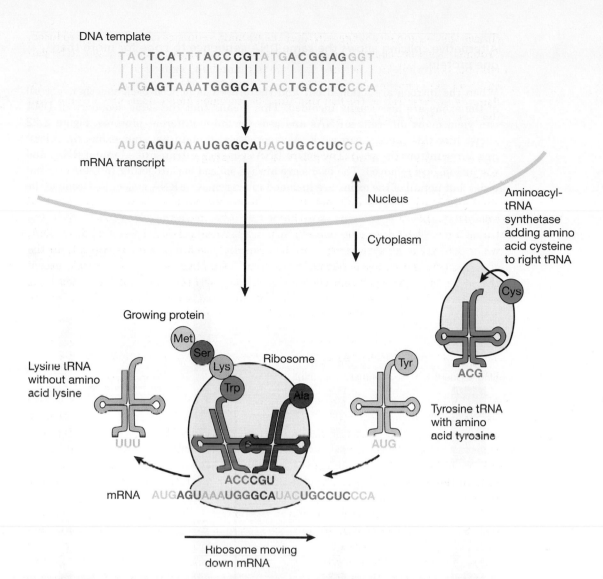

DNA template

TACTCATTTACCCGTATGACGGAGGGT

ATGAGTAAATGGGCATACTGCCTCCCA

AUGAGUAAAUGGGCAUACUGCCUCCCA

mRNA transcript

Nucleus

Cytoplasm

Aminoacyl-tRNA synthetase adding amino acid cysteine to right tRNA

Cys

Growing protein

Met
Ser
Lys
Trp
Ala

Lysine tRNA without amino acid lysine

UUU

Ribosome

Tyr

ACG

Tyrosine tRNA with amino acid tyrosine

AUG

ACCCGU

mRNA   AUGAGUAAAUGGGCAUACUGCCUCCCA

Ribosome moving down mRNA

FIGURE 2.21

The information encoded in DNA determines the structure of proteins in the following way: Inside the nucleus, an mRNA copy is made of the original DNA template (top DNA strand). The mRNA is coded in three-base codons. Here each mRNA codon is given a different color. For example, the sequence AUG codes for the start of a protein and the amino acid methionine, the sequence AGU codes for serine, and the sequence AAA codes for lysine. The mRNA then migrates to the cytoplasm. In the cytoplasm, special enzymes called aminoacyl-tRNA synthetases locate a specific kind of tRNA and attach the amino acid whose mRNA codon will bind to the anticodon on the tRNA. For example, the mRNA codon for cysteine is UGC, and the appropriate anticodon is ACG because U binds to A, G binds to C, and C binds to G. In this diagram, matching mRNA codons and tRNA anticodons are given the same color. The initiation of protein assembly is complicated and involves specialized enzymes. Once the process is started, each codon of the mRNA binds to the ribosome. Then the matching tRNA is bound to the mRNA, the amino acid is transferred to the growing protein, the tRNA is released, the ribosome shifts to the next mRNA codon, and the process is repeated.

sometimes many—noncoding sequences called **introns**. (The protein-coding sequences are called **exons**.) Protein synthesis in eukaryotes includes one additional step not mentioned in our discussion so far: After the entire DNA sequence is copied to make an mRNA molecule in the nucleus, the intron-based parts of the mRNA are snipped out and the mRNA molecule is spliced back together. Then the mRNA is exported out of the nucleus, and protein synthesis takes place.

Alternative splicing allows the same DNA sequence to code for more than one protein.

When the introns are snipped out and the mRNA is spliced back together, not all of the exons are necessarily included. This means that the same sequence of DNA can yield many different mRNAs and code for many different proteins. **Figure 2.22** shows how this process can work. A hypothetical protein has four exons (in colors) and three introns (in gray). The entire DNA sequence is transcribed into mRNA, and the introns are removed. The exons are always spliced back together in their original order, but not all of the exons are included in every new mRNA molecule. Because the final mRNA can include different exons, different proteins can be produced.

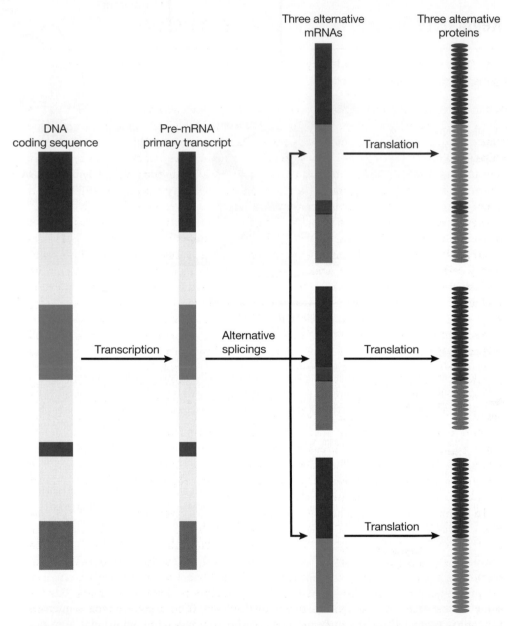

FIGURE 2.22

A hypothetical protein has four exons (*red, green, blue,* and *tan*) and three introns (*gray*). The entire DNA sequence is transcribed into a pre-mRNA molecule. After introns are snipped out, the exons are spliced back together to yield three different mRNAs, and these are translated into three different proteins.

Alternative splicing seems to be important. It has been estimated that more than half of the protein-coding sequences in the human genome yield more than one protein, and some yield as many as 10 proteins. Some biologists think that this flexibility is important in multicellular organisms in which different types of proteins are required for the function of diverse cell types. Adherents of this hypothesis think that introns proliferated at the origin of eukaryotes. Other biologists think that introns are weakly deleterious bits of DNA that are maintained in eukaryotes because their population sizes are much smaller than those of prokaryotes. As we will see in Chapter 3, small populations are subject to a random nonadaptive evolutionary process called genetic drift that usually works against natural selection. In bacterial populations numbering in the hundreds of millions, drift is weak and selection eliminates the introns. In the much smaller eukaryotic populations, drift is strong enough to maintain introns.

Regulatory Sequences Control Gene Expression

The DNA sequence in regulatory genes determines when protein-coding genes are expressed.

Gene regulation in the bacterium *Escherichia coli* provides a good example of how the DNA sequence in regulatory genes interacts with the environment to control gene expression. *Escherichia coli* uses the sugar glucose as its primary source of energy. When glucose runs short, *E. coli* can switch to other sugars, such as lactose, but this switch requires several enzymes that allow lactose to be metabolized. The genes for these lactose-specific enzymes are always present, but they are not expressed when there is plenty of glucose available. It would be wasteful to produce these enzymes if they were not needed. The genes for making enzymes that allow the metabolism of lactose to glucose are expressed only in environments in which doing so is necessary: when glucose is in short supply *and* enough lactose is present. Two regulatory sequences are located near the protein-coding genes that encode the amino acid sequence for three enzymes necessary for lactose metabolism. When there is no lactose in the environment, a **repressor** protein binds to one of the two regulatory sequences. The repressor interferes with RNA polymerase, an enzyme necessary for transcription thereby preventing the protein-coding genes from being transcribed (**Figure 2.23a**). When lactose is present, the repressor protein changes shape and does not bind to the DNA in the regulatory sequence (**Figure 2.23b**). If glucose is present the activator protein does not bind, and the protein-coding genes are expressed at a low rate. However, when glucose is absent the **activator** protein binds to a second regulatory DNA sequence, greatly increasing the rate at which the protein-coding genes are transcribed (**Figure 2.23c**). The specific DNA sequences of the regulatory genes control whether the repressor and activator proteins that control DNA transcription bind to the DNA. This means that the sequence of DNA in regulatory genes affects the phenotype and creates variation. Therefore, regulatory genes are subject to natural selection, just as are protein-coding genes.

In humans and other eukaryotes, the expression of a given protein-coding gene is often affected by many regulatory sequences, which are sometimes located quite far from the coding sequence that they regulate. The proteins that bind to these regulatory genes can interact in complex ways, so that multiple proteins bound to sequences of DNA at widely separated sites may act to activate or repress a particular protein-coding gene. However, the basic principle is usually the same: Under some circumstances, the DNA sequence in the regulatory gene binds to a protein that in turn affects the expression of a particular protein-coding gene.

The existence of multiple regulatory sequences allows for **combinatorial control** of gene expression. In *E. coli,* the combination of a repressor and an activator means that the genes necessary to metabolize lactose are synthesized in an environment that contains lactose *but lacks* glucose. This is a simple example. Combinatorial control of

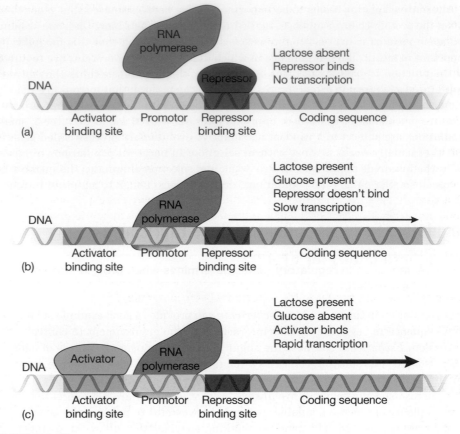

FIGURE 2.23

This diagram shows how regulatory sequences binding activator and repressor proteins control the expression of the enzymes necessary to digest lactose in *E. coli*. (a) When lactose is absent, the repressor protein binds to the regulatory sequence, thus preventing the enzyme RNA polymerase from creating an mRNA from the DNA template. (b) When lactose is present the repressor no longer binds to the DNA. However, if glucose is also present, the activator does not bind and the genes necessary for metabolism of lactose are expressed, but at a low (basal) level. (c) When lactose is present and glucose is absent, the activator protein can bind, greatly increasing the rate at which the RNA polymerase binds and increasing the level of gene expression.

gene expression is often more complex in eukaryotes. There may be dozens of activators that allow gene expression to respond to environmental differences in complex and subtle ways.

Gene regulation allows cell differentiation in complex multicellular organisms such as humans.

Complex, multicellular organisms are made of many kinds of cells, each with its own specific chemical machinery: Cells in the liver and pancreas secrete digestive enzymes, nerve cells carry electrical signals throughout the body, muscle cells respond to these signals by performing mechanical work, and so on. Nonetheless, in a single individual, all cells have the same genetic composition. The cells differ in their function because different genes are activated in different cell types.

The development of the vertebrate nervous system provides an example of how this process works. At a certain point in the development of all vertebrate embryos, cells that are going to give rise to the spinal cord differentiate to form the neural tube. Special cells at one end of this structure secrete a molecule called "Sonic Hedgehog" (after the video-game character of the same name). Cells close to the source of this signaling molecule experience a high concentration of Sonic Hedgehog; those more distant experience a lower concentration. The concentrations of Sonic Hedgehog affect the expression of genes in these future nerve cells. Low concentrations lead to the expression of genes that destine the cells to become motor neurons, which control muscles; higher concentrations lead to the expression of genes that cause cells to become the neurons of the brain and spinal cord.

A single signal can trigger the complex series of events that transform a cell into a liver cell or a nerve cell because the expression of one gene causes a different set of genes to be expressed, and this in turn causes additional sets of genes to be expressed. For example, the PAX6 gene codes for a regulatory protein that is important in the differentiation of the cells in the developing eye. Artificially activating this gene in one cell in a fruit fly antenna leads to the synthesis of a regulatory protein that sets off a cascade of gene expression, which eventually involves the expression of about 2,500 other genes and the development of an extra eye on the fly's antenna.

Not All DNA Codes for Protein

Some DNA sequences code for functional RNA molecules.

Messenger RNA is not the only important form of RNA. Some RNA molecules bind together with proteins to perform a variety of cellular functions. Ribosomes are an example, as are **spliceosomes**, the organelles that splice the mRNA in eukaryotes after the introns have been snipped out. It has long been known that some DNA codes for these RNA molecules, but it was thought that the bulk of the genome (98%) that lies outside of exons was just "junk DNA" with no real function. However, at least half of the DNA that lies in introns and between genes seems to be expressed as **noncoding RNA (ncRNA)**. One form of ncRNA, short RNA segments called **microRNAs (miRNAs)**, plays an important role in regulating the translation of mRNA into protein. When these sequences are transcribed, the resulting miRNAs move outside the nucleus where, after much processing, they bind to complementary mRNA molecules and affect the rate at which the mRNA is translated into protein, providing an alternative mechanism for regulation of gene expression. Much like regulatory sequences, some miRNAs play an important role in regulating development and cell differentiation in complex organisms such as humans. Longer RNA segments called **long noncoding RNAs (lncRNAs)** have a wide variety of functions and may be especially important in regulating the expression of genes during development.

Chromosomes also contain long strings of simple repeated sequences.

Introns are not the only kind of DNA that is not involved in the synthesis of proteins. Chromosomes also contain a lot of DNA that is composed of simple repeated patterns. For example, in fruit flies there are long segments of DNA composed of adenine and thymine in the following monotonous five-base pattern:

...ATAATATAATATAATATAATATAATATAATATAATATAAT...

In all eukaryotes, simple repeated sequences of DNA are found in particular sites on particular chromosomes.

At this point, there is some danger of losing sight of the genes amid the discussion of introns, exons, and repeat sequences. A brief reprise may be helpful.

In summary, chromosomes contain an enormously long molecule of DNA. Genes are short segments of this DNA. After suitable editing, a gene's DNA is transcribed into mRNA, which in turn is translated into a protein whose structure is determined by the gene's DNA sequence. Proteins determine the properties of living organisms by selectively catalyzing some chemical reactions and not others and by forming some of the structural components of cells, organs, and tissues. Genes with different DNA sequences lead to the synthesis of proteins with different catalytic behavior and structural characteristics. Many changes in the morphology or behavior of organisms can be traced back to variations in the proteins and genes that build them. It is important to remember that evolution has a molecular basis. The changes in the genetic constitution of populations that we will explore in Chapter 3 are grounded in the physical and chemical properties of molecules and genes discussed here.

CHAPTER REVIEW

Key Terms

variants (p. 26)
crosses (p. 26)
F_0 generation (p. 26)
F_1 generation (p. 26)
F_2 generation (p. 26)
genes (p. 26)
gametes (p. 27)
independent assortment (p. 27)
chromosome (p. 28)
nucleus (p. 28)
mitosis (p. 28)
diploid (p. 28)
meiosis (p. 29)
homologous chromosomes (p. 29)
haploid (p. 29)
zygote (p. 29)
alleles (p. 30)

homozygous (p. 30)
heterozygous (p. 30)
genotype (p. 31)
phenotype (p. 31)
dominant (p. 31)
recessive (p. 31)
Punnett square (p. 32)
recombination (p. 33)
locus (p. 36)
genome (p. 36)
linked (p. 36)
unlinked (p. 37)
crossing-over (p. 37)
deoxyribonucleic acid (DNA) (p. 37)
bases (p. 38):
    adenine (p. 38)
    guanine (p. 38)

cytosine (p. 38)
thymine (p. 38)
protein-coding genes (p. 39)
enzymes (p. 39)
regulatory genes (p. 40)
biochemical pathways (p. 41)
proteins (p. 42)
amino acids (p. 42)
primary structure (p. 42)
tertiary structure (p. 42)
hemoglobin (p. 43)
sickle-cell anemia (p. 43)
codons (p. 43)
ribonucleic acid (RNA) (p. 44)
uracil (p. 44)
messenger RNA (mRNA) (p. 44)

transfer RNA (tRNA) (p. 44)
anticodon (p. 44)
ribosomes (p. 44)
organelles (p. 44)
prokaryotes (p. 44)
eukaryotes (p. 44)
introns (p. 45)
exons (p. 45)
repressor (p. 47)
activator (p. 47)
combinatorial control (p. 47)
spliceosome (p. 49)
noncoding RNA (ncRNA) (p. 49)
microRNA (miRNA) (p. 49)
long noncoding RNA (lncRNA) (p. 49)

Study Questions

1. Explain why Mendel's principles follow from the mechanics of meiosis.

2. Mendel did experiments in which he kept track of the inheritance of seed texture (wrinkled or smooth). First he created true-breeding lines: Parents with smooth seeds produced offspring with smooth seeds, and parents with wrinkled seeds produced offspring with wrinkled seeds. Then when he crossed wrinkled and smooth peas from these true-breeding lines, all of the offspring were smooth. Which trait was dominant? What happened when he crossed members of the F_1 generation? What would have happened had he backcrossed members of the F_1 generation with individuals from a true-breeding line (like their parents) with smooth seeds? What about crosses between the F_1 generation and the wrinkled-seed true-breeding line?

3. Mendel also did experiments in which he kept track of two traits. For example, he created true-breeding

lines of smooth-green and wrinkled-yellow peas, and then he crossed these lines to produce an F_1 generation. What were the F_1 individuals like? He then crossed the members of the F_1 generation to form an F_2 generation. Assuming that the seed-color locus and the seed-texture locus are on different chromosomes, calculate the ratio of each of the four phenotypes in the F_2 generation. Calculate the approximate ratios, assuming that the two loci are very close together on the same chromosome.

4. Many animals, such as birds and mammals, have physiologic mechanisms that enable them to maintain their body temperature well above the air temperature. They are called *homeotherms*. *Poikilotherms*, such as snakes and other reptiles, regulate body temperature by moving closer to or farther away from sources of heat. Use what you have learned about chemical reactions to develop a hypothesis that explains why animals have mechanisms for controlling body temperature.

5. Why is DNA so well suited to carrying information?

6. Recall the discussion about crossing-over, and try to explain why introns might increase the rate of recombination between surrounding exons.

7. Explain how Mendel's work helped to resolve the problem that Darwin's critics had raised concerning how variation is maintained in populations.

8. Explain three ways in which DNA influences phenotypes.

9. All of the cells in your body carry the same genes, but cells in your lungs work differently from cells in your kidneys or your leg muscles. Briefly explain the mechanisms that make this differentiation possible.

10. How does the information contained in DNA determine the structure of proteins?

Further Reading

Barton, N. H., D. E. G. Briggs, J. A. Eisen, D. B. Goldstein, and N. H. Patel. 2007. *Evolution.* Woodbury, N.Y.: Cold Spring Harbor Press.

Bergstrom, C. and L. A. Dugatkin. 2016. *Evolution.* 2nd ed. New York: Norton.

Maynard Smith, J. 1998. *Evolutionary Genetics.* 2nd ed. New York: Oxford University Press.

Olby, R. C. 1985. *Origins of Mendelism.* 2nd ed. Chicago: University of Chicago Press.

Snustad, D. P., and M. J. Simmons. 2011. *Principles of Genetics.* 6th ed. Hoboken, N.J.: Wiley.

Watson, J. D., T. A. Baker, S. P. Bell, et al. 2013. *Molecular Biology of the Gene.* 7th ed. San Francisco: Pearson/Benjamin Cummings.

Review this chapter with personalized, interactive questions through INQUIZITIVE.

3

THE MODERN SYNTHESIS

CHAPTER OBJECTIVES

By the end of this chapter you should be able to:

A Describe the genetic composition of populations in terms of the frequencies of genes and genotypes.

B Explain how sexual reproduction changes genotypic frequencies, sometimes leading to the Hardy–Weinberg equilibrium.

C Describe how natural selection changes gene frequencies in populations.

D Explain how population genetics explains the maintenance of variation, which is necessary for evolution to occur.

E Explain how natural selection shapes learned behavior.

F Explain why evolution does not always produce adaptations.

Population Genetics

Evolutionary change in a phenotype reflects change in the underlying genetic composition of the population. When we discussed Mendel's experiments in Chapter 2, we briefly introduced the distinction between genotype and phenotype: Phenotypes are the observable characteristics of organisms, and genotypes are the underlying genetic compositions. There need not be a one-to-one correspondence between genotype and phenotype. For example, there were only two seed-color phenotypes in Mendel's pea populations (yellow and green), but there were three genotypes (*AA*, *Aa*, and *aa*; **Figure 3.1**).

From what we now know about the genetic nature of inheritance, it is also clear that evolutionary processes must entail changes in the genetic composition of populations. When evolution alters the morphology of a trait such as the finch's beak, there must be a corresponding change in the distribution of genes that control beak development within the population. To understand how Mendelian genetics solves each of Darwin's difficulties, we need to look more closely at what happens to genes in populations that are undergoing natural selection. This is the domain of **population genetics**.

Genes in Populations

Biologists describe the genetic composition of a population by specifying the frequency of alternative genotypes.

It's easiest to see what happens to genes in populations if we consider a trait that is controlled by one gene operating at a single locus on a chromosome. Phenylketonuria (PKU), a potentially debilitating, genetically inherited disease in humans, is determined by the substitution of one allele for another at a single locus. Individuals who are homozygous for the PKU allele are missing a crucial enzyme in the biochemical pathway that allows people to metabolize the amino acid phenylalanine. If the disease is not treated, phenylalanine builds up in the bloodstream of children with PKU and leads to severe mental retardation. Fortunately, treatment is possible. People with PKU raised on a special low-phenylalanine diet can develop normally and lead normal lives as adults.

How do evolutionary processes control the distribution of the PKU allele in a population? The first step in answering this question is to characterize the distribution of the harmful allele by specifying the **genotypic frequency**, which is simply the fraction of the population that carries that genotype. Using established biochemical methods, population geneticists can determine an individual's genotype for a specific genetic locus. Suppose that we perform a census of a population of 10,000 individuals and determine the number of individuals with each genotype. Let's label the normal allele *A* and the deleterious PKU allele *a*. **Table 3.1** shows the number of individuals with each genotype and the frequencies of each genotype in this hypothetical population. (In most real populations, the frequency of individuals homozygous for the

FIGURE 3.1

Peas were a useful subject for Mendel's botanical experiments because they have several dichotomous traits. For example, pea seeds are either yellow or green but not an intermediate color.

TABLE 3.1

Genotype	Number of Individuals	Frequency of Genotype
aa	2,000	freq(*aa*) = 2,000/10,000 = 0.2
Aa	4,000	freq(*Aa*) = 4,000/10,000 = 0.4
AA	4,000	freq(*AA*) = 4,000/10,000 = 0.4

The distribution of individuals with each of the three genotypes in a population of 10,000.

PKU allele is only about 1 in 10,000. We have used larger numbers here to make the calculations simpler.)

Genotypic frequencies must add up to 1.0 because every individual in the population has to have a genotype. We keep track of the frequencies of genotypes, rather than the numbers of individuals with each genotype, because the frequencies provide a description of the genetic composition of populations that is independent of population size. This makes it easy to compare populations of different sizes.

One goal of evolutionary theory is to determine how genotypic frequencies change through time.

A variety of events in the lives of plants and animals may act to change the frequency of alternative genotypes in populations from generation to generation. Population geneticists categorize these processes into several evolutionary mechanisms, or forces. The most important mechanisms are sexual reproduction, natural selection, mutation, and genetic drift. In the rest of this section, we will see how sexual reproduction and natural selection alter the frequencies of genes and genotypes, and later in the chapter we will return to consider the effects of mutation and genetic drift.

How Random Mating and Sexual Reproduction Change Genotypic Frequencies

The events that occur during sexual reproduction can lead to changes in genotypic frequencies in a population.

First, let's consider the effects of the patterns of inheritance that Mendel observed. Imagine that men and women do not choose their mates according to whether they are afflicted with PKU but mate randomly with respect to the individual's genotype for PKU. It is important to study the effects of random mating because, for most genetic loci, mating is random. Even though humans may choose their mates with care and might even avoid mates with particular genetic characteristics, such as PKU, they cannot choose mates with a particular allele at each locus because there are more than 22,000 genetic loci in humans. Random mating between individuals is equivalent to the random union of gametes. In this sense, it's not really different from oysters shedding their eggs and sperm into the ocean, where chance dictates which gametes will form zygotes.

The first step in determining how sexual reproduction affects genotypic frequencies is to calculate the frequency of the PKU allele in the pool of gametes.

We can best understand how segregation affects genotypic frequencies by breaking the process into two steps. In the first step, we determine the frequency of the PKU allele among all the gametes in the mating population. Remember that the a allele is the PKU allele and the A allele is the "normal" allele. Table 3.1 gives the frequency of each of the three genotypes among the parental generation. We want to use this information to determine the genotypic frequencies among the F_1 generation. First, we calculate the frequencies of the two types of alleles in the pool of gametes. (The frequency of an allele is also referred to as its **gene frequency**.) Let's label the frequency of A as p and the frequency of a as q. (Because there are only two alleles, $p + q = 1$.) If all individuals produce the same number of gametes, then we can calculate q as follows:

$$q = \frac{\text{no. of } a \text{ gametes}}{\text{total no. of gametes}}$$

Note that this is simply the definition of a frequency. We can calculate the values of the numerator and denominator of this fraction from information we already know. Because a gametes can be produced only by aa and Aa individuals, the total number of a gametes is simply the sum of the number of Aa and aa parents multiplied by the number of a gametes that each parent produces. The denominator is the number of gametes per parent multiplied by the total number of parents. Hence

$$q = \frac{\left(\begin{array}{c}\text{no. of } a \\ \text{gametes per} \\ aa \text{ parent}\end{array}\right)\left(\begin{array}{c}\text{no. of } aa \\ \text{parents}\end{array}\right) + \left(\begin{array}{c}\text{no. of } a \\ \text{gametes per} \\ Aa \text{ parent}\end{array}\right)\left(\begin{array}{c}\text{no. of } Aa \\ \text{parents}\end{array}\right)}{(\text{no. of gametes per parent})(\text{total no. of parents})} \tag{3.1}$$

To simplify this equation, let's first examine the terms that involve numbers of individuals. Remember that the population size is 10,000 individuals. This means that the number of aa parents is equal to the frequency of aa parents multiplied by 10,000. Similarly, the number of Aa parents is equal to the frequency of Aa parents multiplied by 10,000. Now we examine the terms that involve numbers of gametes. Suppose each parent produces two gametes. For aa individuals, both gametes contain the a allele, so the number of a gametes per aa parent is 2. For Aa individuals, half of the gametes will carry the a allele and half will carry the A allele, so here the number of a gametes per Aa parent is 0.5×2. Now we substitute all these values into Equation (3.1) to get

$$q = \frac{2[\text{freq}(aa) \times 10{,}000] + (0.5 \times 2)[\text{freq}(Aa) \times 10{,}000]}{2 \times 10{,}000}$$

We can reduce this fraction by dividing the top and bottom by $2 \times 10{,}000$, which yields the following formula for the frequency of the a allele in the pool of gametes:

$$q = \text{freq}(aa) + 0.5 \times \text{freq}(Aa) \tag{3.2}$$

Notice that this form of the formula contains neither the population size nor the average number of gametes per individual in the population. Under normal circumstances, the population size and the average number of gametes per individual do not matter. This means you can use this expression as a general formula for calculating gene frequencies among the gametes produced by any population of individuals, as long as the genetic locus of interest has only two alleles. It is important to keep in mind that the formula results from applying Mendel's laws and from counting the number of a gametes produced.

By using Equation (3.2) and values from Table 3.1, for this particular population we get $q = 0.2 + (0.5 \times 0.4) = 0.4$. Because $p + q$ must sum to 1.0, p (the frequency of the A gametes) must be 0.6. Notice that the frequency of each allele in the pool of gametes is the same as the frequency of the same allele among parents.

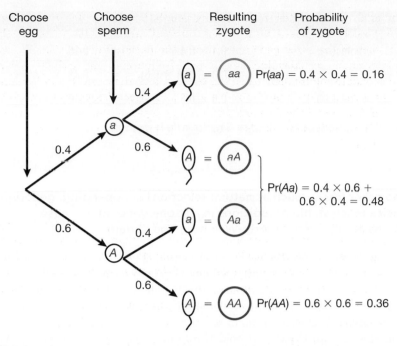

Choose egg   Choose sperm   Resulting zygote   Probability of zygote

a = aa Pr(aa) = 0.4 × 0.4 = 0.16

A = aA

Pr(Aa) = 0.4 × 0.6 + 0.6 × 0.4 = 0.48

a = Aa

A = AA Pr(AA) = 0.6 × 0.6 = 0.36

FIGURE 3.2

This event tree shows how to calculate the frequency of each genotype among the zygotes created by the random union of gametes. In this pool of gametes, the frequency of a is 0.4 and the frequency of A is 0.6. The first node represents the choice of an egg. There is a 40% chance that the egg will carry the a allele and a 60% chance it will carry the A allele. The second node represents the choice of the sperm. Once again, there is a 40% chance of drawing an a-bearing sperm and a 60% chance of drawing an A-bearing sperm. We can calculate the probability of each genotype by computing the probability of taking a path through the tree. For example, the probability of getting an aa zygote is 0.4 × 0.4 = 0.16.

The next step is to calculate the frequencies of all the genotypes among the zygotes.

Now that we have calculated the frequency of the PKU allele among the pool of gametes, we can determine the frequencies of the genotypes among the zygotes. If each zygote is the product of the random union of two gametes, the process of zygote formation can be schematically represented in an event tree (**Figure 3.2**) similar to the one we used to represent Mendel's crosses (see Figure 2.9). First we select a gamete—say, an egg. The probability of selecting an egg carrying the a allele is 0.4 because this is the frequency of the a allele in the population. Now we randomly draw a second gamete, the sperm. Again, the probability of getting a sperm carrying an a allele is 0.4. The chance that these two randomly chosen gametes are both a is 0.4 × 0.4, or 0.16. Figure 3.2 also shows how to compute the probabilities of the other two genotypes. Note that the sum of the three genotypes always equals 1.0, because all individuals must have a genotype.

If we form many zygotes by randomly drawing gametes from the gamete pool, we will obtain the genotypic frequencies shown in **Table 3.2**. Compare the frequencies of each genotype in the parental population (Table 3.1) with the frequencies of each genotype in the F_1 generation. The genotypic frequencies have changed because the processes of independent segregation of alleles into gametes and random mating alter the distribution of alleles in zygotes. As a result, genotypic frequencies between the F_0 generation and the F_1 generation are altered. [Note, however, that the frequencies of the two *alleles*, a and A, have not changed; you can check this out by using Equation (3.2).]

TABLE 3.2

freq(*aa*)	=	0.4×0.4	=	0.16
freq(*Aa*)	=	$(0.4 \times 0.6) + (0.4 \times 0.6)$	=	0.48
freq(*AA*)	=	0.6×0.6	=	0.36

The distribution of genotypes in the population of zygotes in the F_1 generation.

When no other forces (such as natural selection) are operating, genotypic frequencies reach stable proportions in just one generation. These proportions are called the Hardy–Weinberg equilibrium.

If no other processes act to change the distribution of genotypes, the set of genotypic frequencies in Table 3.2 will remain unchanged in later generations. That is, if members of the F_1 generation mate at random, the distribution of genotypes in the F_2 generation will be the same as the distribution of genotypes in the F_1 generation. The fact that genotypic frequencies remain constant was recognized independently by the British mathematician Godfrey Harold Hardy and the German physician Wilhelm Weinberg in 1908, and these constant frequencies are now called the **Hardy–Weinberg equilibrium**. As we will see later in this chapter, the realization that not just sexual reproduction alone alters phenotypic and genotypic frequencies was the key to understanding how variation is maintained.

In general, the Hardy–Weinberg proportions for a genetic locus with two alleles are

$$\text{freq}(aa) = q^2$$
$$\text{freq}(Aa) = 2pq$$
$$\text{freq}(AA) = p^2 \tag{3.3}$$

where q is the frequency of allele a, and p is the frequency of allele A. Using a Punnett square, **Figure 3.3** shows how to calculate these frequencies.

If no other processes act to change genotypic frequencies, the Hardy–Weinberg equilibrium frequencies will be reached after only one generation and will remain

FIGURE 3.3

This Punnett square shows a second way to calculate the frequency of each type of zygote when there is random mating. The horizontal axis is divided according to the proportion of *A* and *a* eggs and is thus divided into fractions *p* and *q*, where $q = 1 - p$. The vertical axis is divided according to the proportion of sperm carrying each allele, and again it is divided into fractions *p* and *q*. The areas of the rectangles formed by the intersection of the vertical and horizontal dividing lines give the proportion of zygotes resulting from each of the four possible fertilization events. The area of the square containing *aa* zygotes is q^2, the area of the square containing *AA* zygotes is p^2, and the total area of the two rectangles containing *Aa* zygotes is $2pq$.

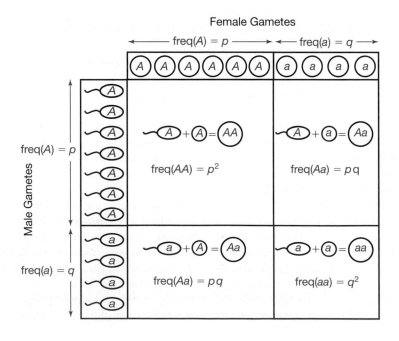

3.1 Genotypic Frequencies after Two Generations of Random Mating

From Equation (3.2) given in the text, we know that the frequency of the gametes produced by adults in the second generation carrying the a allele will be

$$q = \text{freq}(aa) + 0.5 \times \text{freq}(Aa)$$
$$= 0.16 + 0.5 \times 0.48$$
$$= 0.16 + 0.24 = 0.4$$

Because there are only two alleles, the frequency of A is 0.6. As we asserted in the text, the frequencies of the two alleles remain unchanged, and the frequencies of the gametes produced by random mating among members of this generation are the same as in the previous generation.

The calculations for these frequencies are the same as those shown in Table 3.2.

---

unchanged thereafter. Moreover, if the Hardy–Weinberg proportions are altered by chance, the population will return to Hardy–Weinberg proportions in one generation. If this seems an unlikely conclusion, work through the calculations in **A Closer Look 3.1**, showing how to calculate the genotypic frequencies after a second episode of segregation and random mating. As you can see, genotypic frequencies remain constant.

We have seen that sexual reproduction and random mating can change the distribution of genotypes, which will reach equilibrium after one generation. We have also seen that neither process changes the frequencies of alleles. Clearly, sexual reproduction and random mating alone cannot lead to evolution over the long run. We now turn to a process that can produce changes in the frequencies of alleles: natural selection.

How Natural Selection Changes Gene Frequencies

If different genotypes are associated with different phenotypes and those phenotypes differ in their ability to reproduce, then the alleles that lead to the development of the favored phenotype will increase in frequency.

Genotypic frequencies in the population will remain at the Hardy–Weinberg proportions [Equation (3.3)] as long as all genotypes are equally likely to survive and produce gametes. For PKU, this prediction is approximately correct in wealthy countries where the disease can be treated. However, this assumption will not be valid in environments in which PKU is not treated. Suppose a catastrophe interrupted the supply of modern medical care to our hypothetical population just after a new generation of zygotes was formed. In this case, virtually none of the PKU individuals (genotype aa) would survive to reproduce. Suppose the frequency of the PKU allele was 0.4 among zygotes and the population was in Hardy–Weinberg equilibrium. Let's compute q', the frequency of the PKU allele among adults, which is given by

$$q' = \frac{\text{no. of } a \text{ gametes produced by adults in the next generation}}{\text{total no. of gametes produced by adults in the next generation}} \quad (3.4)$$

We have already calculated the frequency of each of the genotypes among zygotes just after conception (see Table 3.2), but if PKU is a lethal disease without treatment, not all of these individuals will survive. Now we need to calculate the frequency of a gametes after selection, which we can do by expanding Equation (3.4) as follows:

$$q' = \frac{\begin{pmatrix} \text{no. of } a \\ \text{gametes per} \\ aa \text{ parent after} \\ \text{selection} \end{pmatrix} \begin{pmatrix} \text{no. of } aa \\ \text{parents} \\ \text{after} \\ \text{selection} \end{pmatrix} + \begin{pmatrix} \text{no. of } a \\ \text{gametes per} \\ Aa \text{ parent after} \\ \text{selection} \end{pmatrix} \begin{pmatrix} \text{no. of } Aa \\ \text{parents} \\ \text{after} \\ \text{selection} \end{pmatrix}}{(\text{no. of gametes per parent})(\text{total no. of parents after selection})} \quad (3.5)$$

Because none of the aa individuals will survive, the first term in the numerator will be equal to zero. If we assume that all of the AA and Aa parents survive, then the number of parents after selection is $10{,}000 \times [\text{freq}(AA) + \text{freq}(Aa)]$, and thus Equation (3.5) can be simplified to

$$q' = \frac{(0.5 \times 2)(0.48 \times 10{,}000)}{2 \times (0.36 + 0.48) \times 10{,}000} = 0.2857$$

This calculation shows that the frequency of the PKU allele in adults is 0.2857, a big decrease from 0.4.

Several important lessons can be drawn from this example:

- Selection cannot produce change unless there is variation in the population. If all the individuals were homozygous for the normal allele, gene frequencies would not change from one generation to the next.

- Selection does not operate directly on genes and does not change gene frequencies directly. Instead, natural selection changes the frequency of different phenotypes. In this case, individuals with PKU cannot survive without treatment. Selection decreases the frequency of the PKU allele because it is more likely to be associated with the lethal phenotype.

- The strength and direction of selection depend on the environment. In an environment with medical care, the strength of selection against the PKU allele is negligible.

It is also important to see that although this example shows how selection can change gene frequencies, it does not yet show how selection can lead to the evolution of new adaptations. Here, all phenotypes were present at the outset; all selection did was change their relative frequency.

The Modern Synthesis

The Genetics of Continuous Variation

When Mendelian genetics was rediscovered, biologists at first thought it was incompatible with Darwin's theory of evolution by natural selection.

Darwin believed that evolution proceeded by the gradual accumulation of small changes. But Mendel and the biologists who elucidated the structure of the genetic system around the turn of the twentieth century were dealing with genes that had a noticeable effect on the phenotype. The substitution of one allele for another in Mendel's peas changed pea color. Genetic substitutions at other loci had similarly visible effects on the shape of the pea seeds and the height of the pea plants. Genetics seemed to prove that inheritance was fundamentally discontinuous, and early-twentieth-century geneticists such as Hugo de Vries and William Bateson argued that this fact could not

[a]

[b]

[c]

FIGURE 3.4

Shown here are the three architects of the modern synthesis, which explained how Mendelian genetics could be used to account for continuous variation: (a) Ronald A. Fisher, (b) J. B. S. Haldane, and (c) Sewall Wright.

be reconciled with Darwin's idea that adaptation occurs through the accumulation of small variations. If one genotype produces short plants and the other two genotypes produce tall plants, then there will be no intermediate types, and the size of pea plants cannot change in small steps. In a population of short plants, tall ones must be created all at once by mutation, not gradually lengthened by selection. Most biologists of the time found these arguments convincing and, consequently, Darwinism was in decline during the early twentieth century.

Mendelian genetics and Darwinism were eventually reconciled, resulting in a body of theory that solved the problem of explaining how variation is maintained.

In the early 1930s, British biologists Ronald A. Fisher and J. B. S. Haldane and American biologist Sewall Wright showed how Mendelian genetics could be used to explain continuous variation (**Figure 3.4**). We will see how their insights led to the resolution of the two main objections to Darwin's theory: (1) the absence of a theory of inheritance and (2) the problem of accounting for how variation is maintained in populations. When the theory of Wright, Fisher, and Haldane was combined with Darwin's theory of natural selection and with modern field studies by biologists such as Theodosius Dobzhansky, Ernst Mayr, and George Gaylord Simpson, a powerful explanation of organic evolution emerged. This body of theory with its supporting empirical evidence is now called the **modern synthesis**.

Continuously varying characters are affected by genes at many loci, each locus having only a small effect on phenotype.

To see how the theory of Wright, Fisher, and Haldane works, let's start with an unrealistic, but instructive, case. Suppose that there is a measurable, continuously varying character such as beak depth, and suppose that two alleles, + and −, operating at a single genetic locus control the character. We'll assume that the gene at this locus influences the production of a hormone that stimulates beak growth and that each allele leads to production of a different amount of the growth hormone. Let's say that each "dose" of the + allele increases the beak depth, whereas a − dose decreases it. Thus + + individuals have the deepest beaks, − − individuals have the shallowest beaks, and + − individuals have intermediate beaks. In addition, suppose the frequency of the + allele in the population is 0.5. Now we use the Hardy–Weinberg rule [Equation (3.3)] to calculate the frequencies of different beak depths in the population. A quarter of the population will have deep beaks (+ +), half will have intermediate beaks (+ −), and the remaining quarter will have shallow beaks (− −; **Figure 3.5**).

This does not look like the smooth, bell-shaped distribution of beak depths that the Grants observed on Daphne Major (see Chapter 1). If beak depth were controlled by a single locus, natural selection could not increase beak depth in small increments. But it turns out that the beak morphology is actually affected by genes at more than one locus. Arkhat Abzhanov, a biologist at Imperial College in London, and the Grants have identified several genes that affect beak morphology in Darwin's finches. The level of expression of one of these genes, BMP4, affects beak depth, and the level of expression of two other genes, BMP2 and BMP7, influences overall beak size. More recently, several other genes have been shown to affect beak size and shape. So, it is reasonable to think about what would happen if genes at a second locus on a different chromosome also affected beak depth. As before, we assume there is a + allele that leads to larger beaks and a − allele that leads to smaller beaks. Using the Hardy–Weinberg proportions and assuming the independent segregation of chromosomes, we can show that there are now more types of genotypes, and the distribution of phenotypes begins to look

FIGURE 3.5

The hypothetical distribution of beak depth, assuming that beak depth is controlled by a single genetic locus with two alleles that occur at equal frequency. The height of each bar represents the fraction of birds in the population with a given beak depth. The heights of the bars are computed by means of the Hardy–Weinberg formula. The birds with the smallest beaks are homozygous for the − allele and have a frequency of $0.5 \times 0.5 = 0.25$. The birds with intermediate beaks are heterozygotes and have a frequency of $2(0.5 \times 0.5) = 0.5$. The birds with the deepest beaks are homozygous for the + allele and thus have a frequency of $0.5 \times 0.5 = 0.25$.

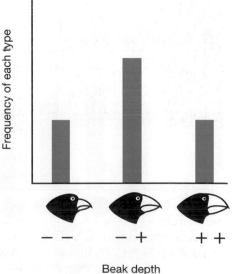

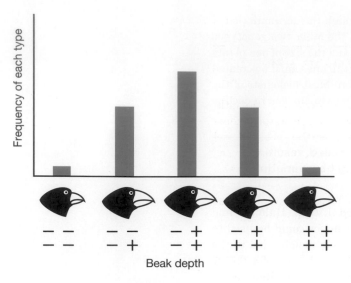

FIGURE 3.6

When beak depth is controlled by two loci, each locus having two alleles that occur with equal frequency, intermediate types are observed. The frequencies are calculated by means of a more advanced form of the Hardy–Weinberg formula, on the assumption that the genotypes at each locus are independent.

FIGURE 3.7

When beak depth is controlled by three loci, each locus having two alleles that occur with equal frequency, the distribution of phenotypes begins to resemble a bell-shaped curve.

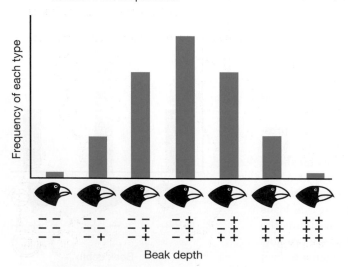

somewhat smoother (**Figure 3.6**). Now imagine that a third locus controls the calcium supply to the growing beak, and that once again there is a + allele leading to larger beaks and a − allele leading to smaller beaks. As **Figure 3.7** shows, the distribution of beak depths is now even more like the bell-shaped distribution seen in nature. However, small gaps still exist in this distribution of beaks. If genes were the only influence on beak depth, we might expect to see a more broken distribution than the Grants actually observed in the Galápagos.

The observed distribution of phenotypic values is a smooth, bell-shaped curve for two reasons. First, the phenotypic expression of all characters, whether affected by one locus or by many loci, depends on the environment in which the organism develops, leading to **environmental variation**. For example, the size of a bird's beak will depend on how well the bird is nourished during its development. When there is only one locus and the effect of an allelic substitution at that locus is large, environmental variation is not important; it is easy to distinguish the phenotypes associated with different genotypes. This was the case for seed color in Mendel's peas. But when there are many loci and each has a small effect on the phenotype, environmental variation tends to blur together the phenotypes associated with different genotypes. You cannot be sure whether you are measuring a + + + − − − bird that matured in a good environment or a + + + + − − bird that grew up in a poor environment. Second, complex characters such as beak depth are probably affected by genes at many more than three loci. As we will see in Chapter 14, new tools have allowed geneticists to identify the loci that affect human height. So far 423 loci have been implicated, and many more are probably involved. Such a large number of loci would lead to a very smooth, bell-shaped distribution of phenotypes.

Darwin's view of natural selection is easily incorporated into the genetic view that evolution typically results from changes in gene frequencies.

Darwin knew nothing about genetics, and his theory of adaptation by natural selection was framed in purely phenotypic terms: There is a struggle for existence, there is phenotypic variation that affects survival and reproduction, and this phenotypic variation is heritable. In Chapter 1, we saw how this theory could explain the adaptive changes in beak depth in a population of Darwin's finches on the Galápagos Islands. As we saw earlier in this chapter, however, population geneticists take the seemingly different view that evolution means changes in allelic frequencies by natural selection. But these two views of evolution are easily reconciled. Suppose that Figure 3.7 gave the distribution of beak depths before the drought on Daphne Major. Remember that individuals with deep beaks were more likely to survive the drought and reproduce than individuals with smaller beaks. Figure 3.7 illustrates that individuals with deep beaks are more likely to have + alleles at the three loci assumed to affect beak depth. Thus at each locus, + + individuals have deeper beaks on average than + − individuals, which in turn have deeper beaks on average than − − individuals. Because individuals with deeper beaks had higher fitness, natural selection would favor the + alleles at each of the three loci affecting beak depth, and the + alleles would increase in frequency.

How Variation Is Maintained

Genetics readily explains why the phenotypes of offspring tend to be intermediate between phenotypes of their parents.

Recall from Chapter 1 that the blending model of inheritance appealed to nineteenth-century thinkers because it explained the fact that for most continuously varying characters, offspring are intermediate between their parents. However, the genetic model developed by Fisher, Wright, and Haldane is also consistent with this fact. To see why, consider a cross between the individuals with the biggest and smallest beaks: $(+ + + + + +) \times (- - - - - -)$. All of the offspring will be $(+ - + - + -)$, intermediate between the parents because, during development, the effects of $+$ and $-$ alleles are averaged. Other matings will produce a distribution of different kinds of offspring, but intermediate types will tend to be the most common.

There is no blending of genes during sexual reproduction.

We know from population genetics that, in the absence of selection (and other factors to be discussed later in this chapter), genotypic frequencies reach equilibrium after one generation and the distribution of phenotypes does not change. Furthermore, we know that sexual reproduction produces no blending in the genes, even though offspring may appear to be intermediate between their parents. This is because genetic transmission involves faithful copying of the genes themselves and their reassembly in different combinations in zygotes. The only blending that occurs takes place at the level of the expression of genes in phenotypes. The genes themselves remain distinct physical entities (**Figure 3.8**).

These facts do not completely solve the problem of the maintenance of variation because selection tends to deplete variation. When selection favors birds with deeper beaks, we might expect $-$ alleles to be replaced at all three loci affecting the trait, leaving a population in which every individual has the genotype $+ + + + + +$. There would still be phenotypic variation due to environmental effects, but without genetic variation there can be no further adaptation.

Mutation slowly adds new variation.

Genes are copied with amazing fidelity, and several molecular repair mechanisms protect their messages from random degradation. Every once in a while, however, when a mistake in copying is made and goes unrepaired, a new allele is introduced into the population. In Chapter 2, we learned that genes are pieces of DNA. Certain forms of ionizing radiation (such as X-rays) and certain kinds of chemicals damage the DNA and alter the message that it carries. These changes are called **mutations**, and they add variation to a population by continuously introducing new alleles, some of which may produce novel phenotypic effects that selection can assemble into adaptations. Although rates of mutation are very low—ranging from 1 in 100,000 to 1 in 10 million per locus per gamete in each generation—this process plays an important role in generating variation.

Low mutation rates can maintain variation because a lot of variation is protected from selection.

For characters that are affected by genes at many loci, low rates of mutation can maintain variation in

FIGURE 3.8

The blending model of inheritance assumes that the hereditary material is changed by mating. When red and white parents are crossed to produce a pink offspring, the blending model posits that the hereditary material has mixed, so that when two pink individuals mate, they produce only pink offspring. According to Mendelian genetics, however, the effects of genes are blended in their *expression* to produce a pink phenotype, but the genes themselves remain unchanged. Thus when two pink parents mate, they can produce white, pink, or red offspring.

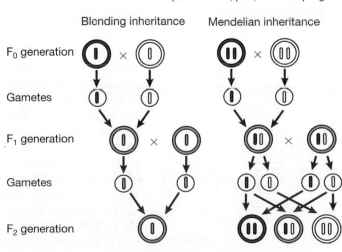

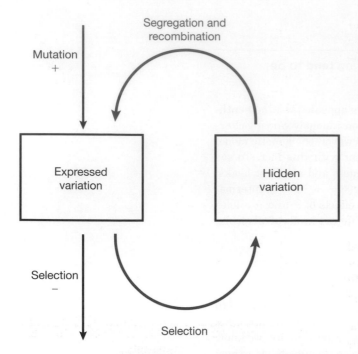

FIGURE 3.9

There are two pools of genetic variation: hidden and expressed. Mutation adds new genetic variation to the pool of expressed variation, and selection removes it. Segregation and recombination shuffle variation back and forth between the two pools within each generation.

populations. This is possible because many genotypes generate intermediate phenotypes that are favored by stabilizing selection. If individuals with a variety of genotypes are equally likely to survive and reproduce, a considerable amount of variation is protected (or hidden) from selection. To make this concept clearer, let's consider the action of stabilizing selection on beak depth in the medium ground finch. Remember that stabilizing selection occurs when birds with intermediate-size beaks have higher fitness than birds with deeper or shallower beaks. If beak depth is affected by genes at three loci, as shown in Figure 3.7, individuals with several genotypes may develop very similar, intermediate-size beaks. For example, individuals that are + + at one locus affecting the trait may be − − at another locus and thus may have the same phenotype as an individual that is + − at both loci. Thus, when many loci affect a single trait, only a small fraction of the genetic variability present in the population is expressed phenotypically. As a result, selection removes variation from the population very slowly. The processes of segregation and recombination slowly shuffle and reshuffle the genome and expose the hidden variation to natural selection in later generations (**Figure 3.9**). This process solves Darwin's other dilemma; because a considerable amount of variation is protected from selection, a very low mutation rate can maintain variation despite the depleting action of selection.

Hidden variation explains why selection can move populations far beyond their initial range of variation.

Recall from Chapter 1 Fleeming Jenkin's argument that Darwin's theory could not explain cumulative evolutionary change because it provided no account of how a population could evolve beyond the initial range of variation present. Selection would cull away all of the small-beaked finches, Jenkin would have argued, but it could never make the average beak bigger than the biggest beak initially present. If this argument were correct, then selection could never lead to cumulative, long-term change.

Jenkin's argument was wrong because it failed to take hidden variation into account. Hidden variation is always present in continuously varying traits. Let's suppose that environmental conditions favor larger beaks. When beak depth is affected by genes at many loci, the birds in a population with the deepest beaks do not carry all + alleles. They carry a lot of + alleles and some − alleles. When the finches with the shallowest beaks die, alleles leading to small beaks are removed from the breeding population. As a result, the frequency of + alleles at every locus increases. But because even the deepest-beaked individuals had some − alleles, a huge amount of variation remains. This variation is shuffled through the process of sexual reproduction. Because + alleles become more common, more of these alleles are likely to be combined in the genotype of a single individual. The greater the proportion of + alleles in an individual, the larger the beak will be. Thus the biggest beak will be larger than the biggest beak in the previous generation. In the next generation, the same thing happens again: The deepest-beaked individuals still carry − alleles, but there are fewer than before. As a result, the biggest beaks can be bigger than any beaks in the previous generation.

This process can go on for many generations. An experiment on oil content in corn begun at the University of Illinois Agricultural Experiment Station in 1896 provides a good example. Researchers selected for high and low oil content in corn and, as **Figure 3.10** shows, each generation showed significant change. In the initial population of 163 ears of corn, oil content ranged from 4% to 6%. After nearly 80 generations of selection, both the high and low values for oil content far exceeded the initial range of variation. Researchers then even reversed the direction of adaptation by taking plants from the high-oil line and selecting for low oil content.

Natural Selection and Behavior

The evolution of mate guarding in the soapberry bug illustrates how flexible behavior can evolve.

So far we have considered the evolution of morphological characters, such as beak depth and eye morphology, that do not change once an individual has reached adulthood. In much of this book, we will be interested in the evolution of the behavior of humans and other primates. Behavior is different from morphology in an important way: It is flexible, and individuals adjust their behavior in response to their circumstances. Some people think natural selection cannot account for flexible responses to environmental contingencies because natural selection acts only on phenotypic variation resulting from genetic differences. Although this view is very common (particularly among social scientists), it is incorrect. To see why, let's consider an elegant empirical study that illustrates exactly how natural selection can shape flexible behavioral responses.

The soapberry bug (*Jadera haematoloma*), a seed-eating insect found in the southeastern United States, has been studied by biologist Scott Carroll of the Institute for Contemporary Evolution. (It's okay to call them "bugs" because they are members of the insect order Hemiptera, the true bugs.) Adult soapberry bugs are bright red and black and 1 to 1.5 cm (0.5 in.) long (**Figure 3.11**). They gather in huge groups near the plants they eat. During mating, males mount females and copulate. The transfer of sperm typically takes about 10 minutes. However, males remain in the copulatory position, sometimes for hours, securely anchored to females by large genital hooks. This behavior is called **mate guarding**. Biologists believe that the function of mate guarding is to prevent other males from copulating with the female before she lays her eggs. When a female mates with several males, they share in the paternity of the eggs that she lays. By guarding his mate and preventing her from mating with other males, a male can increase his reproductive success. However, mate guarding also has a cost: A male cannot find and copulate with other females while guarding a mate. The relative magnitude of the costs and benefits of mate guarding depends on the **sex ratio** (the relative numbers of males and females). When the ratio of males to females is high, males have little chance of finding another female, and guarding is the best strategy. When females are more common than males, the chance of finding an unguarded female increases, so males may benefit more from looking for additional mates than from guarding.

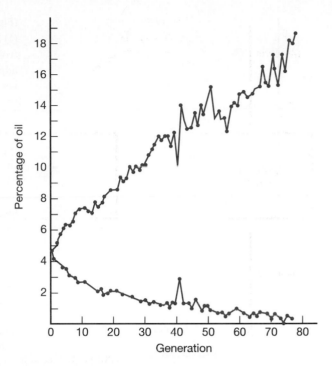

FIGURE 3.10

Selection can carry a population far beyond the original range of variation because at any given time a lot of genetic variation is not expressed as phenotypic variation. The range of variation in the oil content of corn at the beginning of this experiment was 4% to 6%. After about 80 generations, oil content in the line selected for high oil content had increased to 19%, and oil content in the low-oil line had decreased to less than 1%.

FIGURE 3.11

The male soapberry bug on the left is guarding the (larger) female to his right. Biologist Scott Carroll painted numbers on the bugs to identify individuals.

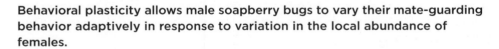

Behavioral plasticity allows male soapberry bugs to vary their mate-guarding behavior adaptively in response to variation in the local abundance of females.

In populations of soapberry bugs in western Oklahoma, the sex ratio is quite variable. In some places, there are equal numbers of males and females; in others, there are twice as many males as females. Males guard their mates more often in situations where females are rare than in situations where they are common

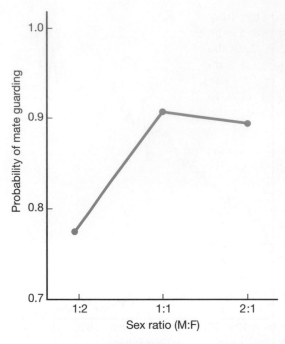

FIGURE 3.12

The probability of mate guarding is a function of sex ratio in this population of the soapberry bug. Males are less likely to guard their mates when females are relatively more common than when they are relatively rare.

(**Figure 3.12**). Two possible mechanisms might produce this pattern: (1) Males in populations with high sex ratios might differ genetically from males in populations with low sex ratios or (2) males might adjust their behavior in response to the local sex ratio. To distinguish between these two possibilities, Carroll brought soapberry bugs into the laboratory and created populations with different sex ratios. Then he watched males mate in each of these populations. If the mate-guarding trait were **canalized** (that is, showed the same phenotype in a wide variety of environments), then a male would behave the same way in each population; if the trait were **plastic**, then a male would adjust his behavior in relation to the local sex ratio. Carroll's data confirmed that male soapberry bugs in Oklahoma have a plastic behavioral strategy that causes them to modify their mating behavior in response to current social conditions.

Evidence suggests that the soapberry bug's plasticity has evolved in response to the variability in conditions in Oklahoma.

Most soapberry bugs live south of Oklahoma, in warmer, more stable habitats such as the Florida Keys, and the sex ratio in these areas is always close to even. (Hence unguarded females are relatively rare.) Carroll subjected male soapberry bugs from the Florida Keys to the same experimental protocol as that for the bugs from Oklahoma. As **Figure 3.13** shows, the Florida males did not change their behavior in response to changes in sex ratio. They guarded their mates about 90% of the time, regardless of the abundance of females. In a stable environment like the Florida Keys, sex ratios do not vary much from time to time, and the

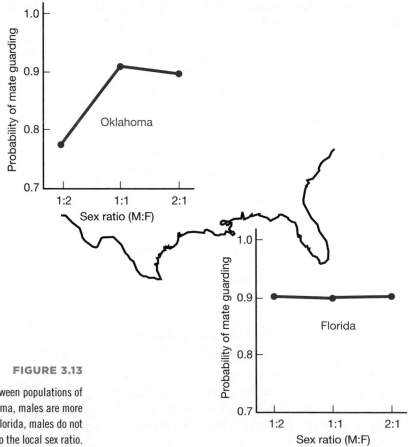

FIGURE 3.13

The average probability of mate guarding varies between populations of soapberry bugs in Oklahoma and Florida. In Oklahoma, males are more likely to guard when females are relatively rare; in Florida, males do not vary their mate-guarding behavior in relation to the local sex ratio.

ability to adjust mate-guarding behavior provides no advantage. Behavioral flexibility is costly in several ways. For example, flexible males must spend time and energy assessing the sex ratio before they mate, males sometimes make mistakes about the local sex ratio and behave inappropriately, and flexibility probably requires a more complex nervous system. Thus a simple fixed behavioral rule is likely to be better in stable environments. In the variable climate of Oklahoma, however, the ability to adjust mate-guarding behavior provides enough fitness benefits to compensate for the costs of maintaining behavioral flexibility.

Behavioral plasticity evolves when the nature of the behavioral response to the environment is genetically variable.

How did behavioral flexibility evolve in the Oklahoma bugs? It evolved like any other adaptation: by the selective retention of beneficial genetic variants. For any character to evolve, (1) the character must vary, (2) the variation must affect reproductive success, and (3) the variation must be heritable. Mate guarding in soapberry bugs satisfies each of these conditions: First, there is variation, and this variation affects fitness. By bringing individual bugs into the laboratory and observing their behavior at different sex ratios, Carroll showed that individual males in Oklahoma have different behavioral strategies. **Figure 3.14** plots the probability of mate guarding for four representative individuals numbered 1 to 4. Males 1 and 4 both have fixed behavioral strategies. Male 1 guards about 90% of the time, and male 4 guards about 80% of the time. Males 2 and 3 have variable behavioral strategies. Male 2 is very sensitive to changes in sex ratio; male 3 is less sensitive. Notice that both the amount of mate guarding and the amount of flexibility vary among the bugs in Oklahoma.

Second, it seems likely that this variation would affect male reproductive success. In Oklahoma, males would experience a range of sex ratios, so it seems likely that males with flexible strategies, such as male 2, would tend to have the most offspring. In Florida, males with inflexible strategies, such as male 1, would have the most offspring.

Third, the character is heritable. By controlling matings in the laboratory, Carroll showed that males tended to have the same strategies as their fathers. Bugs like male 1 had sons with fixed strategies, and bugs like male 2 had sons with flexible strategies. Thus the Oklahoma bugs would come to have a variable strategy, whereas the Florida bugs would come to have a fixed strategy.

Behavior in the soapberry bug is relatively simple. Mate guarding depends on the sex ratio in the local population. The behavior of humans and other primates is much more complex, but the principles that govern the evolution of complex forms of behavior are the same as the principles that govern the evolution of simpler forms of behavior. That is, individuals must differ in how they respond to the environment, these differences must affect their ability to

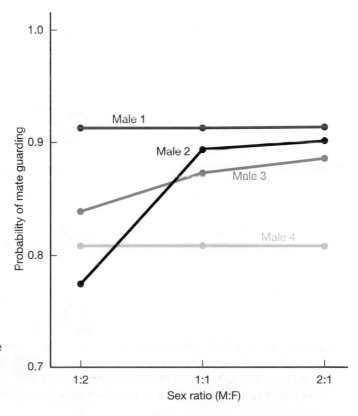

FIGURE 3.14

There is genetic variation in the behavioral rules of individual soapberry bugs, illustrated here by four representative males. There is variation both in the level of mate guarding (for example, male 1 spends more time guarding mates than male 4 does for all sex ratios) and in the amount of plasticity (for example, the behavior of males 1 and 4 does not change, the behavior of male 3 changes a bit, and the behavior of male 2 changes a lot). If this variation is heritable, the rule that works best, averaged over all of the environments of the population, will tend to increase.

survive and reproduce, and at least some of these differences must be heritable. Then individual responses to environmental circumstances will evolve in much the same way that finch beaks and male soapberry bugs' mate-guarding behavior evolve.

Constraints on Adaptation

Natural selection plays a central role in our understanding of evolution because it is the only mechanism that can explain adaptation. However, in this section, we consider five reasons why evolution does not always lead to the best possible phenotype.

Correlated Characters

When individuals that have particular variants of one character also tend to have particular variants of a second character, the two characters are said to be correlated.

So far, we have considered the action of natural selection on only one character at a time. This approach is misleading if natural selection acts on more than one character simultaneously and the characters are nonrandomly associated, or correlated. It's easiest to grasp the meaning and importance of correlated characters in the context of a now familiar example: Darwin's finches. When the Grants and their colleagues captured medium ground finches on Daphne Major, they measured beak depth, beak width, and several other morphological characters. Beak depth is measured as the top-to-bottom dimension of the beak; beak width is the side-to-side dimension. As is common for such morphological characters, the Grants found that beak depth and beak width are positively correlated: Birds with deep beaks also have wide beaks (**Figure 3.15**). Each point in Figure 3.15 represents one individual. Beak width is plotted on the vertical axis, and beak depth is plotted along the horizontal axis. If the cloud of points were round or the points were randomly scattered in the graph, it would mean these characters were uncorrelated. Then information about an individual's beak depth would tell us nothing about its beak width. However, the cloud of points forms an ellipse with the long axis oriented from the lower left to the upper right, so we know that the two characters are **positively correlated**: Deep beaks also tend to be wide. If birds with deep beaks tended to have narrow beaks, and birds with shallow beaks

FIGURE 3.15

Beak depth and beak width are correlated in the medium ground finch on Daphne Major. The vertical axis gives the difference between the individual's beak width and the mean width in the population, and the horizontal axis gives the difference between the individual's beak depth and the mean depth in the population. Each point represents one individual. The data show that birds with deep beaks are likely to have wide beaks, and birds with shallow beaks generally have narrow beaks.

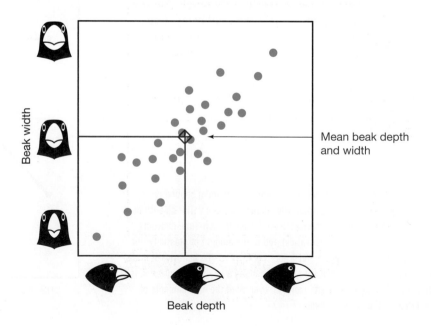

Mean beak depth and width

tended to have wide beaks, the two characters would be **negatively correlated**, and the long axis of the cloud of points would run from the upper left to the lower right.

Correlated characters occur because some genes affect more than one character.

Genes that affect more than one character are said to have **pleiotropic effects**, and most genes probably fall into this category. Often genes that are expressed early in development and affect overall size also influence several other discrete morphological traits. This means individuals carrying mutations that lead to increased expression in the genes that influence overall beak size, BMP2 and BMP7, will have deeper and wider beaks than individuals that do not carry these mutations. The PKU locus provides another good example of pleiotropy. Untreated PKU homozygous individuals are phenotypically different from heterozygous individuals or normal homozygous individuals in several ways. For example, they have lower IQs and different hair color than individuals with other genotypes. Thus the substitution of another PKU allele for the normal allele in a heterozygous individual would affect a wide variety of phenotypic characters.

When two characters are correlated, selection that changes the mean value of one character in the population also changes the mean value of the correlated character.

Returning to the finches, suppose that there is selection for individuals with deep beaks and that beak width has no effect on survival (**Figure 3.16**). As we would expect, the mean value of beak depth increases. Notice, however, that the mean value of beak width also increases, even though beak width has no effect on the probability that an individual will survive. Selection on beak depth affects the mean value of beak width because the two traits are correlated. This effect is called the **correlated response** to selection. It results from the fact that selection increases the frequency of genes that increase beak depth, and individuals carrying those genes also tend to have genes that lead to the development of wider beaks.

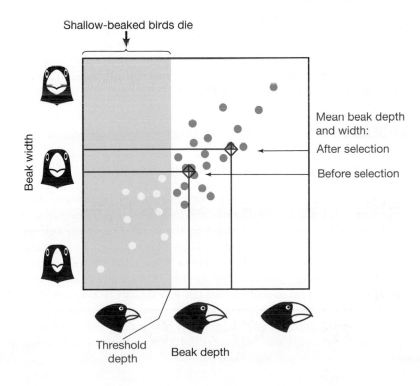

FIGURE 3.16

Selection on one character affects the mean value of correlated characters, even if the other characters have no effect on fitness. The distribution of beak depth and beak width in the population is plotted here, assuming that only birds whose beaks are greater than a threshold depth survive. Beak width is assumed to have no effect on survival. Selection leads to an increase in both average beak depth and average beak width. Mean beak width is increased by the correlated response of selection on beak depth.

FIGURE 3.17

The correlated response to selection can cause less fit phenotypes to become more common. The distribution of beak depth and beak width in the population is plotted here, assuming that all birds whose beaks are less than a threshold depth or greater than a threshold width die. All others survive. Even though there is selection favoring narrower beaks, beak width increases as a result of the correlated response to selection on beak depth.

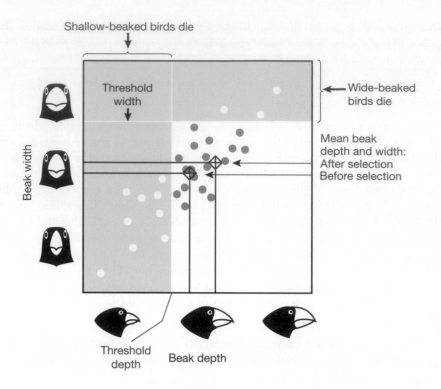

A correlated response to selection can cause other characters to change in a maladaptive direction.

To understand how selection on one character can cause other characters to change in a **maladaptive** (less fit) direction, let's continue with our example. It turns out that beak width did affect survival during the Galápagos drought. By holding beak depth constant, the Grants showed that individuals with *thinner* beaks were more likely to survive during the drought, probably because birds with thinner beaks were able to generate more pressure on the tough seeds that predominated during the drought. If selection were acting only on beak width, then we would expect the mean beak width in the population to decrease. However, the correlated response to selection on beak depth also acts to increase mean beak width. If the correlated response to selection on beak depth were stronger than the effect of selection directly on beak width, mean beak width would increase, even though selection favors thinner beaks. This is exactly what happened on Daphne Major (**Figure 3.17**).

Disequilibrium

Selection produces optimal adaptations only at equilibrium.

In Chapter 1, we saw how natural selection could gradually increase beak depth, generation by generation, until an equilibrium was reached—a point where stabilizing selection maintained the average beak depth in the population at the optimum size. This example illustrates the principle that selection keeps changing a population until an adaptive equilibrium is reached. It is easy to forget that the populations being observed have not necessarily reached equilibrium. If the environment has changed recently, there is every reason to suspect that the morphology or behavior of residents is not adaptive under current conditions. How long does it take for populations to adapt to environmental change? This depends on how quickly selection acts. As we have seen, enormous changes can be created by artificial selection in a few dozen generations.

Disequilibrium is particularly important for some human characters because there have been big changes in the lives of humans during the past 10,000 years. It seems likely that many aspects of the human phenotype have not had time to catch up with recent changes in our subsistence strategies and living conditions. Our diet provides a good example. Ten thousand years ago, most people hunted wild game and gathered wild plants for food (**Figure 3.18**). Typically, people had little access to fat, salt, and sugar, which are essential dietary requirements for the proper functioning of the human body. So it was probably good for people to eat as much of these types of food as they could find. In an adaptive response to human dietary needs, evolution equipped people with a nearly insatiable appetite for fat, salt, and sugar. With the advent of agriculture and trade, however, these substances became readily available, and our evolved appetites for them became problematic. Today, eating too much fat, salt, and sugar is associated with a variety of health problems, including bad teeth, obesity, diabetes, and high blood pressure.

FIGURE 3.18

For most of our evolutionary history, humans have subsisted on wild game and plant foods. Fat, salt, and sugar were in short supply. Here, a Hadza woman digs up a tuber.

Genetic Drift

When populations are small, genetic drift may cause random changes in gene frequencies.

So far, we have assumed that evolving populations are always very large. When populations are small, however, random effects caused by sampling variation can be important. To see what this means, consider a statistical analogy. Suppose that we have a huge urn like the one in **Figure 3.19**. The urn contains 10,000 balls—half of them black and half of them red. Suppose we also have a collection of small urns, each holding 10 balls. We draw 10 balls at random from the big urn to put in each small urn. Not all of the little urns will have five red balls and five black balls. Some will have four red balls, some three red balls, and a few may even have no red balls. The fact that the distribution of black and red balls among the small urns varies is called **sampling variation**.

The same thing happens during genetic transmission in small populations (**Figure 3.20**). Suppose there is an organism that has only one pair of chromosomes with two possible alleles, *A* and *a*, at a particular locus and that selection does not act on this trait. In addition, a population of five individuals of this species is newly isolated from the rest. In this small population (generation 1), each allele has a frequency of 0.5, so five chromosomes carry *A* and five carry *a*. These five individuals mate at random and produce five surviving offspring (generation 2). To keep things simple, assume that the gamete pool is large, so there is no sampling variation. This means that half of the gametes produced by generation 1 will carry *A* and half will carry *a*. However, only 10 gametes will be drawn from this gamete pool to form the next generation of five individuals. These gametes will be sampled in the same way the balls were sampled from the urn (see Figure 3.19). The most likely result is that there are five *A* alleles and five *a* alleles in generation 2, as there were in generation 1. However, just as there is some chance of drawing three, four, six, seven, or even zero black balls from the urn, there is some chance that generation 2 will

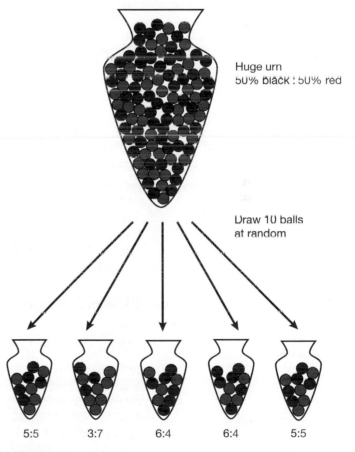

Huge urn
50% black : 50% red

Draw 10 balls
at random

5:5   3:7   6:4   6:4   5:5

FIGURE 3.19

A huge urn contains 10,000 balls, half of them black and half red. Suppose in each of five small urns are placed 10 balls randomly drawn from the large urn. Not all of the small urns will contain five red balls and five black balls. This phenomenon is called sampling variation.

FIGURE 3.20

Sampling variation leads to changes in gene frequency in small populations. Suppose that a population produces half *A* gametes (*black*) and half *a* gametes (*white*), and the next generation consists of just five individuals. It is not unlikely that, by chance, these individuals will carry six *A* gametes and four *a* gametes. This new population will produce 60% *A* gametes. Thus sampling variation can change gene frequencies. Notice here that gene frequencies change again in generation 3, this time in the reverse direction, to 30% *A*.

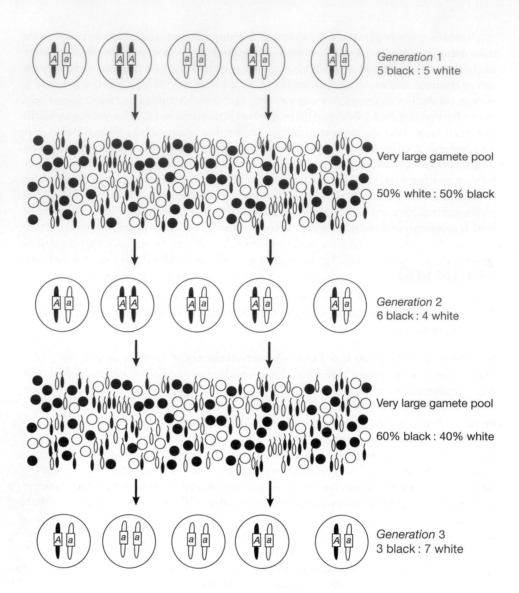

Generation 1
5 black : 5 white

Very large gamete pool

50% white : 50% black

Generation 2
6 black : 4 white

Very large gamete pool

60% black : 40% white

Generation 3
3 black : 7 white

not carry equal numbers of *A* and *a* alleles. Suppose the five individuals in generation 2 have six *A* alleles and four *a* alleles. Because the frequency of the *A* allele is now 0.6, when these individuals mate, the frequency of the *A* allele in the pool of gametes will be 0.6 as well. This means the frequency of the allele in the population has changed by chance alone. In the third generation the gene frequencies change again, this time in the opposite direction. This phenomenon is called **genetic drift**. In small populations, genetic drift causes random fluctuations in genetic frequencies.

The way alleles are sampled in real populations depends on the biology of the evolving species and the nature of the selective forces that the species faces. For instance, population size in some bird species might be limited by nesting sites, and success in obtaining nesting sites might be related to body size. Perhaps the larger the bird, the more chance there would be that it would have success in battling others for a prime nesting spot. Although body size is influenced partly by genes, and selection would favor genes for large body size, many genes do not affect body size, and they may be sampled randomly in this particular population.

The rate of genetic drift depends on population size.

Genetic drift causes more rapid change in small populations than in large ones because sampling variation is more pronounced in small samples. To see why this is true, consider a simple example. Suppose a company is trying to predict whether a new product—low-fat peanut butter—will succeed in the marketplace. The company hires two survey firms to find out how many peanut butter fans would switch to the new product. One firm conducts interviews with five people, four of whom say that they would switch immediately to the low-fat variety. The other company polls 1,000 people, 50% of whom say they would prefer the low-fat peanut butter. Which survey should the company trust? Clearly, the second survey is more credible than the first one. It is easy to imagine that by chance we might find four fans of low-fat peanut butter in a sample of five people, even if the actual frequency of such preferences in the population is only 50%. In contrast, in a sample of 1,000 people, we would be very unlikely to find such a large discrepancy between our sample and the population.

The same principle applies to genetic drift. In a population of five individuals (with 10 chromosomes) in which two alleles are equally common (frequency = 0.5), there is a good chance that in the next generation the frequency of one of the alleles will be greater than or equal to 0.8. But in a population of 1,000 individuals, there is virtually no chance that such a large deviation from the initial frequency will occur.

Genetic drift causes isolated populations to become genetically different from each other.

Genetic drift leads to unpredictable evolution because the changes in gene frequency caused by sampling variation are random. As a result, drift causes isolated populations to become genetically different from one another over time. Results from a computer simulation illustrate how this works for a single genetic locus with two alleles (**Figure 3.21**). Initially, there is a large population in which each allele has a frequency of 0.5. Then four separate populations of 20 individuals (each individual carrying two sets of chromosomes) are created. These four populations are maintained at a constant size and remain isolated from each other during all later generations. During the first generation, the frequency of A in the gamete pool is 0.5 in each population. However, the unpredictable effects of sampling variation alter the frequencies of A among adults in each population. From the first generation, we sample 40 gametes from each of the four populations. In population 1 (indicated by red solid circles), the frequency of A increases dramatically; in populations 2 (blue open triangles) and 3 (green solid triangles), it increases a little; and in population 4 (orange open circles), it decreases. In each case, the change is created by chance alone.

This computer simulation demonstrates that genetic drift causes isolated populations to become genetically different from one another. Four populations are initially formed by drawing 20 individuals from a single large population. In the original population, one gene has two alleles, each of which has a frequency of 0.5. The graph tracks the frequency of one of these alleles in each of these populations over time. Population 1 is in red (*solid circles*), population 2 in blue (*open triangles*), population 3 in green (*solid triangles*), and population 4 in orange (*open circles*). The dotted line shows the global frequency of the A allele. In each population, the frequency of the allele fluctuates randomly under the influence of genetic drift. Notice that populations 2 and 4 eventually reach fixation, meaning that one of the two alleles is lost.

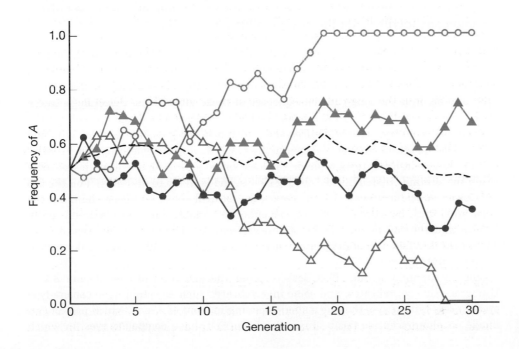

During the next generation, the same process is repeated, except that the gamete pools of each of the four populations now differ. Once again, 40 gametes are sampled from each population. This time the frequency of A in populations 2, 3, and 4 increases, whereas the frequency of A in population 1 decreases substantially. As these random and unpredictable changes continue, the four populations become more and more different from one another. Eventually, one of the alleles is lost entirely in two of our populations (A in population 2 and a in population 4), making all individuals in these populations identical at the locus in question. Such populations are said to have reached **fixation**. In general, the smaller the population, the sooner it reaches fixation. Theoretically, if genetic drift goes on long enough, all populations eventually reach fixation. When populations are large, however, this may take such a long time that the species becomes extinct before fixation occurs. Populations remain at fixation until mutation introduces a new allele.

As you might expect, the rate at which populations become genetically different is strongly affected by their size: Small populations differentiate rapidly; larger populations differentiate more slowly. We will see in Chapter 13 that the relationship between population size and the rate of genetic drift allows us to make interesting and important inferences about the size of human populations tens of thousands of years ago.

Populations must be quite small for drift to lead to significant maladaptation.

Random changes produced by genetic drift will create adaptations only by chance, and we have seen that the probability of assembling complex adaptations in this way is small indeed. This means that genetic drift usually leads to maladaptation. The importance of genetic drift in evolution is a controversial topic. Nearly everyone agrees that populations must be fairly small (say, less than 100 individuals) for drift to have an important effect when it is opposed by strong natural selection. There is a lot of debate about whether populations in nature are usually that small. Most scientists agree, however, that genetic drift is not likely to generate significant maladaptation in traits that vary continuously and are affected by many genetic loci. Thus a trait such as beak size in Darwin's finches is unlikely to be influenced much by genetic drift.

Local versus Optimal Adaptations

Natural selection may lead to an evolutionary equilibrium at which the most common phenotype is not the best possible phenotype.

Natural selection acts to increase the adaptedness of populations, but it does not necessarily lead to the best possible phenotypes. The reason is that natural selection is myopic: It favors small improvements to the existing phenotype, but it does not take into account the long-run consequences of these alterations. Selection is like a mountaineer who tries to climb a peak cloaked in a dense cloud; she cannot see the surrounding country, but she figures she will reach her goal if she keeps climbing uphill. The mountain climber will eventually reach the top of something if she continues climbing, but if the topography is the least bit complicated, she won't necessarily reach the summit. Instead, when the fog clears, she is likely to find herself on the top of a lower, subsidiary peak. In the same way, natural selection keeps changing the population until no more small improvements are possible, but there is no reason to believe that the end product is the optimal phenotype. The phenotype arrived at in such cases is called a *local adaptation*; it is analogous to the subsidiary peak reached by the mountaineer.

For an example of this effect, let's consider the evolution of eyes (**Figure 3.22**). Humans, other vertebrates, and some invertebrates such as octopi have **camera-type eyes**. In this type of eye, there is a single opening in front of a lens, which projects an image on photoreceptive tissue. Insects, by contrast, have **compound eyes**, in which

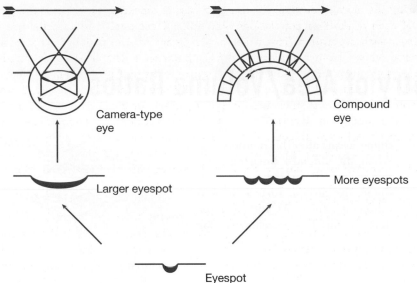

FIGURE 3.22

Researchers believe that different evolutionary pathways led to the development of camera-type eyes and compound eyes. Selection for greater light sensitivity in a simple organism could favor the development of a larger eyespot or of multiple eyespots. The first pathway could lead to camera-type eyes and the second to compound eyes. Images are formed very differently in compound and camera-type eyes, making it virtually impossible for compound eyes to evolve into camera-type eyes.

many very small, separate photoreceptors build up an image composed of a grid of dots, something like a television image. Compound eyes are inferior to camera-type eyes in most ways. For example, compound eyes have lower resolution and less light-gathering power than similarly sized camera-type eyes have. But if compound eyes are indeed inferior to camera-type eyes, why haven't insects evolved camera-type eyes?

The most likely answer is that once a species has evolved complex compound eyes, selection cannot favor intermediate types, even though this might eventually allow superior camera-type eyes to evolve. Consider the early lineages in which eyes were first evolving. There may have come a time when selection favored greater light sensitivity. In the vertebrates and mollusks, greater light sensitivity was achieved by increases in the area of sensitive tissue within each eye. In insects, it seems likely that increased sensitivity was achieved by multiplication of the number of small photoreceptors, each in its own cup. At this early stage, these alternatives yielded equally useful eyes. Once the insect lineage had evolved a visual system based on many small eyes, however, images could not be formed the same way as in camera-type eyes because the camera-type eye inverts images, but the compound eye does not. It does not seem structurally possible to evolve from a compound eye to a camera-type eye in a series of small steps, each favored by selection.

Some local adaptations are called developmental constraints.

Most kinds of organisms begin life as a single fertilized cell, a zygote. As an organism grows, this cell divides many times, giving rise to specialized nerve cells, liver cells, and so on. This process of growth and differentiation is called **development**, and the development of complex structures such as eyes involves many interdependent processes. Developmental changes that would produce desirable modifications are often selected against because these alterations have many other negative effects. For example, as we will see later in the text, there is good reason to believe that it would be adaptive for males in some primate species to be able to produce milk for their young. However, no primate males can do this, and most biologists believe that the developmental changes that would allow males to produce milk would also make them sterile. Thus developmental processes constrain evolution, but such constraints are not absolute. They result from the particular phylogenetic history of the lineage. If primate reproductive biology had evolved along a different pathway so that the development of mammary glands and other elements of the primate reproductive system were independent, then there would be no constraints on the evolution of male lactation.

3.2 The Geometry of Area/Volume Ratios

When an animal becomes larger and does not otherwise change shape, the ratio of any fixed-area measurement to its volume decreases. This is easiest to understand if we compute how the ratio of an animal's entire surface area to its volume changes as the animal becomes larger. To see why, suppose that the animal is a cube x centimeters on a side. Then its volume is x^3 and its surface area is $6x^2$. Thus the ratio of its surface area to its volume is

$$\frac{\text{surface area}}{\text{volume}} = \frac{6x^2}{x^3} = \frac{6}{x}$$

This means an animal measuring 1 cm on a side has 6 cm^2 of surface area for each cubic centimeter of volume. An animal 2 cm on a side has only 3 cm^2 of surface area for each cubic centimeter of volume. Of course, there aren't any cubic animals, but it turns out that the shape doesn't alter this relationship. When an animal's linear dimension is doubled without a change in shape, the ratio of surface area to volume is halved. We will see later that this fact has consequences for how temperature affects the size of animals.

The same geometric principle governs the relationship between any area measurement and volume. Suppose the cubic animal has a vertical bone running through its center that supports the weight of the rest of the animal. For simplicity, we make the bone square in cross section and assume that it has the dimensions $1/2x$ by $1/2x$, as shown in **Figure 3.23**. The cross-sectional area of this bone, then, is $1/2x \times 1/2x = 1/4x^2$. Thus the ratio of the cross-sectional area of the bone to the volume of the animal is

$$\frac{\frac{1}{4}x^2}{x^3} \times \frac{1}{4x}$$

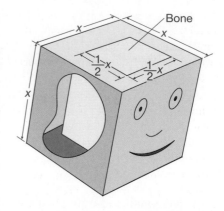

FIGURE 3.23

A cutaway diagram of the cubic animal discussed here.

So if the linear dimensions of an animal are doubled, its weight will increase eightfold, but the cross-sectional area of its bones—and therefore their strength—will increase only fourfold.

Other Constraints on Evolution

Evolutionary processes are also constrained by the laws of physics and chemistry.

The laws of physics and chemistry place additional constraints on the kinds of adaptations that are possible. For example, the laws of mechanics predict that the strength of bones is proportional to their cross-sectional area. This fact constrains the evolution of morphology. To see why, suppose that an animal's size (that is, its *linear* dimensions) doubles as the result of natural selection. This means that the animal's weight (which is proportional to its *volume*) will increase by a factor of 8 (see **A Closer Look 3.2**). The strength to bear an animal's weight comes from its muscles and bones and is determined largely by their cross-sectional areas. If the linear dimensions of bones and muscles double, however, their cross-sectional areas will increase by a factor of only 4. This means that the bones and muscles will be only half as strong in relation to the animal's weight as they were before. Thus if the bones are to be as strong as they were before, they must be more than four times greater in cross section, which means that they must become proportionally thicker. Of course, thicker bones are heavier, and this greater weight imposes other constraints on the animal. If it is heavier, it may not be able to move as fast or suspend itself from branches. This constraint (and a closely related one that governs the strength of muscles) explains why big animals such as elephants are typically slow and ponderous, whereas smaller animals such as squirrels move quickly and are often agile leapers. It also explains why all flying animals are relatively light. Natural selection might well favor an elephant that could run like a cheetah, vault obstacles like an impala, and climb like a monkey. But because of the trade-offs imposed by the laws of physics, there are no mutant pachyderms whose bones are both light and strong enough to permit these behaviors (**Figure 3.24**).

Such constraints are closely related to genetic correlations. Selection eliminates mutants with heavier, weaker bones; the trade-offs inherent in the design of structures mean that mutant animals that have lighter, stronger bones are also not possible. The genotypes that remain in the population lead to the development of either lighter bones or stronger bones, but not both. Thus bone weight and bone strength typically show a positive genetic correlation.

FIGURE 3.24

It might be advantageous for animals to be large enough to be relatively invulnerable to predators but agile enough to leap considerable distances. But physical constraints limit evolution; not all things are possible.

CHAPTER REVIEW

Key Terms

population genetics (p. 54)
genotypic frequency (p. 54)
gene frequency (p. 56)
Hardy–Weinberg equilibrium (p. 58)
modern synthesis (p. 61)

environmental variation (p. 62)
mutations (p. 63)
mate guarding (p. 65)
sex ratio (p. 65)
canalized (p. 66)

plastic (p. 66)
positively correlated (p. 68)
negatively correlated (p. 69)
pleiotropic effects (p. 69)
correlated response (p. 69)
maladaptive (p. 70)

sampling variation (p. 71)
genetic drift (p. 72)
fixation (p. 74)
camera-type eyes (p. 74)
compound eyes (p. 74)
development (p. 75)

Study Questions

1. In human populations in Africa, two common alleles affect the structure of hemoglobin, the protein that carries oxygen on red blood cells. What is regarded as the normal allele (the allele common in European populations) is usually labeled A, and the sickle-cell allele is designated S. There are three hemoglobin genotypes, and the phenotypes associated with these genotypes can be distinguished in a variety of ways. In one African population of 10,000 adults, for example, there are 3,000 AS individuals, 7,000 AA individuals, and 0 SS individuals.

   (a) Suppose that these individuals were to mate at random. What would be the frequency of the A and S alleles among the gametes that they produced?

(b) What would be the frequency of the three genotypes that resulted from random mating?

(c) Is the original population of adults in Hardy–Weinberg equilibrium?

2. The three common genotypes at the hemoglobin locus have very different phenotypes: *SS* individuals suffer from severe anemia, *AS* individuals have a relatively mild form of anemia but are resistant to malaria, and *AA* individuals have no anemia but are susceptible to malaria. The frequency of the *S* allele among the gametes produced by the first generation of a central African population is 0.2.

(a) Assuming that mating occurs at random, what are the frequencies of the three genotypes among zygotes produced by this population?

(b) In this area, no *SS* individuals survive to adulthood, 70% of the *AA* individuals survive, and all of the *AS* individuals survive. What is the frequency of each of the three genotypes among the second generation of adults?

(c) What is the frequency of the *S* allele among gametes produced by these adults?

3. Tay–Sachs disease is a lethal genetic disease controlled by two alleles: the Tay–Sachs allele *T* and the normal allele *N*. Children who have the *TT* genotype suffer from mental deterioration, blindness, paralysis, and convulsions, and they die sometime between the ages of 3 and 5 years. The proportion of infants afflicted with Tay–Sachs disease varies in different ethnic groups. The frequency of the Tay–Sachs allele is highest among descendants of central European Jews. Suppose that in a population of 5,000 people descended from central European Jews the frequency of the Tay–Sachs allele is 0.02.

(a) Assuming that members of this population mate at random and each adult has four children on average, how many gametes from each person are successful at being included in a zygote?

(b) Calculate the frequency of each of the three genotypes among the population of zygotes.

(c) Assume that no *TT* zygotes survive to become adults, and that 50% of the *TN* and *NN* individuals survive to become adults. What is the frequency of the *TT* and *TN* genotypes among these adults?

(d) When the adult individuals resulting from zygotes in (b) produce gametes, what is the frequency of the *T* allele among their gametes? Compare your answer with the frequency of the *T* allele among gametes produced by their parents. Think about your result, given that these numbers are roughly realistic and

that there have been no substantial medical breakthroughs in the treatment of Tay–Sachs disease. What is the paradox here?

4. Consider a hypothetical allele that is lethal, like the Tay–Sachs allele, but is dominant rather than recessive. Assume that this allele has a frequency of 0.02, like the Tay–Sachs allele. Recalculate the answers to 3a, 3b, and 3c. Is selection stronger against the recessive allele or the dominant one? Why? Assuming that mutation to the deleterious allele occurs at the same rate at both loci, which allele will occur at a higher frequency? Why?

5. Consider the Rh blood group. For this example we will assume that there are only two alleles, *R* and *r*, of which *r* is recessive. Homozygous *rr* individuals produce certain nonfunctional proteins and are said to be Rh-negative (Rh⁻), whereas *Rr* and *RR* individuals are Rh-positive (Rh⁺). When an Rh⁻ female mates with an Rh⁺ male, their offspring may suffer serious anemia while still in the uterus. The frequency of the *r* allele in a hypothetical population is 0.25. Assuming that the population is in Hardy–Weinberg equilibrium, what fraction of the offspring in each generation will be at risk for this form of anemia?

6. The blending model of inheritance was attractive to nineteenth-century biologists because it explained why offspring tend to be intermediate in appearance between their parents. This model fails, however, because even though it predicts a loss in variation, variation is maintained. How does Mendelian genetics explain the intermediate appearance of offspring without the loss of variation? Be sure to explain why the properties of the Hardy–Weinberg equilibrium are an important part of this explanation.

7. In a particular species of fish, egg size and egg number are negatively correlated. Draw a graph that illustrates this fact. What will happen to egg size if selection favors larger numbers of eggs?

8. The compound eyes of insects provide poor image clarity in comparison with the images produced by the camera-type eyes of vertebrates. Explain why insects do not evolve camera-type eyes.

9. Why does natural selection produce adaptations only at equilibrium?

10. When selection makes animals larger overall, their bones also get thicker. The increase in bone thickness could be due to the correlated response to selection or to independent selection for thicker bones or to some combination of the two factors. How could you determine the relative importance of these two processes?

Further Reading

Bergstom, C. T. and L. A. Dugatkin. 2019. *Evolution.* 2nd ed. New York: Norton.

Dawkins, M. S. 1995. *Unraveling Animal Behavior.* 2nd ed. New York: Wiley.

Dawkins, R. 2015. *The Blind Watchmaker: Why the Evidence of Evolution Reveals a Universe without Design.* New York: Norton.

Hedrick, P. W. 2009. *Genetics of Populations.* 4th ed. Boston: Jones & Bartlett.

Lane, N. 2009. *Life Ascending, Ten Great Inventions of Evolution.* New York: Norton.

Walsh, B. and M. Lynch, 2018. *Evolution and Selection of Quantitative Traits.* Oxford: Oxford University Press.

 Review this chapter with personalized, interactive questions through INQUIZITIVE.

 This chapter also features a Guided Learning Exploration on continuous traits.

4

SPECIATION AND PHYLOGENY

CHAPTER OBJECTIVES

By the end of this chapter you should be able to:

A Describe how species are defined.

B Explain how new species arise through the process of evolution.

C Explain why speciation causes organisms to be organized hierarchically and how this pattern can be described with a phylogenetic tree.

D Assess why reconstructing phylogenies is important.

E Reconstruct phylogenies using patterns of variation in living species.

What Are Species?

Microevolution refers to how populations change under the influence of natural selection and other evolutionary forces; macroevolution refers to how new species and higher taxa are created. So far, we have focused on how natural selection, mutation, and genetic drift cause populations to change through time. These are mechanisms of **microevolution**, and they affect the morphology, physiology, and behavior of particular species in particular environments. For example, microevolutionary processes are responsible for variation in the size and shape of the beaks of medium ground finches on Daphne Major.

However, this is not all there is to evolution. Darwin's major work was titled *On the Origin of Species* because he was interested in how new species are created as well as how natural selection operates within populations. Evolutionary theory tells us how new species, genera, families, and higher groupings come into existence.

These processes are mechanisms of **macroevolution**. Macroevolutionary processes play an important role in the story of human evolution. To properly interpret the fossil record and reconstruct the history of the human lineage, we need to understand how new species and higher groupings are created and transformed over time.

Species can usually be distinguished by their behavior and morphology.

Organisms cluster into distinct types called species. The individual organisms that belong to a species are similar to each other and are usually quite distinct from the members of other species. For example, in Africa some tropical forests house two species of apes: chimpanzees and gorillas. These two species are similar in many ways: Both are tailless, both bear weight on their knuckles when they walk, and both defend territories. Nonetheless, these two species can easily be distinguished on the basis of their morphology: Chimpanzees are smaller than gorillas; male gorillas have tiny **testes** (singular *testis;* the organs that produce sperm), whereas male chimpanzees have quite large ones; and gorillas have a fin of bone on their skull, whereas chimpanzees have more rounded skulls. Gorillas and chimpanzees also differ in their behavior: Chimpanzees make and use tools in foraging, whereas gorillas do not; male gorillas beat their chests when they perform displays, whereas male chimpanzees flail branches and charge about; and gorillas live in smaller groups than chimpanzees do. These two species are easy to distinguish because no animals are intermediate between them; there are no "gimps" or "chorillas" (**Figure 4.1**).

Species are not abstractions created by scientists; they are real biological categories. People all over the world name the plants and animals around them, and biologists use the same kinds of phenotypic characteristics to sort animals into species that other people use. For the most part, there is little problem in identifying any particular specimen from its phenotype.

Although nearly everyone agrees that species exist and can recognize species in nature, biologists are much less certain about how species should be defined. This uncertainty arises from the fact that evolutionary biologists do not agree about *why* species exist. There is now a considerable amount of controversy about the processes that give rise to new species and the processes that maintain established ones. Although there are many different views on these topics, we will concentrate on two of the most widely held points of view: the biological species concept and the ecological species concept.

[a]

[b]

FIGURE 4.1

(a) Chimpanzees and (b) gorillas sometimes occupy the same forests and share certain traits. However, these two species are readily distinguishable because no animals are intermediate between them.

The Biological Species Concept

The biological species concept defines a species as a group of interbreeding organisms that are reproductively isolated from other organisms.

Most zoologists believe in the **biological species concept**, which defines a biological species as a group of organisms that interbreed in nature and are reproductively isolated. **Reproductive isolation** means that members of a given group of organisms do not mate successfully with organisms outside the group. For example, there is just one species of gorilla, *Gorilla gorilla*, and this means all gorillas can mate with one another but do not breed with any other kinds of animals in nature. According to adherents of the biological species concept, reproductive isolation is the reason that there are no gimps or chorillas.

The biological species concept defines a species in terms of the ability to inter breed because successful mating leads to **gene flow**, the movement of genetic material from one population to another or from one part of a population to another. Gene flow tends to maintain similarities among members of the same species. To see how gene flow preserves homogeneity within species, consider the hypothetical situation diagrammed in **Figure 4.2a**. Imagine a population of finches living on a small island in which there are two habitats: wet and dry. Natural selection favors different sizes of beaks in each habitat. Large beaks are favored in the dry habitat, and small beaks are favored in the wet habitat. Because the island is small, however, birds fly back and forth between the two environments and mate at random, so there is a lot of gene flow between habitats. Unless selection is very strong, gene flow will swamp its effects. On average, birds in both habitats will have medium-size beaks, a compromise between the optimal phenotypes for each habitat. In this way, gene flow tends to make the members of a species evolve as a unit.

Now suppose that there are finches living on two different islands, one wet and one dry, and that the islands are far enough apart that the birds cannot fly from one island to the other (**Figure 4.2b**). This means there will be no interbreeding, and no gene flow will occur. With no genetic exchange between the two independent groups to counter the effects of selection, the two populations will diverge genetically and become less similar phenotypically. Birds on the dry island will develop large beaks, and birds on the wet island will develop small beaks.

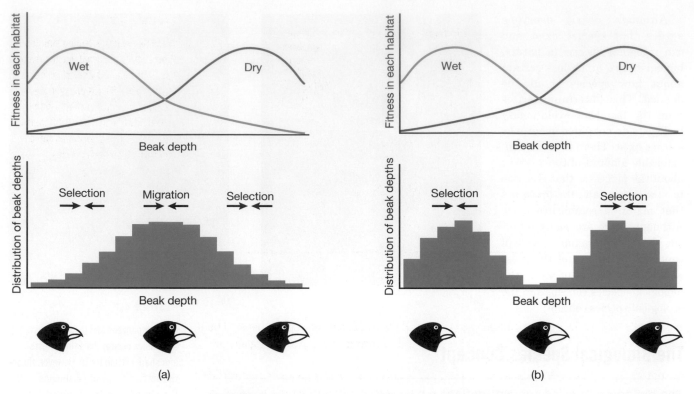

FIGURE 4.2

Gene flow among populations destroys differences between them. (a) Suppose a population of finches lives on an island with both wet and dry habitats. In the dry habitat, selection favors deep beaks, but selection in the wet habitat favors shallow beaks. Because interbreeding leads to extensive gene flow between populations in the two habitats, the population beak size responds to an average of the selection in the two environments, and so beaks are of intermediate size. (b) Now suppose two populations of finches live on different islands, one wet and one dry, with no gene flow between the two islands. In this scenario, deep beaks are common on the dry island, and shallow beaks are common on the wet island.

Reproductive isolation prevents species from genetically blending.

Reproductive isolation is the flip side of interbreeding. Suppose chimpanzees and gorillas could and did interbreed successfully in nature. The result would be gene flow between the two kinds of apes, and this phenomenon would produce animals that were genetically intermediate between chimpanzees and gorillas. Eventually, the two species would merge. There are no chorillas or gimps in nature because chimpanzees and gorillas are reproductively isolated.

Reproduction is a complicated process, and anything that alters the process can act as an isolating mechanism. Even subtle differences in activity patterns, courtship behavior, or appearance may prevent individuals of different types from mating. Moreover, even if a mating among individuals of different types does take place, the egg may not be fertilized or the zygote may not survive.

The Ecological Species Concept

The ecological species concept emphasizes the role of selection in maintaining species boundaries.

Critics of the biological species concept point out that lack of gene flow is neither necessary nor sufficient to maintain species boundaries in every case, and they argue that

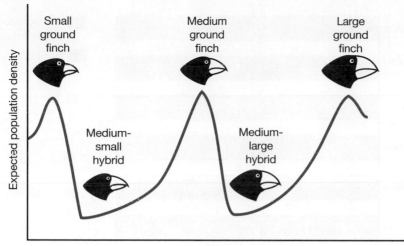

Beak depth

FIGURE 4.3

In the Galápagos, selection maintains three species of ground finches, even though there is substantial gene flow among them. The red line represents the amount of food available in the environment for birds with different-size beaks. Each peak in this curve represents a different species.

selection plays an important role in preserving the boundaries between species. The view that emphasizes the role of natural selection in creating and maintaining species is called the **ecological species concept**.

In nature, species boundaries are often maintained even when there are substantial amounts of gene flow between species. For example, the medium ground finch readily breeds with the large ground finch on islands where they coexist. Peter Grant and his colleagues have estimated that approximately 10% of the time, medium ground finches mate with large ground finches, leading to considerable gene flow between these two species. Yet the species have not merged. Grant and his colleagues have concluded that the medium ground finch and the large ground finch have remained distinct because these two species represent two of the three optimal beak sizes for ground finches (**Figure 4.3**). These three optimal sizes are based on the availability of seeds of different size and hardness and the ability of birds with different-size beaks to harvest these seeds. According to the calculations of Grant's group, the three optimal beak sizes for ground finches correspond to the average beak sizes of the small (*Geospiza fuliginosa*), medium (*G. fortis*), and large (*G. magnirostris*) ground finch species (**Figure 4.4**). These researchers suggest that hybrids are selected against because their beaks fall in the "valleys" between these selective "peaks." The kind of interbreeding seen in the finches of the Galápagos Islands is not particularly unusual. A survey of 114 plant species and 170 animal species by Loren Rieseberg and his collaborators at Indiana University indicates that a substantial fraction of species is not reproductively isolated (**Figure 4.5**).

FIGURE 4.4

(a) The small ground finch, *Geospiza fuliginosa*; (b) the medium ground finch, *G. fortis*; (c) the large ground finch, *G. magnirostris*.

[a]

[b]

[c]

FIGURE 4.5

About half of a broad sample of plant and animal species is reproductively isolated, and about half is not. Members of species that are reproductively isolated do not produce viable offspring when mated to other species.

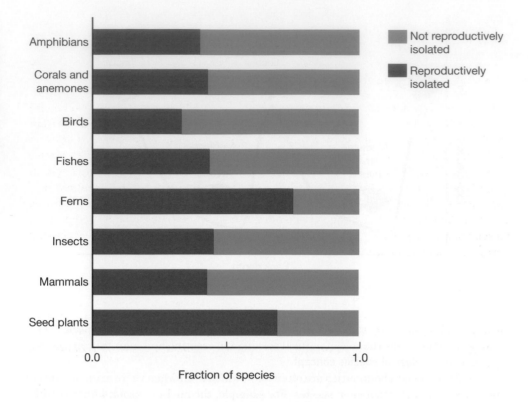

FIGURE 4.6

Checkerspot butterflies living in localized populations throughout California are all members of the same species, but there is probably very little gene flow among different populations.

In addition, several species have maintained their coherence with no gene flow between isolated subpopulations. For example, the checkerspot butterfly is found in scattered populations throughout California (**Figure 4.6**). Members of different populations are similar morphologically and are all classified as members of the species *Euphydryas editha*. However, careful studies by Stanford University biologist Paul Ehrlich have shown that these butterflies rarely disperse more than 100 m (about 110 yd.) from their place of birth. Given that populations are often separated by several kilometers and sometimes by as much as 200 km (about 124 mi.), it seems unlikely that there is enough gene flow to unify the species.

The existence of strictly asexual organisms provides additional evidence that species can be maintained without gene flow. Asexual species, which reproduce by budding or fission, simply replicate their own genetic material and produce exact copies of themselves. Thus gene flow cannot occur in asexual organisms, and it does not make sense to think of them as being reproductively isolated. Nonetheless, many biologists who study purely asexual organisms contend that it is just as easy to classify them into species as it is to classify organisms that reproduce sexually.

What maintains the coherence of species in these cases? Most biologists think the answer is natural selection. To see why, let's return to the Galápagos once more. Imagine a situation in which selection favored finches with small beaks in one habitat and finches with large beaks in another habitat but did not favor birds with medium beaks in either habitat. If birds with medium-size beaks consistently failed to survive or reproduce successfully, natural selection could maintain the difference between the two bird species even if they interbred freely. For the checkerspot butterflies and asexual species, we can imagine the opposite scenario: Selection favors organisms with the same morphology and physiology and thus maintains their similarity even without gene flow.

Today, most biologists concede that selection maintains species boundaries in a few odd cases like Darwin's finches, but they generally insist that reproductive isolation plays the major role in most instances. However, a growing minority of researchers contends that selection plays an important role in maintaining virtually all species boundaries.

The Origin of Species

Speciation is difficult to study empirically.

The species is one of the most important concepts in biology, so it would be helpful to know how new species come into existence. Despite huge amounts of hard work and many heated arguments, however, there is still uncertainty about what Darwin called the "mystery of mysteries"—the origin of species. The mystery persists because speciation is difficult to study empirically. Unlike microevolutionary change within populations, which researchers can sometimes study in the field or laboratory, new species usually evolve too slowly for any single individual to study the entire process. And on the flip side, speciation usually occurs much too rapidly to be detected in the fossil record. Nonetheless, biologists have compiled a substantial body of evidence that provides important clues about how new species arise.

Allopatric Speciation

If geographic or environmental barriers isolate part of a population, and selection favors different phenotypes in these regions, then a new species may evolve.

Allopatric speciation occurs when a population is divided by some type of barrier, and different parts of the population adapt to different environments. The following hypothetical scenario captures the essential elements of this process. Mountainous islands in the Galápagos contain dry habitats at low elevations and wet habitats at higher elevations. Suppose that a finch species that lives on mountainous islands has a medium-size beak, which represents a compromise between the best beak for survival in the wet areas and the best beak for survival in the dry areas. Further suppose that several birds are carried on the winds of a severe storm to another island that is uniformly dry, like one of the small, low-lying islands in the Galápagos archipelago. On this new dry island, only large-beaked birds are favored because large beaks are best for processing the large, hard seeds that predominate there. As long as there is no competition from some other small bird adapted to the dry habitat and there is very little movement between the two islands, the population of finches on the dry island will rapidly adapt to their new habitat, and the birds' average beak size will increase (**Figure 4.7**).

Now let's suppose that after some time, finches from the small, dry island are blown back to the large island from which their ancestors came. If the large-beaked newcomers successfully mate with the medium-beaked residents, then gene flow between the two populations will rapidly eliminate the differences in beak size between them, and the recently created large-beaked variety will disappear. If, on the other hand, large-beaked immigrants and small-beaked residents cannot successfully interbreed, then the differences between the two populations will persist. As we noted earlier, a variety of mechanisms can prevent successful interbreeding, but it seems that the most common obstacle to interbreeding is that hybrid progeny are less viable than other offspring. In this case, we might imagine that when the two populations of finches became isolated, they diverged genetically because of natural selection and genetic drift. The longer they remain isolated, the greater the genetic difference between the populations becomes. When the two distinct types then come into contact, hybrids may have reduced viability, either because genetic incompatibilities have arisen during their isolation or because hybrid birds are unable to compete successfully for food. If these processes cause complete reproductive isolation, then a new species has been formed.

Even if there is some gene flow after the members of the two populations initially come back into contact, two additional processes may increase the degree of

FIGURE 4.7

A likely sequence of events is depicted in a hypothetical allopatric speciation event in the Galápagos. Initially, one population of finches occupies an island with both wet and dry habitats. Because there is extensive gene flow between habitats, these finches have intermediate-size beaks. By chance, some finches disperse to a dry island, where they evolve larger beaks. Then, again by chance, some of these birds are reintroduced to the original island. If the two populations are reproductively isolated, then a new species has been formed. Even if there is some gene flow, competition between the two populations may cause the beaks of the residents and immigrants to diverge further, a process called character displacement.

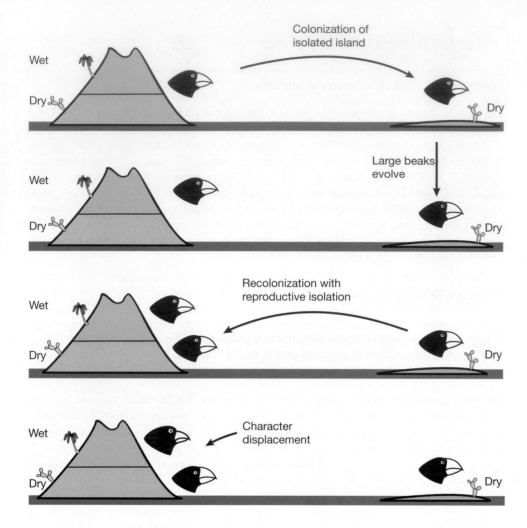

reproductive isolation and facilitate the formation of a new species. The first process, **character displacement**, may occur if competition over food, mates, or other resources increases the morphological differences between the immigrants and the residents. In our example, the large-beaked immigrants will be better suited to the dry parts of the large island and will be able to outcompete the original residents of these areas. Resident birds will be better off in the wet habitats, where they face less competition from the immigrants. Because small beaks are advantageous in wet habitats, residents with smaller-than-average beaks will be favored by selection. At the same time, because large beaks are advantageous in dry habitats, natural selection will favor increased beak size among the immigrants. This process will cause the beaks of the competing populations to diverge. There is good evidence that character displacement has played an important role in shaping the morphology of Darwin's finches (**Figure 4.8**).

A second process, called **reinforcement**, may act to reduce the extent of gene flow between the populations. Because hybrids have reduced viability, selection will favor behavioral or morphological adaptations that prevent matings between members of the two populations. This process will further increase the reproductive isolation between the two populations. Thus character displacement and reinforcement may amplify the initial differences between the populations and lead to two new species.

Allopatric speciation requires a physical barrier that initially isolates part of a population, interrupts gene flow, and allows the isolated subpopulation to diverge from the original population under the influence of natural selection. In our example, the physical barrier is the sea, but mountains, rivers, and deserts can also restrict

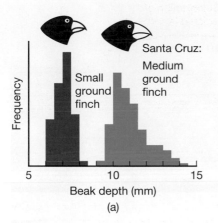 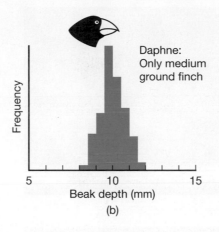 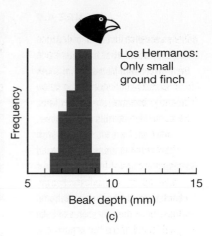

FIGURE 4.8

The distribution of beak sizes of the small ground finch (*G. fuliginosa*) and the medium ground finch (*G. fortis*) on three of the Galápagos Islands illustrates the effects of character displacement. (a) The distribution of beak depth for each species on Santa Cruz Island, where the two species compete. (b, c) The distributions of beak depth for the small and medium ground finches on islands where they do not compete with each other (*G. fortis* on Daphne Major and *G. fuliginosa* on Los Hermanos). The beaks of the two species are the most different where there is direct competition. Careful measurements of seed density and other environmental conditions suggest that competition, not environmental differences, is responsible for the difference between the populations.

movomont and intorrupt gonc flow. Character displacement and reinforcement may work to increase differences when members of the two populations renew contact. However, these processes are not necessary elements in allopatric speciation. Many species become completely isolated while they are separated by a physical barrier.

Parapatric and Sympatric Speciation

New species may also form if there is strong selection that favors two different phenotypes.

Biologists who endorse the biological species concept usually argue that allopatric speciation is the most important mechanism for creating new species in nature. For these scientists, gene flow "welds" a species together, and species can be split only if gene flow is interrupted. Biologists who think that natural selection plays an important role in maintaining species often contend that selection can lead to speciation even when there is interbreeding. There are two versions of this hypothesis. One version, called **parapatric speciation**, holds that selection alone is not sufficient to produce a new species, but new species can be formed if selection is combined with partial genetic isolation. For example, baboons range from Saudi Arabia to the Cape of Good Hope and occupy a diverse array of environments. Some baboons live in moist tropical forests, some live in arid deserts, and some live in high-altitude grasslands (**Figure 4.9**). Different behaviors and morphological traits may be favored in each of these environments, and this variation may cause baboons in different regions to vary. At the boundaries between these various habitats, baboons that come from different habitats and have different characteristics may mate and create a **hybrid zone**. Studies of hybrid zones for a wide variety of species suggest that hybrids are usually less fit than nonhybrids. When this is the case, selection should favor behavior or morphology that prevents mating between species members from different habitats. If such reinforcement does occur, gene flow will be reduced, and eventually two new, reproductively isolated species will evolve.

FIGURE 4.9

Baboons are distributed all over Africa and occupy a variety of habitats. (a) In the Drakensberg Mountains in South Africa, winter temperatures drop below freezing, and snow sometimes falls. (b) In the Matopo Hills of Zimbabwe, woodlands are interspersed with open savanna areas. (c) Amboseli National Park lies at the foot of Mount Kilimanjaro just inside the Kenyan border. (d) Gombe Stream National Park lies on the hilly shores of Lake Tanganyika in Tanzania.

[a]

[b]

[c]

[d]

The other version of the hypothesis, called **sympatric speciation**, contends that strong selection favoring different phenotypes can lead to speciation even when there is no geographic separation and, therefore, initially there is extensive gene flow among individuals in the population. Sympatric speciation is theoretically possible, but it is uncertain how often it occurs in nature. The three speciation mechanisms are diagrammed in **Figure 4.10**.

FIGURE 4.10

There are three mechanisms of speciation. (a) Allopatric speciation occurs when a species is divided into two reproductively isolated populations that later diverge. If the divergent populations regain contact, they cannot interbreed, and two new species exist. (b) Parapatric speciation occurs when a species experiences different environments in different parts of its geographic range. Natural selection causes different populations of the species to diverge while adapting to these different surroundings, and eventually reproductive isolation is achieved. (c) Sympatric speciation occurs when selection strongly favors different adaptations to a similar environment within a single species.

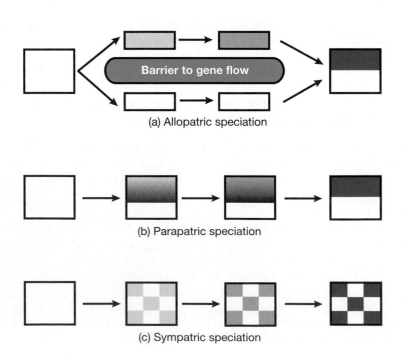

(a) Allopatric speciation

(b) Parapatric speciation

(c) Sympatric speciation

[a]

[b]

[c]

FIGURE 4.11

When the first finches arrived in the Galápagos 500,000 years ago, there were many empty niches. (a) Some finches became cactus eaters, (b) others became seed eaters, and (c) some became predators on insects and other arthropods.

Adaptive radiation occurs when there are many empty niches.

Ecologists use the term **niche** to refer to a particular way of "making a living," which includes the kinds of food eaten as well as when, how, and where the food is acquired. One consequence of all three models of speciation is that the rate of speciation depends on the number of available ecological niches. Once again, Darwin's finches provide a good example. About half a million years ago when the first finches migrated to the Galápagos from the mainland of South America (or perhaps from the Cocos Islands), all of the niches for small birds in the Galápagos were empty. There were opportunities to make a living as a seed eater, a cactus eater, and so on. The finches' ancestors diversified to fill all of these ecological niches, and eventually they became 14 distinct species (**Figure 4.11**). An even more spectacular example of this process occurred at the end of the Cretaceous period. Dinosaurs dominated the earth during the Cretaceous period but then disappeared suddenly 65 Ma. The mammals that coexisted with the dinosaurs were mostly small, nocturnal, and insectivorous. But when the dinosaurs became extinct, these small creatures diversified to fill a variety of ecological niches, evolving into elephants, killer whales, buffalo, wolves, bats, gorillas, humans, and other kinds of mammals. When a single kind of animal or plant diversifies to fill many available niches, the process is called **adaptive radiation** (**Figure 4.12**)

The Tree of Life

Organisms can be classified hierarchically on the basis of similarities. Many such similarities are unrelated to adaptation.

In Chapter 1, we saw how Darwin's theory of evolution explains the existence of adaptation. Now we can show how the same theory explains why organisms can be classified into a hierarchy on the basis of their similarities, a fact that puzzled nineteenth-century biologists much more than the existence of adaptations did. Richard Owen, a nineteenth-century anatomist who was one of Darwin's principal opponents in the debates after the publication of *On the Origin of Species*, used dugongs (aquatic mammals much like manatees), bats, and moles as an example of this phenomenon. All three of these creatures have the same kind and number of bones in their forelimbs, even though the shapes of these bones are quite different (**Figure 4.13**). The forelimb of the bat is adapted to flying and that of the dugong to paddling (**Figure 4.14**). Nonetheless, the basic structure of a bat's forelimb is much more similar to that of a dugong than to the forelimb of a swift (a type of bird), even though the swift's forelimb is also designed for flight.

FIGURE 4.12

Darwin's finches are not the only example of adaptive radiations that occur when immigrants encounter an array of empty ecological niches. In the Hawaiian archipelago, the adaptive radiation of honeycreeper finches, shown here, produced an even greater diversity of species.

FIGURE 4.13

Organisms show patterns of similarity that have nothing to do with adaptation. Richard Owen's drawings of the forelimbs of three mammalian species illustrate this fact. (a) The dugong, (b) the mole, and (c) the bat all have the same basic bone structure in their forelimbs; dugongs use their forelimbs to swim in the sea, moles use their forelimbs to burrow into the earth, and bats use their forelimbs to fly.

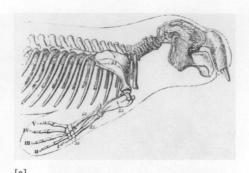

[a]

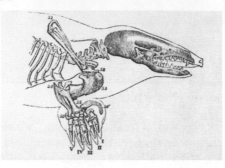

[b]

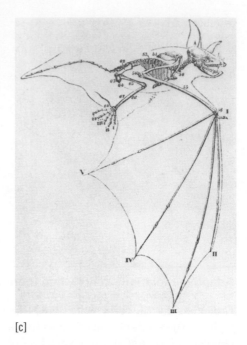

[c]

Such patterns of similarity make it possible to cluster species hierarchically—like a series of nested boxes (**Figure 4.15**). This remarkable property of life is the basis of the system for classifying plants and animals devised by the eighteenth-century Swedish biologist Carolus Linnaeus. All species of bats share many similarities and are grouped in one box. Bats can be clustered with dugongs and moles in a larger box that contains all of the mammals. Mammals in turn are classified together with birds, reptiles, and amphibians in an even larger box that contains all of the animals. Sometimes the similarities that lead us to classify animals together are functional similarities. Many of the features shared by different species of bats are related to the fact that they fly at night. However, many shared features bear little relation to adaptation; bats, tiny aerial acrobats, for example, are grouped with the enormous placid dugong and not with the small acrobatic swift.

Speciation explains why organisms can be classified hierarchically.

FIGURE 4.14

The dugong is an aquatic animal. Its forelimb is adapted for paddling through the water.

The fact that new species derive from existing species accounts for the existence of the patterns of nonadaptive similarity that allow organisms to be classified hierarchically. Clearly, new species originate by splitting off from older ones, and we can arrange a group of species that share a common ancestor into a family tree, or **phylogeny**. **Figure 4.16** shows the family tree of the **hominoids**—the superfamily that includes apes and humans—and **Figure 4.17** shows what some of these present-day apes look like. At the root of the tree is an unknown ancestral species from which all hominoids evolved. Each branch represents a speciation event in which one species split into two daughter species, and one or both of the new daughter species diverged in morphology or behavior from the parent species.

It is important to realize that when two daughter species diverge, they do not differ in all phenotypic details. A few traits may differ while most others retain their original form. For example, the hominoids share many common features, including an unspecialized digestive system, five toes on each foot, and no tail. At the same time, these animals have diverged in several ways. Gorillas usually live in groups that contain one adult male

Vertebrates

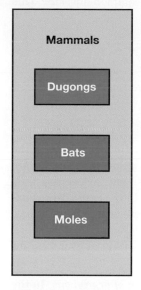

Mammals

Dugongs

Bats

Moles

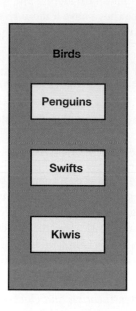

Birds

Penguins

Swifts

Kiwis

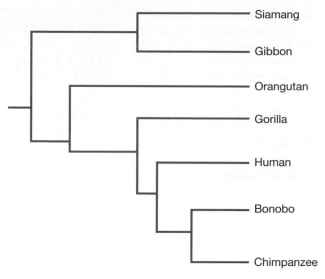

Siamang

Gibbon

Orangutan

Gorilla

Human

Bonobo

Chimpanzee

FIGURE 4.15

Patterns of similarity allow organisms to be classified hierarchically into a series of nested boxes. All mammals share more similarities with one another than they do with birds, even though some mammals, such as bats, must solve the same adaptive problems that birds face.

FIGURE 4.16

A phylogeny for the hominoids. The names of living species are given at the ends of the branches. The branching pattern of the phylogenetic tree reflects the ancestry of extant (still living) lineages. For example, chimpanzees and bonobos have a more recent common ancestor than do chimpanzees and orangutans.

and several adult females; orangutans are mainly solitary. Male and female gibbons are about the same size, but among the other apes, males are larger than females. In general, we expect to see the greatest divergence in those traits that are related to making a living in different habitats and to choosing mates.

Now let's return to the family tree. Each time one species splits to become two new species, the new daughter species will differ in some way. They will continue to diverge through time because once speciation has occurred, the two lineages evolve independently. Some differences will arise because the two species adapt to different habitats; other differences may result from random processes such as genetic drift. In general, species that have recently diverged will have more characteristics in common with one another than with species that diverged in the more distant past.

FIGURE 4.17

(a) Gibbons and siamangs are called *lesser apes* and are classified in the family Hylobatidae. (b) Orangutans, (c) gorillas, (d) chimpanzees, and humans are classified in the family Hominidae. Orangutans, gorillas, and chimpanzees are called *great apes* classified in the sub-family Pongidae.

[a]

[b]

[c]

[d]

For example, chimpanzees and gorillas share more traits with each other than they do with orangutans or gibbons because they share a more recent common ancestor. This pattern of sharing traits is the source of the hierarchical nature of life.

Why Reconstruct Phylogenies?

Phylogenetic reconstruction plays three important roles in the study of organic evolution.

FIGURE 4.18

Knuckle walking among the great apes. (a) Chimpanzees and gorillas are knuckle walkers; this means that when they walk, they bend their fingers under their palms and bear weight on their knuckles. (b) In contrast, orangutans are not knuckle walkers; when they walk, they bear their weight on their palms.

[a]

[b]

We have seen that descent with modification explains the hierarchical structure of the living world. Because new species always evolve from existing species and species are reproductively isolated, all living organisms can be placed on a single phylogenetic tree, which we can then use to trace the ancestry of all living species. In the rest of this chapter, we will see how the pattern of similarities and differences observed in living things can be used to construct phylogenies and to help establish the evolutionary history of life.

Reconstructing phylogenies plays an important role in the study of evolution for three reasons:

1. *Phylogeny is the basis for the identification and classification of organisms.* In the latter part of this chapter we will see how scientists use phylogenetic relationships to name organisms and arrange them into hierarchies. This endeavor is called **taxonomy**.

2. *Knowing phylogenetic relationships often helps explain why a species evolved certain adaptations and not others.* Natural selection creates new species by modifying existing body structures to perform new functions. To understand why a new organism evolved a particular trait, it helps to know what kind of organism it evolved from, and this is what phylogenetic trees tell us. The phylogenetic relationships among the apes provide a good example of this point. Most scientists used to believe that chimpanzees and gorillas shared a more recent common ancestor than either of them shared with humans. This view influenced their interpretation of the evolution of locomotion, or forms of movement, among the apes. All of the great apes are **quadrupedal**, which means that they walk on their hands and feet. However, gorillas and chimpanzees curl their fingers over their palms and bear weight on their knuckles, a form of locomotion called **knuckle walking**, whereas orangutans bear weight on their palms (**Figure 4.18**). Humans, of course, stand upright on two legs. Knuckle walking involves distinctive modifications of the anatomy of the hand, and because human hands show none of these anatomic features, most scientists believed that humans did not evolve from a knuckle-walking species. Because both chimpanzees and gorillas are knuckle walkers, it was generally assumed that this trait evolved in their common ancestor (**Figure 4.19**). However, more recent measurements of genetic similarity have now convinced most scientists that humans and chimpanzees are more closely related to each other than either species is to the gorilla. If this is correct, then the old account of the evolution of locomotion in apes must be wrong. Two accounts are consistent with the new phylogeny. It is possible that the common ancestor of humans, chimpanzees, and gorillas was a knuckle walker and that knuckle walking was retained in the common ancestor of humans and chimpanzees. This would mean that knuckle walking evolved only once and that humans are descended from a knuckle-walking species (**Figure 4.20a**). Alternatively, the common ancestor of humans, chimpanzees, and other apes may not have been a knuckle walker. If this were the case, then knuckle walking evolved independently in chimpanzees and gorillas (**Figure 4.20b**). Each of these scenarios raises interesting questions about the evolution of locomotion in humans and apes. If chimpanzees and gorillas evolved knuckle walking independently, then a close examination should reveal subtle differences in their morphology. If humans evolved from a knuckle walker, then perhaps more careful study will reveal the

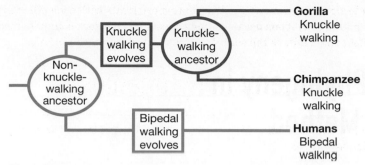

Knuckle walking evolved in the common ancestor of chimpanzees and gorillas.

FIGURE 4.19

If chimpanzees are more closely related to gorillas than to humans, as morphological evidence suggests, then this is the most plausible scenario for the evolution of locomotion in gorillas, chimpanzees, and humans. The fossil record suggests that the ancestor of chimpanzees, gorillas, and humans was neither knuckle walking nor bipedal. The simplest account of the current distribution of locomotion is that knuckle walking evolved once in the common ancestor of chimpanzees and gorillas.

traces of our former mode of locomotion. As we will see in Chapter 10, such traces can be found in the wrists of one of our putative ancestors.

3. *We can deduce the function of morphological features or behaviors by comparing the traits of different species.* This technique is called the **comparative method**. As we will see in Part Two, most primates live in groups. Some scientists have argued that **terrestrial** (ground-dwelling) primates live in larger groups than **arboreal** (tree-dwelling) primates do because terrestrial species are more vulnerable to predators, and animals are safer in larger groups. To test the relationship between group size and terrestriality by using the comparative method, we would collect data on group size and lifestyle (arboreal/terrestrial) for many primate species. However, most biologists believe that only independently evolved cases should be counted in comparative analyses, so we must take the phylogenetic relationships among species into account. **A Closer Look 4.1** provides a hypothetical example of how phylogenetic information can alter our interpretations of comparative data.

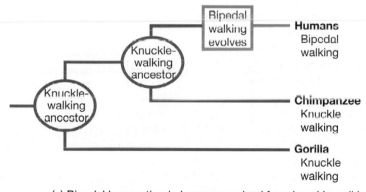

(a) Bipedal locomotion in humans evolved from knuckle walking.

FIGURE 4.20

If humans are more closely related to chimpanzees than to gorillas, as the genetic data suggest, then there are two possible scenarios for the evolution of locomotion in these three species. (a) If the common ancestor of chimpanzees and gorillas was a knuckle walker, then it follows that human bipedal locomotion evolved from knuckle walking. (b) If the common ancestor of all three species was not a knuckle walker, then knuckle walking must have evolved independently in chimpanzees and gorillas.

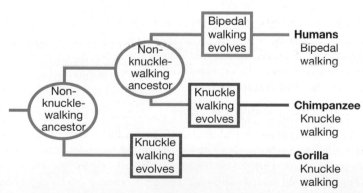

(b) Knuckle walking evolved independently in chimpanzees and gorillas.

Why Reconstruct Phylogenies? ■ **95**

4.1 The Role of Phylogeny in the Comparative Method

To understand why it is important to take phylogeny into account when using the comparative method, consider the phylogeny illustrated in **Figure 4.21**, which shows the pattern of relationships for eight hypothetical primate species.

As you can see, three terrestrial species live in large groups and three arboreal species live in small groups. Only one terrestrial species lives in small groups, and only one arboreal species lives in large groups. Thus, if we based our determinations on living species, we would conclude that there is a statistical relationship between group size and lifestyle. If we count independent evolutionary events, however, we get a very different answer. Large group size and terrestriality are found together only in species B and its descendants: B1, B2, and B3. This combination evolved only once, although we now observe this combination in three living species. There is also only one case of selection creating an arboreal species that lives in small groups (species C and its descendants: C1, C2, and C3). Note that each of the other possible combinations (large arboreal groups and small terrestrial groups) has also evolved once. When we tabulate independent evolutionary events, we find no consistent relationship between lifestyle and group size. Clearly, phylogenetic information is crucial to making sense of the patterns we see in nature.

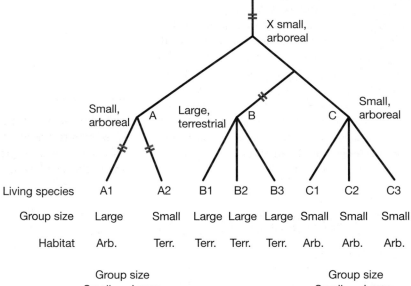

Living species	A1	A2	B1	B2	B3	C1	C2	C3
Group size	Large	Small	Large	Large	Large	Small	Small	Small
Habitat	Arb.	Terr.	Terr.	Terr.	Terr.	Arb.	Arb.	Arb.

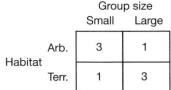

	Group size	
	Small	Large
Habitat Arb.	3	1
Terr.	1	3

Counting living species implies that habitat predicts group size.

	Group size	
	Small	Large
Habitat Arb.	1	1
Terr.	1	1

Counting evolutionary events implies that habitat does *not* predict group size.

FIGURE 4.21

The phylogenetic relationships among eight hypothetical primate species are shown here. Living species lie at the ends of branches, and their ancestors are identified at the branching points of the tree. These species vary in group size (small or large) and lifestyle (arboreal or terrestrial). The lineages in which novel associations between group size and lifestyle first evolved are marked with a double red bar. The left-hand matrix below tallies the association between group size and lifestyle among living species and suggests that, whereas arboreal species live in small groups, terrestrial species live in large ones. The right-hand matrix tallies the number of times the association between group size and lifestyle changed during the evolution of these species. In this case, no relationship between group size and lifestyle is evident. This example shows why it is important to keep track of independent evolutionary events in comparative analyses.

For many years, scientists constructed phylogenies only for classification. The terms *taxonomy* and *systematics* were used interchangeably to refer to the construction of phylogenies and to the use of such phylogenies for naming and classifying organisms. With the recent realization that phylogenies have other important uses such as the ones just described, there is a need for terms that distinguish phylogenetic construction from classification. Here we adopt the suggestion of the Field Museum of Natural History anthropologist Robert Martin to employ the term **systematics** to refer to the construction of phylogenies and the term *taxonomy* to mean the use of phylogenies in naming and classification. Although this distinction may not seem important now, it will become more relevant as we proceed.

How to Reconstruct Phylogenies

We reconstruct phylogenies on the assumption that species with many phenotypic similarities are more closely related than are species with fewer phenotypic similarities.

To see how systematists reconstruct a phylogeny, consider the following example. We begin with three species—named for the moment A, B, and C—whose myoglobin (a cellular protein involved in oxygen metabolism) differs in the ways shown in **Table 4.1**. Remember from Chapter 2 that proteins are long chains of amino acids; the letters in Table 4.1 stand for the amino acids present at various positions along the protein chain.

A systematist wants to find the pattern of descent that is most likely to have produced these data. The first thing she would notice is that all three types of myoglobin have the same amino acid at many positions, exemplified here by the positions 1 (all G), 2 (all L), and 3 (all S). At positions 5, 9, 13, 30, and 34, species A and B have the same amino acids, but species C has a different set. At position 48, species B and C have the same amino acid; and at position 59, A and C have the same amino acid. At position 66, all three species have different amino acids. The systematist would see that there are fewer differences between species A and B than between A and C or between B and C. She would infer that fewer genetic changes have accumulated since A and B shared a common ancestor sometime in the past than since A and C or B and C shared a common ancestor. Because fewer evolutionary changes are necessary to convert A to B than A to C or B to C, A and B are assumed to have a more recent common ancestor than either species shares with C (**Figure 4.22**). In other words, species that are more similar are assumed to be more closely related.

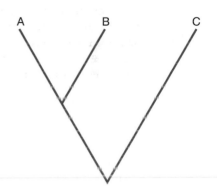

FIGURE 4.22

The phylogenetic tree for species A, B, and C derived from their myoglobin amino acid sequences, portions of which are shown in Table 4.1.

TABLE 4.1										

	Amino Acid Number									
Species	**1**	**2**	**3**	**5**	**9**	**13**	**30**	**34**	**48**	**59**	**66**
A	G	L	S	G	L	V	I	K	H	E	A
B	G	L	S	G	L	V	I	K	G	A	I
C	G	L	S	Q	Q	I	M	H	G	E	Q

The amino acid sequence for the protein myoglobin is shown for three species. The numbers refer to positions on the myoglobin chain, and the letter in each cell stands for the particular amino acid found at that position in each kind of myoglobin. All three species have the same amino acids at all of the positions not shown here, as well as at positions 1 to 3. At eight positions, however, there is at least one discrepancy among the three species (shaded).

FIGURE 4.23

Duck-billed platypuses illustrate the importance of distinguishing between derived and ancestral traits. These creatures lay eggs and have horny bills like birds, but they lactate like mammals.

Let's end the suspense and reveal the identities of the three species. Species A is our own species, *Homo sapiens*; species B is the duck-billed platypus (**Figure 4.23**); and species C is the domestic chicken. The phylogeny shown in Figure 4.22 suggests that humans and duck-billed platypuses (A and B) are more closely related to each other than either of them is to chickens (C). However, just one character (such as myoglobin) doesn't provide nearly enough data to establish a conclusive phylogeny. To be convinced of the relationships among these three organisms, we would need to gather and analyze data on many more characters and generate the same tree each time. And, indeed, many other characters *do* show the same pattern as myoglobin for these organisms. Humans and duck-billed platypuses, for example, have hair and mammary glands (**Figure 4.24a**), but chickens lack these structures. The phylogeny in Figure 4.22 is the currently accepted tree for these three species.

Problems Due to Convergence

In constructing phylogenies, we must avoid basing decisions on characters that are similar because of convergent evolution.

Despite the evidence just discussed, not everything about the phylogeny for humans, platypuses, and chickens is hunky-dory. Two amino acid positions, 48 and 59 (see Table 4.1), are not consistent with the phylogeny shown in Figure 4.22. Moreover, certain other characters do not fit this tree. For example, both platypuses and humans lactate, but chickens do not (Figure 4.24a); both humans and chickens are bipedal, but platypuses are not (**Figure 4.24b**); and both platypuses and chickens lay eggs (**Figure 4.24c**), have a feature of the gut called a cloaca, and sport horny bills, but humans do not. Why don't these characters fit neatly into our phylogeny?

One reason for these anomalies is convergent evolution. Sometimes traits shared by two species are not the result of common ancestry. Instead, they are separate adaptations independently produced by natural selection. Chickens and humans are bipedal *not* because they are descended from the same bipedal ancestor, but rather because they *each* evolved this mode of locomotion independently. Similarly, the horny bills of chickens and platypuses are not similarities due to descent; they are independently derived characters. Systematists say that characters similar because of convergence

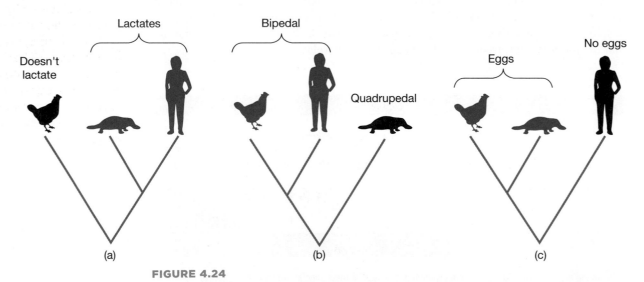

FIGURE 4.24

(a) Many traits, such as lactation, generate the same pattern of relationships among humans, chickens, and platypuses that myoglobin does (see Figure 4.22). (b) Other characters, such as bipedal locomotion, suggest a closer relationship between humans and chickens than between humans and platypuses or between platypuses and chickens. (c) Some traits, such as egg laying, suggest a closer relationship between chickens and platypuses than between chickens and humans or between platypuses and humans.

are **analogous**, whereas characters whose similarity is due to descent from a common ancestor are **homologous**. It is important to avoid using convergent traits in reconstructing phylogenetic relationships. Although it is fairly obvious that bipedal locomotion in humans and chickens is not homologous, convergence is sometimes very difficult to detect.

Problems Due to Ancestral Characters

It is also important to ignore similarity based on ancestral characters, traits that also characterized the common ancestor of the species being classified.

There are also homologous traits that do not fit neatly into correct phylogenies. For example, chickens and platypuses reproduce by laying eggs, whereas humans do not. It seems likely that both chickens and platypuses lay eggs because they are descended from a common egg-laying ancestor. But if these characters are homologous, why don't they allow us to generate the correct phylogeny? (See Figure 4.24c.)

Egg laying is an example of what systematists call an **ancestral trait**, one that characterized the common ancestor of chickens, platypuses, and humans. Egg laying has been retained in the chicken and the platypus but lost in humans. It is important to avoid using ancestral characters when constructing phylogenies. Only **derived traits**—features that have evolved since the time of the last common ancestor of the species under consideration—can be used in constructing phylogenies.

To see why we need to distinguish between ancestral and derived traits, consider the three hypothetical species of "cooties" pictured in **Figure 4.25**. At first glance the red-eyed cootie and the blue-necked cootie seem more similar to each other than either is to the orange-spotted cootie, and a careful count of characters will show that they do share more traits with each other than with the orange-spotted variety. But consider **Figure 4.26**, which shows the phylogeny for the three species. You can see that

Red-eyed cootie Blue-necked cootie Orange-spotted cootie

FIGURE 4.25

The red-eyed cootie and the blue-necked cootie share more traits with each other (such as green legs, blue body, and green antennae) than either of them shares with the orange-spotted cootie. If phylogenetic reconstruction were based on overall similarity, we would conclude that the blue-necked cootie and the red-eyed cootie are more closely related to each other than either is to the orange-spotted cootie.

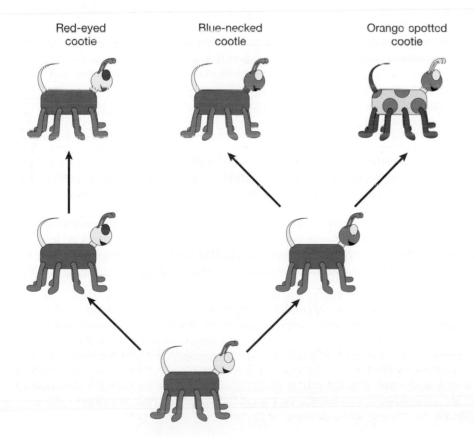

FIGURE 4.26

The phylogenetic history of the three cootie species shown in Figure 4.25 indicates that the blue-necked cootie is actually more closely related to the orange-spotted cootie than to the red-eyed cootie. This is because the blue-necked cootie and the orange-spotted cootie have a more recent common ancestor with each other than with the red-eyed cootie.

FIGURE 4.27

In this phylogeny, only derived characters are shown: red eyes in the red-eyed cootie; orange face in the last common ancestor of blue-necked and orange-spotted cooties; blue neck and orange face in the blue-necked cootie; and red legs and tail, yellow antennae, orange face, light orange body, and orange spots in the orange-spotted species. The correct phylogeny is based on similarity in shared, derived characters.

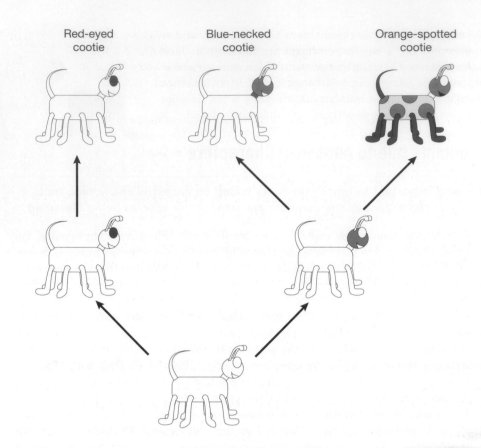

Red-eyed cootie   Blue-necked cootie   Orange-spotted cootie

the blue-necked cootie is actually more closely related to the orange-spotted cootie because they share a more recent common ancestor. The blue-necked cootie seems more similar to the red-eyed cootie because they share many ancestral characters, but the orange-spotted cootie has undergone a period of rapid evolution that has eliminated most ancestral characters. Looking at ancestral traits will not generate the correct phylogeny if rates of evolution differ among species.

If we base our assessment of similarity only on the number of derived characters that each species displays (as shown in **Figure 4.27**), then the most similar species are the ones most closely related. The blue-necked cootie and orange-spotted cootie share one derived character (an orange face) with each other, but they share no derived characters with the red-eyed cootie. Thus if we avoid ancestral characters, we can construct the correct phylogeny.

Systematists distinguish between ancestral and derived characters by using the following criteria: Ancestral characters (1) appear earlier in organismal development, (2) appear earlier in the fossil record, and (3) are seen in out-groups.

It is easy to see that distinguishing ancestral characters from derived characters is important, but it is hard to see how to do it in practice. If you can observe only living organisms, how can you tell whether a particular character in two species is ancestral or derived? There is no surefire solution to this problem, but biologists use three rules of thumb:

1. The development of a multicellular organism is a complex process, and it seems logical that modifications occurring early in the process are likely to be more disruptive than modifications occurring later. As a result, evolution often (but not always) proceeds by modifying the *ends* of existing developmental pathways rather than by modifying earlier stages. To the extent that this generalization is true, characters that occur early in development are ancestral. For example, humans and other apes do not have tails, but the fact that a tail appears in the development of the human

embryo and then disappears is evidence that the tail is an ancestral character (**Figure 4.28**). We conclude from this that the absence of a tail in humans is derived. This reasoning will not be helpful if ancestral traits have been completely lost during development or if traits (such as egg laying) are expressed only in adults.

2. Fossils often provide information about the ancestors of modern species. If we see in the fossil record that all of the earliest likely ancestors of apes had tails, and that primates without tails appear only later in the fossil record, then it is reasonable to infer that having a tail is an ancestral character. This criterion may fail if the fossil record is incomplete and, therefore, a derived character appears in the fossil record before an ancestral one. Finally, we can determine which characters are ancestral in a particular group by looking at neighboring groups, or **out-groups**. Suppose we are trying to determine whether having a tail is ancestral in the primates. We know that monkeys have tails and that apes do not, but we do not know which state is ancestral. To find out, we look at neighboring mammalian groups, such as insectivores or carnivores. Because the members of these out-groups typically have tails, it is reasonable to infer that the common ancestor of all primates also had a tail.

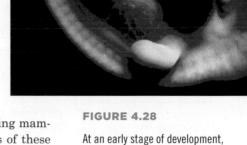

FIGURE 4.28

At an early stage of development, the human embryo has a tail. This feature disappears as development continues.

Using Genetic-Distance Data to Date Phylogenetic Events

Genetic distance measures the overall genetic similarity of two species.

Biologists and anthropologists often find it useful to compute a measure of overall genetic similarity, called **genetic distance**, between pairs of species. There are several ways to measure genetic distance, but here we will focus on genetic-distance estimates based on DNA sequence data. For example, to compute the genetic distance between humans and chimpanzees, the biologist first identifies homologous DNA segments in the two species. This means the segments are descended from the same DNA sequence in the common ancestor of humans and chimpanzees. These DNA segments are then sequenced. The number of nucleotide sites at which the two sequences differ is used to compute the genetic distance between the two species. The mathematical formulas used for these calculations are beyond the scope of this book, but suffice it to say that the more nucleotide differences, the bigger the genetic distance.

Genetic-distance data are often consistent with the hypothesis that genetic distance changes at an approximately constant rate.

The genetic distances among noncoding sequences of the DNA of humans, chimpanzees, gorillas, and orangutans are shown in **Table 4.2**. Notice that gorillas, humans,

TABLE 4.2				
	Human	**Chimpanzee**	**Gorilla**	**Orangutan**
Human	—	1.24	1.62	3.08
Chimpanzee		—	1.63	3.12
Gorilla			—	3.09
Orangutan				—

Genetic distances among humans and the three great ape species, based on sequence divergence in noncoding regions of DNA.

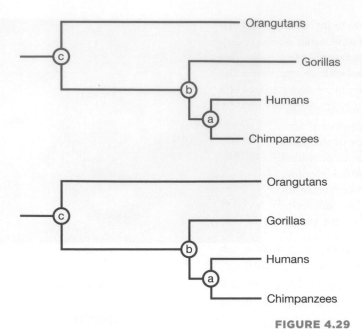

FIGURE 4.29

Trees representing the evolution of genetic distance when the rate of evolution is not constant (top) and is constant (bottom). The last common ancestor of humans and chimpanzees is labeled a; the last common ancestor of humans, chimpanzees, and gorillas is labeled b; and the last common ancestor of all four primate species is labeled c. The amount of genetic distance between two species is represented by the horizontal distance traversed when following the line connecting these species. For example, the genetic distance between humans and gorillas is given by the horizontal distance between humans and a plus the distance from a to b plus the distance from b to gorillas. Because rates of evolution vary in the top graph, the distance between humans and gorillas is not the same as the distance between gorillas and chimpanzees. In the bottom graph the rates are constant and the distances fit the data in Table 4.2.

and chimpanzees are essentially the same genetic distance from orangutans. Also notice that the distance between humans and gorillas is approximately the same as the distance between chimpanzees and gorillas. This pattern is evidence that genetic distance changes at an approximately constant rate. To see why, examine **Figure 4.29** representing the evolution of genetic distance in four species of primates. The letter a labels the last common ancestor of humans and chimpanzees; b labels the last common ancestor of humans, chimpanzees, and gorillas; and c labels the last common ancestor of all four species.

The lengths of the horizontal lines represent the amount of genetic change between any two species. For example, the horizontal distance between c and orangutans represents the change that occurred between the time that orangutans split off from the lineage leading to the other three species and the present. The horizontal distance between b and a represents the amount of genetic change that occurred in the lineage leading to humans and chimpanzees between the time the lineage leading to gorillas split off and the time the lineages leading to humans and chimpanzees diverged. The genetic distance between two living species is just the sum of the distances from the last common ancestor of those two species. The top figure gives a hypothetical example in which evolution proceeds at different rates in different lineages. For example, it is slower in the lineage leading to orangutans and more rapid in the lineage leading to gorillas. You can see that this tree does not fit the genetic-distance data given in Table 4.2. The horizontal distances from c to modern species are not equal, nor are the distances between b and humans and b and chimpanzees. The bottom figure assumes that the rate of genetic change in all lineages is constant. Now the horizontal distances closely fit the data given in the table. Because the time that elapsed since branching off from the last common ancestor is the same for all these species, it follows that the rate of change of genetic distance along both of these paths through the phylogeny must be approximately the same and that the genetic distance between two species is a measure of the time elapsed since both had a common ancestor. Evolutionists refer to genetic distances with a constant rate of change as **molecular clocks** because genetic change acts like a clock that measures the time since two species shared a common ancestor. Data from many other groups of organisms suggest that genetic distances often have this clocklike property, but there are also important exceptions.

Most biologists agree that as long as genetic distances are not too big or too small, the molecular-clock assumption is a useful approximation. Some biologists believe drift and mutation are the key factors influencing rates of change, whereas others argue the molecular clock is controlled more by natural selection.

If the molecular-clock hypothesis is correct, then knowing the genetic distance between two living species allows us to estimate how long ago the two lineages diverged.

The data from contemporary species indicate that genetic distance changes at a constant rate, but the data do not provide a clue as to what that rate might be. However, dated fossils allow us to estimate when the splits between lineages occurred. By dividing the known genetic distance between a pair of species by the time since the last common ancestor, we can estimate the rate at which genetic distance changes through time. For example, fossil evidence indicates that the last common ancestor of orangutans and humans lived about 14 Ma and the genetic distance between these two species is 3.08. Dividing the genetic distance between humans and orangutans by the time since their last common ancestor indicates that genetic distance accumulates at a rate of 0.22 unit per million years. The rate of change can also be estimated by directly estimating the rate of mutation. This method yields somewhat lower estimates of the rate of change of genetic distance.

Once we have an estimate of the rate at which genetic distance changes through time, the molecular-clock hypothesis can be used to date the divergence times for lineages, even when we do not have any fossils. For example, to estimate the last common ancestor of humans and chimpanzees, we divide the genetic distance, 1.24, by the estimated rate, 0.22. This calculation indicates that the last common ancestor of these two species lived about 5.6 Ma.

In practice, scientists use several divergence dates to calibrate the rate at which genetic distance changes through time. Each divergence date produces a slightly different estimate of the rate of change of genetic distance, and this estimate in turn generates different estimates of the divergence times. Thus, divergence dates for humans and chimpanzees range from 8 Ma to 5 Ma.

Taxonomy: Naming Names

The hierarchical pattern of similarity created by evolution provides the basis for the way science classifies and names organisms.

Putting names on things is, to some extent, arbitrary. We could give species names such as Sam or Ruby, or perhaps use numbers like the Social Security system does. This is, in fact, the way common names work. The word *lion* is an arbitrary label, as is "Charles" or "550-72-9928." The problem for scientists is that the number of animals and plants is very large, encompassing far too many species for any individual to keep track of. One way to cope with this massive complexity is to devise a system in which organisms are grouped in a hierarchical system of classification. Once again, there are many possible systems. For example, we could categorize organisms alphabetically, grouping alligators and apricots with the *A*s, barnacles and baboons with the *B*s, and so on. This approach has little to recommend it because knowing how an animal is classified tells us nothing about the organism besides its location in the alphabet. An alternative approach would be to adopt a system of classification that groups organisms with similar characteristics, analogous to the Library of Congress system used by libraries to classify books. The *Q*s might be predators, the *QH*s aquatic predators, the *QP*s aerial predators, and so on. Thus knowing that the scientific name of the red-tailed hawk is QP604.4 might tell you that the hawk is a small aerial predator that lives in North America. The problem with such a system is that not all organisms would fall into a single category. Where would you classify animals that eat both animals and plants, such as bears, or amphibious predators, such as frogs?

The scientific system for naming animals is based on the hierarchy of descent: Species that are closely related are classified together. Closely related species are grouped together in the same **genus** (plural *genera*; **Figure 4.30**). For example, the genus *Pan* contains two closely related species of chimpanzees: the common chimpanzee, *Pan troglodytes*; and

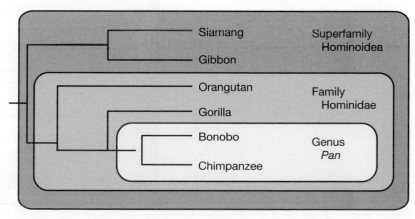

the bonobo, *Pan paniscus*. Closely related genera are usually grouped in a higher unit, often the **family**. Chimpanzees are in the family Hominidae along with orangutans (*Pongo pygmaeus*), gorillas (*Gorilla gorilla*), and humans (*Homo sapiens*). Closely related families are then grouped in a more inclusive unit, often a **superfamily**. The superfamily Hominoidea includes all of the apes.

Taxonomists disagree about whether overall similarity should also be used in classifying organisms.

Most taxonomists agree that descent should play a major role in classifying organisms. However, taxonomists vehemently disagree about whether descent should be the *only* factor used to classify organisms. The members of a relatively new school of thought, called **cladistic taxonomy** (or sometimes *cladistic systematics*), argue that only descent should matter. Adherents to an older school of taxonomy, called **evolutionary taxonomy** (or *evolutionary systematics*), believe that classification should be based both on descent *and* on overall similarity. To understand the difference between these two philosophies, consider **Figure 4.31**, in which humans have been added to the phylogeny of the apes shown in Figure 4.30. Evolutionary taxonomists would say that humans

FIGURE 4.31

Cladistic and evolutionary taxonomic schemes generate two phylogenies for the hominoids. (a) The evolutionary classification classifies humans in a different family from other apes because apes are more similar to one another than they are to humans. (b) According to the cladistic classification, humans must be classified in the same family as the other great apes because they share a common ancestor with these creatures.

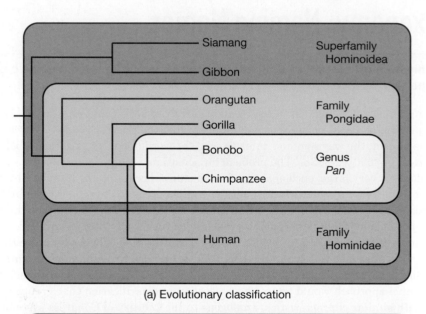

(a) Evolutionary classification

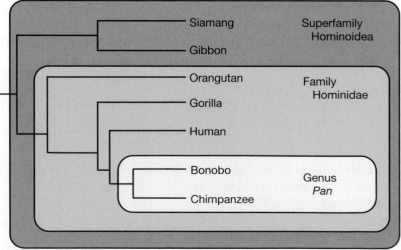

(b) Cladistic classification

are qualitatively different from other apes and so deserve to be distinguished at a higher taxonomic level (Figure 4.31a). Accordingly, these taxonomists classify humans in a family of their own, the Hominidae, and the other great apes would be in the family Pongidae. For a cladist, that approach is unacceptable because humans are descended from the same common ancestor as orangutans, gorillas, and chimpanzees. This means humans *must* be classified in the same family as orangutans, gorillas, and chimpanzees (Figure 4.31b). It is not just chauvinism about our own place in the primate phylogeny that causes discrepancies between these classification schemes. The same problem arises in many other taxa. For example, it turns out that crocodiles and birds share a more recent common ancestor than either does with lizards. For a cladist, that approach means that birds and crocodiles must be classified together, and lizards must be classified separately. Evolutionary taxonomists argue that birds are obviously distinctive and deserve a separate taxonomic grouping.

In theory, cladistic taxonomy is both informative and unambiguous. It is informative because knowing an organism's name and its position in the hierarchy of life tells us how it is related to other organisms. It is unambiguous because the position of each organism is given by the actual pattern of descent. Once you are confident that you understand the phylogenetic relationships within a group, there is no doubt about how any organism in that group should be classified. Cladists believe that evolutionary taxonomy is ambiguous because judgments of overall similarity are necessarily subjective. On the other hand, evolutionary taxonomists complain that the advantages of cladistics are mainly theoretical. In real life, they argue, uncertainty about phylogenetic relationships introduces far more ambiguity and instability in classification than do judgments about overall similarity. For example, until modern high-resolution genetic data became available, it was not clear whether chimpanzees were more closely related to humans or to gorillas. Morphological data suggested that chimpanzees are more closely related to gorillas, and the limited genetic data were not definitive. Given such uncertainty, how would cladists name and classify these species?

It is important to keep in mind that this controversy is not about what the world is like or even about how evolution works; rather, it is a debate about how we should name and classify organisms. Thus no experiment or observation can prove either school right or wrong. Instead, scientists must determine which system is more useful in practice. We have adopted a cladistic taxonomy in this text.

CHAPTER REVIEW

Key Terms

microevolution (p. 82)
macroevolution (p. 82)
testes (p. 82)
biological species concept (p. 83)
reproductive isolation (p. 83)
gene flow (p. 83)
ecological species concept (p. 85)
allopatric speciation (p. 87)

character displacement (p. 88)
reinforcement (p. 88)
parapatric speciation (p. 89)
hybrid zone (p. 89)
sympatric speciation (p. 90)
niche (p. 91)
adaptive radiation (p. 91)
phylogeny (p. 92)
hominoids (p. 92)

taxonomy (p. 94)
quadrupedal (p. 94)
knuckle walking (p. 94)
comparative method (p. 95)
terrestrial (p. 95)
arboreal (p. 95)
systematics (p. 97)
analogous (p. 99)
homologous (p. 99)
ancestral trait (p. 99)

derived trait (p. 99)
out-groups (p. 101)
genetic distance (p. 101)
molecular clocks (p. 102)
genus (p. 103)
family (p. 104)
superfamily (p. 104)
cladistic taxonomy (p. 104)
evolutionary taxonomy (p. 104)

Study Questions

1. If you visit a botanical garden, you may see plants from certain arid regions of Africa and South America that look very much alike.
   (a) What are two explanations for the similarity of these plants?
   (b) It is also known that these continents have been separated for at least 30 million years. Which of the two explanations given in part (a) is consistent with this fact?

2. Chimpanzees and gorillas more closely resemble each other anatomically than either resembles humans. For example, the hands of chimpanzees and gorillas are structurally similar and quite different from human hands. Genetic-distance data suggest, however, that humans and chimpanzees are more closely related to each other than either is to gorillas. Assuming that the genetic-distance data are correct, give two explanations for the observed anatomic similarity between chimpanzees and gorillas.

3. Use the genetic-distance matrix that follows to establish the taxonomic relationships between the species listed. (*Hint:* Draw a phylogenetic tree to illustrate these taxonomic relationships.)

	A	B	C	D
B	4.8	—		
C	0.7	5.0	—	
D	3.6	4.7	3.6	—

4. According to the biological species concept, what is a species? Why do some biologists define species in this way?

5. What is the ecological species concept? Why have some biologists questioned the biological species concept?

6. Molecular methods allow biologists to measure the amount of gene flow among populations that make up a species. When such methods first became available, systematists were surprised to find that many morphologically indistinguishable populations seem to be reproductively isolated from each other and thus, according to the biological species concept, are considered entirely different species. Is it possible to account for the existence of such cryptic species by allopatric speciation? by parapatric speciation? by sympatric speciation?

7. New plant species are sometimes formed by the hybridization of existing species. A new species retains all of the genes of each parent. For example, the variety of wheat used to make bread is a hybrid of three grass species. Explain how such hybridization affects the family tree of these plants.

8. In some areas of Africa, the ranges of different baboon species overlap, and individuals from one species sometimes mate with members of the other species and form a hybrid zone. How can these two species remain distinct if there is interbreeding between them?

9. What is the difference between ancestral traits and derived traits? Why is it important to make this distinction?

10. In this book, you will often see phylogenetic diagrams like the one in Figure 4.16, which can be based on morphological or genetic data. Why do trees based on genes provide more powerful insights than trees based on morphology?

Further Reading

Bergstrom, C. T. and L. A. Dugatkin. 2019. *Evolution.* 2nd ed. New York: Norton.

Darwin, C., and P. Appleman, ed. 2002. *Origin of Species.* New York: Norton.

Dawkins, R. 2015. *The Blind Watchmaker: Why the Evidence of Evolution Reveals a Universe without Design.* New York: Norton, Ch. 11.

Ridley, M. 1986. *Evolution and Classification: The Reformation of Cladism.* New York: Longman.

 Review this chapter with personalized, interactive questions through INQUIZITIVE.

 This chapter also features a Guided Learning Exploration on speciation and ancestral/derived traits.

PART
TWO

PRIMATE ECOLOGY
AND BEHAVIOR

5

PRIMATE DIVERSITY AND ECOLOGY

CHAPTER OBJECTIVES

By the end of this chapter you should be able to:

A Identify the complex of traits that defines the primate order.

B Show where primates live in the world.

C Describe the major characteristics that differentiate one kind of primate from another.

D Describe how primates cope with primary ecological challenges: finding food and avoiding predation.

E Identify what kinds of groups primates form.

F Discuss major factors that threaten the status of wild primate populations.

Two Reasons to Study Primates

The chapters in Part Two focus on the behavior of living nonhuman primates. Studies of nonhuman primates help us understand human evolution for two complementary but distinct reasons. First, closely related species tend to be similar morphologically because, as we saw in Chapter 4, they share traits acquired through descent from a common ancestor. For example, **viviparity** (bearing live young) and lactation are traits that all placental and marsupial mammals share, and these traits distinguish mammals from other taxa, such as reptiles. The existence of such similarities means that studies of living primates often give us more insight into the behavior of our ancestors than do studies of other organisms. This approach is called "reasoning by homology." The second reason we study primates is based on the idea that natural selection favors similar adaptations in similar environments. By assessing the patterns of diversity in the behavior and morphology of organisms in relation to their environments, we can see how evolution shapes adaptation in response to different selective pressures. This approach is called "reasoning by analogy."

Primates Are Our Closest Relatives

Because humans and other primates share many characteristics, other primates provide valuable insights about early humans.

Humans are more closely related to nonhuman primates than to any other animal species. The anatomic similarities among monkeys, apes, and humans led the Swedish naturalist Carolus Linnaeus to place us in the order Primates in the first scientific taxonomy, *Systema Naturae*, published in 1735. Later, naturalists such as Georges Cuvier and Johann Blumenbach placed us in our own order because of our distinctive mental capacities and upright posture. In *The Descent of Man*, however, Charles Darwin firmly advocated reinstating humans in the order Primates; he cited the biologist Thomas Henry Huxley's essay listing the many anatomical similarities between us and apes, and he mused that "if man had not been his own classifier, he would never have thought of founding a separate order for his own reception." Modern systematics unambiguously confirms that humans are more closely related to other primates than to any other living creatures.

Because we are closely related to other primates, we share with them many aspects of morphology, physiology, and development. For example, like other primates, we have well-developed vision and grasping hands and feet. We share features of our life history with other primates as well, including an extended period of juvenile development and relatively long life span, as well as larger brains in relation to body size than that found in members of other taxonomic groups. Homologies between humans and other primates also extend to behavior because the physiologic and cognitive structures that underlie human behavior are more similar to those of other primates than to members of other taxonomic groups. The existence of this extensive array of homologous traits, the product of the common evolutionary history of the primates, means that nonhuman primates provide useful models for understanding the evolutionary roots of human morphology and for unraveling the origins of human nature.

Primates Are a Diverse Order

Diversity within the primate order helps us understand how natural selection shapes behavior.

During the past 50 years, hundreds of researchers from a variety of academic disciplines have spent thousands of hours observing many species of nonhuman primates in the wild, in captive colonies, and in laboratories. All primate species have evolved adaptations that enable them to meet the basic challenges of life, such as finding food, avoiding predators, obtaining mates, rearing young, and coping with competitors. At the same time, there is great morphological, ecological, and behavioral diversity among species within the primate order. For example, primates range in size from the pygmy mouse lemur, which weighs about 30 g (about 1 oz.), to the male gorilla, which weighs about 260 times more—160 kg (350 lb.). Some species live in dense tropical forests; others are at home in open woodlands and savannas. Some subsist almost entirely on leaves; others rely on an omnivorous diet of fruits, leaves, flowers, seeds, gum, nectar, insects, and small animal prey. Some species are solitary, and others are highly gregarious. Some are active at night (**nocturnal**); others are active during daylight hours (**diurnal**). One primate, the fat-tailed dwarf lemur, enters a torpid state and sleeps for 6 months each year. Some species actively defend territories from incursions by other members of their own species (**conspecifics**); others do not. In some species, only females provide care of their young; in others, males participate actively in this process.

This variety is inherently interesting. However, evidence of diversity among closely related organisms living under somewhat different ecological and social conditions also helps researchers understand how evolution shapes behavior. Animals that are closely related to one another phylogenetically tend to be very similar in morphology, physiology, life history, and behavior. Thus, differences observed among closely related species are likely to represent adaptive responses to specific ecological conditions. At the same time, similarities among more distantly related creatures living under similar ecological conditions are likely to be the product of convergence.

This approach, sometimes called the *comparative method*, has become an important form of analysis as researchers attempt to explain the patterns of variation in morphology and behavior observed in nature. The same principles have been borrowed to reconstruct the behavior of extinct hominins, early members of the human lineage. Because behavior leaves little trace in the fossil record, the comparative method provides one of our only objective means of testing hypotheses about the lives of our hominin ancestors. For example, the observation that substantial differences exist in male and female body size—a phenomenon called **sexual dimorphism**—in species in which males compete over access to females and form groups that contain one male and multiple females or multiple males and multiple females suggests that highly dimorphic hominins may have lived in similar groups. In Part Three, we will see how the data and theories about behavior produced by primatologists have played an important role in reshaping our ideas about human origins.

Features That Define the Primates

Members of the primate order are characterized by several shared, derived characters, but not all primates share all of these traits.

The animals pictured in **Figure 5.1** are all members of the primate order. These animals are similar in many ways: They are covered with a thick coat of hair, they have four limbs, and they have five fingers on each hand. They give birth to live young, and mothers suckle their offspring. However, they share these ancestral features with all mammals. Beyond these ancestral features, it is hard to see what the members of this group of animals have in common that makes them distinct from other mammals. What distinguishes a ring-tailed lemur from a mongoose or a raccoon? What features link the elegant leaf monkey and the bizarre aye-aye?

In fact, primates are a rather nondescript mammalian order that cannot be unambiguously characterized by a single derived feature shared by all members. In his extensive treatise on primate evolution, however, biologist Robert Martin of the Field Museum of Natural History in Chicago defines the primate order in terms of the derived features listed in **Table 5.1**.

The first two traits in Table 5.1 are related to the flexible movement of hands and feet. Primates can grasp with their hands and feet (**Figure 5.2a**), and most monkeys

FIGURE 5.1

All of these animals are primates: (a) aye-aye, (b) ring-tailed lemur, (c) leaf monkey, (d) howler, and (e) gelada monkey. Primates are a diverse order and do not possess a suite of traits that unambiguously distinguishes them from other animals.

[a]   [b]   [c]

[d]   [e]

TABLE 5.1

1. The big toe on the foot is **opposable**, and hands are **prehensile**. This means primates can use their feet and hands for grasping. The opposable big toe has been lost in humans.

2. There are flat nails on the hands and feet in most species, instead of claws, and there are sensitive tactile pads with "fingerprints" on the fingers and toes.

3. Locomotion is **hind-limb dominated**, meaning that the hind limbs do most of the work, and the center of gravity is nearer to the hind limbs than to the forelimbs.

4. There is an unspecialized **olfactory** (smelling) apparatus that is reduced in diurnal primates.

5. The visual sense is highly developed. The eyes are large and moved forward in the head, providing stereoscopic vision.

6. Females have small litters, and gestation and juvenile periods are longer than in other mammals of similar size.

7. The brain is larger than the brains of similarly sized mammals, and it has several unique anatomic features.

8. The **molars** are relatively unspecialized, and there is a maximum of two **incisors**, one **canine**, three **premolars**, and three molars on each side of the upper and lower jaws.

9. There are several other subtle anatomical characteristics that are useful to systematists but are hard to interpret functionally.

Definition of the primate order. See the text for more complete descriptions of these features.

[a]

[b]

[c]

FIGURE 5.2

(a) Primates have grasping feet, which they use to climb, cling to branches, hold food, and scratch themselves. (b) Primates can oppose the thumb and forefinger in a precision grip—a feature that enables them to hold food in one hand while they are feeding, to pick small ticks and bits of debris from their hair while grooming, and (in some species) to use tools. (c) Most primates, like this squirrel monkey, have flat nails on their hands and sensitive tactile pads on the tips of their fingers.

[a]

[b]

[c]

FIGURE 5.3

Primates display a variety of different forms of locomotion: (a) vertical clinging and leaping, (b) walking on top of branches, (c) leaping from one support to another, (d) climbing, and (e) swinging by their arms underneath branches.

[d]

[e]

FIGURE 5.4

In most primates, the eyes are moved forward in the head. The field of vision of the two eyes overlaps, creating binocular, stereoscopic vision.

and apes can oppose their thumb and forefinger in a precision grip (**Figure 5.2b**). The flat nails, distinct from the claws of many animals, and the tactile pads on the tips of primate fingers and toes further enhance their dexterity (**Figure 5.2c**). These traits enable primates to use their hands and feet differently from the ways most other animals do. Primates can grasp fruit, squirming insects, and other small items in their hands and feet, and they can grip branches with their fingers and toes. During grooming sessions, a primate delicately parts its partner's hair and uses its thumb and forefinger to remove small bits of debris from the skin.

The third trait in Table 5.1 is related to how primates move about their environments. All nonhuman primates spend at least part of their time in trees, and most move quadrupedally on relatively small supports. Some primates walk or run along branches (and along the ground); some leap from one horizontal branch to another; some cling to a vertical support (such as a tree trunk) and launch themselves through the air to another vertical support, a form of locomotion called vertical clinging and leaping; and humans move upright along the ground (**Figure 5.3**). Powerful hind limbs play an important role in walking, running, jumping, and climbing. Some primates, such as apes, swing along underneath branches using a form of locomotion called **brachiation**.

Traits 4 and 5 in Table 5.1 are related to a shift in emphasis among the sense organs. Most primates are characterized by a greater reliance on visual stimuli and less reliance on olfactory stimuli than other mammals. Many primate species can perceive color, and their eyes are set forward in the head, providing them with binocular, stereoscopic vision (**Figure 5.4**). **Binocular vision** means that the fields of vision of the two eyes overlap so that both eyes perceive the same image. **Stereoscopic vision** means that each eye sends a signal of the visual image to both hemispheres in the brain to create an image with depth. These trends are not uniformly expressed within the primate order; for example, olfactory cues play a more important role in the lives of **strepsirrhine** primates than in the lives of **haplorrhine** primates. As we will explain shortly, the strepsirrhine primates include the lorises and lemurs, and the haplorrhine primates include tarsiers, monkeys, and apes.

Features 6 and 7 in Table 5.1 result from the distinctive life history of primates. As a group, primates have longer pregnancies, mature at later ages, live longer, and have larger brains than other animals of similar body size. These features reflect a progressive trend toward increased dependence on complex behavior, learning, and behavioral flexibility within the primate order. As the late primatologist Alison Jolly pointed out, "If there is an essence of being a primate, it is the progressive evolution of intelligence as a way of life." As we will see in the chapters that follow, these traits profoundly affect mating and parenting strategies and the patterns of social interaction within primate groups.

The eighth feature in Table 5.1 concerns primate dentition. Teeth play a very important role in the lives of primates and in our understanding of their evolution. The utility of teeth to primates themselves is straightforward: Teeth are necessary for processing food and are also used as weapons in conflicts with other animals. Teeth are also useful features for researchers who study living and fossil primates. Primatologists sometimes rely on tooth wear to gauge the age of individuals, and they use features of the teeth to assess the phylogenetic relationships among species. As we will see, paleontologists often rely on teeth, which are hard and preserve well, to identify the phylogenetic relationships of extinct creatures and to make inferences about their developmental patterns, their dietary preferences, and their social structure. **A Closer Look 5.1** describes primate dentition in greater detail.

Although these traits are generally characteristic of primates, you should keep two points in mind. First, none of the traits makes primates unique. Dolphins, for example, have large brains and extended periods of juvenile development, and their social behavior may be just as complicated and flexible as that of any nonhuman primate (**Figure 5.5**) Second, not every primate possesses all of these traits. Humans have lost the grasping big toe that characterizes other primates, some strepsirrhine primates have claws on some of their fingers and toes, and not all monkeys have color vision.

FIGURE 5.5

A high degree of intelligence characterizes some animals besides primates. Dolphins, for example, have very large brains in relation to their body size, and their behavior is quite complex.

Primate Biogeography

Primates are restricted mainly to tropical regions of the world.

The continents of Asia, Africa, and South America and the islands that lie near their coasts are home to most of the world's nonhuman primates (**Figure 5.6**). A few species remain in Mexico and Central America. Nonhuman primates were once found in southern Europe, but no natural population survives there now. There are no natural populations in Australia or Antarctica, and none occupied these continents in the past.

Nonhuman primates are found mainly in tropical regions, where the fluctuations in temperature from day to night greatly exceed fluctuations in temperature during the year. In

FIGURE 5.6

The distribution of living and fossil nonhuman primates. Primates are now found in Central America, South America, Africa, and Asia. They are found mainly in tropical regions of the world. Primates were formerly found in southern Europe and northern Africa. There have never been indigenous populations of primates in Australia or Antarctica.

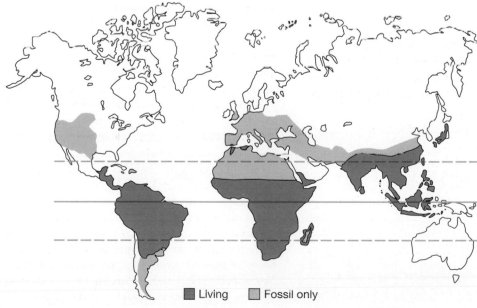

Living   Fossil only

the tropics, the distribution of resources that primates rely on for subsistence is affected more strongly by seasonal changes in rainfall than by seasonal changes in temperature. Some species extend their ranges into temperate areas of Africa and Asia, where they manage to cope with substantial seasonal fluctuations in environmental conditions.

Nonhuman primate species occupy an extremely diverse set of habitats, including all types of tropical forests, savanna woodlands, mangrove swamps, grasslands, high-altitude plateaus, and deserts. Almost all species, however, are found in forested areas, where they travel, feed, socialize, and sleep in the safety of trees.

A Taxonomy of Living Primates

Scientists classify primates into two suborders: Strepsirrhini and Haplorrhini (**Table 5.2**). Many of the primates included in the suborder Strepsirrhini are nocturnal, and, like some of the earliest primates that lived 50 Ma, they have many adaptations

TABLE 5.2

Suborder	Infraorder	Superfamily	Family	Subfamily	Examples
Strepsirrhini	Lemuriformes	Lemuroidea	Cheirogaleidae		Dwarf lemurs, mouse lemurs
			Daubentoniidae		Aye-ayes
			Indriidae		Indris, sifakas
			Lemuridae		Lemurs
			Lepilemuridae		Sportive lemurs
	Lorisiformes	Lorisoidea	Galagidae	Galaginae	Galagos
			Lorisidae	Lorisinae	Lorises
				Perodicticinae	Pottos
Haplorrhini	Tarsiiformes	Tarsioidea	Tarsiidae	Tarsinae	Tarsiers
	Platyrrhini	Ceboidea	Atelidae	Alouattinae	Howler monkeys
				Atelinae	Spider monkeys
			Cebidae	Aotinae	Owl monkeys
				Callitrichinae	Marmosets, tamarins
				Cebinae	Capuchins
				Saimirinae	Squirrel monkeys
			Pitheciidae	Callicebinae	Titi monkeys
				Pitheciinae	Sakis, uakaris
	Catarrhini	Cercopithecoidea	Cercopithecidae	Cercopithecinae	Mangabeys, macaques, vervets, baboons
				Colobinae	Langurs, colobus, leaf monkeys
		Hominoidea (apes, humans)	Hylobatidae		Gibbons, siamang
			Hominidae	Ponginae	Orangutans
				Homininae	Gorillas, chimpanzees, humans

Taxonomy of the living primates.

to living in darkness, including a well-developed sense of smell, large eyes, and independently movable ears. By contrast, monkeys, apes, and humans, which make up the suborder Haplorrhini, evolved adaptations more suited to a diurnal lifestyle early in their evolutionary history. In the Haplorrhini, traits related to increased complexity of behavior, including large brains and longer life spans, are most fully developed. Haplorrhine monkeys are generally larger than strepsirrhines, are active during the day, are more fully dependent on vision than smell, and live in bigger and more complex social groups.

The classification of the primates that we have adopted here reflects the pattern of descent within the order. Tarsiers are included in the haplorrhines because genetic data indicate that they are more closely related to monkeys and apes than to the strepsirrhines. However, like many of the strepsirrhines, they are small-bodied and nocturnal. A cladistic classification places tarsiers within the Haplorrhini, but an evolutionary taxonomy would group tarsiers with strepsirrhines because of their overall similarity in morphology and behavior.

Primate Diversity

The Strepsirrhines

The strepsirrhine primates are divided into two infraorders: Lemuriformes and Lorisiformes.

The **infraorder** Lemuriformes includes lemurs, which are found only on Madagascar and the Comoro Islands, off the southeastern coast of Africa. These islands have been separated from Africa for 120 million years. The primitive primates that reached Madagascar evolved in total isolation from primates elsewhere in the world and from many of the predators and competitors that primates confront in other places. Faced with a diverse set of available ecological niches, the lemurs underwent a spectacular adaptive radiation. When humans first colonized Madagascar about 2,000 years ago, there were approximately 44 species of lemurs, some as small as mouse lemurs and others as big as gorillas. In the next few centuries, all of the larger lemur species became extinct, probably the victims of human hunters or habitat loss. The extant lemurs are mainly small- or medium-size arboreal residents of forested areas (**Figure 5.9a**). They travel quadrupedally or by vertical clinging and leaping (**Figure 5.9b**). Activity patterns of lemurs are quite variable: About half are primarily diurnal, others are nocturnal, and some are active during both day and night. One of the most interesting aspects of lemur behavior is that females routinely dominate males. In most lemur species, females can supplant males from desirable feeding sites; and in some lemur species, females regularly defeat males in aggressive encounters. Although such behavior may seem unremarkable in our own liberated times, female dominance is rare in other primate species.

The infraorder Lorisiformes is composed of small, nocturnal, arboreal residents of the forests of Africa and Asia. These animals include two families with different locomotion and activity patterns. Galagos are active and agile, leaping through the trees and running quickly along the tops of branches (**Figure 5.9c**). The lorises and pottos move with ponderous deliberation, and their wrists and ankles have a specialized network of blood vessels that allows them to remain immobile for long periods. These traits may be adaptations that help them avoid detection by predators. Traveling alone, the Lorisiformes generally feed on fruit, gum, and insect prey. They leave their dependent offspring in nests built in the hollows of trees or hidden in masses of tangled vegetation. During the day, females sleep, nurse their young, and groom, sometimes in the company of mature offspring or familiar neighbors.

5.1 Teeth and Guts: You Are What You Can Chew and Digest

For various reasons, biological anthropologists spend a lot of time thinking about teeth. Teeth are useful markers for taxonomic identity because various primates have different numbers of teeth. Teeth are also useful because they tell us things about what kinds of food primates eat. If we can detect a relationship between dental morphology and diet, we can apply these insights to the fossil record. This is particularly handy because teeth are the most commonly preserved parts of the body. Finally, teeth and gut morphology provide examples of how natural selection has created adaptations that enable animals to cope with their environments more effectively.

Dental Formula

To appreciate the basic features of primate dentition, you can consult **Figure 5.7** or you can simply look in a mirror because your teeth are much like those of other primates. Teeth are rooted in the jaw. The jaw holds four kinds of teeth: In order, they are, first, the incisors at the front; then come the canines, premolars, and the molars in the rear. All primates have the same kinds of teeth, but species vary in how many of each kind of tooth they have. For convenience, these combinations are expressed in a standard format called the **dental formula**, which is commonly written in the following form:

$$\frac{2.1.3.3}{2.1.3.3}$$

Reading from left to right, the numerals tell us how many incisors, canines, premolars, and molars a particular species has (or had) on one side of its jaws. The top line of numbers represents the teeth on one side of the upper jaw (**maxilla**), and the bottom line represents the teeth on the corresponding side of the lower jaw (**mandible**). Usually, but not always, the formula is the same for both upper and lower jaws. Like most other parts of the body, our dentition is **bilaterally symmetrical**, which means that the left side is identical to the right side.

The dental formulas among living primates vary (**Table 5.3**). The lorises, pottos, galagos, and several lemurids have $\frac{2.1.3.3}{2.1.3.3}$ dental formula, but other strepsirrhine taxa have lost incisors, canines, or premolars. Tarsiers have lost one incisor on the mandible but have retained two on the maxilla. All of the platyrrhine monkeys, except the marmosets and tamarins, have retained the same dental formula as most of the strepsirrhines; the marmosets and tamarins have lost one molar. The catarrhine monkeys, apes, and humans have only two premolars.

Dental Morphology

Primates who rely heavily on gum for food tend to have large and prominent incisors, which they use to gouge holes in the bark of trees (**Figure 5.8**). In some strepsirrhine species, the incisors and canines are projected forward in the jaw and are used to scrape hardened gum off the surface

TABLE 5.3		
Primate Taxa		**Dental Formula**
Strepsirrhines	Lorises, pottos, galagos, dwarf lemurs, mouse lemurs, true lemurs	$\frac{2.1.3.3}{2.1.3.3}$
	Indris	$\frac{2.1.2.3}{2.0.3.3}$
	Aye-ayes	$\frac{1.0.1.3}{1.0.0.3}$
Haplorrhines	Tarsiers	$\frac{2.1.3.3}{1.1.3.3}$
	Platyrrhine monkeys (most species)	$\frac{2.1.3.3}{2.1.3.3}$
	Marmosets, tamarins	$\frac{2.1.3.2}{2.1.3.2}$
	Catarrhine monkeys, apes, humans	$\frac{2.1.2.3}{2.1.2.3}$

Primates vary in the numbers of each type of tooth that they have. The dental formulas listed here give the number of incisors, canines, premolars, and molars on each side of the upper jaw (maxilla) and lower jaw (mandible).

FIGURE 5.7

The upper jaw (left) and lower jaw (right) are shown here for a male colobus monkey (a) and a male gorilla (b). In catarrhine monkeys, the prominent anterior and posterior cusps of the lower molars form two parallel ridges. In apes, the five cusps of the lower molar form a Y-shaped pattern.

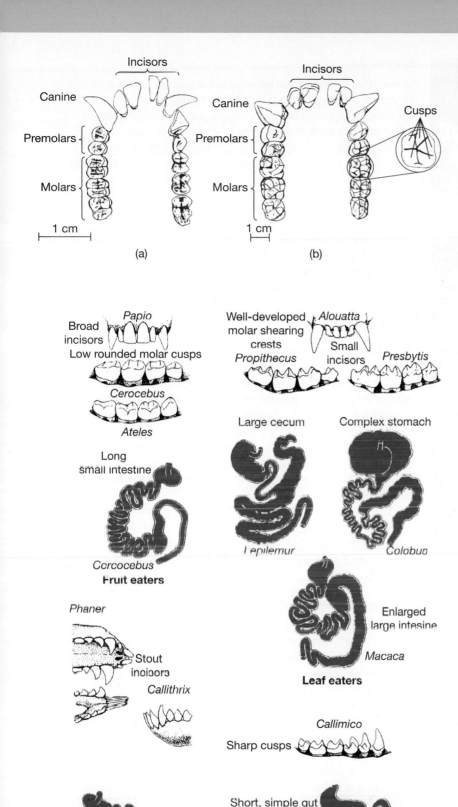

of branches and tree trunks. Dietary specializations are also reflected in the size and shape of the molars. Primates who feed mainly on insects and leaves have molars with well-developed shearing crests that permit them to cut their food into small pieces when they chew. Insectivores tend to have higher and more pointed cusps on their molars, which are useful for puncturing and crushing the bodies of their prey. The molars of frugivores tend to have flatter, more rounded cusps, with broad and flat areas used to crush their food. Primates who rely on hard seeds and nuts have molars with very thick enamel that can withstand the heavy chewing forces needed to process these types of food.

Guts

Primates who feed principally on insects or animal prey have relatively simple digestive systems that are specialized for absorption. They generally have a simple small stomach, a small cecum (a pouch located at the upper end of the large intestine), and a small colon in relation to the rest of the small intestine. Frugivores also tend to have simple digestive systems, but frugivorous species with large bodies have capacious stomachs to hold large quantities of the leaves they consume along with the fruit in their diet. Folivores have the most specialized digestive systems because they must deal with large quantities of cellulose and secondary plant compounds. Because primates cannot digest cellulose or other structural carbohydrates directly, folivores maintain colonies of microorganisms in their digestive systems that break down these substances. In some species, these colonies of microorganisms are housed in an enlarged cecum; in other species, the colon is enlarged for this purpose. Colobines, for example, have an enlarged and complex stomach divided into several sections where microorganisms help process cellulose.

FIGURE 5.8

The dentition and digestive tracts of fruit-eating (frugivorous), leaf-eating (folivorous), gum-eating (gummivorous), and insect-eating (insectivorous) primates typically differ.

[a]

[b]

[c]

FIGURE 5.9

(a) Ring-tailed lemurs, with their distinctive striped tails, live in social groups and are active during daylight hours. In several lemur species, females are dominant over males. (b) Sifakas use their powerful legs to jump in an upright posture, a form of locomotion known as vertical clinging and leaping. (c) Galagos are small, arboreal, nocturnal animals that can leap great distances. They are mainly solitary, though residents of neighboring territories sometimes rest together during the day.

The Haplorrhines

The suborder Haplorrhini contains three infraorders: Tarsiiformes, Platyrrhini, and Catarrhini.

The infraorder Tarsiiformes includes tarsiers, which are enigmatic primates that live in the rain forests of Borneo, Sulawesi, and the Philippines (**Figure 5.10**). Like many of the strepsirrhine primates, tarsiers are small, nocturnal, and arboreal, and they move by vertical clinging and leaping. Some tarsiers live in pair-bonded family groups, but many groups have more than one breeding female. Female tarsiers give birth to infants that weigh 25% of their own weight; mothers leave their bulky infants behind in safe hiding places when they forage for insects. Tarsiers are unique among primates because they are the only primates that rely exclusively on animal matter for food, feeding on insects and small vertebrate prey.

The two other infraorders, Plattyrrhini and Catarrhini, correspond to the geographic regions of the world in which their members are found. Platyrrhine monkeys are found in South and Central America and southern Mexico, whereas catarrhine monkeys and apes are found in Africa and Asia. This geographic dichotomy breaks down with humans, however: We are catarrhine primates, but we are spread over the globe.

The infraorder Platyrrhini is divided into three families: Atelidae, Cebidae, and Pitheciidae. Although these monkeys encompass considerable diversity in size, diet, and social organization, they do share some basic features. All but those in one genus are diurnal, all live in forested areas, and all are mainly arboreal. Platyrrhine monkeys range in size from the 600-g (21-oz.) squirrel monkey to the 9.5-kg (21-lb.) muriqui (**Figure 5.11**). Most platyrrhine monkeys are quadrupedal, moving along the tops of branches and jumping between adjacent trees. Some species in the family Atelidae can suspend themselves by their hands, feet, or tail and can move by swinging by their arms beneath branches. Although many people think that all monkeys can swing by their tails, prehensile tails are actually restricted to the largest species of platyrrhine monkeys.

The family Atelidae is composed of howler monkeys, spider monkeys, woolly monkeys, and muriquis. Howler monkeys are named for their long-distance roars in intergroup interactions. They live in one-male or small multimale groups, defend their home ranges, and feed mainly on leaves. Spider monkeys, woolly monkeys, and muriquis subsist mainly on fruit and leaves, and they live in multimale, multifemale groups of 15 to 25. Spider monkeys (Figure 5.11b) rely heavily on ripe fruit, which is often found in patches that are not big enough to feed the entire group, so groups often break up into temporary subgroups for feeding, a pattern that is called fission–fusion social

FIGURE 5.10

Tarsiers are small, insectivorous primates that live in Asia. Some tarsiers form pair bonds.

[a]

[b]

[c]

[d]

FIGURE 5.11

Portraits of some platyrrhine monkeys. (a) Muriquis, or woolly spider monkeys, are large-bodied and arboreal. They are extremely peaceful creatures, rarely fighting or competing over access to resources. (b) Spider monkeys rely heavily on ripe fruit and travel in small parties. They have prehensile tails that they can use much like an extra hand or foot. (c) Capuchin monkeys have a larger brain in relation to their body size than any of the other nonhuman primates. (d) Squirrel monkeys form large multimale, multifemale groups. In the mating season, males gain weight and become "fatted" and then compete actively for access to receptive females.

organization. Muriquis and spider monkeys (Figure 5.11a and b) are unusual among primates because females disperse from their natal (birth) groups when they reach sexual maturity, whereas males remain in their natal groups for life. As we will see, female dispersal and fission–fusion social organization also characterize chimpanzees.

The family Cebidae includes capuchins, owl monkeys, squirrel monkeys, marmosets, and tamarins. Owl monkeys, which form pair bonds and defend territories, are the only nocturnal haplorrhine primates. Capuchin monkeys (Figure 5.11c) are notable, in part, because they have very large brains in relation to their body size (see Chapter 8). They display several behavioral traits that play an important role in thinking about human origins, including tool use, social learning, and the development of behavioral traditions. Capuchins and squirrel monkeys (Figure 5.11d) live in multimale, multifemale groups of 10 to 50 individuals and forage for fruit, leaves, and insects.

The marmosets and tamarins, which belong to the subfamily Callitrichinae, share several morphological features that distinguish them from other haplorrhine primate species: They are extremely small, the largest weighing less than 1 kg (2.2 lb.); they have claws instead of nails; they have only two molars, whereas all other monkeys have three; and they often give birth to twins (**Figure 5.12**). Marmosets and tamarins are also notable for their domestic arrangements: they are **cooperative breeders**. This means there is a single breeding female, and all the other group members help rear the offspring.

FIGURE 5.12

Marmosets are small-bodied South American monkeys that form pair-bonded or polyandrous social groups. Males and older offspring actively participate in the care of infants.

The family Pitheciidae includes the titi monkey, which lives in pair-bonded family groups. This family also includes the uakaris and sakis, which are not yet very well studied in the wild. Whereas most primates that eat fruit swallow or spit out the seeds, which are rich in lipids, the sakis are specialized seed eaters.

The infraorder Catarrhini contains the monkeys and apes of Africa and Eurasia and humans.

As a group, the catarrhine primates share several anatomic and behavioral features that distinguish them from the platyrrhine primates. For example, most catarrhine monkeys and apes have narrow nostrils that face downward, whereas platyrrhine monkeys have round nostrils. Catrrhine primates have two premolars on each side of the upper and lower jaws; platyrrhine monkeys have three. Most catarrhine primates are larger than most platyrrhine species, and catarrhine monkeys and apes occupy a wider variety of habitats than platyrrhine species do.

The catarrhine primates are divided into two superfamilies: Cercopithecoidea (catarrhine monkeys) and Hominoidea (apes and humans). Cercopithecoidea contains one extant (still living) family, which is further divided into two subfamilies of monkeys: Cercopithecinae and Colobinae.

The superfamily Cercopithecoidea encompasses great diversity in social organization, ecological specializations, and biogeography.

Members of the subfamily Colobinae, which includes the colobus monkeys of Africa and the langurs and leaf monkeys of Asia, may be the most elegant of the primates (**Figure 5.13**). They have slender bodies, long legs, long tails, and often beautifully colored coats. The guereza colobus monkey, for example, has a white ring around its black face, a striking white cape on its black back, and a bushy white tail that flies out behind as it leaps from tree to tree. These monkeys are mainly leaf and seed eaters, and most species spend much of their time in trees. They have complex stomachs, almost like the chambered stomachs of cows, which allow them to maintain bacterial colonies that

FIGURE 5.13

(a) African colobines, such as these guereza colobus monkeys, are arboreal and feed mainly on leaves. These animals are sometimes hunted for their spectacular coats. (b) Gray langurs, also known as Hanuman langurs, are native to India and have been the subject of extensive study during the past four decades. In some areas, gray langurs form one-male, multifemale groups, and males engage in fierce fights over membership in bisexual groups. In these groups, infanticide often follows when a new male takes over the group.

[a]

[b]

[a] [b] [c]

FIGURE 5.14

Some representative cercopithecines: (a) Bonnet macaques are one of several species of macaques that are found throughout Asia and North Africa. Like other macaques, bonnet macaques form multimale, multifemale groups, and females spend their entire lives in their natal (birth) groups. (b) Vervet monkeys are found throughout Africa. Like macaques and baboons, females live among their mothers, daughters, and other maternal kin. Males transfer to nonnatal groups when they reach maturity. Vervets defend their ranges against incursions by members of other groups. (c) Blue monkeys live in one-male, multifemale groups. During the mating season, however, one or more unfamiliar males may join bisexual groups and mate with females.

facilitate the digestion of cellulose. Colobines, langurs, and leaf monkeys are most often found in groups composed of one adult male and several adult females. As in many other vertebrate taxa, the replacement of resident males in one-male groups is often accompanied by lethal attacks on infants by new males. Infanticide under such circumstances is believed to be favored by selection because it improves the relative reproductive success of infanticidal males. This issue is discussed more fully in Chapter 6.

Most cercopithecine monkeys are found in Africa, though one particularly adaptable genus (*Macaca*) is widely distributed through Asia and part of northern Africa (**Figure 5.14**). The cercopithecines occupy a wide variety of habitats and are quite variable in body size and dietary preferences. The social behavior, reproductive behavior, life history, and ecology of several cercopithecine species (particularly baboons, macaques, and vervets) have been studied extensively and will figure prominently in the discussions of mating strategies and social behavior in Chapters 6 and 7. Cercopithecines typically live in medium or large one-male or multimale groups. Females typically remain in their natal groups (the groups into which they are born) throughout their lives and establish close and enduring relationships with their maternal kin; males leave their natal groups and join new groups when they reach sexual maturity.

The superfamily Hominoidea includes two families of apes: Hylobatidae (gibbons) and Hominidae (orangutans, gorillas, chimpanzees, and humans).

The hominoids differ from the cercopithecoids in several ways. The most readily observed difference between apes and monkeys is that apes lack tails. But there are many other more subtle differences between apes and monkeys. For example, the apes share some derived traits, including broader noses, broader palates, and larger brains; and they retain some primitive traits, such as relatively unspecialized molars. In catarrhine monkeys the prominent anterior and posterior cusps are arranged to form two parallel ridges. In apes, the five cusps on the lower molars are arranged to form a side-turned, Y-shaped pattern of ridges.

[a]

[b]

FIGURE 5.15

(a) Gibbons and (b) siamangs live in pair-bonded groups and actively defend their territories against intruders. They have extremely long arms, which they use to propel themselves from one branch to another as they swing hand over hand through the canopy, a form of locomotion called brachiation. Siamangs and gibbons are confined to the tropical forests of Asia. As with other residents of tropical forests, their survival is threatened by the rapid destruction of tropical forests.

The family Hylobatidae, sometimes called lesser apes, includes gibbons and siamangs, and its living members are now found in Asia. The family Hominidae includes the larger-bodied great apes (orangutans, gorillas, bonobos, chimpanzees, and humans). Orangutans are found in Asia, whereas chimpanzees, bonobos, and gorillas are restricted to Africa.

The lesser apes are slightly built creatures with extremely long arms in relation to their body size (**Figure 5.15**). Gibbons and siamangs are strictly arboreal, and they use their long arms to perform spectacular acrobatic feats, moving through the canopy with grace, speed, and agility. Gibbons and siamangs are the only true brachiators among the primates, propelling themselves by their arms alone, and are in free flight between handholds. (The monkey bars in your elementary school playground should really be called ape bars.) Gibbons and siamangs typically live in pair-bonded family groups, vigorously defend their home ranges (the areas they occupy), and feed on fruit, leaves, flowers, and insects. Siamang males play an active role in caring for young, often carrying them during the day; male gibbons are less attentive fathers. In territorial displays, mated pairs of siamangs perform coordinated vocal duets that can be heard over long distances.

Orangutans are the only extant great apes outside Africa. One species, *Pongo pygmaeus*, lives in the forests of Borneo, and another species, *Pongo abelli*, lives in Sumatra. In 2017, scientists discovered a third species in northern Sumatra, *Pongo taunuliensis*. Orangutans are among the largest and most solitary species of primates (**Figure 5.16**). They have been studied since the 1970s by Birutė Galdikas in Tanjung Puting, Borneo. Long-term studies of orangutans have also been conducted at Cabang Panti in Borneo and at Ketambe and Suaq Balimbing in Sumatra. Orangutans feed primarily on fruit, but they also eat some leaves and bark. Adult females associate mainly with their own infants and immature offspring and do not often meet or interact with other orangutans. Adult males spend most of their time alone. A single adult male may defend a home range that encompasses the home ranges of several adult females; other males wander over larger areas and mate opportunistically with receptive females. When resident males encounter these nomads, fierce and noisy encounters may take place.

Gorillas, the largest of the apes, existed in splendid isolation from Western science until the middle of the nineteenth century (**Figure 5.17**). Gorillas are divided into two

FIGURE 5.16

(a) Orangutans are large, ponderous, and mostly solitary creatures. Male orangutans often descend to the ground to travel; lighter females often move through the tree canopy. (b) Today, orangutans are found only on the islands of Borneo and Sumatra, in tropical forests like this one.

[a]

[b]

[a]

[b]

FIGURE 5.17

(a) Gorillas are the largest of the primates. Mountain gorillas usually live in one-male, multifemale groups, but some groups contain more than one adult male. (b) Most behavioral information about gorillas comes from observations of mountain gorillas living in the Virunga Mountains of central Africa, pictured here. The harsh montane habitat may influence the nature of social organization and social behavior in these animals, and the behavior of gorillas living at lower elevations may differ.

species, the western gorilla (*Gorilla gorilla*) and the eastern gorilla (*Gorilla berengei*), which are separated by about 1,000 km across central Africa. The eastern gorillas are divided into two subspecies, mountain gorillas and lowland gorillas. George Schaller pioneered studies of gorilla social organization and ecology in the 1950s. In the 1960s, Diane Fossey began studying mountain gorillas in the Virunga Mountains of Rwanda and was the first to habituate gorillas to the presence of observers and collect detailed data on known individuals. It has been more difficult to study western gorillas, partly because they live in closed canopy forests where it is hard to find and follow them. However, it is becoming clear that there are a number of differences between western and eastern gorillas. Nearly all western gorilla groups contain only one adult male, while about 40% of eastern gorilla groups contain more than one adult male. Western and eastern lowland gorillas also eat more fruit, travel farther each day, and have larger home ranges than mountain gorillas do. Meanwhile, dispersal by both sexes is common in both eastern and western gorilla species.

As humankind's closest living relatives, chimpanzees (*Pan troglodytes*; **Figure 5.18a**) have played a uniquely important role in the study of human evolution. Whether reasoning by homology or by analogy, researchers have found observations about chimpanzees to be important bases for hypotheses about the behavior of early hominins.

Detailed knowledge of chimpanzee behavior and ecology comes from several long-term studies conducted at sites across Africa. In the 1960s, Jane Goodall began her well-known study of chimpanzees at the Gombe Stream National Park on the shores of Lake Tanganyika in Tanzania (**Figure 5.18b**). About the same time, a study was initiated by the late Toshisada Nishida at a site in the Mahale Mountains not far from Gombe. These studies are now moving into their sixth decade. Other important study sites have been established in the Taï Forest of Ivory Coast and at two sites in the Kibale Forest of Uganda: Kanyawara and Ngogo.

Bonobos (*Pan paniscus*; **Figure 5.18c**), the other member of the genus *Pan*, live in inaccessible places and have not been studied as extensively as chimpanzees. However, important field studies on bonobos have been conducted at several sites in the Democratic Republic of the Congo, including Wamba, Lomako, and LuiKotale.

Chimpanzees and bonobos form large multimale, multifemale communities. Female chimpanzees and bonobos usually disperse from their natal groups when they reach sexual maturity, whereas males remain in their natal groups throughout their lives. The members of chimpanzee communities are rarely found together in a unified group. Like spider monkeys, they split up into smaller parties that vary in size and composition from day to day. Bonobo groups are generally more cohesive than chimpanzee groups. In chimpanzees, the strongest social bonds among adults are formed among males, whereas bonobo females form stronger bonds with one another and with

[a]

[b]

[c]

FIGURE 5.18

(a) Chimpanzees live in multimale, multifemale social groups. In this species, males form the core of the social group and remain in their natal groups for life. Many researchers believe that chimpanzees and bonobos are our closest living relatives. (b) Like other apes, chimpanzees are found mainly in forests such as this area on the shores of Lake Tanganyika in Tanzania. However, chimpanzees sometimes range into more open areas as well. (c) Bonobos are members of the same genus as chimpanzees and are similar in many ways. Bonobos are sometimes called "pygmy chimpanzees," but this is a misnomer because bonobos and chimpanzees are about the same size. This infant bonobo is sitting in a patch of terrestrial herbaceous vegetation, one of the staples of the bonobo's diet.

their adult sons than males do. Chimpanzees display a number of behaviors that are of particular interest for researchers who study human origins. First, they regularly hunt vertebrate prey, particularly red colobus monkeys. Second, chimpanzees often share meat with other group members. Third, chimpanzees make and use tools for various tasks.

Primate Ecology

Much of the day-to-day life of primates is driven by two concerns: getting enough to eat and avoiding being eaten. Food is essential for growth, survival, and reproduction, and it should not be surprising that primates spend much of every day finding, processing, consuming, and digesting a wide variety of foods (**Figure 5.19**). At the same time, primates must always be on guard against predators such as lions, pythons, and eagles that hunt them by day and leopards that stalk them by night. Access to food and safety from predators drive the evolution of primate social organization and this, in turn, influences the evolution of mating systems and social behavior. It is important to understand these relationships because the same ecological factors are likely to have influenced the social organization and behavior of our earliest ancestors.

The Distribution of Food

Food provides energy that is essential for growth, survival, and reproduction.

Like all other animals, primates need energy to maintain normal metabolic processes; to regulate essential body functions; and to sustain growth, development, and

FIGURE 5.19

A female baboon feeds on corms in Amboseli, Kenya.

reproduction. The total amount of energy that an animal requires depends on four components:

1. *Basal metabolism.* **Basal metabolic rate** is the rate at which an animal expends energy to maintain life when at rest. As **Figure 5.20** shows, large animals have higher basal metabolic rates than small animals have. However, large animals require relatively fewer calories *per unit* of body weight.

2. *Active metabolism.* When animals become active, their energy needs rise above baseline levels. The number of additional calories required depends on how much energy the animal expends. The amount of energy expended, in turn, depends on the size of the animal and how fast it moves. In general, to sustain a normal range of activities, an average-size primate like a baboon or macaque requires enough energy per day to maintain a rate about twice its basal metabolic rate.

3. *Growth rate.* Growth imposes further energetic demands on organisms. Infants and juveniles, which are gaining weight and growing in stature, require more energy than would be expected from their body weight and activity levels alone.

4. *Reproductive effort.* For female primates, the energetic costs of reproduction are substantial. During the latter stages of their pregnancies, for example, primate females require about 25% more calories and, during lactation, about 50% more calories than usual.

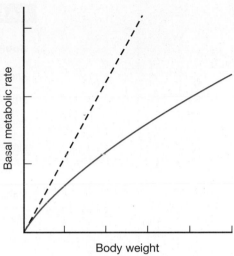

FIGURE 5.20

Average basal metabolism is affected by body size. The dashed line represents a direct linear relationship between body weight and basal metabolic rate. The solid line represents the actual relationship between body weight and basal metabolic rate. The fact that the curve bends means that larger animals use relatively less energy per unit of body weight.

A primate's diet must satisfy the animal's energy requirements, provide specific types of nutrients, and minimize exposure to dangerous toxins.

The food that primates eat provides them with energy and essential nutrients, such as amino acids and minerals, that they cannot synthesize themselves. Proteins are essential for virtually every aspect of growth and reproduction and to regulate many body functions. As we saw in Chapter 2, proteins are composed of long chains of amino acids. Primates cannot synthesize amino acids from simpler molecules, so to build many essential proteins, they must ingest foods that contain sufficient amounts of several amino acids. Fats and oils are important sources of energy for animals and provide about twice as much energy as equivalent volumes of **carbohydrates**. Vitamins, minerals, and trace amounts of certain elements play an essential role in regulating many of the body's metabolic functions. Although specific vitamins, minerals, and trace elements are needed in only small amounts, deficiencies of these nutrients can significantly impair normal body function. For example, trace amounts of iron and copper are important to synthesize hemoglobin, vitamin C is essential for growth and healing of wounds, and sodium regulates the quantity and distribution of body fluids. Primates cannot synthesize any of these compounds and must acquire them from the foods they eat. Further, water is the major constituent of the bodies of all animals and most plants. For survival, most animals must balance their water intake with their water loss; moderate dehydration can be debilitating, and significant dehydration can be fatal.

At the same time that primates obtain nourishment from food, they must also take care to avoid **toxins**, substances in the environment that are harmful to them. Many plants produce toxins called **secondary compounds** to protect themselves from being eaten. Thousands of these secondary compounds have been identified: Caffeine and morphine are among the secondary compounds most familiar to us. Some secondary compounds, such as **alkaloids**, are toxic to consumers because they pass through the stomach into various types of cells, where they disrupt normal metabolic functions. Common alkaloids include capsaicin (the compound that brings tears to your eyes when you eat red peppers) and chocolate. Other secondary compounds, such as

TABLE 5.4

Source	Protein	Carbohydrates	Fats and Oils	Vitamins	Minerals	Water
Animals	×	(×)	×	×	×	×
Fruit		×				×
Seeds	×		×	×		
Flowers		×				×
Young leaves	×			×	×	×
Mature leaves	(×)					
Woody stems	×					
Sap		×			×	×
Gum	×	(×)			×	
Underground parts	×	×				×

Sources of nutrients for primates. (×) indicates that the nutrient content is generally accessible only to animals that have specific digestive adaptations.

tannins (the bitter-tasting compound in tea), act in the consumer's gut to reduce the digestibility of plant material. Secondary compounds are particularly common among tropical plant species and are often concentrated in mature leaves and seeds. Young leaves, fruit, and flowers tend to have lower concentrations of secondary compounds, making them relatively more palatable to primates.

Primates obtain nutrients from many sources.

Primates obtain energy and essential nutrients from a variety of sources (**Table 5.4**). Carbohydrates are obtained mainly from the simple sugars in fruit, but animal prey, such as insects, also provides a good source of fats and oils. **Gum**, a substance that plants produce in response to physical injury, is an important source of carbohydrates for some primates, particularly galagos, marmosets, and tamarins. Primates get most of their protein from insect prey or from young leaves. Some species have special adaptations that facilitate the breakdown of cellulose, enabling them to digest more of the protein contained in the cells of mature leaves. Although seeds provide a good source of vitamins, fats, and oils, many plants package their seeds in husks or pods that shield their contents from seed predators. Many primates drink daily from streams, water holes, springs, or puddles of rainwater (**Figure 5.21**). Primates can also obtain water from fruit, flowers, young leaves, animal prey, and the underground storage parts (roots and tubers) of various plants. These sources of water are particularly important for arboreal animals that do not descend from the canopy and for terrestrial animals during times of the year when surface water is scarce. Vitamins, minerals, and trace elements are obtained in small quantities from many sources.

Although primates display considerable diversity in their diet, some generalizations are possible:

1. All primates rely on at least one type of food that is high in protein and another that is high in carbohydrates. Strepsirrhines generally obtain protein from insects and

FIGURE 5.21

These savanna baboons are drinking from a pool of rainwater. Most primates must drink every day.

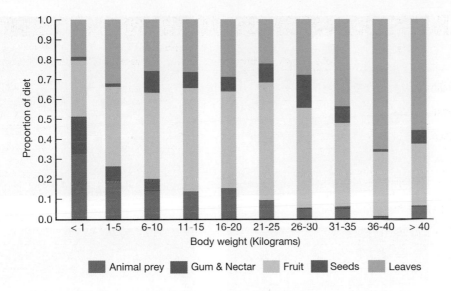

FIGURE 5.22

Diet and body size are related in primates. Most species eat some of various different types of foods, but relative amounts of different types of foods varies. The smallest species tend to eat a lot of animal prey, mainly insects and other invertebrates, and gum and nectar. Medium-sized primates eat a lot of fruit. The largest species eat a lot of other types of plant material, particularly leaves.

carbohydrates from gum and fruit. Monkeys and apes usually obtain protein from insects or young leaves and carbohydrates from fruit.

2. Most primates rely more heavily on some types of foods than on others. Chimpanzees, for example, feed mainly on ripe fruit throughout their range from Tanzania to Ivory Coast. Scientists use the terms **frugivore**, **folivore**, **insectivore**, and **gummivore** to refer to primates who rely most heavily on fruit, leaves, insects, and plant gum, respectively. A Closer Look 5.1 examines some of the morphological adaptations among primates with different diets.

3. In general, insectivores are smaller than frugivores, and frugivores are smaller than folivores (**Figure 5.22**). These differences in size are related to differences in energy requirements; small animals have relatively higher energy requirements than larger animals do, and they require relatively small amounts of high-quality foods that can be processed quickly. Larger animals are less constrained by the quality of their food than by the quantity because they can afford to process lower-quality foods more slowly.

The nature of dietary specializations and the challenge of foraging in tropical forests influence ranging patterns.

Nonhuman primates do not have the luxury of shopping in supermarkets, where abundant supplies of food are concentrated in a single location and are constantly replenished. Instead, the availability of their preferred foods varies widely in space and time, making their food sources patchy and often unpredictable. Most primate species live in tropical forests. Although such forests, with their dense greenery, seem to provide abundant supplies of food for primates, appearances can be deceiving. Tropical forests contain many tree species, and individual trees of any particular species are few in number.

Primates with different dietary specializations confront different foraging challenges (**Figure 5.23**). Plants generally produce more leaves than flowers or fruit, and they bear leaves for a longer period during the year than they bear flowers and fruit. As a result, foliage is normally more abundant than fruit or flowers at a given time during the year, and mature leaves are more abundant than young leaves. Insects and other suitable prey animals occur at even lower densities than plants. This means that folivores can generally find more food in a given area than frugivores or insectivores can. However, the high concentration of toxic secondary compounds in mature leaves complicates the foraging strategies of folivores. Some leaves must be avoided altogether, and others can be eaten only in small quantities. Nonetheless, the food supplies of folivorous species

[a] [b] [c]

[d] [e] [f]

FIGURE 5.23

(a) Some primates feed mainly on leaves, though many leaves contain toxic secondary plant compounds. The monkey shown here is a red colobus monkey in the Kibale Forest of Uganda. (b) Some primates include a variety of insects and other animal prey in their diet. This tamarin is eating a grasshopper. (c) Mountain gorillas are mainly vegetarians. Like this male eating leaves in Volcanoes National Park in Rwanda, they consume vast quantities of plant material. (d) This vervet monkey is feeding on grass stems. (e) Although many primates feed mainly on one type of food, such as leaves or fruit, no primate relies exclusively on one type of food. For example, the main bulk of the muriqui diet comes from fruit, but muriquis also eat leaves, as shown here. (f) Langurs are folivores. Here, gray langurs in Ramnagar, Nepal, forage for water plants.

are generally more uniform and predictable in space and time than the food supplies of frugivores or insectivores. Thus it is not surprising to find that folivores generally have smaller home ranges than those of frugivores or insectivores.

Activity Patterns

Primate activity patterns show regularity in seasonal and daily cycles.

Primates spend most of their time feeding, moving around their home ranges, and resting (**Figure 5.24**). Relatively small portions of each day are spent grooming, playing, fighting, or mating (**Figure 5.25**). The proportion of time devoted to various activities is influenced to some extent by ecological conditions. For primates living in seasonal habitats, for example, the dry season is often a time of scarce resources, and it is harder to find enough of the appropriate types of food. In some cases, this means the proportion of time spent feeding and traveling increases during the dry season, while the proportion of time spent resting decreases.

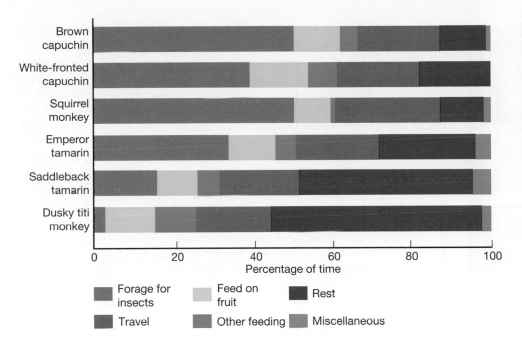

FIGURE 5.24

The amount of time that animals devote to various types of activities is called a *time budget*. Time budgets of different species vary considerably. These six monkey species all live in a tropical rain forest in Manú National Park in Peru.

Chart showing Percentage of time (x-axis 0 to 100) for species: Brown capuchin, White-fronted capuchin, Squirrel monkey, Emperor tamarin, Saddleback tamarin, Dusky titi monkey.

Legend:
- Forage for insects
- Travel
- Feed on fruit
- Other feeding
- Rest
- Miscellaneous

Primate activity also shows regular patterns during the day. When primates wake up, their stomachs are empty, so one of the first tasks of the day is to visit a feeding site. Much of the morning is spent eating and moving between feeding sites. As the sun moves directly overhead and the temperature rises, most species settle down in a shady spot to rest, socialize, and digest their morning meals. Later in the afternoon they resume feeding. Before dusk they move to the night's sleeping site; some species sleep in the same trees every night, and others have multiple sleeping sites within their ranges.

Ranging Behavior

All primates have home ranges, but only some species are territorial—defending their home range against incursions by other members of their species.

In all primate species, groups range over a relatively fixed area, and members of a given group can be consistently found in a particular area over time. These areas are called home ranges, and they contain all of the resources that group members exploit in feeding, resting, and sleeping. However, the extent of overlap among adjacent home ranges and the nature of interactions with members of neighboring groups or strangers vary considerably among species. Some primate species, such as gibbons, maintain exclusive access to fixed areas called **territories**. Territory residents regularly advertise their

FIGURE 5.25

(a) All diurnal primates, like this red colobus, spend some part of each day resting. (b) Immature monkeys spend much of their free time playing. These vervet monkeys are play wrestling. (c) Gorillas often rest near other group members and socialize during a midday rest period.

[a]

[b]

[c]

FIGURE 5.26

Siamangs and gibbons perform complex vocal duets as part of territorial defense.

presence by vocalizing, and they aggressively protect the boundaries of their territories from encroachment by outsiders (**Figure 5.26**). Although some territorial birds defend only their nest sites, primate territories contain all of the sites at which the residents feed, rest, and sleep, and the areas in which they travel. Thus, among territorial primates, the boundaries for the territory are essentially the same as for their home range, and territories of neighboring groups do not overlap. Nonterritorial species, such as squirrel monkeys and long-tailed macaques, establish home ranges that overlap considerably with those of neighboring groups. When members of neighboring nonterritorial groups meet, they may fight, avoid one another, or, more rarely, mingle peacefully. It has been hard for primatologists to study intergroup encounters because unhabituated groups may avoid groups that are being followed by human observers. Groups may avoid groups that they hear in the distance, and observers may miss these encounters entirely. To avoid these problems, Margaret Crofoot (now at the Max Planck Institute of Animal Behavior in Konstanz, Germany) used an automated radio telemetry system (ARTS) to monitor the movements of six capuchin groups on Barro Colorado Island in the Panama Canal. She fitted one monkey in each group with a transmitter attached to a collar that sent radio signals at regular intervals to a set of antennas mounted on towers located throughout the forest. When she and her colleagues analyzed the ARTS data, they found that the ranges of the groups overlapped considerably, and intergroup encounters were concentrated in areas of range overlap (**Figure 5.27**). The outcome of these encounters was influenced by two factors: the relative size of the two groups and the location of the encounter. Larger groups usually defeated smaller groups, but

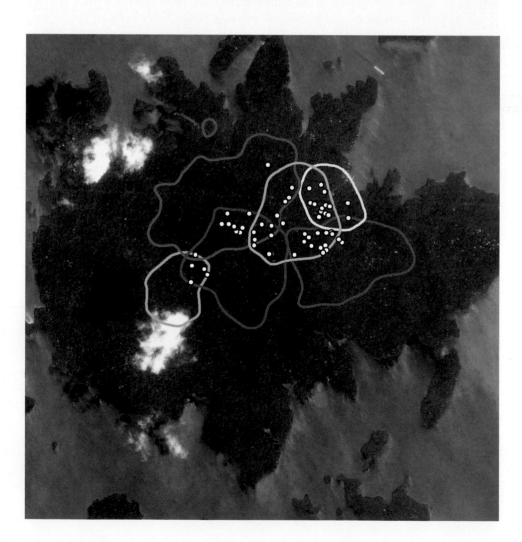

FIGURE 5.27

Overlapping home ranges of capuchin monkey groups on Barro Colorado Island, Panama. Sites of intergroup encounters are marked with a dot. (Satellite image by DigitalGlobe.)

smaller groups that were closer to the center of their home ranges were able to defeat larger groups (**Figure 5.28**).

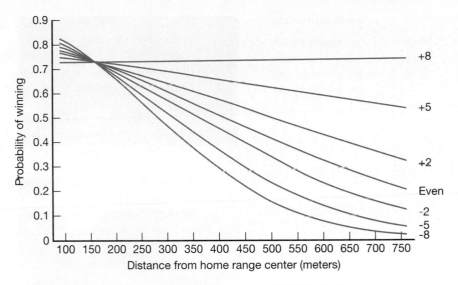

The two main functions suggested for territoriality are resource defense and mate defense.

To understand why some primate species defend their home ranges from intruders and others do not, we need to think about the costs and benefits associated with defending resources from conspecifics. Costs and benefits are measured in terms of the impact on the individual's ability to survive and reproduce successfully. Territoriality is beneficial because it prevents outsiders from exploiting the limited resources within a territory. At the same time, however, territoriality is costly because the residents must be constantly vigilant against intruders, regularly advertise their presence, and be prepared to defend their ranges against encroachment. Territoriality is expected to occur only when the benefits of maintaining exclusive access to a particular piece of land outweigh the costs of protecting these benefits.

When will the benefits of territoriality exceed the costs? The answer to this question depends in part on the kinds of resources individuals need to survive and reproduce successfully and in part on the way these resources are distributed spatially and seasonally. For reasons we will discuss more fully in Chapter 6, the reproductive strategies of mammalian males and females generally differ. Usually, female reproductive success depends mainly on getting enough to eat for themselves and their dependent offspring, and male reproductive success depends mainly on their ability to mate with females. As a consequence, females are more concerned about access to food, and males are more interested in access to females. Thus territoriality has two functions. Sometimes females defend food resources or males defend food resources on their behalf. Other times, males defend groups of females against incursions by other males. In primates, both resource defense and mate defense seem to have influenced the evolution of territoriality.

Predation

Predation is believed to be a significant source of mortality among primates, but direct evidence of predation is difficult to obtain.

Primates are hunted by a variety of predators, including pythons, raptors, crocodiles, leopards, lions, tigers, and humans (**Figure 5.29**). In Madagascar, large lemurs are preyed upon by fossas, pumalike carnivores. Primates are also preyed upon by other primates. Chimpanzees, for example, hunt red colobus monkeys, and baboons sometimes prey on vervet monkeys.

The estimated rates of predation vary from less than 1% of the population per year to more than 15%. The available data suggest that small-bodied primates are more vulnerable to predation than larger ones are, and immature primates are generally more susceptible to predation than adults are. These data are not very solid, however, because systematic information about predation is quite hard to come by. Most predators avoid close contact with humans, and some predators, such as leopards, generally hunt at night, when most researchers are asleep. Usually predation is inferred when a healthy animal that is unlikely to have left the group abruptly vanishes (**Figure 5.30**). Such inferences are, of course, subject to error.

FIGURE 5.28

In capuchin monkeys, groups are likely to win encounters when they have more members than their opponents. (The relative size of focal groups and their opponents is shown on the right side of the graph.) However, when groups are close to the center of their own home range, the effect of relative group size largely disappears.

FIGURE 5.29

Primates are preyed upon by a variety of predators, including the (a) python, (b) lion, (c) leopard, (d) crowned hawk eagle, and (e) crocodile.

[a]

[b]

[d]

[c]

[e]

FIGURE 5.30

Researchers can sometimes confirm predation. Here, an adult female baboon in the Okavango Delta, Botswana, was killed by a leopard. You can see (a) the depression in the sand that was made when the leopard dragged the female's body out of the sleeping tree and across a small sandy clearing, (b) the leopard's footprints beside the drag marks, and (c) the remains of the female the following morning—her jaw, bits of her skull, and clumps of hair.

Another approach is to study the predators, not their prey. Crowned hawk eagles are the only large raptors that live in the tropical rain forests of Africa. They are formidable predators; although they weigh only 3 to 4 kg (6.6 to 8.8 lb.), they have powerful legs and large talons and can take prey weighing up to 20 kg (44 lb.). Crowned hawk eagles carry their prey, intact or after dismembering them on the ground, back to their nests and discard the bones. By sorting through the remains under crowned hawk eagle nests, researchers can figure out what they eat. Analyses of nest remains in the Kibale Forest of Uganda and the Taï Forest in Ivory Coast indicate that crowned hawk eagles prey on all of the primates in these forests except chimpanzees. Monkeys make up 60% to 80% of the crowned hawk eagles' diets at these sites, and the eagles kill a sizable fraction (2% to 16%) of the total populations of various primate species in these forests each year.

[a]

[b]

[c]

Susanne Shultz at the University of Manchester and her colleagues compared the characteristics of mammalian prey taken by crowned hawk eagles, leopards, and chimpanzees in the Taï Forest (**Figure 5.31**). In general, terrestrial species are more vulnerable than arboreal species, and species that live in small groups are more vulnerable than animals that live in large groups. Thus arboreal monkeys that live in large groups face the lowest risks. Shultz and her colleagues suggest that these results may explain some aspects of the distribution of terrestrial primates in Africa, Asia, and the neotropics. In Africa, with crowned hawk eagles and at least two large predatory felids present, terrestrial primates are large-bodied or live in large groups. In Asia, where there are no large forest raptors and few large felids, there are several semiterrestrial macaque species. And in South and Central America, where there are several species of large felids and forest raptors, there are no terrestrial monkeys at all.

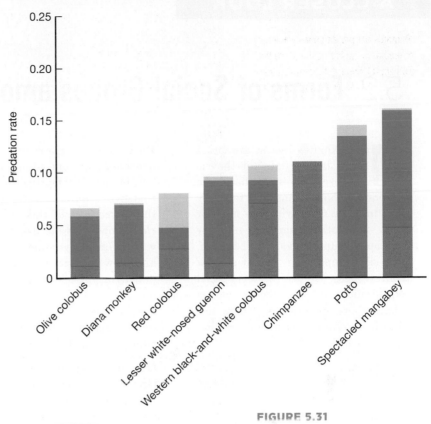

FIGURE 5.31

The rate of predation by leopards (*orange*), eagles (*green*), and chimpanzees (*blue*) in the Taï Forest on different primate species is shown here. Note that the preferred prey of chimpanzees is the red colobus monkey, and chimpanzees' only predator is the leopard.

Primates have evolved an array of defenses against predators.

Many primates give alarm calls when they sight potential predators, and some species have specific vocalizations for particular predators. Vervet monkeys, for example, give different calls when they are alerted to the presence of leopards, small carnivores, eagles, snakes, baboons, and unfamiliar humans. In many species, the most common response to predators is to flee or take cover. Small primates sometimes try to conceal themselves from predators; larger ones may confront potential predators. When slow-moving pottos encounter snakes, for example, they fall to the ground, move a short distance, and freeze. At some sites, adult red colobus monkeys aggressively attack chimpanzees that stalk their infants.

Another antipredator strategy that some primates adopt is to associate with members of other primate species. In the Taï Forest, several monkey species share the canopy and form regular associations with one another. For example, groups of red colobus monkeys spend approximately half their time with groups of Diana monkeys. Interspecific associations may enhance predator detection if each species occupies a different portion of the canopy and is oriented toward different predators. In addition, by associating with members of different species, monkeys may increase group size without increasing levels of competition from conspecifics that have similar dietary preferences.

Primate Sociality

Sociality has evolved in primates in response to ecological pressures.
Social life has both costs and benefits.

Nearly all primates live in social groups of one kind or another. Sociality has evolved in primates because there are important benefits associated with living in groups. As we saw earlier, grouping offers safety from predators because groups

5.2 Forms of Social Groups among Primates

Most primates live in groups. A group is a social unit that is composed of animals that share a common home range or territory and interact more with one another than with other members of their species. Groups can vary in their size, age–sex composition, and degree of cohesiveness. We use the term **social organization** to describe variation along these dimensions. There are four basic types of social organization among primates (**Figure 5.32**):

Solitary: Females maintain separate home ranges or territories and associate mainly with their dependent offspring. Males establish their own territories or home ranges, which may encompass the ranges of one or more adult females. All of the **solitary** primates are strepsirrhines, except for orangutans.

Pairs: Groups are composed of one adult male, one adult female, and immature offspring. Species that live in pairs usually defend the boundaries of their territories. Gibbons live in pairs, along with a few platyrrhine monkeys and a few strepsirrhines. In some pair-living species, males and females remain close together, but in others, they may travel independently within their territories much of the time.

One male, multiple females: Groups are composed of several adult females, one resident adult male, and immature offspring. Males compete vigorously over residence in these kinds of groups, and males may band together to oust established residents. This form of social organization is characteristic of howler monkeys, some langurs, and gelada baboons.

Multiple males, multiple females: Groups are composed of several adult females, several adult males, and immature offspring. This form of social organization is characteristic of macaques, baboons, capuchin monkeys, squirrel monkeys, and some colobines. Some species, such as chimpanzees and spider monkeys, living in these kinds of groups regularly divide up into smaller temporary parties (fission–fusion groups).

Primates also vary in their **mating systems**, the pattern of mating activity and reproductive outcomes. There is a close, but not perfect, relationship between social organization and mating systems. There are

FIGURE 5.32

The major types of social groups that primates form. When males and females share their home ranges, their home ranges are drawn here in grey. When the ranges of the two sexes differ, male home ranges are drawn in blue and female home ranges are drawn in red. The sizes of the male and female symbols reflect the degree of sexual dimorphism among males and females. Bold symbols represent adults.

Solitary

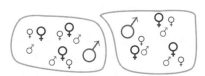

One male, multiple females (polygyny)

Pairs (pair bonding)

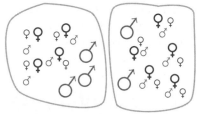

Multiple males, multiple females (polygynandry)

four main forms of mating systems in primates:

Monogamy/pair bonding: In a strictly monogamous mating system, each male and female will mate with only one member of the opposite sex. Most primates that live in pairs mate mainly with each other, but matings with outside males have been observed in several species. Thus the term **pair bonding** may be a more accurate description of the mating system of most pair-living primates than monogamy.

Cooperative breeding/polyandry: Marmosets and tamarins live in groups that may contain one or more adult females, one or more adult males, and offspring from one or more litters. Typically, only one female breeds, but more than one male may share paternity of infants. **Polyandry**, in which a single female forms a stable pair bond with two males at the same time, is rare among mammals.

Polygyny: Males mate with multiple females, but each female mates with a single male. This mating system characterizes most of the species that live in one-male, multifemale groups. **Polygyny** generates considerable skew in male reproductive success, as resident males largely control access to receptive females. However, in some of these species, including blue monkeys, males from outside the group sometimes enter groups and mate with females.

Polygynandry: Both males and females mate with more than one partner. This mating system is generally associated with species that live in multimale, multifemale groups and might also characterize some solitary species. In most species that form multimale, multifemale groups, males compete over access to mating females, and there is considerable skew in male reproductive success.

These classifications of social organization and mating systems represent idealized descriptions of residence and mating patterns. The reality is inevitably more complicated. Not all groups of a particular species may have the same social organization or mating system. For example, some groups of tarsiers are composed of a single mated pair, whereas others include additional females. Hamadryas and gelada baboons form one-male, multifemale units, but several of those units collectively belong to larger aggregations.

provide the "three *D*s": detection, deterrence, and dilution. Animals in groups are more likely to detect predators because there are more pairs of eyes on the lookout for predators. Animals in groups are also more effective in deterring predators by actively mobbing or chasing them away. Finally, the threat of predation to any single individual is diluted when predators strike at random. If there are two animals in a group, and a predator strikes, each animal has a 50% chance of being eaten. If there are 10 individuals, the individual risk is decreased to 10%. Primates that live in groups may be better able to acquire and control resources. Animals that live in groups can chase away lone individuals from feeding trees and can protect their own access to food and other resources against smaller numbers of intruders.

Although there are important benefits associated with sociality, there are equally important costs. Animals that live in groups may encounter more competition over access to food and mates, become more vulnerable to disease, and face various hazards from conspecifics (such as cannibalism, cuckoldry, inbreeding, or infanticide). The size and composition of the groups that we see in nature (**A Closer Look 5.2**) are expected to reflect a compromise between the costs and benefits of sociality for individuals. The magnitude of these costs and benefits is influenced by both social and ecological factors.

Primatologists are divided over whether predation or competition for food is the primary factor favoring sociality among primates.

It is not entirely clear whether predation or competition for resources was the primary factor favoring the evolution of sociality in primates. However, many primatologists are convinced that the nature of resource competition affects the behavioral strategies of primates, particularly females, and influences the composition of primate groups (A Closer Look 5.2). Females come first in this scenario because their fitness depends mainly on their nutritional status: Well-nourished females grow faster, mature earlier, and have higher fertility rates than do poorly nourished females. In contrast, males' fitness depends primarily on their ability to obtain access to fertile females, not on their nutritional status. Thus ecological pressures influence the distribution of females, and males distribute themselves to maximize their access to females. (We will discuss male and female reproductive strategies more fully in Chapter 6.)

Primate Conservation

Many species of primates are in real danger of extinction in the wild.

Sadly, no introduction to the primate order would be complete without noting that the prospects for the continued survival of many primate species are grim. Figures released by the International Union for the Conservation of Nature in 2019 show that the populations of about 75% of all primate species are declining, and about 60% of all primate species are threatened with extinction. Two primate species have already become extinct. The most recent list of the 25 most endangered primates includes Bornean orangutans, one subspecies of western gorillas, the ring-tailed lemur, and two species of spider monkeys. Species in all primate families are at risk (**Figure 5.33**).

Primates are threatened because their habitats are disappearing. Most primates live in the tropics, and they are directly affected by the widespread destruction of the world's forests. Between 2000 and 2017, approximately 1,790,000 square kilometers of forest have been lost in South and Central America, Africa, and Asia; this is an area about 4.5 times the size of California. In response to global demand for palm oil, which supplies about 30% of the world's vegetable oil and is becoming an important biofuel source, huge swaths of forest have been converted to palm oil plantations, particularly in Indonesia and Malaysia. This has contributed to major declines of orangutan populations. Between 1999 and 2015, 100,000 orangutans disappeared from Borneo, and these losses are likely to continue (**Figure 5.34**). Large areas of forest have been converted to production of natural rubber in Southeast Asia and for soybeans and beef production in South America.

Trade in forest products (such as lumber and charcoal), fossil fuels, and mineral extraction also pose threats to primates. When forests are logged, the climate of the

FIGURE 5.33

A substantial fraction of species from all primate families are vulnerable (purple), endangered (green), or critically endangered (blue).

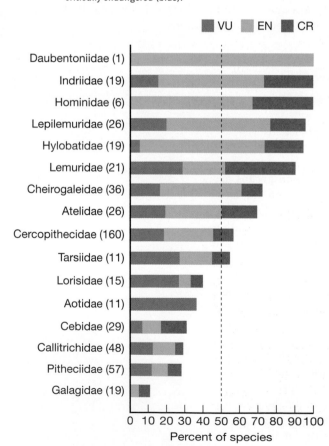

Orangutan density (individuals per square kilometer)

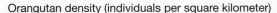

0.01–0.5	0.5–1.0	1.0–2.0	2.0–5.0	5.0–10.9

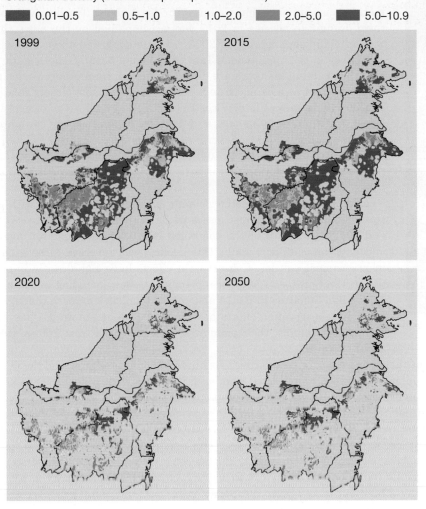

FIGURE 5.34

The population of Bornean orangutans has dropped dramatically over the past two decades and is expected to decline further in the future.

forest changes. Humidity levels drop, and this increases tree mortality and the risk of fires. Mining degrades the environment by poisoning soil and polluting groundwater.

Habitat loss is not the only threat to primates. In many areas of the world, primates and other forest animals have traditionally been hunted for subsistence. However, there is a growing commercial demand for bushmeat, which includes primates. The bushmeat trade has been a major contributor to primate population declines in Africa and Southeast Asia. In a number of areas, primate populations have also been severely affected by outbreaks of epidemic disease. Several hundred gorillas, belonging to more than 100 groups, regularly foraged in a swampy clearing in Odzala-Kokoua National Park in the Republic of the Congo. During a 2-year period, 95% of these gorillas died from Ebola. Anthropogenic climate change also threatens primates, as both cyclones and droughts are expected to increase in frequency and intensity. Sixteen percent of all primate species live in areas vulnerable to cyclones, many in Madagascar, and 22% live in areas that are vulnerable to droughts.

FIGURE 5.35

The kapunji monkey was first discovered by scientists in 2005.

It is particularly disturbing that some of the most endangered primate species are ones that we know the least about. The discovery of the third orangutan species in 2017 was accompanied by the news that it was critically endangered. In 2005, scientists learned about a new kind of monkey in the highlands of Tanzania (**Figure 5.35**). Genetic data indicate that kapunji monkeys are different enough from other species to be placed in their own genus, *Rungwecebus*, and are most closely related to baboons. This species is now restricted to two small areas of evergreen forest, 350 km (230 mi.) apart, and the total population was estimated to be only 1,100 individuals. Several new primate species have been identified recently in Madagascar, and as the forests in Madagascar shrink, there is a real possibility that others will become extinct before we know that they ever existed.

Efforts to save endangered primate populations have met with some success. For example, forest losses in Brazil declined 70% between 2005 and 2013 as a result of the combined efforts of governmental agencies and environmentalists; sadly, this decline was reversed after a change in government policies. Conservation efforts have significantly improved the survival prospects of several primate species, including muriquis and golden lion tamarins in Brazil and golden bamboo lemurs in Madagascar. Intensive conservation efforts have helped save mountain gorillas in the Virunga Mountains; their population has increased from 250 in 1981 to about 600 in 2018.

But there is no room for complacency. As human populations grow and the demand for commodities grown or extracted from land that primates live on increases, pressure on primate populations will become more severe. Several strategies to conserve forest habitats and preserve animal populations are on the table. These include land-for-debt swaps in which foreign debts are forgiven in exchange for commitments to conserve natural habitats, to develop ecotourism projects, and to promote sustainable development of forest resources. But as conservationists study these solutions and try to implement them, the problems facing the world's primates become more pressing. More and more forests disappear each year, and many primates are lost, perhaps forever.

CHAPTER REVIEW

Key Terms

viviparity (p. 110)	canine (p. 113)	bilaterally symmetrical (p. 118)	frugivore (p. 129)
nocturnal (p. 111)	premolars (p. 113)	cooperative breeders (p. 121)	folivore (p. 129)
diurnal (p. 111)	brachiation (p. 114)		insectivore (p. 129)
conspecifics (p. 111)	binocular vision (p. 114)	basal metabolic rate (p. 127)	gummivore (p. 129)
sexual dimorphism (p. 111)	stereoscopic vision (p. 114)	carbohydrates (p. 127)	territories (p. 131)
opposable (p. 113)	strepsirrhine (p. 115)	toxins (p. 127)	social organization (p. 136)
prehensile (p. 113)	haplorrhine (p. 115)	secondary compounds (p. 127)	solitary (p. 136)
hind-limb dominated (p. 113)	infraorder (p. 117)	alkaloids (p. 127)	mating systems (p. 136)
olfactory (p. 113)	dental formula (p. 118)	gum (p. 128)	pair bonding (p. 137)
molars (p. 113)	maxilla (p. 118)		polyandry (p. 137)
incisors (p. 113)	mandible (p. 118)		polygyny (p. 137)

Study Questions

1. What is the difference between homology and analogy? What evolutionary processes correspond to these terms?

2. Suppose that a group of extraterrestrial scientists lands on Earth and enlists your help in identifying animals. How do you help them recognize members of the primate order?

3. What kinds of habitats do most primates occupy? What are the features of this kind of environment?

4. Large primates often subsist on low-quality food such as leaves; small primates specialize in high-quality foods such as fruit and insects. Why is body size associated with dietary quality in this way?

5. For folivores, tropical forests seem to provide an abundant and constant supply of food. Why is this not an accurate assessment?

6. Territorial primates do not have to share access to food, sleeping sites, mates, and other resources with members of other groups. Given that territoriality reduces the extent of competition over resources, why are not all primates territorial?

7. Most primates specialize in one type of food, such as fruit, leaves, or insects. What benefits might such specializations have? What costs might be associated with specialization?

8. Nocturnal primates are smaller, more solitary, and more arboreal than diurnal primates. What might be the reason(s) for this pattern?

9. Sociality is a relatively uncommon feature in nature. What are the potential advantages and disadvantages of living in social groups? Why are virtually all primates social?

10. The future of primates, and other occupants of tropical forests, is precarious. What are the major hazards that primates face?

Further Reading

Campell, C. J., A. Fuentes, K. C. MacKinnon, M. Panger, and S. K. Bearder, eds. 2007. *Primates in Perspective*. New York: Oxford University Press.

Estrada, A., P. A. Garber, A. B. Rylands, et al. (2017). "Impending Extinction Crisis of the World's Primates: Why Primates Matter." *Science Advances*, 3(1): e1600946.

Kappeler, P. M., and M. Pereira, eds. 2003. *Primate Life Histories and Socioecology*. Chicago: University of Chicago Press.

Mitani, J., J. Call, P. Kappeler, R. Palombit, and J. B. Silk, eds. 2012. *The Evolution of Primate Societies*. Chicago: University of Chicago Press.

Strier, K. B. 2010. *Primate Behavioral Ecology*. 4th ed. Boston: Allyn & Bacon.

 Review this chapter with personalized, interactive questions through INQUIZITIVE.

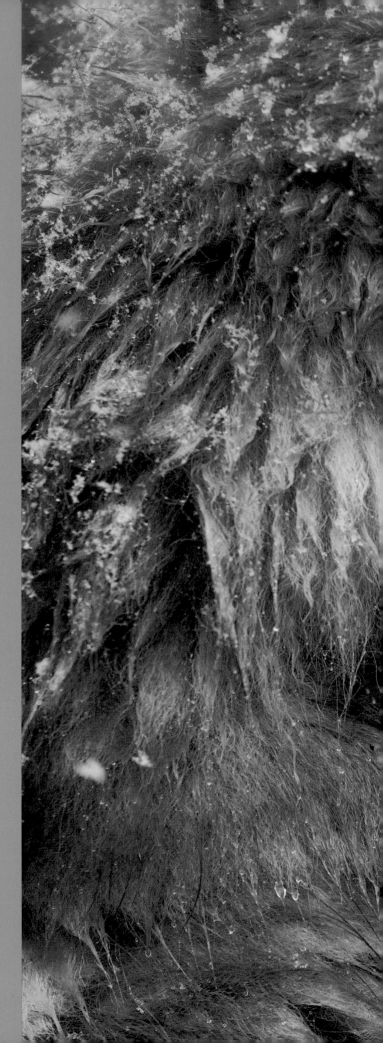

6

PRIMATE REPRODUCTIVE STRATEGIES

CHAPTER OBJECTIVES

By the end of this chapter you should be able to:

A Explain why reproduction is the central act of all living things.

B Describe how mammalian reproductive biology influences the reproductive tactics of primate females.

C Discuss the factors that influence female reproductive success.

D Describe the process of sexual selection and explain why it favors traits that would not be favored by normal natural selection.

E Describe how competition among males over access to females influences male reproductive tactics.

F Explain why infanticide is an adaptive tactic for male primates in some circumstances.

Reproduction is the central act in the life of every living thing. Primates perform a dizzying variety of behaviors: Gibbons fill the forest with their haunting duets, baboons threaten and posture in their struggle for dominance over other members of their group, and chimpanzees use carefully selected stone hammers to crack open tough nuts. But all of these behaviors evolved for a single ultimate purpose: to enhance reproduction. According to Darwin's theory, complex adaptations exist because they evolved step by step through natural selection. At each step, only those modifications that increased reproductive success were favored and retained in later generations of offspring. Thus, each morphological feature and every behavior exist only because they were part of an adaptation that contributed to reproduction in ancestral populations. As a consequence, mating systems (the way animals find mates and care for offspring) play a crucial role in our understanding of primate societies.

Understanding the diverse reproductive systems of nonhuman primates illuminates human evolution because we share many elements of our reproductive physiology with other species of primates.

To understand the evolution of primate reproductive systems, we must take into account that the reproductive strategies of living primates are influenced by their phylogenetic heritage as mammals. Mammals reproduce sexually. After conception, mammalian females carry their young internally. Then females give birth and suckle their young for an extended period (**Figure 6.1**). The mammalian male's role in the reproductive process is more variable than that of the female. In some species, males' only contribution to their offspring is the sperm they provide at the moment of conception. In other species, males defend territories; provide for their mates; and feed, carry, and protect their offspring.

Although mammalian physiology constrains primate reproductive tactics, there is still considerable diversity in the patterns of reproduction, mating, and parenting among primates. In some primate species, male reproductive success is determined mainly by success in competition with other males over access to mates; in others, it is strongly influenced by the quality of care that they provide for their offspring. In most primate species, female reproductive success depends mainly on their own qualities, such as their age, experience, and ability to gain access to resources; in others, it depends on efforts of other group members who help carry, protect, and provide food for their offspring.

The amount of time, energy, and resources that the males and females of a species invest in mating activities and offspring care has profound consequences for the evolution of virtually every aspect of their social behavior and many aspects of their morphology. The selection pressures that affect males and females in species with equal parental investment are very different from the selection pressures that affect males and females in species in which females invest much more than males do. Although this variation is inherently interesting, it is also useful for thinking about the evolutionary foundations of human mating and parenting

FIGURE 6.1

In all primate species, females nurse their young. In baboons and many other species, females provide most of the direct care that infants receive.

systems. We share many aspects of our reproductive physiology and behavior with other primates, and so it is important to understand how and why mating systems and parenting behavior differ among species.

The Language of Adaptive Explanations

In evolutionary biology, the term *strategy* is used to refer to a set of traits that are designed to solve a particular adaptive problem, such as finding food, avoiding predation, or rearing offspring. Strategies may encompass a range of *tactics* that organisms use in a particular functional context. For example, the potto's anti-predation tactic is to remain very still and avoid detection, while the baboon's anti-predation tactics include vigilance, alarm calling, and fleeing.

When evolutionary biologists use these terms, they mean something very different from what we normally mean when we think about a general's military maneuvers or a baseball manager's decision about when to put in a pinch hitter. In common usage, *strategy* and *tactics* imply a conscious plan of action to achieve a known goal. Evolutionary biologists don't think that other animals consciously decide to defend their territories against intruders, wean their offspring at a particular age, monitor their ingestion of secondary plant compounds, and so on. The behavioral tactics that we observe are the product of natural selection acting on individuals to shape their motivations, reactions, capacities, and decisions. Tactics that led to greater reproductive success in ancestral populations have been favored by natural selection and represent adaptations.

Cost and benefit refer to how particular behavioral tactics affect reproductive success.

Different behaviors have different impacts on an animal's genetic fitness. Behaviors are said to be beneficial if they increase the genetic fitness of individuals and costly if they reduce the genetic fitness of individuals. For example, we argued in Chapter 5 that ranging behavior involves a trade-off between the benefits of exclusive access to a particular area and the costs of defending territorial boundaries. Ultimately, benefits and costs matter because they influence lifetime reproductive success, but it is often very difficult to evaluate the links between morphological or behavioral traits and fitness directly, particularly in long-lived animals such as primates. Instead, researchers rely on indirect measures, such as foraging efficiency (measured as the quantity of nutrients obtained per unit time) and assume that, all other things being equal, behavioral tactics that increase foraging efficiency also enhance genetic fitness and will be favored by natural selection. We will encounter many other examples of this type of reasoning in this chapter and the chapters that follow.

The Evolution of Reproductive Strategies

Reproduction can be divided into two components: mating effort and parenting effort.

Mating effort refers to all of the activities that are related to conception, including the effort required to find mates and gain access to them. This may involve courtship displays, competition with rivals, or establishing a territory. **Parenting effort** includes all of the activities that are related to offspring care after conception occurs, such as sitting on the nest after eggs are laid or nursing infants after birth. All other things being equal, we expect evolution to favor traits that increase success in both mating and

FIGURE 6.2

Mammalian mothers protect their offspring from predators and defend them from aggression from conspecifics. Here, a female lion defends her cub.

FIGURE 6.3

In most non-pair-bonded species, males have relatively little contact with infants. Although males such as this bonnet macaque are sometimes quite tolerant of infants, they rarely carry, groom, feed, or play with them.

parenting efforts. If time, energy, and other resources were unlimited, then animals would invest heavily in both mating effort and parenting effort. In real life, though, time, energy, and material resources are always in short supply. The effort that an individual devotes to parenting diverts time and energy away from mating effort. Natural selection will favor individuals that allocate effort among these competing demands so as to maximize the number of surviving offspring they produce.

In most mammalian species, females invest more in parenting effort than in mating effort, and males invest more in mating effort than in parenting effort. Females nurse their young, and in many species females also keep their infants warm, carry them, groom them, protect them from predators, and defend them from aggression by conspecifics (**Figure 6.2**). It is less common for females to fight over access to males or compete for membership in social groups. In contrast, competition among males over access to mates is relatively common, and there are relatively few species in which males invest much in parental care (**Figure 6.3**). These general patterns also characterize primates.

The amount of time, energy, and resources that the males and females of a species invest in offspring care has profound consequences for the evolution of virtually every aspect of their social behavior and many aspects of their morphology. The selection pressures that affect males and females in species with equal parental investment are very different from the selection pressures that affect males and females in species in which females invest much more than males do. Thus, it is important to understand how and why the amounts and patterns of parental investment differ among species.

In this chapter, we will discuss the reproductive behavior of males and females separately, although selective pressures acting on males and females are not entirely independent. For example, females' distribution in time and space has important effects on the evolution of male mating and parenting behavior. There are also cases in which males' reproductive tactics are costly for females, and females evolve counterstrategies to minimize the impact of male tactics.

The mammalian reproductive system commits primate females to investing in their offspring.

Why aren't there mammalian species in which males do all the work and females compete with each other for access to males? This is not simply a theoretical possibility: Female sea horses deposit their eggs in their mate's brood pouch and then swim away and look for a new mate (**Figure 6.4**). There are whole families of fish in which male parental care is more common than female parental care, and in several species of birds—including rheas, spotted sandpipers, and jacanas—females abandon their clutches after the eggs are laid, leaving their mates to feed and protect their young.

In primates and other mammals, selection tends to favor high female investment because females become pregnant and lactate and males do not. Pregnancy and lactation commit mammalian females to invest in their young and limit the benefits of male investment in offspring. Because offspring depend on their mothers for nourishment during pregnancy and after birth, mothers cannot abandon their young without greatly reducing their offspring's chances of surviving. On the other hand, mammalian males are never capable of rearing their offspring without help from a female and can generally benefit more from investing in mating effort than in parenting effort. Therefore, when only one sex invests in offspring, it is invariably the female.

You may be wondering why selection has not produced males that can lactate. This would seem like a highly desirable adaptation because it would enable males to make important contributions to offspring care and protect infants from the consequences of maternal mortality. But, as we noted in Chapter 3, most biologists believe that the developmental changes that would enable males to lactate would also make them sterile. This is an example of a developmental constraint.

Uncertainty about paternity also shapes the benefits derived from male parenting effort.

Characteristics of mammalian reproductive systems, including internal fertilization and gestation, make it much easier for females to identify their own offspring than for males to do so. Females are able to guarantee that their care is directed to their own offspring. Even in species that live in pairs or form one-male groups, there is some possibility that infants will be sired by outside males. And in species that form multi-male groups, females may mate with more than one male near the time of conception. If males run some risk of directing care to other males' offspring, this will dilute the fitness benefits they gain from investment in parenting effort.

Female Reproductive Tactics

Female primates invest heavily in each of their offspring.

Pregnancy and lactation are time-consuming and energetically expensive activities for all female primates, including humans. The duration of pregnancy ranges from 59 days in the tiny mouse lemur to 255 days in the hefty gorilla. In primates, as in most other animals, larger animals tend to have longer pregnancies than do smaller animals (**Figure 6.5**), but primates have considerably longer pregnancies than we would expect on the basis of their body sizes alone. The extended duration of pregnancy in primates is related to the fact that brain tissue develops very slowly. Primates have very large brains in relation to their body sizes, so extra time is needed for fetal brain growth and development during pregnancy. Primates also have an extended period of dependence after birth, further increasing the amount of care mothers must provide. Throughout this period, mothers must meet not only their own nutritional requirements but also those of their growing infants.

FIGURE 6.4

Male sea horses carry fertilized eggs in a special pouch and provide care for their offspring as the young grow.

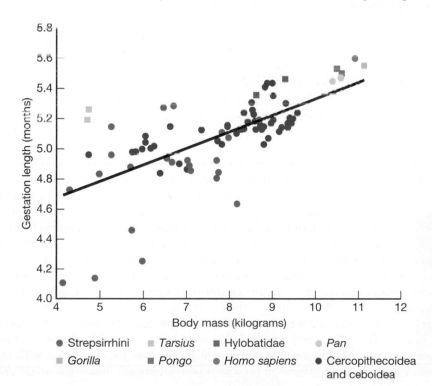

FIGURE 6.5

As in other mammalian taxa, maternal body size is correlated with gestation length. Great apes have the longest pregnancies, and small-bodied strepsirrhine primates have relatively short pregnancies.

In some species, offspring may weigh as much as 30% of their mother's body weight at the time of weaning.

The energy costs of pregnancy and lactation impose important constraints on female reproductive behavior. Because it takes so much time and energy to produce an infant, each female can rear only a relatively small number of surviving infants during her lifetime (**Figure 6.6**). For example, a female baboon that gives birth for the first time when she is about 6 or 7 years old and lives to a ripe old age of 30 would give birth to about a dozen offspring during her life. Of course, not all females will survive to old age, and nearly half of all infants will die before they reach reproductive age. Thus most baboon females produce a relatively small number of surviving infants over a lifetime, and each infant represents a substantial proportion of a female's lifetime fitness. Therefore, we would expect mothers to be strongly committed to the welfare of each of their offspring.

A female's reproductive success depends on her ability to obtain enough resources to support herself and her offspring.

In most species of primates, including humans, females must achieve a minimum nutritional level to ovulate and to conceive. For animals living in the wild, without 24-hour grocery stores and Grubhub, getting enough to eat each day can be a serious challenge. Female reproductive success is often limited by the availability of resources within the local habitat. The most convincing evidence of the effects of food availability on female reproduction come from situations in which animals suddenly gain access to exceptionally abundant sources of food. For example, at a number of "monkey parks" in Japan, monkeys were fed to encourage them to visit areas where they could be observed by tourists (**Figure 6.7**). When the monkeys were being intensively provisioned, groups grew very rapidly because the females grew faster, matured earlier, and had higher fertility rates. Similarly, in the Amboseli basin of Kenya, female members of a baboon group that foraged on food scraps from a nearby tourist lodge were fatter than females in neighboring groups that subsisted on wild foods, and they also matured earlier, had shorter intervals between successive births (**interbirth intervals**), and their infants were more likely to survive. Similar, but less pronounced, changes were observed when the wild-feeding groups shifted their home ranges to an area with more abundant natural food resources (**Figure 6.8**).

Female Reproductive Careers

There are a number of different components of female reproductive success.

The lifetime fitness of females is based on the number of surviving infants that they produce over the course of their lives. Female lifetime fitness will be influenced by their age at first birth, the length of interbirth intervals, infant survival, and life span. All other things being equal, females that begin to reproduce at earlier ages have shorter interbirth intervals, higher survivorship among their offspring, live longer, and are likely to have higher lifetime fitness than other females. However, there may be important trade-offs between these components of fitness. For example, females could shorten their interbirth intervals by weaning their infants earlier, but this might jeopardize infant survival.

First-time (**primiparous**) mothers provide a good example of these kinds of trade-offs. When females begin to reproduce, they are often not yet fully grown. Their energetic investment in their infants competes with their energetic investment in their own skeletal growth and development. This

FIGURE 6.6

This female bonnet macaque produced twins, but only one survived beyond infancy. Twins are common among marmosets and tamarins but are otherwise uncommon among monkeys and apes.

FIGURE 6.7

At several locations in Japan, indigenous monkeys are fed regularly. The size of these artificially fed groups has risen rapidly, indicating that population growth is limited by the availability of resources.

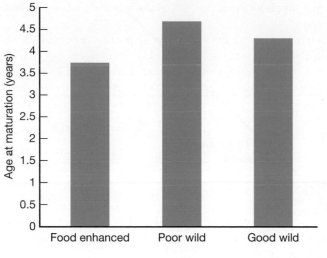

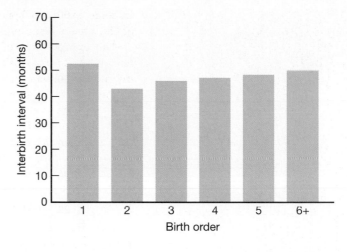

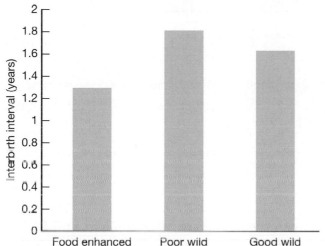

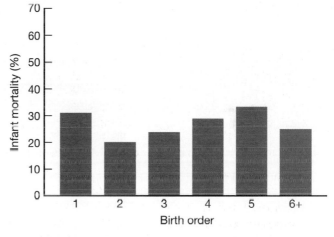

FIGURE 6.8

Food availability influences the age at first birth and length of interbirth intervals in baboon groups. In Amboseli National Park in Kenya, females that lived in groups had access to food scraps at a tourist lodge (food enhanced), matured earlier, and had shorter interbirth intervals than females that subsisted on wild foods. In addition, differences in wild-feeding groups reflected habitat quality.

FIGURE 6.9

The infants of primiparous mountain gorillas are less likely to survive than the offspring of multiparous females, and their interbirth intervals are also longer than the interbirth intervals of multiparous females.

is part of the reason why the offspring of primiparous mothers are less likely to survive than the offspring of **multiparous** females who have already produced offspring, and why it takes longer for primiparous females to recover from raising their infants. For example, the normal interbirth interval for well-fed captive rhesus macaques that produce surviving infants is 1 year, but primiparous mothers of surviving infants are twice as likely to skip a year between births as multiparous females are. Primiparous mountain gorillas' infants are less likely to survive than the offspring of multiparous females, and the interbirth intervals of primiparous mountain gorillas are longer than those of multiparous females (**Figure 6.9**).

In marked contrast to humans, most primate females continue to reproduce throughout their lives. Susan Alberts of Duke University led a team of researchers who examined the survivorship and reproductive activity of females in several well-studied populations of primates as varied as sifakas and gorillas. Their analyses

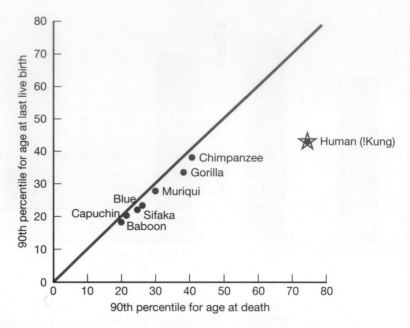

FIGURE 6.10

Female longevity is plotted against the age at last birth for a variety of primates. The 90th percentile for age at death represents the age by which 90% of the females in the sample have died (in other words, 10% of the females live past this age). The 90th percentile for age at last live birth represents the age at which 90% of the females in the sample produced their last infant. The close association between age at death and age at last infant means that most primate females continue to reproduce throughout their lives.

show that the age of last reproduction is close to the age at death for most females (**Figure 6.10**). The postreproductive period represents only 1% to 6% of the life span in these species but 43% of the life span in a representative population of human foragers, the !Kung.

Sources of Variation in Female Reproductive Performance

Variation in access to resources is likely to be reflected in variation in female reproductive success.

As we explained in Chapter 5, primates often compete for access to food resources. In nature, resource competition often leads to the formation of dominance hierarchies, which are kind of like rankings of players in tennis or chess (**A Closer Look 6.1**). High-ranking animals tend to have priority of access to food resources, particularly when resources are clumped and can be monopolized. If female reproductive success is influenced by female nutritional status, and dominance rank is related to food access, it seems reasonable to expect that rank would be associated with reproductive success. And, in fact, high rank does confer reproductive advantages on females in a number of primate populations.

Among female chimpanzees at the Gombe Stream National Park in Tanzania, daughters of ranking females mature at younger ages than daughters of lower-ranking females do (**Figure 6.12a**). High-ranking females also have shorter interbirth intervals than lower-ranking females do (**Figure 6.12b**). Recent analyses of long-term data on female mountain gorillas by Martha Robbins of the Max Planck Institute for Evolutionary Anthropology and her colleagues also show that high-ranking females have substantially shorter interbirth intervals than those of low-ranking females

6.1 Dominance Hierarchies

In many animal species as varied as crickets, chickens, and chimpanzees, competitive encounters within pairs of individuals are common. The outcome of these contests may be related to the participants' relative size, strength, experience, or willingness to fight. In many species, for example, larger and heavier individuals regularly defeat smaller individuals. If there are real differences in power (based on size, weight, experience, or aggressiveness) between individuals, then we would expect the outcomes of dominance contests to be about the same from day to day. This is often the case. When dominance interactions between two individuals have predictable outcomes, we say that a **dominance** relationship has been established.

When dominance interactions have predictable outcomes, we can assign dominance rankings to individuals. Consider the four hypothetical females in **Figure 6.11a**, which we will call Blue, Turquoise, Green, and Purple. Blue always beats Turquoise, Green, and Purple. Turquoise never beats Blue but always beats Green and Purple. Green never beats Blue or Turquoise but always beats Purple. Poor Purple never beats anybody. We can summarize the outcome of these confrontations between pairs of females in a **dominance matrix** such as the one in **Figure 6.11b**, and we can use the data to assign numerical ranks to the females. In this case, Blue ranks first, Turquoise second, Green third, and Purple fourth. When females can defeat all the females ranked below them and none of the females ranked above them, dominance relationships are said to be **transitive**. When the relationships within all sets of three individuals (trios) are transitive, the hierarchy is linear.

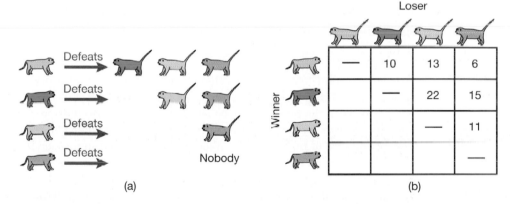

FIGURE 6.11

(a) Suppose that four hypothetical females—named Blue, Turquoise, Green, and Purple—have the following transitive dominance relationships: Blue defeats the other three in dominance contests. Turquoise cannot defeat Blue but can defeat Green and Purple. Green loses to Blue and Turquoise but can defeat Purple. Purple can't defeat anyone. (b) The results of data such as those in part a are often tabulated in a dominance matrix, with the winners listed down the left side and the losers across the top. The value in each cell of the matrix represents the number of times one female defeated the other. Here, Blue defeated Turquoise 10 times and Green defeated Purple 11 times. There are no entries below the diagonal because females were never defeated by lower-ranking females.

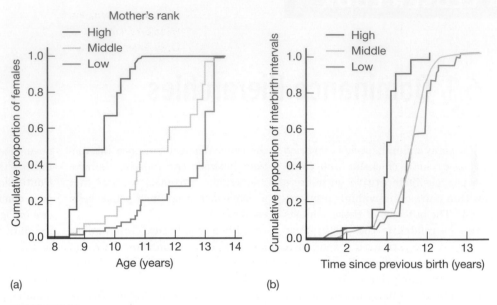

Mother's rank
— High
— Middle
— Low

(a)

(b)

FIGURE 6.12

In (a), the cumulative proportion of females reaching sexual maturity at a given age is plotted. Half of all the daughters of high-ranking females reach sexual maturity before the age of 10, while less than 20% of the daughters of middle- and low-ranking females reach sexual maturity by that age. In (b), the cumulative proportion of interbirth intervals lasting a given length of time is plotted. Among high-ranking females, about half of all interbirth intervals last less than 4 years, while among middle- and low-ranking females about 20% of all interbirth intervals are that short.

(**Figure 6.13**). Similarly, studies of gray langurs near Jodhpur, India, conducted by a group of primatologists, including Carola Borries (at Stony Brook University in New York) and Volker Sommer (at University College London), have shown that young, high-ranking females reproduce more successfully than do older, lower-ranking females (**Figure 6.14**).

It is important to consider whether the advantages of high rank generate variation in lifetime fitness among females. If female rank changes over the course of their lives, then the reproductive benefits that females derive from being high ranking may be offset by the disadvantages that they incur when they are low ranking. The advantages of high rank are likely to be magnified in species in which female ranks remain relatively stable over the course of their lives. For example, at Gombe Stream National Park, ranks of female chimpanzees are relatively stable, and high-ranking females have higher lifetime fitness than that of low-ranking females. Female baboons establish dominance hierarchies that typically remain stable over decades, and in Mikumi, Tanzania, higher-ranking female baboons produced more surviving offspring over the course of their lives than lower-ranking females did (**Figure 6.15**).

The quality of social bonds may also influence female reproductive success.

Dominance rank is not the only factor that contributes to variation in female reproductive success. A growing body of evidence from studies of baboons and macaques suggests that the quality of females' social bonds may also affect their fitness. In these species, females spend a considerable amount of time sitting near other group members and grooming them. These interactions seem to be important to females

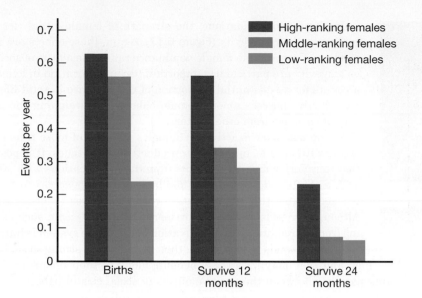

FIGURE 6.14

Female gray langurs reproduce more successfully when they are young and hold higher rank than when they are older and have lower rank. The three bars on the left represent the proportion of females of each rank category who give birth each year. The other sets of bars represent the proportion of females in each rank category who give birth each year to infants who survive to 12 months and 24 months. Dominance rank influences both the likelihood of giving birth and the likelihood that infants will survive.

because they save time for socializing, even when times are tough and food is scarce (**Figure 6.16**). For example, female baboons are forced to spend more time foraging and moving between feeding sites in the dry season than in the wet season. During the dry season, they cut down on the amount of time that they spend resting, but they preserve time for socializing. Social bonds seem to matter to females. Anne Engh of Kalamazoo College and her colleagues found that female baboons who lose close companions to predators experience substantial increases in cortisol levels, a hormonal indicator of stress. You might think that females are simply stressed about living through a predator attack, but females who were present in the group and didn't lose close associates were less strongly affected than those that did.

Data derived from long-term studies of two baboon populations show that females who have strong social bonds reproduce more successfully than females that have weaker social bonds. Females who spent more time grooming and near other group members had more surviving infants than other females did, and these effects were independent of the females' dominance rank. In fact, sociality seems to insulate females from some of the costs of low rank. The most sociable low-ranking females reproduce as successfully as the most sociable high-ranking females.

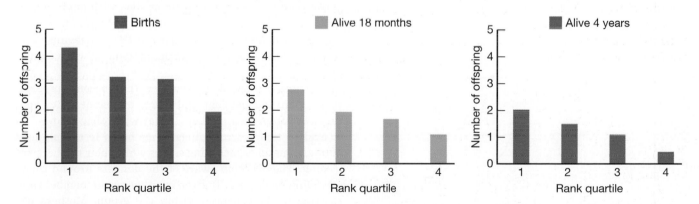

FIGURE 6.15

Among yellow baboons in Tanzania, female dominance rank is associated with lifetime reproductive success. In this graph, females are divided into quartiles, with 1 representing the top-ranked quarter of females and 4 representing the lowest-ranking ones. High-ranking females produced more offspring over the course of their lives and also produced more offspring that survived to the ages of 18 months and 4 years.

FIGURE 6.16

In many primate species, females form close and lasting ties to selected partners, often their own mothers and daughters. Here, two female chacma baboons groom in the Moremi Reserve of the Okavango Delta in Botswana.

FIGURE 6.17

This graph illustrates the relationship between social connectedness and longevity in female baboons. The blue line represents females at the 75th percentile, and the red line represents females in the 25th percentile. Slightly more than half of females in the 75th percentile live to at least 20 years of age, while less than 20% of females in the 25th percentile live that long.

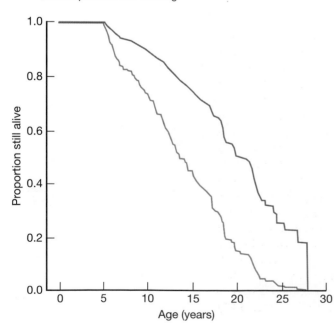

In these two populations, the strength of female social ties also influences their longevity (**Figure 6.17**). Again, these effects are independent of variation in female dominance rank. The effects of sociality on longevity are particularly important because variation in longevity accounts for a substantial proportion of the variance in total lifetime reproductive success among female baboons, outweighing the influence of variance from other factors.

We do not know exactly why females who spend more time interacting with others enjoy these reproductive advantages. It is possible that they derive material benefits from their associations with others, such as better protection from predators or better access to food. It is also possible that social contact reduces females' levels of stress. Although stress is an adaptive response to acute danger, such as fleeing from a predator, chronic activation of the stress system is harmful. So, if social bonds reduce stress, then females with strong social bonds may be healthier, and they may pass these benefits along to their offspring. There are intriguing parallels between the positive effects of social ties in baboons and in humans, although we do not yet know whether these parallels are based on similar mechanisms.

Trade-Offs between Quantity of Offspring and Quality of Care

Females must make a trade-off between the number of offspring they produce and the amount of care that they provide to each of their offspring.

Just as both males and females must allocate a limited effort to parental investment and mating, females must apportion resources among their offspring. All other things being equal, natural selection will favor individuals that can convert effort into offspring most efficiently. Because mothers have a finite amount of effort to devote to offspring, they cannot maximize both the quality and the quantity of the offspring that they produce. If a mother invests great effort in one infant, she must reduce her investment in others. If a mother produces many offspring, she will be unable to invest very much in any of them.

In nature, maternal behavior reflects this trade-off when a mother modifies her investment in relation to an offspring's needs. Initially, infants spend virtually all of their time in contact with their mothers. The very young infant depends entirely on its mother for food and transportation and cannot anticipate or cope with environmental hazards. At this stage, mothers actively maintain close contact with their infants (**Figure 6.18**), retrieving them when they stray too far and scooping them up when danger arises.

As infants grow older, however, they become progressively more independent and more competent. They venture away from their mothers to play with other infants and to explore their surroundings. They begin to sample food plants, sometimes mooching scraps while their mothers are feeding. They become aware of the dangers around them, attending to alarm calls given by other group members and reacting to disturbances within the group. Mothers use a variety of tactics to actively encourage their infants to become more independent. They may subtly resist their infants' attempts to suckle. They may also encourage their infants to travel independently. Nursing is gradually limited to briefer and more widely spaced bouts that may

FIGURE 6.18

A female chimpanzee sits beside her youngest infant in Gombe Stream National Park in Tanzania.

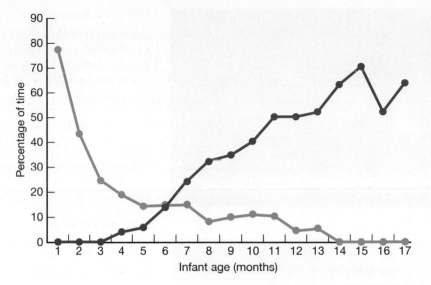

FIGURE 6.19

The *y* axis shows the proportion of time that free-ranging baboon infants spend suckling (*turquoise circles*) and feeding on their own (*red circles*). As infants get older, they spend less time suckling and more time feeding independently. These changes reflect changes in the costs and benefits of infant care for mothers.

eventually provide the infant more with psychological comfort than physical nourishment. At this stage, infants are carried only when they are ill, injured, or in great danger.

The changes in maternal behavior reflect the shifting balance between the requirements of the growing infant and the energy costs to the mother of catering to her infant's needs. As infants grow older, they become heavier to carry and require more nourishment, imposing substantial burdens on mothers. However, as infants grow older they also become more capable of feeding themselves and of traveling independently, and this means that mothers can gradually limit investment in their older infants without jeopardizing their welfare (**Figure 6.19**). Mothers can conserve resources that can be allocated to infants born later.

Sexual Selection and Male Mating Tactics

Sexual selection leads to adaptations that allow males to compete more effectively with other males for access to females.

So far, we have seen that primate females invest heavily in each of their young and produce relatively few offspring during their lives. Moreover, in most primate species, females can raise their offspring without much help from males. Female reproductive success is mainly limited by access to food, not access to mates. From the males' point of view things look quite different. Conception opportunities are relatively scarce because of the females' long interbirth intervals. At the same time, one male can potentially sire the offspring of many females. As a result, males in many species compete for access to sexually receptive females. Characteristics that increase male success in competition for mates will spread as a result of what Darwin called **sexual selection**.

It is important to understand the distinction between natural selection and sexual selection. Most kinds of natural selection favor phenotypes in both males and females that enhance their ability to survive and reproduce. Many of these traits are related to resource acquisition, predator avoidance, and offspring care. Sexual selection is a special category of natural selection that favors traits that increase success in competition for mates, and it will be expressed most strongly in the sex whose access to members of the opposite sex is most limited. Sexual selection may favor traits that increase mating success, even if those traits reduce the ability of the animal to survive or acquire resources—outcomes not usually favored by natural selection (**Figure 6.20**).

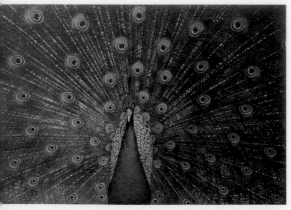

[a]

[b]

FIGURE 6.20

Sexual selection can favor traits not favored by natural selection. (a) The peacock's tail hinders his ability to escape from predators, but it enhances his attractiveness to females. Female peahens are attracted to males that have the most eyespots in their trains. (b) Male red deer use their antlers when they fight with other males. Red deer antlers are a good example of a trait that has been favored by sexual selection.

There are two types of sexual selection: (1) Intrasexual selection, which results from competition among males, and (2) intersexual selection, which results from female choice.

Sexual selection is typically divided into two categories: intrasexual selection and intersexual selection. In species in which females cannot choose their mates, access to females will be determined by competition among males. In such species, **intrasexual selection** favors traits that enhance success in male–male competition. In species in which females can choose their mates, selection favors traits that make males more attractive to females. This is called **intersexual selection**. Intersexual selection is responsible for some spectacular traits, such as the peacock's tail (see Figure 6.20a) and the lion's thick mane. There is not much evidence that intersexual selection plays an important role in primates, although it is tempting to hypothesize that this is the explanation for the proboscis monkey's big nose (**Figure 6.21**).

Sexual selection is often much stronger than ordinary natural selection.

In mammalian males, sexual selection can affect behavior and morphology more than other forms of natural selection can because male reproductive success usually varies much more than female reproductive success. Data from long-term studies of lions conducted by Craig Packer of the University of Minnesota and Anne Pusey of Duke University show that the lifetime reproductive success of the most successful males is often much greater than that of even the most successful females (**Figure 6.22**). The same pattern is likely to hold for most non-pair-bonded primates. Analyses of parentage in a large population of rhesus macaques on Cayo Santo Island in Puerto Rico show that the most successful male in the population sired 47 infants, while the most successful female produced only 16, and the variance in number of offspring produced over the life span is about four times higher for males than that for females (**Figure 6.23**). Because the strength of selection depends on how much variation in fitness there is among individuals, sexual selection acting on male primates can be much stronger than selective forces acting on female primates. (Incidentally, in species such as sea horses, in which males invest in offspring and females do not, the entire pattern is reversed: Sexual selection acts much more strongly on females than on males.)

FIGURE 6.21

These are proboscis monkeys, named for the prominence of their noses. Males have considerably larger and more bulbous noses than females.

Intrasexual Selection

Competition among males for access to females favors large body size, large canine teeth, and other weapons that enhance male competitive ability.

For primates and most other mammals, intrasexual competition is most intense among males. In the most basic form of male–male competition, males simply drive other males away from females. Males who regularly win such fights have higher

[a]

[b]

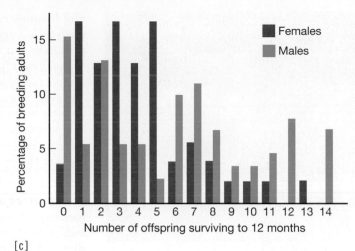

[c]

FIGURE 6.22

The reproductive success of (a) male lions is considerably more variable than that of (b) female lions. (c) In Serengeti National Park and the Ngorongoro Crater of Tanzania, few female lions fail to produce any surviving cubs, but most females produce fewer than six surviving cubs during their lives. Many males fail to produce any cubs, and a few males produce many cubs.

reproductive success than those who lose. Thus intrasexual selection favors features such as large body size, horns, tusks, antlers, and large canine teeth that enable males to be effective fighters. For example, male gorillas compete over access to groups of females, and males weigh twice as much as females and have longer canine teeth.

As explained in Chapter 5, when the two sexes consistently differ in size or appearance, they are said to be sexually dimorphic (**Figure 6.24**). The body sizes of males

FIGURE 6.23

These graphs plot the number of offspring produced by (a) male and (b) female rhesus macaques that lived to different ages. There is much more variation in these values for males than for females.

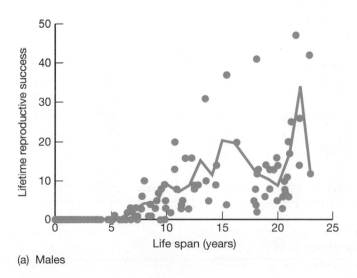

(a) Males

(b) Females

and females represent compromises among many competing selective pressures. Larger animals are better fighters and are less vulnerable to predation, but they also need more food and take longer to mature. Males compete over females; females compete over resources but generally do not compete over mates. The effect of intrasexual competition among males, however, is quantitatively greater than the effect of competition among females because the fitness payoff to a very successful male is greater than it is to a very successful female. Therefore, sexual selection is much more intense than ordinary natural selection. As a result, intrasexual selection leads to the evolution of sexual dimorphism.

The fact that sexual dimorphism is greater in primate species forming one-male, multifemale groups than in pair-bonded species indicates that intrasexual selection is the likely cause of sexual dimorphism in primates.

FIGURE 6.24

Adult male yellow baboons are nearly twice the size of adult females. The degree of sexual dimorphism in body size is most pronounced in species with the greatest amount of competition over access to females.

If sexual dimorphism among primates is the product of intrasexual competition among males over access to females, then we should expect to see the most pronounced sexual dimorphism in the species in which males compete most actively over access to females. One indirect way to assess the potential extent of competition among males is to consider composition of social groups. In general, male competition is expected to be most intense in social groups in which males are most outnumbered by females. At first, this prediction might seem paradoxical because we might expect to have more competition when more males are present. The key to resolving this paradox is to remember that in most natural populations, there are approximately equal numbers of males and females at birth. In species that form one-male groups, there are many **bachelor males** (males who don't belong to social groups) who exert constant pressure on resident males. In species that form pair bonds, each male is paired with a single female, reducing the intensity of competition among males over access to females.

Comparative analyses originally conducted by Paul Harvey of the University of Oxford and Tim Clutton-Brock of the University of Cambridge have shown that the extent of sexual dimorphism in primates corresponds roughly to the composition of the groups in which the males live (**Figure 6.25**). There is little difference in body weight or canine size between males and females in species that typically form pair bonds, such as gibbons and titi monkeys. At the other extreme, the most pronounced dimorphism is found in species that live in one-male, multifemale groups, such

FIGURE 6.25

The degree of sexual dimorphism is a function of the ratio of males to females in social groups. (a) Relative canine size (male canine length divided by female canine length) and (b) body size dimorphism (male body weight divided by female body weight) are greater in species that form one-male, multifemale groups than in species that form multimale, multifemale groups or pair-bonded groups.

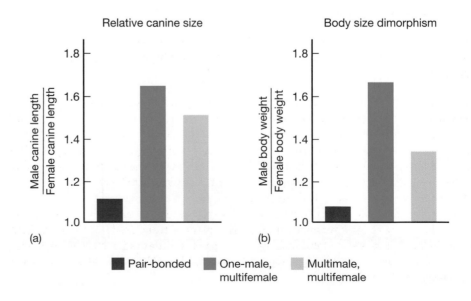

as gorillas and black-and-white colobus monkeys. And in species that form multimale, multifemale groups, the extent of sexual dimorphism is generally intermediate between these extremes. Thus sexual dimorphism is most pronounced in the species in which the ratio of females to males within groups is highest.

In multimale, multifemale groups, in which females mate with several males during a given estrous period, sexual selection favors increased sperm production.

In most primate species, as with mammals in general, the female is receptive to mating mainly during the portion of her reproductive cycle when fertilization is possible. That period is called **estrus**. In primate species that live in multimale, multifemale groups, females can often mate with several males during a single estrous period. In such species, sexual selection favors increased sperm production because males who deposit the most sperm in the female reproductive tract have the greatest chance of impregnating them. Competition in the quantity of sperm is likely to be relatively unimportant in pair-bonded species because females mate mainly with the resident male. Similarly, competition in sperm quantity probably does not play a very important role in species that form one-male, multifemale groups. In these species, male fitness may depend more on their ability to establish residence in groups than on their ability to produce large quantities of sperm.

Group composition is associated with testis size, much as we would expect. Males with larger testes typically produce more sperm than do males with smaller testes, and males who live in multimale groups have much larger testes in relation to their body size than do males who live in either pair-bonded or one-male, multifemale groups (**Figure 6.26**).

Male Reproductive Tactics

As we noted earlier, there is considerable variation in the mating strategies and parenting behavior of primate males. In some primate species, males compete fiercely over access to females, while in others, there is relatively little direct competition. In some species, males invest substantial effort in caring for offspring, and in others, males provide no direct care for offspring. As we will see below, males' mating and parenting strategies are often linked.

Pair-Bonding Species

Pair bonding is generally associated with relatively high levels of paternal investment.

In species that live in pairs, males do not compete directly over access to females. In these species, males' reproductive success depends mainly on their ability to establish territories, find mates, and rear surviving offspring. Males can be relatively confident that they are the sires of the offspring of their mates and that their care will be directed to their own genetic offspring. As we noted earlier, extra pair-matings have been observed in a number of pair-bonded primate species, but the resident male sires a larger proportion of his mates' offspring than do outside males.

In some pair-bonded species, like owl monkeys, males play an active role in infant care. They carry them, groom them, and protect them from predators (**Figure 6.27**). Male siamangs are also helpful fathers, carrying older infants for long periods every day. However, there are some pair-bonded species, like gibbons, in which males provide little care for infants.

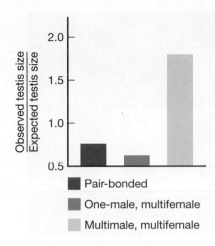

FIGURE 6.26

The average size of testes in species that typically form pair-bonded and one-male, multifemale groups is relatively smaller than the average size of testes in multimale, multifemale groups. Here, observed testis weight is divided by the expected testis weight to produce relative testis size. The expected testis weight is derived from analyses that correct for the effects of body size.

Legend:
- Pair-bonded
- One-male, multifemale
- Multimale, multifemale

FIGURE 6.27

A pair of white-handed gibbons. Gibbons form pair bonds, but males and females sometimes mate with individuals from outside the group.

Cooperative Breeding

Some cooperatively breeding species are monogamous and others are polyandrous.

Cooperatively breeding species typically live in groups that include more than one adult of each sex (**Figure 6.28**). It has been clear for some time that normally only one female breeds in these groups, but genetic data were needed to work out whether infants were all sired by one male (monogamy) or sired by multiple males (polyandry). Current evidence suggests that the dominant male sires almost all of the infants in marmoset groups, but multiple males may sire infants in tamarin groups. Research conducted by Samuel Diaz-Munoz of the University of California, Davis, suggests that infants from the same litter are sometimes sired by different males, and sometimes different litters in the same group are sired by different males. He points out that if different litters are sired by different males, then short-term studies that sample only one litter per group may underestimate the extent of polyandry.

Helpers, particularly males, play an active role in child care, often carrying infants, grooming them, and sharing food with them. The contribution of helpers makes it possible for marmosets and tamarins to sustain unusually high fertility rates. While most monkeys and apes produce single young and have interbirth intervals that last at least a year, marmosets and tamarins typically produce twins, and females sometimes produce more than two litters per year. The presence of multiple males in the group increases the likelihood of infant survival in a population of golden-headed lion tamarin infants in Brazil (**Figure 6.29**). Infants raised in groups with more than one adult male also grew faster and had higher adult weights than infants raised in groups with only one male.

Polygyny

In species with polygynous mating systems, the reproductive success of males depends mainly on their ability to establish and maintain residence in social groups.

Most species that live in one-male, multifemale groups have polygynous mating systems. This means the resident male mates with multiple females, but females mainly mate with a single male. In this situation, male mating success depends mainly on gaining access to groups of females. This can lead to intense conflicts among males. Among gray langurs, dispersing males join all-male bands that collectively attempt to oust resident males from their groups. Once they succeed in driving out the resident male, the members of the all-male band compete among themselves for sole access to the group of females. In the highlands of Ethiopia, male geladas that are

FIGURE 6.28

Some marmoset and tamarin groups contain more than one adult male and a single breeding female. In at least some of these groups, mating activity is limited to the group's dominant male, even though all the males participate in the care of offspring.

FIGURE 6.29

In tamarin groups, infant survival is higher when there is more than one adult male in the group (blue line) than when there is only one male in the group (red line). In groups that contain more than one adult male, about 80% of infants survive at least 100 days, while in groups with only one male, survival to 100 days is about 40%.

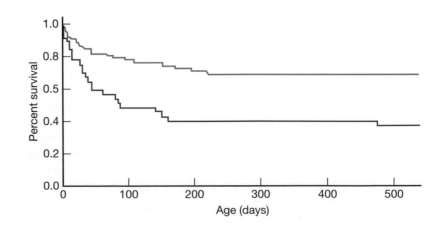

not associated with groups of females form "bachelor" groups. Bachelor males regularly challenge resident males, leading to fierce confrontations that may last for several days (**Figure 6.30**). Males are sometimes killed or seriously injured in contests for control of one-male groups.

Competitive pressures affect males' ability to monopolize conceptions within their groups.

As we saw earlier, in species that form one-male groups, males have relatively small testes. This implies that resident males are generally able to monopolize conceptions within their groups, and there is relatively little sperm competition. In species for which we have genetic evidence, the data indicate that most, but not necessarily all, infants are sired by the resident male. For example, all of the infants born in one-male groups of gray langurs, red howlers, and geladas are sired by the resident male. However, in some species, resident males are not able to prevent other males from associating with the group and mating with females. In blue monkeys and patas monkeys, influxes of nonresident males are concentrated during the mating season and may involve several males. Genetic data from a long-term study of blue monkeys in the Kakamerga Forest of Kenya directed by Marina Cords of Columbia University found that resident males sire on average 60% of the infants born in their groups.

Competitive pressures from outside males may sometimes lead males to tolerate the presence of secondary males (or "followers"). For example, in gelada baboons, approximately one-third of all social units include a leader and one or more followers. Follower males are sometimes former leaders who remain in the group after being deposed in a takeover. In other cases, bachelor males team up to take over social units; afterward, one male becomes the leader and one or more other males may remain in the group as followers. Noah Snyder-Mackler, now at Arizona State University, and his colleagues Jacinta Beehner and Thore Bergman at the University of Michigan have found that followers tend to be found in units containing larger numbers of females and that the presence of additional males increases leader males' tenure by 30%, reduces the rate of takeover attempts, and shortens females' interbirth intervals (**Figure 6.31**). Leader males sire about 83% of the infants born within their groups, and followers sire the rest. Similar patterns have been observed in red howler monkeys and hamadryas baboons. In these cases, the benefits of tolerating secondary males (longer tenure, larger female group size, higher female fertility) may outweigh the costs (lost conceptions).

FIGURE 6.30

Most gelada groups contain only one male. Males sometimes attempt to take over groups and oust the resident male, and this can result in fierce battles among males.

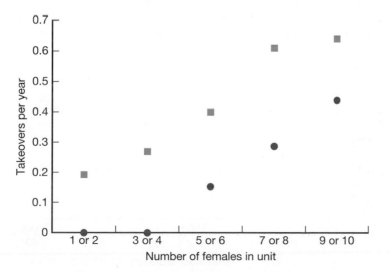

FIGURE 6.31

The rate of takeover attempts rises as the number of females in social units increases. However, units with more than one male (*circles*) experience fewer takeover attempts than do units with only one male (*squares*).

In most species with polygynous mating systems, males generally provide little care for infants. Gorillas provide an interesting exception to the general rule.

As we explained before, in species that live in one-male groups, males are expected to invest more in mating effort than in parenting effort. Males in these species do not often provide direct care for infants. Male mountain gorillas represent an interesting exception to this pattern. Males are remarkably tolerant of infants, and infants and juveniles cluster around them (**Figure 6.32**). Males sometimes groom infants and juveniles and defend them in disputes with other group members. The majority of mountain gorilla groups contain only one male, but some groups contain multiple males. In these groups, the top-ranked male sires more offspring than do other males but does not completely monopolize conceptions. Research led by Stacy Rosenbaum, now at the University of Michigan, and her colleagues shows that infants in multimale groups are particularly attracted to top-ranked males, who are not necessarily their own fathers, and males do not treat their own infants differently than they treat other males' infants. In addition, Rosenbaum and her colleagues have found that males who spend the most time caring for infants have substantially higher lifetime fitness than that of males who spend less time caring for infants, and this effect is not simply due to the fact that high-ranking males sire more offspring and associate more with infants (**Figure 6.33**).

Polygynandry

In species with polygynandrous mating systems, competition focuses on access to sexually receptive females.

Most species that live in multimale, multifemale groups have a polygynandrous mating system, which means that both males and females mate with multiple partners. However, that does not mean that all males have equal opportunities to mate. Males often compete over access to receptive females. Sometimes males attempt to drive other males away from females, to interrupt copulations, or to prevent other males from approaching or interacting with females. More often, however, male–male competition is mediated through dominance relationships that reflect competitive abilities. These relationships are generally established in contests that can involve threats

FIGURE 6.32

Adult male gorillas are remarkably tolerant of infants and juveniles, who often cluster around them. (Photo printed with permission from the Dian Fossey Gorilla Fund International.)

FIGURE 6.33

In this graph, adult male mountain gorillas are divided into three levels on the basis of how much time they spent affiliating with immatures. Males that spent the most time affiliating with immatures had the highest lifetime fitness. (The orange line shows all males, and the blue line shows males with a dominance rank of beta or lower.)

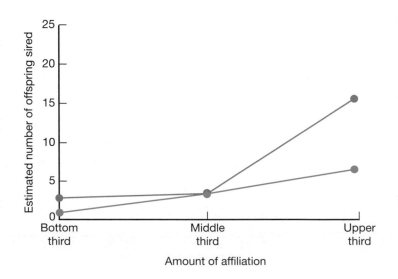

and stereotyped gestures but that can also lead to escalated conflicts in which males chase, wrestle, and bite one another (**Figure 6.34**). Male fighting ability and dominance rank are generally closely linked to physical condition. In many species, there is an ∩-shaped relationship between male age and dominance rank, as male condition and dominance rank peak in early adulthood (**Figure 6.35**).

In many species that form multimale groups, male dominance rank influences reproductive success. Figure 6.35 shows that in yellow baboons, paternity success tracks male dominance rank. A team led by Susan Perry of the University of California, Los Angeles, showed that in white-faced capuchin groups, the top-ranking male sired 38% to 70% of all infants. Analyses of paternity in multimale groups of hanuman langurs, long-tailed macaques, howler monkeys, patas monkeys, and chimpanzees also show that high-ranking males reproduce more successfully than do other males.

Although there is often a positive correlation between dominance rank and reproductive success, there is considerable variation in the extent of reproductive skew within populations, across populations, and across species. In Amboseli, top-ranking males monopolize access to receptive females most effectively when there are relatively few other males present in the group and when there are relatively few estrous females present at the same time. In chimpanzees, the mean dominance rank of sires ranges from about 0.9 in the Mahale Mountains of Tanzania (a score of 1.0 would mean that all infants were sired by the top-ranking male) to just over 0.6 in one chimpanzee community in the Taï Forest (**Figure 6.36**).

FIGURE 6.34

Male baboons sometimes fight directly over access to estrous females. Here a male baboon challenges another male who is mate-guarding a sexually receptive female. If the challenger is successful, he will begin mate-guarding the female himself.

There are some species with polygynandrous mating systems that confound our expectations about male reproductive tactics.

Although most species that live in multimale, multifemale groups conform to the pattern we have described, there are a few intriguing exceptions. A long-term project on muriquis in Brazil led by Karen Strier of the University of Wisconsin has shown that male muriquis (**Figure 6.37**) rarely fight, and it is not possible to rank them in a dominance hierarchy. Genetic analyses indicate that the extent of reproductive skew among males is very low. The most successful male in Strier's study population sired only 18% of the infants; in contrast, the most successful baboon and capuchin males sire about 80% of the infants in their groups. Muriquis have relatively large testes for the body size, suggesting that sperm competition may play an important role.

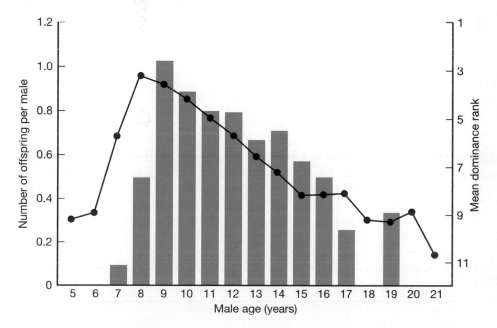

FIGURE 6.35

Among yellow baboons in the Amboseli basin, both male rank (*black dots*) and paternity success (green bars) are closely linked to male age and physical condition. Males reach their highest ranks and sire the most infants, on average, when they are about 8 years old, and then gradually fall in rank and paternity success as they age.

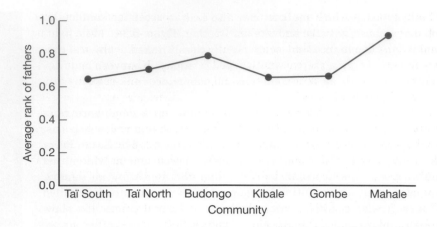

FIGURE 6.36

Here, the average rank of sires is plotted for six chimpanzee communities. In the Mahale community, the mean rank of sire is about 0.9 (maximum value = 1.0), which means that the top-ranking male sired most of the infants. The rank of sires in Taï South is about 0.65, which means that a smaller fraction of infants was sired by the top-ranked male.

FIGURE 6.37

Muriquis live in multimale, multifemale groups, but males rarely fight and do not form a dominance hierarchy. Mating is fairly evenly distributed among males in the group.

In species with polygynandrous mating systems, males sometimes interact selectively with their own offspring.

We once thought that there would be little parental care in species that form multimale groups because males would not be able to distinguish between their own infants and other males' infants. But this does not always seem to be the case. For example, male baboons selectively support their own offspring in disputes with other group members (**Figure 6.38**), and male rhesus macaques and chimpanzees associate more with their own infants than with unrelated infants. Pregnant and lactating female baboons often form close ties to one or sometimes two adult males in their groups. Females' primary associates are often the sires of their infants. Although male baboons do not provide much direct care for infants, their presence may represent a form of passive protection for females and their infants. Nga Nguyen of California State University, Fullerton, who led a study of male–female relationships in Amboseli, describes these relationships as a form of "joint parental care." It is not entirely clear how males know which infants they have fathered. In baboons, mate guarding near the time of conception is closely related to paternity success, and males may use mate-guarding activity as a proxy for paternity.

Infanticide

Infanticide is a sexually selected male reproductive strategy.

We have seen that high-ranking males can monopolize access to receptive females, and this generates fierce competition over residence in one-male groups and competition for high-ranking positions within multimale groups. Males have a limited window of time in which they have priority of access to sexually receptive females. Sarah Blaffer Hrdy, now retired from the University of California, Davis, was the first to see that these circumstances might favor the evolution of infanticide as a male reproductive tactic. Her reasoning was based on the following logic: When a female monkey gives birth to an infant, she nurses it for several months and does not become pregnant again for a considerable time. After the death of an infant, lactation ends abruptly and females resume cycling. Thus, the death of nursing infants hastens the resumption of maternal receptivity, particularly in species that do not breed seasonally. A male who takes over a group or rises to the top-ranking position may benefit from killing nursing infants because their deaths cause their mothers to become sexually receptive much sooner than they would otherwise.

This hypothesis, which has become known as the **sexual selection infanticide hypothesis**, was initially controversial. There were no direct observations of males killing infants, and some researchers found it hard to believe that this form of violence was an evolved strategy. However, infanticide by males has now been documented in approximately 40 primate species (and many nonprimate species, such as lions). Researchers have witnessed dozens of infanticidal attacks in the wild and have recorded many nonlethal attacks on infants by adult males. There are many more instances in which healthy infants have disappeared after takeovers or changes in male rank. Infanticide occurs in species that typically form one-male groups and in multimale groups of chacma baboons, langurs, capuchins, and Japanese macaques.

This body of data enables researchers to test several predictions derived from Hrdy's hypothesis. If infanticide is a male reproductive strategy, then we would expect that (1) infanticide would be associated with changes in male residence or status; (2) males should kill infants whose deaths hasten their mothers' resumption of cycling; (3) males should kill other males' infants, not their own; and (4) infanticidal males should achieve reproductive benefits.

All of these predictions have been supported. Carel van Schaik of the University of Zurich compiled information about 55 infanticides in free-ranging groups that were actually witnessed by observers. He found that nearly all infanticides (85%) followed changes in male residence or dominance rank. He also found that most infanticides involve unweaned infants, whose deaths will have the greatest impact on female sexual receptivity. Males largely avoid killing related infants. Only 7% of infanticides were committed by males who were sexually active in the group at the time the infants were conceived. Finally, in at least 45%—and possibly as much as 70%—of these cases, the infanticidal male later mated with the mother of the infant that he killed.

FIGURE 6.38

Male yellow baboons selectively support their own offspring in conflicts. Here a juvenile takes refuge beside his father when he is challenged by an older juvenile. The male then threatens the older juvenile and subsequently chases him away.

Infanticide is sometimes a substantial source of mortality for infants.

Among mountain gorillas in the Virunga Mountains of Rwanda, savanna baboons in the Moremi Game Reserve in Botswana, gray langurs in Ramnagar in Nepal, and red howlers in Venezuela, approximately one-third of all infant deaths are due to infanticide. Among the gelada baboons of the Simien Mountains in Ethiopia, approximately 10% of infants are killed during each takeover.

Females have evolved a battery of responses to infanticidal threats.

Although infanticide may enhance male reproductive success, it can have a disastrous effect on females who lose their infants. Thus we should expect females to evolve counterstrategies to infanticidal threats. The most obvious counterstrategy would be for females to try to prevent males from harming their infants. However, females' efforts to defend their infants are unlikely to be effective. Remember that males are generally larger than females in species without pair bonds, and the extent of sexual dimorphism is most pronounced in species that form one-male groups.

Instead, females may try to confuse males about paternity. As we have seen already, males seem to kill infants when there is no ambiguity about their paternity; if females can increase uncertainty about paternity, they may reduce the risk of infanticide. Females might confuse males about paternity by obscuring information about their reproductive state, by mating with multiple males when they are sexually receptive, and by mating with males at times when they are not likely to conceive. All of these strategies have been documented among primates. A group of researchers led by Michael Heistermann of the German Primate Center has examined the patterns of sexual behavior and ovulatory status among female gray langurs in multimale groups. They found that females are sexually receptive for about 9 days, on average, and that they can ovulate anytime within that period (**Figure 6.39a**). Females sexually solicit males throughout this period, and male mating behavior is not concentrated around the day of ovulation (**Figure 6.39b**), suggesting that males can't tell when females are likely to conceive. Female langurs also mate with multiple males and sometimes solicit males after takeovers when they are already pregnant.

Gelada females have adopted a different strategy to reduce the impact of infanticide. Eila Roberts of Michigan State University and her colleagues used hormonal data to monitor the reproductive status of females before and after takeovers. They discovered that females abruptly terminate pregnancies in the days that follow takeovers (**Figure 6.40**). Although this may seem like a very costly strategy for females, it may be advantageous for females to terminate their investment in an infant that is very likely to be killed by the new leader male and reallocate maternal effort to another reproductive attempt. It is important to understand that pregnancy termination reflects an

FIGURE 6.39

In some species, females have evolved counterstrategies to infanticide. Gray langurs seem to obscure information about the timing of ovulation. (a) Females are sexually receptive for about 9 days, and ovulation can occur anytime within that period. Asterisks indicate receptive periods in which conceptions occurred. (b) Sexual activity is not concentrated around the time of ovulation. Females solicit males, and males copulate with females at fairly consistent rates throughout the receptive period.

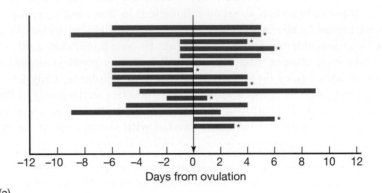

(a)

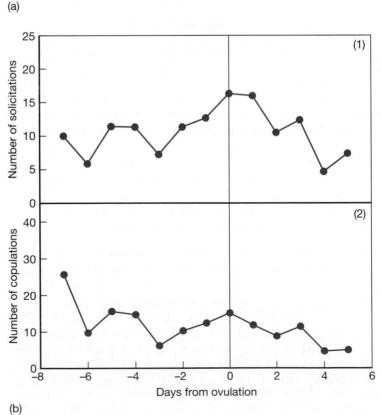

(b)

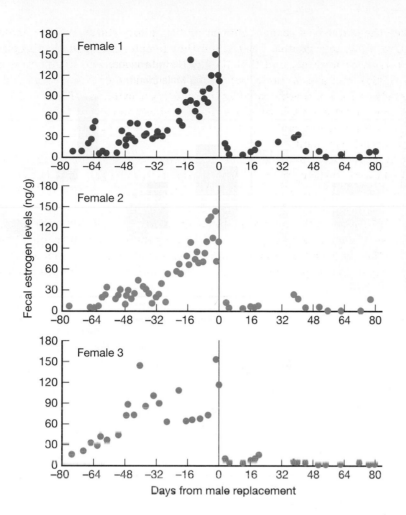

FIGURE 6.40

Gelada females terminate pregnancies immediately after takeovers. Levels of estrogen fall precipitously after takeovers, indicating that pregnancies have ended. Before takeovers, the hormone levels of females who terminated their pregnancies cannot be distinguished from the hormone levels of females who carry pregnancies to term.

adaptive, evolved response to male takeovers and is not based on conscious decisions made by individuals.

Female baboons have evolved a different strategy to reduce their vulnerability to infanticide. Lactating females often form close ties with one or sometimes two adult males (**Figure 6.41**). These males are often, but not always, the fathers of their infants. In the Moremi Reserve of Botswana, infanticide is common, and female baboons are agitated in the presence of new males. Jacinta Beehner and her colleagues have found that females' cortisol levels, which provide a physiologic index of stress, rise sharply when immigrant males enter the group (**Figure 6.42a**). But not all lactating females are equally concerned about the arrival of new males: Those who have established close ties with males have lower cortisol levels than those who have not established close ties to males (**Figure 6.42b**).

If the data on infanticide are so consistent, why is the idea so controversial?

When Hrdy first proposed the idea that infanticide is an evolved male reproductive strategy, there was plenty of room for skepticism and dispute. Direct observations of infanticide were very limited, and it was hard to exclude alternative explanations. Now, however, we have good evidence that the patterning of infanticidal attacks fits predictions derived from the sexual selection infanticide hypothesis. But controversy still lingers. Volker Sommer of University College London, whose own work on infanticide in gray langurs

FIGURE 6.41

Lactating female baboons often form close ties to one or two adult males. Here, a female baboon and her infant sit with the mother's male associate.

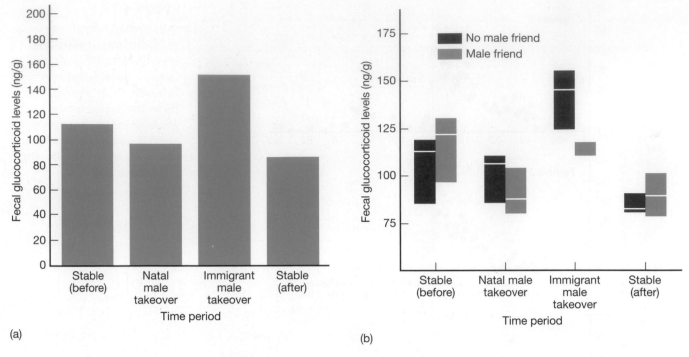

FIGURE 6.42

(a) The cortisol levels of female baboons increased when a new male immigrated into the group and took over the alpha position. (b) This effect was most pronounced for females who did not have a close male associate.

has been attacked by critics of Hrdy's hypothesis, believes that the criticism comes from a tendency to commit what is called the *naturalistic fallacy*, the assumption that what we observe in nature is somehow right, just, and inevitable. Critics are concerned that if we accept the idea that infanticide is an adaptive strategy for langurs or baboons, it will justify similar behavior in humans. However, it is misguided to try to extract moral meaning from the behavior of other animals. As we will discuss in Chapter 16, human societies rely on culturally evolved moral norms, and this makes the naturalistic fallacy an erroneous form of reasoning.

CHAPTER REVIEW

Key Terms

mating effort (p. 145)
parenting effort (p. 145)
interbirth interval (p. 148)
primiparous (p. 148)

multiparous (p. 149)
dominance (p. 151)
dominance matrix (p. 151)
transitive (p. 151)

sexual selection (p. 155)
intrasexual selection
 (p. 156)
intersexual selection (p. 156)

bachelor males (p. 158)
estrus (p. 159)
sexual selection infanticide
 hypothesis (p. 165)

Study Questions

1. Explain why reproductive success is a critical element of evolution by natural selection. When biologists use the terms *cost* and *benefit*, what currency are they trying to measure?

2. What is the difference between polygyny and polyandry? It seems likely that females might prefer polyandry over polygyny, while males would favor polygyny. Explain why males and females might prefer different mating systems. If this conflict of interest occurs, why is polygyny more common than polyandry?

3. In many primate species, reproduction is highly seasonal. Some researchers have suggested that reproductive seasonality has evolved as a means for females to manipulate their reproductive options. How would reproductive seasonality alter females' options? Why do you think this strategy might be advantageous for females?

4. Imagine that you came upon a species in which males and females were the same size, but males had very large testes in relation to their body size. What would you infer about their social organization? Now suppose you found another species in which males were much larger than females but had relatively small testes. What would you deduce about their social system? Why do these relationships hold?

5. Among mammalian species, male fitness is typically more variable than female fitness. Explain why this is often the case. What implications does this have for evolution acting on males and females?

6. What factors influence the reproductive success of females? How do these factors contribute to variance in female reproductive success?

7. Biologists use the term *investment* to describe parental care. What elements of the selective forces acting on parental strategies does this term capture?

8. Explain the logic underlying the sexual selection infanticide hypothesis. What predictions follow from this hypothesis? List the predictions, and explain why they follow from the hypothesis.

9. In general, infanticide seems to be more common in species that form one-male groups than in species that form multimale, multifemale groups or pair-bonded groups. Explain why this might be the case.

10. Why is the naturalistic fallacy considered a problematic way to think about the behavior of other animals?

Further Reading

Alberts, S. C. (2019). Social influences on survival and reproduction: Insights from a long-term study of wild baboons. *Journal of Animal Ecology*, 88(1), 47–66.

Altmann, J. 2001. *Baboon Mothers and Infants*. Chicago: University of Chicago Press.

Fernandez Duque, E., Huck, M., Van Belle, S., & Di Fiore, A. (2020). The evolution of pair living, sexual monogamy, and cooperative infant care: Insights from research on wild owl monkeys, titis, sakis, and tamarins. *American Journal of Physical Anthropology* 171: 118–173.

Kappeler, P. M., and C. P. van Schaik, eds. 2004. *Sexual Selection in Primates: New and Comparative Perspectives*. New York: Cambridge University Press.

Mitani, J., J. Call, P. Kappeler, R. Palombit, and J. B. Silk, eds. 2012. *The Evolution of Primate Societies*. Chicago: University of Chicago Press.

Van Schaik, C. P., and C. H. Janson, eds. 2000. *Infanticide by Males and Its Implications*. New York: Cambridge University Press.

Westneat, D. F., and C. W. Fox, eds. 2010. *Evolutionary Behavioral Ecology*. Section V. Oxford: Oxford University Press.

 Review this chapter with personalized, interactive questions through INQUIZITIVE.

 This chapter also features a Guided Learning Exploration on mating systems and sexual dimorphism.

THE EVOLUTION OF COOPERATION

CHAPTER OBJECTIVES

By the end of this chapter you should be able to:

A Explain why altruism is unlikely to evolve in most circumstances.

B Describe how evolution can favor altruism through the processes of kin selection and reciprocal altruism.

C Discuss the mechanisms that allow primates to recognize their relatives.

D Explain how kinship influences the distribution of altruism in primate groups.

E Evaluate arguments about the importance of reciprocal altruism in primate groups.

Altruism: A Puzzle

So far, we have explained the evolution of morphology and behavior in terms of individual reproductive success. Natural selection favored deeper beaks in Darwin's finches during the drought because deeper beaks allowed individuals to crack tougher seeds. It favors infanticide by male langurs and lions because it allows them to sire more offspring. However, primates (and many other creatures) also perform **altruistic behaviors** that are beneficial to others but costly to themselves. For example, virtually all social primates groom other group members, removing parasites, cleaning scabs, and picking debris from their hair (**Figure 7.1**). Grooming other individuals consumes time that could be spent looking for food, courting prospective mates, caring for offspring, or scanning for predators. The recipient, meanwhile, gets a thorough cleaning of parts of her body that she might find difficult to reach and may enjoy a period of pleasant relaxation. And grooming isn't the only altruistic behavior; primates warn others about the presence of predators, even though doing so makes them more conspicuous. They come to the aid of group members involved in aggressive conflicts, and in some species, individuals share food. If natural selection favors individually advantageous traits, how can we explain the evolution of such altruistic behaviors?

The answer to this question is one of the triumphs of evolutionary biology. Beginning in the 1960s and 1970s with the work of the late William D. Hamilton and Robert Trivers, biologists have developed a rich theory that explains why selection sometimes favors altruistic behavior and why it often does not. This theory has transformed our understanding of the evolution of social behavior. In this chapter, we show how natural selection can favor the evolution of altruistic behavior and describe how it explains the form and pattern of cooperation in primate groups. Cooperation plays an important role in the lives of other primates and plays an even more important role in human societies. In Chapter 16 you will see that the same kinds of processes that shape cooperation in monkeys and apes also influence human cooperation, but human cooperation also extends beyond the patterns that we see in other primates.

FIGURE 7.1

Gray langurs groom one another. Grooming is usually considered altruistic because the groomer expends time and energy when it grooms another animal, and the recipient benefits from having ticks removed from its skin, wounds cleaned, and debris removed from its hair.

Mutualism

Sometimes helping others benefits the actor as well as the recipient. Such behaviors are mutualistic.

Mutualistic interactions provide benefits to both participants. This seems like a win–win proposition, and you might think these kinds of interactions would be very common in nature. But there is a catch. To see what it is, think back to those group projects you worked on in grade school. The problem with group projects is that if someone doesn't do their share, the rest of the group has to take up the slack. You could punish the slacker, but that just compounds the problem because that

FIGURE 7.2

The pyramid will collapse if anyone slacks off.

FIGURE 7.3

Two male baboons (on the right) jointly challenge a third male over access to a receptive female.

would be a lot of trouble. So, you grumble and finish the project yourself. The general lesson here is that mutualistic cooperation is fragile when slacking is profitable for individuals. Mutualism is most likely to work in situations in which slacking off isn't profitable for any of the participants. Imagine that you are entering a human-pyramid contest (**Figure 7.2**). The group that can construct the largest pyramid wins a big prize. If someone doesn't hold up their end, the pyramid collapses, and everyone loses. So no one is motivated to slack off.

Coalitions among male baboons may be an example of mutualism. In yellow and olive baboons, the highest-ranking male usually attempts to monopolize access to females (mate guard) on the days when females are most likely to conceive. Two males may jointly challenge the mate-guarding male and try to gain control of the female (**Figure 7.3**). These interactions can escalate to energetically costly chases and physical confrontations. The challengers sometimes succeed in driving the mate-guarding male away, and one of them begins to mate guard the female. Males that hold middle-ranking positions are most likely to form coalitions. These males have very little chance of gaining access to receptive females on their own, but two middle-ranking males are a formidable force when they work together. As long as each male has some probability of ending up with the female, it may be profitable for both to participate in the coalition. There is no incentive to slack off because slacking off guarantees failure.

The Problem with Group-Level Explanations

Altruistic behaviors cannot be favored by selection just because they are beneficial to the group as a whole.

You might suppose that if the average effect of an act on all members of the group is positive, then it would be beneficial for all individuals to perform it. For example, suppose that when one monkey gives an alarm call, the other members of the

(a) Altruist gives alarm call to group.

(b) Nonaltruist doesn't give alarm call to group.

FIGURE 7.4

Two groups of monkeys are approached by a predator. (a) In one group, one individual (*pink*) has a gene that makes her call in this context. Giving the call lowers the caller's fitness but increases the fitness of every other individual in the group. Like the rest of the population, one out of four of these beneficiaries also carries the gene for calling. (b) In the second group, the female who detects the predator does not carry the gene for calling and remains silent. This lowers the fitness of all members of the group a certain amount because they are more likely to be caught unaware by the predator. Once again, one out of four is a caller. Although members of the caller's group are better off on average than members of the noncaller's group, the gene for calling is not favored, because callers and noncallers in the caller's group both benefit from the caller's behavior but callers incur some costs. Callers are at a disadvantage in comparison with noncallers. Thus calling is not favored, even though the group as a whole benefits.

group benefit, and the total benefits to all group members exceed the actual cost of giving the call. Then, if every individual gave the call when a predator was sighted, all members of the group would be better off than if no warning calls were ever given. You might think that alarm calling would be favored because every individual in the group benefits.

This inference is wrong because it confuses the effect on the group with the effect on the individual that performs the altruistic act. In most circumstances, the fact that alarm calls are beneficial to those hearing them doesn't affect whether the trait of alarm calling evolves; all that matters is how giving the alarm call affects the caller. To see why this is true, imagine a hypothetical monkey species in which some individuals give alarm calls when they are the first to spot a predator. Monkeys who hear the call have a chance to flee. Suppose that one-fourth of the population ("callers") give the call when they spot predators, and three-fourths of the individuals ("noncallers") do not give an alarm call in the same circumstance. (These proportions are arbitrary; we chose them because it's easier to follow the reasoning in examples with concrete numbers.) Let's suppose that in this species the tendency to give alarm calls is genetically inherited.

Now we compare the fitness of callers and noncallers. Because everyone in the group can hear the alarm calls and take appropriate action, alarm calls benefit everyone in the group to the same extent (**Figure 7.4**). Calling does not affect the *relative* fitness of callers and noncallers because, on average, one-fourth of the beneficiaries will be callers and three-fourths of the beneficiaries will be noncallers—the same proportions we find in the population as a whole. Calling reduces the risk of mortality for everyone who hears the call, but it does not change the frequency of callers and noncallers in the population because everyone gains the same benefits. However, callers are conspicuous when they call, so they are more vulnerable to predators. Although all individuals benefit from hearing alarm calls, callers are the only ones who suffer the costs from calling. This means that, on average, noncallers will have a higher fitness than callers. Thus, genes that cause alarm calling will not be favored by selection,

7.1 Group Selection

Group selection was once thought to be the mechanism for the evolution of altruistic interactions. In the early 1960s, the British ornithologist V. C. Wynne-Edwards contended that altruistic behaviors such as those we have been considering here evolved because they enhanced the survival of whole groups of organisms. Thus, individuals gave alarm calls, despite the costs of becoming more conspicuous to predators, because calling protected the group as a whole from attacks. Wynne-Edwards reasoned that groups containing a larger number of altruistic individuals would be more likely to survive and prosper than groups containing fewer altruists, and the frequency of the genes leading to altruism would increase.

Wynne-Edwards's argument is logical because Darwin's postulates logically apply to groups as well as individuals. However, group selection is not an important force in nature because there is generally not enough genetic variation among groups for selection to act on. Group selection can occur if groups vary in their ability to survive and to reproduce and if that variation is heritable. Then group selection may increase the frequency of genes that increase group survival and reproductive success. The strength of selection among groups depends on the amount of genetic variation among groups, just as the strength of selection among individuals depends on the amount of genetic variation among individuals. However, when individual selection and group selection are opposed and group selection favors altruistic behavior while individual selection favors selfish alternatives, individual selection has a tremendous advantage. This is because the amount of variation among groups is much smaller than the amount of variation among individuals, unless groups are very small or there is very little migration among them. Thus, individual selection favoring selfish behavior will generally prevail over group selection, making group selection an unlikely source of altruism in nature.

even if the cost of giving alarm calls is small and the benefit to the rest of the group is large. Instead, selection will favor genes that suppress alarm calling because noncallers have a higher fitness than callers. (See **A Closer Look 7.1** for more about the role of group selection in nature.)

Kin Selection

Natural selection can favor altruistic behavior if altruistic individuals are more likely to interact with each other than chance alone would dictate.

If altruistic behaviors can't evolve by ordinary natural selection or by group selection, then how do they evolve? A clear answer to this question did not come until 1964, when a young biologist named William D. Hamilton published a landmark paper. This paper was the first of a series of fundamental contributions that Hamilton made to our understanding of the evolution of behavior.

The argument made in the previous section contains a hidden assumption: Altruists and nonaltruists are equally likely to interact with one another. We supposed that callers give alarm calls when they hear a predator, no matter who is nearby. Hamilton's insight was to see that any process that causes altruists to be more likely

(a) Altruist gives alarm call to siblings.

(b) Nonaltruist doesn't give alarm call to siblings.

FIGURE 7.5

Two groups of monkeys are approached by a predator. Each group is composed of nine sisters. (a) In one group there is a caller (*pink*), an individual with a gene that makes her call in this context. Her call lowers her own fitness but increases the fitness of her sisters. (b) In the second group, the female who detects the predator is not a caller and does not call when she spots the predator. As in Figure 7.4, calling benefits the other group members but imposes costs on the caller. However, there is an important difference between the situations portrayed here and in Figure 7.4. Here the groups are made up of sisters, so five of the eight recipients of the call also carry the calling gene. In any pair of siblings, half of the genes are identical because the siblings inherited the same gene from one of their parents. Thus, on average, half of the caller's siblings also carry the calling allele because they inherited it from their mother or father. The remaining four siblings carry genes inherited from the other parent; and, like the population as a whole, one out of four of them is a caller. The same reasoning shows that in the group with the noncaller, there is only one caller among the beneficiaries of the call. Half are identical to their sister because they inherited the same noncalling gene from one of their parents; one of the remaining four is a caller. In this situation, callers are more likely to benefit from calling than noncallers are, and so calling alters the relative fitness of callers and noncallers. Whether the calling behavior actually evolves depends on whether these benefits are big enough to compensate for the reduction in the caller's fitness.

to interact with other altruists than they would by chance could facilitate the evolution of altruism.

To see why this is such an important insight, let's modify the previous example by assuming that our hypothetical species lives in groups composed of full siblings, offspring of the same mother and father. The frequencies of the calling and noncalling genes don't change, but their distribution will be affected by the fact that siblings live together (**Figure 7.5**). If an individual is a caller, then, by the rules of Mendelian genetics, there is a 50% chance that the individual's siblings will share the genes that cause calling behavior. This means that the frequency of the genes for calling will be higher in groups that contain callers than in the population as a whole, and, therefore, more than one-fourth of the beneficiaries will be callers themselves. When a caller gives an alarm call, the audience will contain a higher fraction of callers than the population at large does. Thus, the caller raises the average fitness of callers with respect to noncallers. Similarly, because the siblings of noncallers are more likely to be noncallers, then chance alone would dictate callers are less likely to be present in such groups than in the population at large. Therefore, the absence of a warning call lowers the fitness of noncallers more significantly relative to callers.

When individuals interact selectively with genetic relatives, callers are more likely to benefit than noncallers, and, all other things being equal, the benefits of calling will favor the genes for calling. However, we must remember that calling is costly, and this will tend to reduce the fitness of callers. Calling will be favored by natural selection only if its benefits are sufficiently greater than its costs. The exact nature of this trade-off is specified by what we call Hamilton's rule.

Hamilton's Rule

Hamilton's rule predicts that altruistic behaviors will be favored by selection if the costs of performing the behavior are less than the benefits discounted by the coefficient of relatedness between actor and recipient.

Hamilton's theory of **kin selection** is based on the idea that selection could favor altruistic alleles if animals interacted selectively with their genetic relatives. Hamilton's theory also specifies the quantity and distribution of help among individuals. According to **Hamilton's rule**, an act will be favored by selection if

$$rb > c$$

where

      r = the average coefficient of relatedness between the actor and the recipients
      b = the sum of the fitness benefits to all individuals affected by the behavior
      c = the fitness cost to the individual performing the behavior

    The **coefficient of relatedness**, **r**, measures the genetic relationship between interacting individuals. More precisely, r is the probability that two individuals will acquire the same allele through descent from a common ancestor. **Figure 7.6** shows how these probabilities are derived in a simple genealogy. Female A obtains one allele at a given locus from her mother and one from her father. Her half sister, female B, also obtains one allele at the same locus from each of her parents. We obtain the probability that both females receive the same allele from their mother by multiplying the probability that female A obtains the allele (0.5) by the probability that female B obtains the same allele (0.5); the result is 0.25. Thus half sisters have, on average, a 25% chance of obtaining the same allele from their mothers. Now consider the relatedness between female B and her brother, male C. In this case, note that female B and male C are full siblings: They have the same mother and the same father. The probability that both siblings will acquire the same allele from their mother is still 0.25, but female B and male C might also share an allele that they acquired from their father. The probability of this event is also 0.25. Thus, the probability that female B and male C share an allele is equal to the sum of 0.25 and 0.25, or 0.5. This basic reasoning can be extended to calculate the degrees of relatedness among various categories of kin (**Table 7.1**).

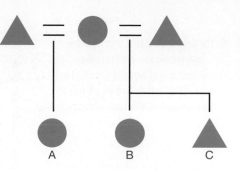

FIGURE 7.6

This genealogy shows how the value of r is computed. Triangles represent males, circles represent females, and the equals sign represents mating. The relationships between individuals labeled in the genealogy are described in the text.

TABLE 7.1

Relationship	r
Parent and offspring	0.5
Full siblings	0.5
Half siblings	0.25
Grandparent and grandchild	0.25
First cousins	0.125 or 0.0625
Unrelated individuals	0

The value of r for selected categories of relatives. (When cousins are offspring of full siblings, they are related by 0.125, but when they are the offspring of half siblings, they are related by 0.0625.)

FIGURE 7.7

As the degree of relatedness (*r*) between two individuals declines, the value of the ratio of benefits to costs (*b:c*) required to satisfy Hamilton's rule for the evolution of altruism rises rapidly.

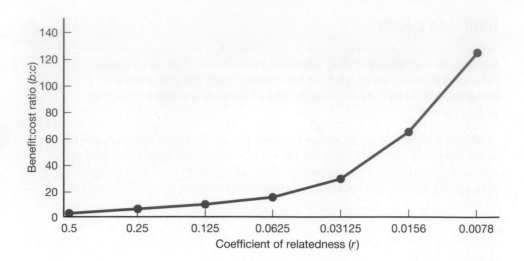

Hamilton's rule leads to two important insights: (1) Altruism is limited to kin and (2) closer kinship facilitates more costly altruism.

If you reflect on Hamilton's rule for a while, you will see that it produces two predictions about the conditions that favor the evolution of altruistic behaviors. First, altruism is not expected to be directed toward nonkin because the coefficient of relatedness, *r*, between nonkin is 0. The condition for the evolution of altruistic traits will be satisfied only for interactions between kin when $r > 0$. Thus, altruists are expected to be nepotistic, showing favoritism toward kin.

Second, close kinship is expected to facilitate altruism. If an act is particularly costly, it is most likely to be restricted to close kin. **Figure 7.7** shows how the benefit:cost ratio scales with the degree of relatedness among individuals. Compare what happens when $r = 1/16$ (or 0.0625) and when $r = 1/2$ (or 0.5). When $r = 1/16$, the benefits must be more than 16 times as great as the costs for Hamilton's inequality $rb > c$ to be satisfied. When $r = 1/2$, the benefits need to be just over twice as large as the costs. All other things being equal, Hamilton's rule will be easier to satisfy for close kin than for distant kin, and altruism will be more common among close relatives than among distant ones.

Kin Recognition

Primates may use contextual cues to recognize maternal relatives.

In order for kin selection to provide an effective mechanism for the evolution of cooperative behavior, animals must be able to distinguish relatives from nonrelatives and close relatives from distant ones. Some organisms can recognize their kin by their likeness to themselves. This is called **phenotypic matching**. Others learn to recognize relatives by using contextual cues—such as familiarity and proximity—that are consistently associated with kinship (**Figure 7.8**). In the past, primatologists assumed that primates relied solely on contextual cues to identify their relatives, but new data suggest that phenotypic matching may also play a role in primate kin recognition.

Mothers seem to make use of contextual cues to recognize their own infants. After they give birth, females repeatedly sniff and inspect their newborns. By the time their infants are a few weeks old, mothers can distinguish between their own infants and unrelated infants. After this, females of most species nurse only their own infants and respond selectively to their own infants' distress calls. Primate mothers don't really need innate means of recognizing their young because young infants spend virtually all of their time in physical contact with their mothers. Thus, mothers have time to get to know their infants and are unlikely to confuse their own newborn with another.

[a]   [b]

FIGURE 7.8

(a) Even mothers must learn to recognize their own infants. Here, a female bonnet macaque peers into her infant's face. (b) Primates apparently learn who their relatives are by observing patterns of association among group members. Here, a female inspects another female's infant.

Monkeys and apes may learn to recognize other maternal kin through contact with their mothers. Offspring continue to spend considerable amounts of time with their mothers even after their younger siblings are born. Thus, they have many opportunities to watch their mothers interact with their new brothers and sisters. Similarly, the newborn infant's most common companions are its mother and siblings (**Figure 7.9**). Because adult females continue to associate with their mothers, infants also become familiar with their grandmothers, aunts, and cousins.

Contextual cues may play some role in paternal kin recognition as well.

Primatologists were once quite confident that most primates could not recognize their paternal kin. This conclusion was based on the following reasoning: First, pair bonds are uncommon in most primate species, so patterns of association between males and females do not provide accurate cues of paternal kinship. Second, females may mate with several males near the time of conception, creating confusion about paternity. Even in pair-bonded species, such as gibbons and titi monkeys, females sometimes mate with males from outside their groups. But as we pointed out in the previous chapter, the conventional wisdom is at least sometimes wrong. In some species that live in multimale groups, males *are* able to distinguish between their own infants and other males' infants and may rely on their memory of associations with females near the time of conception.

New evidence from field studies on macaques and baboons suggests that these monkeys use contextual cues to assess paternal kinship. Jeanne Altmann of Princeton University pointed out that age may provide a good proxy measure of paternal kinship in species in which a single male typically dominates mating activity within the group. When this happens, all infants born at about the same time are likely to have the same father. Recent studies suggest that Altmann's logic is correct—monkeys born about the same time are likely to be paternal siblings, and monkeys use age to identify paternal kin. Female baboons in the population that Altmann studied distinguish between paternal half sisters and unrelated females, and they seem to rely on closeness in age to make these discriminations. Anja Widdig of the University of Leipzig in Germany and her colleagues have also found that females show strong affinities for paternal half sisters (**Figure 7.10**). Their affinities for paternal kin seem to be based partly on strong preferences for interacting with age-mates. However, females also distinguished *among* their age-mates, preferring paternal half sisters over unrelated females of the same age.

[a]

[b]

FIGURE 7.9

Monkey and ape infants grow up surrounded by various relatives. (a) These adult baboon females are mother and daughter. Both have young infants. (b) An adolescent female bonnet macaque carries her younger brother while her mother recovers from a serious illness.

FIGURE 7.10

Female macaques can identify paternal siblings. Females groom far more often with maternal half-siblings (*blue bar*) than with paternal siblings, but they groom more often with paternal siblings than with nonkin. Age similarity seems to provide a cue for paternal kinship: Females groom more often with paternal half-sibling peers (*red hatched bar*) than with paternal half-siblings that are not close in age (*solid red bar*). But note that females also distinguish among peers, preferring half-sibling peers over unrelated peers (*purple hatched bar*). Similar patterns are found for spatial associations.

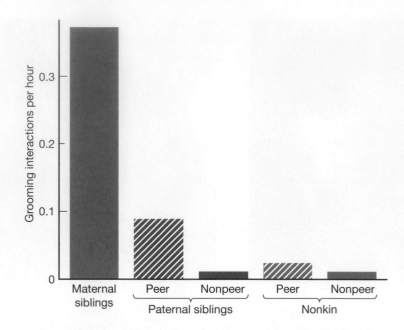

It is not entirely clear what cues macaques might use to distinguish their paternal half sisters from other age-mates. However, research by Anja Widdig and her colleagues suggests that phenotypic cues might play some role. Rhesus macaques respond differently when they are presented with photographs of unfamiliar paternal half-siblings and unfamiliar nonrelatives (**Figure 7.11**), and also respond differently to tape recordings of the vocalizations of these monkeys. Similarly, a study led by Marina Cords of Columbia University showed that female blue monkeys in the Kakamega Forest of Kenya interact preferentially with their paternal half-sisters, but their preferences are not based on age difference.

Kin Biases in Behavior

A considerable body of evidence suggests that the patterns of many forms of altruistic interactions among primates are largely consistent with predictions derived from Hamilton's rule. Here we consider several examples.

FIGURE 7.11

To study kin recognition, researchers present monkeys with pairs of photographs and videotape their responses.

[a]

[b]

[c]

[d]

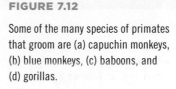

FIGURE 7.12

Some of the many species of primates that groom are (a) capuchin monkeys, (b) blue monkeys, (c) baboons, and (d) gorillas.

Grooming is more common among kin than nonkin.

Social **grooming** plays an important role in the lives of most gregarious primates (**Figure 7.12**). Grooming is likely to be beneficial to the participants in at least two ways: First, grooming serves hygienic functions because bits of dead skin, debris, and parasites are removed and wounds are kept clean and open. Second, grooming may provide a means for individuals to establish relaxed, **affiliative** (friendly) contact and to reinforce social relationships with other group members (**A Closer Look 7.2**). Grooming is also costly because the actor expends both time and energy in performing these services. Moreover, Marina Cords has shown that blue monkeys are less vigilant when they are grooming, perhaps exposing themselves to some risk of being captured by predators.

Grooming is more common among kin, particularly mothers and their offspring, than among nonkin. For example, Ellen Kapsalis and Carol Berman of the University at Buffalo documented the effect of maternal relatedness among rhesus macaques on Cayo Santiago. In this population, females groom close kin at higher rates than those for their grooming of nonkin, and close kin are groomed more often than distant kin (**Figure 7.13**). As relatedness declined, the differences in the proportions of time spent grooming kin and nonkin were essentially eliminated. This may mean that monkeys cannot recognize more distant kin or that the conditions of Hamilton's rule ($rb > c$) are rarely satisfied for distant kin.

Primates most often form coalitions with close kin.

Most disputes in primate groups involve two individuals. Sometimes, however, several individuals jointly attack another individual or one individual comes to the support of another individual involved in an ongoing dispute (**Figure 7.14**). We call these kinds of interactions **coalitions** or **alliances**.

FIGURE 7.13

Rhesus monkeys on Cayo Santiago groom close relatives more often than they groom distant relatives or nonkin.

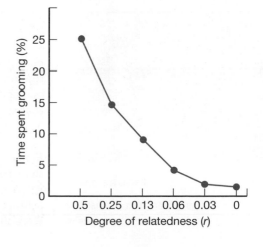

Time spent grooming (%) vs *Degree of relatedness (r)*

7.2 How Relationships Are Maintained

Conflict and competition are fundamental features of social life for many primates: Females launch unprovoked attacks on unsuspecting victims, males battle over access to receptive females, subordinates are supplanted from choice feeding sites, and dominance relationships are clearly defined and frequently reinforced. Although violence and aggression are not prevalent in all primates (muriquis, for example, are so peaceful that dominance hierarchies cannot be detected), many primates can be charitably characterized as contentious. This raises an intriguing question: How is social life sustained in the face of such relentless conflict? After all, it seems inevitable that aggression and conflict will drive animals apart, disrupt social bonds, and reduce the cohesiveness of social groups.

Social relationships matter to primates. They spend a considerable portion of every day grooming other group members. Grooming is typically focused on a relatively small number of partners and is often reciprocated. Robin Dunbar of the University of Oxford contends that in catarrhine primates, grooming has transcended its original hygienic function and now serves as a means to cultivate and maintain social bonds. Social bonds may have real adaptive value to individuals. For example, grooming is sometimes exchanged for support in coalitions, and grooming partners may be allowed to share access to scarce resources.

When tensions do erupt into violence, certain behavioral mechanisms may reduce the disruptive effects of conflict on social relationships. After conflicts end, victims often flee from their attackers—an understandable response. In some cases, however, former opponents make peaceful contact in the minutes that follow conflicts. For example, chimpanzees sometimes kiss their former opponents, female baboons grunt quietly to their former victims, and golden monkeys may embrace or groom their former adversaries. The swift transformation from aggression to affiliation prompted Frans de Waal of Emory University to suggest that these peaceful postconflict interactions are a form of reconciliation, a way to mend relationships that were damaged by conflict. Inspired by de Waal's work, several researchers have documented reconciliatory behavior in a variety of primate species.

Peaceful postconflict interactions seem to have a calming effect on former opponents. When monkeys are nervous and anxious, rates of certain self-directed behaviors, such as scratching, increase. Thus, self-directed behaviors are a good behavioral index of stress. Filippo Aureli of Liverpool John Moores University and his colleagues at Utrecht University and Emory

FIGURE 7.14

Two baboons form an alliance against an adult female.

Support is likely to be beneficial to the individual who receives aid because support alters the balance of power among the original contestants. The beneficiary may be more likely to win the contest or less likely to be injured in the confrontation. At the same time, however, intervention may be costly to the supporter, who expends time and energy and risks defeat or injury by becoming involved. Hamilton's rule predicts that support will be preferentially directed toward kin and that the greatest costs will be expended on behalf of close relatives.

Many studies have shown that support is selectively directed toward close kin. Female macaques and baboons defend their offspring and close kin more often than they defend distant relatives or unrelated individuals (**Figure 7.16**). Females run some risk when they participate in coalitions, particularly when they are allied against higher-ranking individuals. Coalitions against high-ranking individuals are more likely to result in retaliatory attacks against the supporter than are coalitions

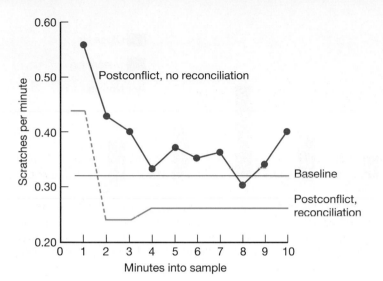

FIGURE 7.15

Rates of scratching, an observable index of stress, by victims of aggression are elevated over normal levels in the minutes that follow aggressive encounters. If some form of affiliative contact (reconciliation) between former opponents occurs during the postconflict period, however, rates of scratching drop rapidly below baseline levels. If there is no reconciliatory contact during the first few minutes of the postconflict period, rates of scratching remain elevated above baseline levels for several minutes. These data suggest that affiliative contact between former opponents has a calming effect. Similar effects on aggressors have also been detected.

University have studied how fighting and reconciliation affect the rate of self-directed behaviors. They found that levels of self-directed behavior, and presumably stress, rise sharply above baseline levels after conflicts. Both victims and aggressors seem to feel the stressful effects of conflicts. If former opponents interact peacefully in the minutes that follow conflicts, rates of self-directed behavior fall rapidly to baseline levels (**Figure 7.15**). If adversaries do not reconcile, rates of self-directed behavior remain elevated above baseline levels for several minutes longer. If reconciliation provides

a means to preserve social bonds, then we would expect primates to reconcile selectively with their closest associates. In several groups, former opponents who have strong social bonds are most likely to reconcile. Kin also reconcile at high rates in some groups, even though some researchers have argued that kin have little need to reconcile because their relationships are unlikely to be frayed by conflict.

Reconciliation may also play a role in resolving conflicts among individuals who do not have strong social bonds. Like many other primates, female baboons are strongly attracted

to newborn infants and make persistent efforts to handle them. Mothers reluctantly tolerate infant handling, but they do not welcome the attention. Female baboons reconcile at particularly high rates with the mothers of young infants, even when they do not have close relationships with them. Reconciliation greatly enhances the likelihood that aggressors will be able to handle their former victims' infants in the minutes that follow conflicts. Thus, in this case, reconciliation seems to be a means to an immediate end but not a means to preserve long-term relationships.

against lower-ranking individuals. Female macaques are much more likely to intervene against higher-ranking females on behalf of their own offspring than on behalf of unrelated females or juveniles. Thus, macaque females take the greatest risks on behalf of their closest kin.

Kin-based support in conflicts affects the social structure of macaque, vervet, and baboon groups.

Maternal support in macaques, vervets, and baboons influences the outcome of aggressive interactions and dominance contests. Initially, an immature monkey can defeat older and larger juveniles only when its mother is nearby. Eventually, regardless of their age or size, juveniles can defeat everyone their mothers can defeat, even when the mother is some distance away. Maternal support contributes

FIGURE 7.16

In five groups of wild baboons, rates of coalitionary support provided by close kin (mothers, daughters, and sisters), more distant maternal kin, and all others are shown. In all five groups, support is biased toward close kin.

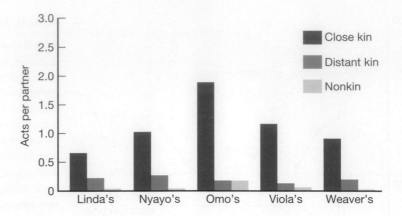

directly to several remarkable properties of dominance hierarchies within these species:

- Maternal rank is transferred with great fidelity to offspring, particularly daughters. In a group of baboons at Gilgil, Kenya, for example, maternal rank is an almost perfect predictor of the daughter's rank (**Figure 7.17**).

- Maternal kin occupy adjacent ranks in the dominance hierarchy, and all the members of one **matrilineage** (maternal kin group) rank above or below all members of other matrilineages.

- Ranking within matrilineages is often quite predictable. Usually, mothers outrank their daughters, and younger sisters outrank their older sisters.

- Female dominance relationships are amazingly stable, often remaining the same over years and sometimes decades.

Kin-biased support plays an important role in the reproductive strategies of red howler males.

Behavioral and genetic studies of red howlers in Venezuela conducted by Teresa Pope have shown that kinship influences howler males' behavior in important ways. Red

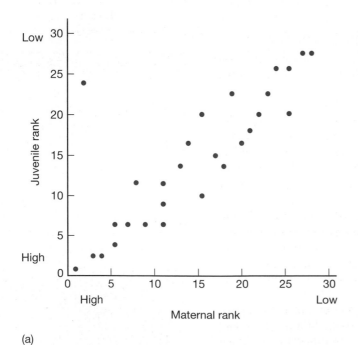

(a)

(b)

FIGURE 7.17

Juvenile female baboons acquire ranks very similar to their mother's rank. (a) The anomalous point at maternal rank 2 belongs to a female whose mother died when she was an infant. (b) Here, the dominant female of a baboon group is flanked by her two daughters, ranked 2 and 3.

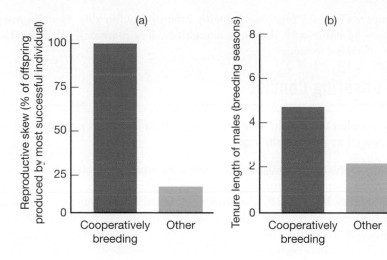

FIGURE 7.18

Cooperatively breeding species are characterized by (a) higher male reproductive skew and (b) longer male tenures than other species that do not breed cooperatively.

howlers live in groups that contain two to four females and one or two males. Males sometimes join up with migrant females and help them establish new territories. Once such groups have been established, resident males must defend their position and their progeny from infanticidal attacks by alien males. When habitats are crowded, males can gain access to breeding females only by taking over established groups and evicting male residents. This is a risky endeavor because males are often injured in takeover attempts. Moreover, as habitats become more saturated and dispersal opportunities become more limited, males tend to remain in their groups longer. Maturing males help their fathers defend their groups against takeover attempts. Collective defense is crucial to males' success because single males cannot defend their group against incursions by rival males.

This situation leads to a kind of arms race because migrating males also form coalitions and cooperate in efforts to evict residents. After they have established residence, males collectively defend the group against incursions by extra-group males. Cooperation among males is beneficial because it helps deter rivals. But it also involves clear fitness costs because, as behavioral and genetic data have demonstrated, only one male succeeds in siring offspring within the group. Not surprisingly, kinship influences the duration and stability of male coalitions. Coalitions that are made up of related males last nearly four times as long as coalitions composed of unrelated males. Coalitions composed of related males are also less likely to experience rank reversals. In this case, the costs of cooperation may be balanced by gains in inclusive fitness.

Kin selection plays an important role in cooperatively breeding primate groups.

In marmosets and tamarins, all group members help care for offspring. In tamarins, males sometimes share paternity of infants. Kin selection helps to explain why helpers help and why males tolerate other males' breeding. Cooperative breeding has evolved several times independently in different groups of mammals, including canids, rodents, and primates. Dieter Lukas of the Max Planck Institute for Evolutionary Anthropology in Leipzig, Germany, and Tim Clutton-Brock of Cambridge University compared characteristics of cooperatively breeding mammalian taxa with characteristics of mammalian taxa that did not breed cooperatively. In cooperatively breeding species, male reproductive skew is considerably higher and the tenure of resident males is considerably longer than in other species. Both these features will increase the degree of relatedness within groups (**Figure 7.18**). In the tamarin groups that Samuel Diaz-Muñoz studied, adult males were related by 0.34 on average, placing them midway between half and full siblings.

Kinship seems to reduce the extent of competition among females to some degree. When subordinate females produce infants, the dominant female often kills them. However, in golden lion tamarins (**Figure 7.19**), there are some groups in which

FIGURE 7.19

Golden lion tamarins are cooperative breeders, but conflict sometimes arises over breeding opportunities.

females temporarily share reproduction with subordinate females. Females are most likely to share breeding with their own daughters, less commonly with sisters, and rarely with unrelated females.

Parent–Offspring Conflict

Kin selection helps to explain why there is conflict between parents and offspring as well as among siblings.

As we explained in the previous chapter, mothers must wean their infants so that they can conserve energy for infants born later. As mothers begin to curtail investment, their infants often resist, sometimes vigorously. Chimpanzee infants throw full-fledged tantrums when their mothers rebuff their efforts to nurse, and baboons whimper piteously when their mothers refuse to carry them. These weaning conflicts arise from a fundamental asymmetry in the genetic interests of mothers and their offspring. Mothers are equally related to all of their offspring ($r = 0.5$), but offspring are more closely related to themselves ($r = 1.0$) than to their siblings ($r = 0.5$ or 0.25). This phenomenon was labeled **parent–offspring conflict** by Rutgers biologist Robert Trivers, who was the first to recognize the evolutionary rationale underlying the conflict between parents and their offspring.

To understand why there is parent–offspring conflict, imagine a mutation that increases the amount of maternal investment in the current infant by a small amount, thereby reducing investment in future infants by the same amount. According to Hamilton's rule, selection will favor the expression of this gene in mothers if

$$0.5 \times \text{(increase in fitness of current infant)}$$
$$> 0.5 \times \text{(decrease in fitness of future offspring)}$$

Because the mother shares half of her genes with each of her offspring, 0.5 appears on both sides of the inequality. The inequality tells us that selection will increase investment in the current offspring until the benefits to the current offspring are equal to the costs to future offspring. The result is quite different if the genes expressed in the current infant control the amount of maternal investment. This time, consider a gene expressed in the current infant that increases the investment the infant receives by a small amount. Once again, we use Hamilton's rule, this time from the perspective of the current infant:

$$1.0 \times \text{(increase in fitness of current fetus)}$$
$$> 0.5 \times \text{(decrease in fitness of future offspring)}$$

In this case, the infant is related to itself by 1.0 and to its full sibling by 0.5. Now selection will increase the amount of maternal investment until the incremental benefit of another unit of investment in the current infant is twice the cost to future brothers and sisters of the fetus (and four times for half siblings). Thus, genetic asymmetries lead to a conflict of interest between mothers and their offspring. Selection will favor mothers who provide less investment than their infants desire, and selection will favor offspring who demand more investment than their mothers are willing to give. This conflict of interest plays out in weaning tantrums and sibling rivalries.

Reciprocal Altruism

Altruism can also evolve if altruistic acts are reciprocated.

The theory of **reciprocal altruism** relies on the basic idea that altruism among individuals can evolve if altruistic behavior is balanced between partners (pairs of interacting individuals) over time. In reciprocal relationships, individuals take turns being actor and recipient—giving and receiving the benefits of altruism (**Figure 7.20**). Reciprocal altruism is favored because over time the participants in reciprocal acts obtain benefits that outweigh the costs of their actions. This theory was first formulated by Robert Trivers and later amplified and formalized by others.

Three conditions occurring together favor the development of reciprocal altruism: Individuals must (1) have an opportunity to interact often, (2) be able to keep track of support given and received, and (3) provide support only to those who help them. The first condition is necessary so that individuals will have the opportunity for their own altruism to be reciprocated. The second condition allows individuals to balance altruism given to and received from particular partners. The third condition produces the nonrandom interaction necessary for the evolution of altruism. If individuals are unrelated, initial interactions will be randomly distributed to altruists and nonaltruists. However, reciprocators will quickly stop helping those who do not help in return, while continuing to help those who do. Thus, as with kin selection, reciprocal altruism can be favored by natural selection because altruists receive a disproportionate share of the benefits of altruistic acts. Note that altruistic acts need not be exchanged in kind; it is possible for one form of altruism (such as grooming) to be exchanged for another form of altruism (such as coalitionary support).

FIGURE 7.20

Two old male chimpanzees groom each other. Reciprocity can involve taking turns or interacting simultaneously. Male chimpanzees remain in their natal communities throughout their lives and develop close bonds with one another.

In primates, the conditions for the evolution of reciprocal altruism probably are satisfied often, and there is some evidence that it occurs.

Most primates live in social groups that are fairly stable, and they can recognize all of the members of their groups. We do not know whether primates have the cognitive capacity to keep track of support given and received from various partners, but we do know that they are very intelligent and can solve complex problems. Thus, primates provide a good place to look for examples of contingent forms of reciprocity.

In several species of macaques, baboons, vervet monkeys, and chimpanzees, individuals tend to spend the most time grooming those who spend the most time grooming them, and they most often support those from whom they most often receive support. In some cases, monkeys seem to exchange grooming for support. Sometimes monkeys switch roles during grooming bouts so that the amount of grooming given and received during each grooming bout is balanced; sometimes grooming is balanced across bouts.

Among male chimpanzees, social bonds seem to be based on reciprocal exchanges in many currencies (see Figure 7.20). For example, John Mitani of the University of Michigan and David Watts of Yale University have found that male chimpanzees at Ngogo, a site in the Kibale Forest of Uganda, share meat selectively with males who share meat with them and with males who regularly support them in agonistic (combative) interactions. Males who hunt together also tend to groom one another selectively, support one another, and participate in border patrols together. Notably, close associates are often not maternal or paternal kin, suggesting that males' relationships are based on reciprocity, not kinship.

These correlational findings are consistent with predictions derived from the theory of reciprocal altruism, but they do not demonstrate that altruism is contingent on reciprocation.

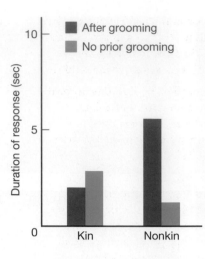

FIGURE 7.21

Vervet monkeys responded more strongly to recruitment calls played from a hidden speaker if the nonkin caller had previously groomed them than if the caller had not.

A few studies suggest that primates keep track of these contingencies, at least over short periods.

Robert Seyfarth and Dorothy Cheney conducted the first study to examine the contingent nature of altruistic exchanges. Like most other monkeys, vervets spend much of their free time grooming. Vervets also form coalitions and use specific vocalizations to recruit support. In this experiment, Seyfarth and Cheney played tape-recorded recruitment calls to individuals in two situations. Vervet A's recruitment call was played to vervet B from a hidden speaker (1) after A had groomed B and (2) after a fixed time during which A and B had not groomed. It was hypothesized that if grooming were associated with support in the future, then B should respond more strongly to A's recruitment call after being groomed. And that's just what the vervets did when they had interacted with nonrelatives. In contrast, when the vervets interacted with their relatives, their responses were not consistently affected by prior grooming (**Figure 7.21**). Seyfarth and Cheney have replicated these results with baboons, adding several controls that help rule out alternative explanations for the subjects' responses to the playbacks.

Although several naturalistic experiments suggest that primates respond to previous help in a contingent way, more controlled experiments conducted in the laboratory have been largely unsuccessful. For example, Alicia Melis, now at the University of Warwick, and her colleagues conducted an experiment in which one chimpanzee needed help from a second chimpanzee to get into a locked room. Each subject was paired with two helpers, one who provided help and another who did not provide help. Then the roles were reversed, and the individual who had needed help was able to provide help to the previously helpful and unhelpful partners. The chimpanzees were as likely to help the unhelpful partner as they were to help the helpful partner.

At this point it is not entirely clear how to interpret the data. Some researchers think that primates selectively help those from whom they have previously received help. Such researchers focus on the correlational evidence and the naturalistic experiments. Others emphasize the shortcomings of correlational studies, such as the lack of evidence from carefully controlled studies in the laboratory, and speculate that primates may not have the cognitive ability to keep track of help given and received from multiple partners over extended periods. However, most researchers would agree on one point: Kin selection plays a more important role in regulating the distribution of altruism in primate groups than does reciprocity.

CHAPTER REVIEW

Key Terms

altruistic behaviors (p. 172)

kin selection (p. 177)

Hamilton's rule (p. 177)

coefficient of relatedness (*r*) (p. 177)

phenotypic matching (p. 178)

grooming (p. 181)

affiliative (p. 181)

coalitions (p. 181)

alliances (p. 181)

matrilineage (p. 184)

parent–offspring conflict (p. 186)

reciprocal altruism (p. 186)

Study Questions

1. Consider the accompanying kinship diagram. What is the kinship relationship (for example, mother, aunt, or cousin) and degree of relatedness (such as 0.5 or 0.25) for each pair of individuals?

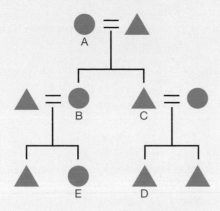

2. In biological terms, explain the difference between the following situations: (a) A male monkey sitting high in a tree gives alarm calls when he sees a lion at a distance. (b) A female monkey abandons a desirable food patch when she is approached by another female.

3. In documentaries about animal behavior, animals are often said to do things "for the good of the species." For example, when low-ranking animals do not reproduce, they are said to give up reproducing to prevent the population from becoming too numerous and exhausting its resource base. What is wrong with this line of reasoning?

4. Why is some sort of nonrandom interaction among altruists necessary for altruism to be maintained?

5. Suppose that primates cannot recognize paternal kin, as many primatologists have assumed. What would that tell you about how natural selection produces adaptations?

6. Data from several studies indicate that primates are more likely to behave altruistically toward kin than toward nonkin. However, many of the same studies show that rates of aggression toward kin and nonkin are basically the same. How does this fit with what you have learned about kin selection? Why are monkeys as likely to fight with kin as with nonkin?

7. There are relatively few good examples of reciprocal altruism in nature. Why is reciprocal altruism uncommon? Why might we expect reciprocal altruism to be more common among primates than among other kinds of animals?

8. Seyfarth and Cheney found that vervet monkeys tended to respond more strongly to the calls of the animals who had groomed them earlier in the day than to the calls of animals who had not groomed them. However, this effect held only for unrelated animals, not for kin. The vervets responded as strongly to the calls of grooming relatives as to those of nongrooming relatives. How might we explain kinship's influence on these results?

9. In addition to kin selection and reciprocal altruism, a third mechanism leading to nonrandom interaction of altruists has been suggested. Suppose altruists had an easily detected phenotypic trait, perhaps a green beard. Then they could use the following rule: "Do altruistic acts only for individuals who have green beards." Once the allele became common, most individuals carrying green beards would not be related to one another, so this would not be a form of kin selection. However, there is a subtle flaw in this reasoning. Assuming that the genes controlling beard color are at different genetic loci from the genes controlling altruistic behavior, explain why green beards would not evolve.

10. Explain why punishment does not provide a ready solution to the problem of cheating in mutualistic interactions.

Further Reading

Chapais, B., and C. M. Berman, eds. 2004. *Kinship and Behavior in Primates*. New York: Oxford University Press.

Kappeler, P. M., and C. P. van Schaik, eds. 2006. *Cooperation in Primates and Humans: Mechanisms and Evolution*. New York: Springer.

Mitani, J., J. Call, P. Kappeler, R. Palombit, and J. B. Silk, eds. 2012. *The Evolution of Primate Societies*. Chicago: University of Chicago Press.

Strier, K. B. 2016. *Primate Behavioral Ecology*, Fifth Edition, Routledge.

Westneat, D. F., and C. W. Fox, eds. 2010. *Evolutionary Behavioral Ecology*. Oxford: Oxford University Press.

 Review this chapter with personalized, interactive questions through INQUIZITIVE.

8

PRIMATE LIFE HISTORIES AND THE EVOLUTION OF INTELLIGENCE

CHAPTER OBJECTIVES

By the end of this chapter you should be able to:

A Explain how life history theory helps us understand why certain features, such as fertility and longevity, are correlated.

B Assess how the evolution of big brains has shaped primate life history strategies.

C Explain why primatologists think natural selection has favored large brains in monkeys and apes.

D Describe what primates know about their physical environment.

E Describe what primates know about their social world.

Big Brains and Long Lives

Large brains and long life spans are two of the features that define the primate order (**Figure 8.1**). Compared with most other animals, primates rely more heavily on learning to acquire the knowledge and skills that they need to survive and reproduce successfully. The complex behavioral strategies that we have explored in the past few chapters depend on primates' ability to respond flexibly in novel situations. Primates also have long periods of development and long life spans. Cognitive complexity and longevity have become even more exaggerated among modern humans, who live longer than any other primates and have relatively larger brains than any other creatures on the planet. It is not a coincidence that primates have both big brains and long lives; these traits are correlated across mammalian species. As we complete our discussion of the behavior and ecology of contemporary primates and turn our attention to the history of the human lineage, it is important to consider the forces that have shaped the evolution of large brains and long life spans in the primate order.

Selection for larger brains generates selection for long lives.

Correlations tell us that two traits are related but not why this relationship exists. In this case, we have some reason to think that the causal arrow goes from brain size to life span, not vice versa. We come to this conclusion because brains are expensive organs to maintain. Our brains account for just 2% of our total body weight, but they consume about 20% of our metabolic energy.

Natural selection does not maintain costly features such as the brain unless they confer important adaptive advantages. Moreover, the extent of investment that organisms make in a particular feature will be linked to the benefit that is derived from the investment. This is the same reason that you are usually willing to spend more for something that you will use for a long time than for something you will use only once. Animals that live for a long time will derive a greater benefit from the energy they expend on building and maintaining their brains than will animals that live for only a short time.

Life History Theory

Life history theory focuses on the evolutionary forces that shape trade-offs between the quantity and quality of offspring and between current and future reproduction.

Birth and death mark the beginning and end of every individual's life cycle. Between these two end points, individuals grow, reach sexual maturity, and begin to reproduce. Natural selection has generated considerable variation around this basic scheme.

FIGURE 8.1

Primates are intelligent and long-lived. The first ape in space was a 4-year-old chimpanzee named Ham, who was trained to perform a variety of tasks while hurtling into space. In May 1961, 3 months after Ham's flight, Alan Shepard followed the chimp into space. Ham was one of several dozen chimpanzees that NASA used to test the safety of space travel for humans.

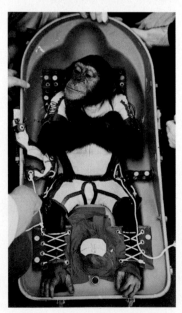

For example, Pacific salmon are hatched in freshwater but spend their adult lives in the open ocean. After years in the sea, they return to the streams where they were hatched to lay or fertilize their eggs; they die soon after they complete this journey. Opossums, the only North American marsupial, produce their first litter at the age of 1 year. Females have one to two litters per year and live less than 3 years (**Figure 8.2**). Lion females produce litters of up to six cubs at 2-year intervals and live into their teens. Elephants conceive for the first time at 10 years of age, have 22-month pregnancies, and give birth to single infants at 4- to 9-year intervals, and may live into their seventies.

If natural selection favors increased reproductive success, why doesn't it extend the opossum's life span, reduce the lion's interbirth interval, or increase the elephant's litter size? The answer is that all organisms face trade-offs that constrain their reproductive options. As we explained in Chapter 6, investment in one infant limits investment in other offspring, so parents must make trade-offs between the quality and quantity of offspring that they produce. Organisms also face trade-offs between current and future reproduction. All other things being equal, fast maturation and early reproduction are advantageous because they increase the length of the reproductive life span and reduce generation time. However, energy devoted to current reproduction diverts energy from growth and maintenance. If growth enhances reproductive success, then it may be advantageous to grow large before beginning to reproduce. Thus many kinds of organisms have a juvenile phase in which they do not reproduce at all. They do not become sexually mature until they reach a size at which the payoffs of allocating energy to current reproduction exceed the payoffs of continued growth. The same kind of argument applies to maintenance. Energy that is diverted from current reproduction to maintenance enables individuals to survive and reproduce successfully in the future.

Aging and death result from trade-offs between reproduction at different ages and survivorship.

Like humans, other primates age; as they get older, their physical abilities deteriorate. They don't run as fast, jump as high, or react as quickly (**Figure 8.3**). Their teeth wear down, making it harder for them to chew their food, and their joints deteriorate. Although humans are the only primates to experience menopause, the fertility of female primates declines when they reach old age. Males in most taxa reach peak physical condition in early adulthood and then decline.

At first glance, aging and death seem to be the inevitable effects of wear and tear on bodies. Organisms are complicated machines, like cars or computers. A machine has many components that must function together for it to work. It seems logical that the components in animals' bodies simply wear out and break down, like a worn clutch or faulty hard disk. But this explanation of aging is flawed because the analogy between organisms and machines is not really apt. Every cell in an organism contains all of the genetic information necessary to build a complete new body, and this genetic information can be used to repair damage. Wounds heal and bones mend, and some organisms, such as frogs, can regenerate entire limbs. Some organisms that reproduce asexually by budding or fission do not experience senescence at all.

If senescence is not inevitable, why doesn't natural selection do away with it? The answer has to do with the relative magnitude of the benefits that animals can derive from current reproduction or from living longer. Organisms could last longer if they were built better. A Lexus is of higher quality than a Subaru and is thus not expected to break down as often, but it also costs much more to build. The same trade-off applies to organisms. Our teeth would last longer if they were protected by a thicker covering of enamel, but building stronger teeth would require more nutrients, particularly calcium. Building higher-quality organisms consumes time and resources, thus reducing the organism's growth rate and early fertility.

FIGURE 8.2

The Virginia opossum, *Didelphis virginiana*, is the only marsupial mammal in North America. Females produce many tiny fetuses, which make their way into the mother's pouch, attach themselves to a nipple, and nurse for 2 to 3 months. Then they emerge from the pouch and cling to their mother's back.

FIGURE 8.3

Virtually all organisms experience senescence (aging). When this photograph was taken, this old male chimpanzee, named Hugo, was missing a lot of hair on his shoulders and back, he had lost a considerable amount of weight, and his teeth had been worn down to the gums.

FIGURE 8.4

Elephants are an example of a species at the slow/long end of the life history continuum. They are very large (approximately 6,000 kg, or 13,200 lb.) and can live up to 70 years in the wild. Females mature at about 10 years of age, have a 22-month gestation period, and give birth to single offspring at 4- to 9-year intervals.

The trade-off between survival and reproduction is strongly biased against characteristics that prolong life at the expense of early survival or reproduction.

Senescence is at least partly the consequence of genes that increase fitness at early ages and decrease it at later ages. Aging is favored by selection because traits that increase fertility at young ages are favored at the expense of traits that increase longevity.

The key to understanding this idea is to realize that selective pressures are much weaker on traits that affect only the old. To see why, think about the fate of two mutant alleles. One allele kills individuals before they reach adulthood, and the other kills individuals late in their lives. Carriers of the allele that kills infants and juveniles have a fitness of zero because none survives long enough to transmit the gene to their descendants. Therefore, there will be strong selection against alleles with deleterious effects on the young. In contrast, selection will have much less impact on a mutant allele that kills animals late in their lives. Carriers of a gene that kills them late in life will have already produced offspring before the effects of the gene are felt. Thus a mutation that affects the old will have limited effects on reproductive performance, and there will be little or no selection against it. This means that genes with pleiotropic effects that enhance early fertility but reduce fitness at later ages may be favored by natural selection because they increase individual fitness.

The trade-offs between current and future reproduction and between the quantity and quality of offspring generate constellations of interrelated traits.

Animals that begin to reproduce early also tend to have small body sizes, small brains, short gestation times, large litters, high rates of mortality, and short life spans. Animals that begin to reproduce at later ages tend to have larger body sizes, larger brains, longer gestation times, smaller litters, lower rates of mortality, and longer life spans (**Figure 8.4**). Life history traits are clustered together in this way because of the inherent trade-offs between current and future reproduction and between the quantity and quality of offspring. Animals that begin to reproduce early divert energy from growth and remain small. Animals that have small litters can invest more in maintenance and extend their life spans. These clusters of correlated traits create a continuum of life history strategies that runs from fast to slow or from short to long. Opossums fall somewhere along the fast/short end of the continuum; elephants lie at the slow/long end.

The trade-off between current and future reproduction depends on ecological factors that influence survival rates.

It makes little sense to divert energy to future reproduction if the prospects for surviving into the future are slim. For example, selection is likely to favor fast/short life histories in species that experience intense predation pressure. If predators are abundant and the prospects of surviving from one day to another are low, it makes little sense to postpone reproducing. In this situation, individuals that mature quickly and begin reproducing at early ages are likely to produce more surviving offspring than those that mature more slowly, so selection will favor faster/shorter life history strategies. Other kinds of ecological factors may favor slower/longer life histories. Suppose that there is severe competition for access to the resources that animals need to reproduce successfully and that larger animals are more successful in competitive encounters than are smaller animals. In this situation, smaller animals will be at a competitive

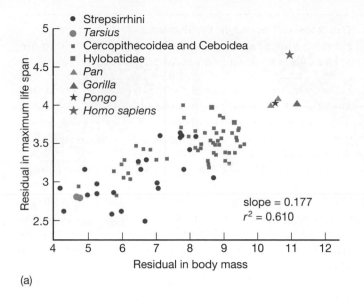

(a)

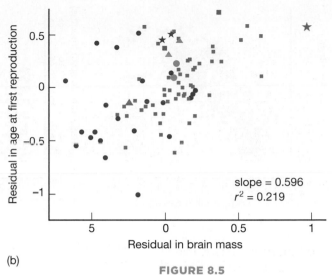

(b)

(a) Body weight is plotted against maximum life span for various primate species. In general, larger-bodied species live longer than smaller-bodied species. Note that humans live considerably longer than expected for their body size. (b) Brain size (corrected for body size) is plotted against age at first reproduction (corrected for body size). Species that have relatively large brains for their body size mature more slowly and reproduce for the first time at older ages than species that have smaller brains for their body size. Humans produce their first infant at about the age expected based on brain size.

disadvantage, and it may be profitable to invest more energy in growth, even if such an investment delays maturation. The life history strategies that characterize organisms reflect the net effects of these kinds of ecological pressures.

Natural selection shifts life history traits in response to changes in environmental conditions.

Natural selection adjusts life history traits in response to changes in prevailing conditions. Because life history traits are tightly clustered together (**Figure 8.5**), selection pressures acting on one trait often influence the value of other traits as well. For example, Steven Austad, a biologist at the University of Texas, compared the effect of predation on the life histories of two populations of opossums. One population lived on the mainland and was vulnerable to a variety of predators. Another population lived on an island that had very few predators for several thousand years. Opossums on the island aged more slowly, lived longer, and had smaller litters than opossums on the mainland. In this case, reduction of predation pressure favored a decelerated life history.

In some cases, organisms adjust their life histories in relation to current ecological conditions. Recall from Chapter 6 that female monkeys mature more quickly and reproduce at shorter intervals when food is abundant. This is not simply an inevitable response to the availability of food; it is an evolved capacity to adjust development in response to local conditions.

Primates fall toward the slow/long end of the life history continuum.

As a group, primates tend to delay reproduction and grow to relatively large sizes, and they also have relatively long gestation times, small litters, low fertility rates, long life spans, and large brains in relation to their body size (**Figure 8.6**). However, there is also variation within the primate order: Monkeys have relatively larger brains and slower life histories than strepsirrhines, and great apes have larger brains and slower life histories than monkeys.

As in other taxa, there is some evidence that ecological conditions influence life history variables in primates. Orangutans live in tropical rain forests on the islands of Borneo and Sumatra. Soil quality is higher in Sumatra than Borneo, and Sumatran forests have more fruit than Bornean forests. In addition, there is less temporal

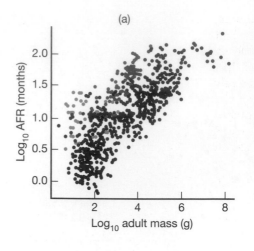

(a)

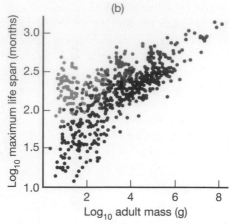

(b)

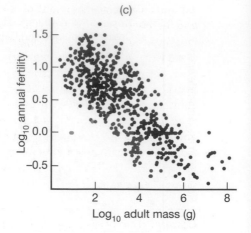

(c)

FIGURE 8.6

Body size is related to life history variables, including (a) the age at first reproduction, (b) maximum life span, and (c) fertility rate. Large animals mature later, live longer, and have slower reproductive rates than smaller animals.

variation in fruit availability in Sumatra than in Borneo. Sumatran orangutans feed mainly on fruit throughout the year, whereas orangutans living in the relatively unproductive forests of eastern and northeastern Borneo experience long periods in which fruit is scarce and they must rely on lower-quality foods, such as bark. Andrea Taylor of Duke University and Carel van Schaik of the University of Zürich suggest that these conditions have favored the evolution of relatively smaller brains and substantially shorter interbirth intervals in Bornean orangutans than in Sumatran orangutans.

Selective Pressures Favoring Large Brains in Monkeys and Apes

Social or ecological pressures may have favored cognitive evolution in monkeys and apes.

Many primatologists believe that the enlargement and reorganization of the brain in monkeys and apes is linked to the competitive pressures produced by sociality. In social groups, animals compete for food, mates, grooming partners, and other valuable resources. They also form social bonds that influence their participation in coalitions, exchange networks, access to resources, and so on (**Figure 8.7**). The larger a group becomes, the more difficult it gets to sustain social bonds and keep track of relationships within the group. The ability to operate effectively in this complicated social world may reward greater flexibility in behavior and favor expansion of the parts of the brain that are linked to learning and planning. This idea is called the **social intelligence hypothesis**.

An alternative set of hypotheses links increased brain size to ecological challenges, behavioral flexibility, innovation, and social learning capacities. Simon Reader of Utrecht University and Kevin Laland of the University of St. Andrews propose that natural selection has favored changes in the primate brain that enhance behavioral flexibility and enable animals to invent appropriate solutions to novel problems and to learn new behaviors from conspecifics (**Figure 8.8**). The benefits derived from innovation and social learning generated selective pressures that favored

FIGURE 8.7

Many primates live in complex social groups. Geladas form one-male units, which combine to form large bands composed of hundreds of individuals.

expansion and development of the parts of the brain linked to learning, planning, and behavioral flexibility. The ability to innovate and learn from others might enhance animals' ability to cope with ecological challenges. Monkeys feed mainly on plants and include many plant species in their diets. They must evaluate the ripeness, nutritional content, and toxicity of food items. Moreover, some primates rely heavily on **extracted foods** that require complex processing techniques. For example, chimpanzees and capuchin monkeys eat hard-shelled nuts that must be cracked open with stones or smashed against a tree trunk; baboons dig up roots and tubers; and aye-ayes extract insect larvae from underneath tree bark (**Figure 8.9**). Extracted foods are valuable elements in primate diets because they tend to be rich sources of protein and energy. However, they require complicated, carefully coordinated techniques to process.

Comparative analyses provide some support for both types of hypotheses about cognitive evolution in primates.

These models of the evolution of cognitive complexity generate specific predictions about the pattern of variation in the brains and cognitive abilities of living primates. For example, the social intelligence hypothesis predicts a link between social complexity and cognitive complexity, and the behavioral flexibility hypothesis predicts that innovations and social learning will be linked to brain size.

To test these hypotheses, we need a reliable measure of cognitive ability. Unfortunately, it is difficult to assess cognitive ability in other species. Instead, most work in this area has relied on measurements of the size or organization of particular parts of the brain. Researchers focus on the development of the forebrain, particularly the **neocortex**, because this is the site of the most substantial evolutionary changes in size and complexity (**Figure 8.10**). Moreover, the neocortex seems to be the part of the brain most closely associated with problem solving and behavioral flexibility. Neocortex size alone is not a very useful measure because larger animals generally have larger brains (and larger neocortexes) than smaller animals. Thus researchers make use of measures that control for these effects. Robin Dunbar, for example, measures the **neocortex ratio**, the ratio between the volume of the neocortex and the volume of the rest of the brain. Dunbar's analyses indicate that animals living in larger groups have larger neocortex ratios than animals living in smaller groups.

There is also evidence that behavioral flexibility and social learning are linked to brain evolution. Reader and Laland surveyed the primate literature for information about three measures of behavioral flexibility: (1) reports of behavioral innovation,

FIGURE 8.8

Wolfgang Köhler was one of the first scientists to systematically study cognitive abilities in captive chimpanzees. He hung a bunch of bananas out of the chimpanzees' reach and put several wooden crates in the room. Eventually, one individual, named Sultan, managed to stack the crates, clamber onto the precarious tower, and grab the bananas.

[a]   [b]

FIGURE 8.9

Primates sometimes exploit foods that are difficult to extract. Here (a) a male chimpanzee pokes a long twig into a hole in a termite mound and extracts termites, and (b) a green monkey punctures an eggshell and extracts the contents.

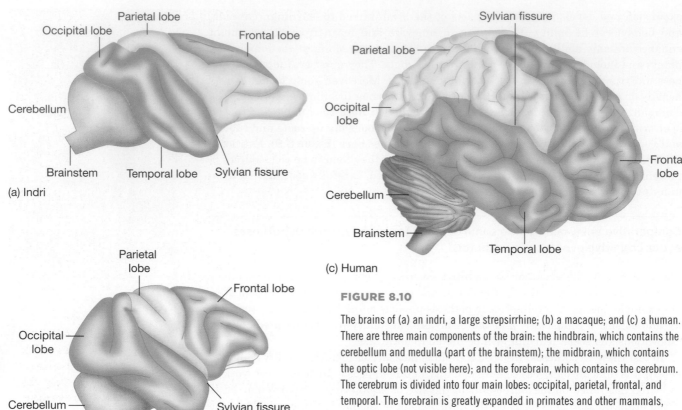

(a) Indri

(b) Macaque

(c) Human

FIGURE 8.10

The brains of (a) an indri, a large strepsirrhine; (b) a macaque; and (c) a human. There are three main components of the brain: the hindbrain, which contains the cerebellum and medulla (part of the brainstem); the midbrain, which contains the optic lobe (not visible here); and the forebrain, which contains the cerebrum. The cerebrum is divided into four main lobes: occipital, parietal, frontal, and temporal. The forebrain is greatly expanded in primates and other mammals, and much of the gray matter (which is made up of cell bodies and synapses) is located on the outside of the cerebrum in a layer called the cerebral cortex. The neocortex is a component of the cerebral cortex, and in mammals the neocortex covers the surface of virtually the entire forebrain.

which they defined as novel solutions to ecological or social problems; (2) examples of social learning, the acquisition of skills and information from others; and (3) observations of tool use. They demonstrated that these measures of behavioral flexibility are closely linked to the executive brain ratio (the size of the executive brain in relation to the brainstem). Primates with relatively large executive brains are more likely to innovate, learn from others, and use tools than primates with relatively small executive brains. In addition, primates seem to be more flexible in their foraging behavior than in their social behavior. In the long list of examples of innovations and socially learned behaviors that Reader and Laland compiled, foraging innovations predominate. In these comparative analyses, there is no consistent relationship between social learning and group size.

Thus, it is not entirely clear whether social or ecological challenges were primary factors favoring the evolution of big brains in primates. And these two hypotheses are not mutually exclusive. Primates may have derived benefits both from being able to cope more effectively with social challenges and from being able to master ecological challenges. Alternatively, cognitive abilities that evolved for one purpose may be applied in other contexts.

Great apes do not fit the social intelligence hypothesis very well.

Great apes have larger brains in relation to their body size than monkeys do, but they live in smaller groups than many monkeys do. Although chimpanzees and bonobos may live in communities that include as many as 50 to 100 individuals, gorillas live in smaller groups, and orangutans are largely solitary. It is possible that group size does

[a]

[b]

not capture all of the critical elements of sociality. This obviously poses a problem for the social intelligence hypothesis.

Richard Byrne points out that great apes make use of more complicated foraging techniques than other primates, enabling them to feed on some foods that other primates cannot process. For example, virtually all plant foods that mountain gorillas rely on are well defended by spines, hard shells, hooks, and stingers. Consuming each of their food items requires a particular routine—a complicated sequence of steps structured in a particular way. Many of the foods that orangutans feed on are also difficult to process. Great apes sometimes use tools to obtain access to certain foods that are not otherwise available to them. For example, chimpanzees poke twigs into holes of termite mounds and anthills, use leaves as sponges to mop up water from deep holes, and employ stones as hammers to break open hard-shelled nuts (**Figure 8.11**). Sumatran orangutans use sticks to probe for insects and to pry seeds out of the husks of fruit. Capuchins, who also have very large brains for their body size, also use complicated foraging procedures and are skilled tool users.

What Do Monkeys Know about One Another?

Although the selective forces that favored the evolution of large brains and slow life histories among primates are not fully understood, it is clear that primates know a lot about the other members of their groups. One of the most striking things about primates is the interest they take in one another. Newborns are greeted and inspected with interest (**Figure 8.12**). Adult females are sniffed and visually inspected regularly during their estrous cycles. When a fight breaks out, other members of the group watch attentively. As we have seen in previous chapters, monkeys know a considerable amount about their own relationships to other group members. A growing body of evidence suggests that monkeys also have some knowledge of the nature of relationships among other individuals, or **third-party relationships**. Monkeys'

knowledge of social relationships may enable them to form effective coalitions, compete successfully, and manipulate other group members to their own advantage.

Monkeys and apes know something about kinship relationships among other members of their groups.

One of the first indications that monkeys understand the nature of other individuals' kinship relationships came from a playback experiment on vervet monkeys conducted by Dorothy Cheney and Robert Seyfarth in Kenya's Amboseli National Park. Several female vervets heard a tape-recorded scream of a juvenile vervet piped from a hidden speaker. When the call was played, the mother of the juvenile stared in the direction of the speaker longer than other females did. This response suggests that mothers recognized the call of their own offspring. Even before the mother reacted, however, other females in the vicinity looked directly at the juvenile's mother. This response suggests that other females understood which monkey the juvenile belonged to and that they were aware that a special relationship existed between the mother and her offspring.

Monkeys may also have broader knowledge of kinship relationships (**Figure 8.13**). The evidence for this claim also comes from Cheney and Seyfarth's work on vervet monkeys. When monkeys are threatened or attacked, they often respond by threatening or attacking a lower-ranking individual who was not involved in the original incident—a phenomenon we call **redirected aggression**. Vervets selectively redirect aggression toward the maternal kin of the original aggressor. So, if female A threatens female B, then B threatens AA, a close relative of A. If monkeys were simply blowing off steam or venting their aggression, they would choose a target at random. Thus, the monkeys seem to know that certain individuals are somehow related.

Monkeys probably understand rank relationships among other individuals.

Because kinship and dominance rank are major organizing principles in most primate groups, it makes sense to ask whether monkeys also understand third-party rank relationships. The most direct evidence that monkeys understand third-party rank relationships comes from two playback experiments conducted on chacma baboons in the Okavango Delta of Botswana by Seyfarth, Cheney, and their colleagues. In this group, dominance relationships were stable, and females never responded submissively toward lower-ranking females.

In one experiment that Seyfarth and Cheney designed, females listened to a recording of a female's grunt followed by another female's submissive fear barks. Female baboons responded more strongly when they heard a higher-ranking female responding submissively to a lower-ranking female's grunt than when they heard a lower-ranking female responding submissively to a higher-ranking female's grunt. Thus, females were more attentive when they heard a sequence of calls that did not correspond to their knowledge of dominance rank relationships among other females. Control experiments excluded the possibility that females were reacting simply to the fact that they had not heard a particular sequence of calls before. The pattern of responses suggests that females knew the relative ranks of other females in their group and were particularly interested in the anomalous sequence of calls. Studies of wild vervet monkeys produce very similar patterns of responses.

FIGURE 8.14

Complex fitness calculations may be involved in decisions about whether to join a coalition. By helping the victim against the aggressor, the ally increases the fitness of the victim and decreases the fitness of the aggressor.

In a second experiment, Thore Bergman and Jacinta Beehner collaborated with Seyfarth and Cheney to probe the baboons' knowledge of the hierarchical nature of rank relationships in groups with matrilineal dominance ranks. Using the same basic experimental paradigm, researchers played sequences of vocalizations that simulated rank reversals within lineages and rank reversals between lineages. As young female baboons mature, they often rise in rank above their older sisters and other female kin. Thus, changes in the relative rank of females in the same lineage are part of the normal course of rank acquisition. However, changes in the relative ranks of unrelated females are much less common. The baboons reacted much more strongly to simulated rank reversals between lineages than to simulated rank reversals within lineages. Again, the researchers were careful to control for confounding variables, such as rank distance and novelty. This result suggests that the females understood the relative ranks of other females and that they understood that changes in rank relationships within lineages are not the same as changes in rank relationships between lineages.

Participation in coalitions probably draws on sophisticated cognitive abilities.

Even the simplest coalition is a complex interaction. When coalitions are formed, at least three individuals are involved, and several kinds of interactions are going on simultaneously (**Figure 8.14**). Consider the case in which one monkey (the aggressor) attacks another monkey (the victim). The victim then solicits support from a third party (the ally), and the ally intervenes on behalf of the victim against the aggressor. The ally behaves altruistically toward the victim, giving support to the victim at some potential cost to itself. At the same time, however, the ally behaves aggressively toward the aggressor, imposing harm or energy costs on the aggressor. Thus, the ally simultaneously has a positive effect on the victim and a negative effect on the aggressor. Under these circumstances, decisions about whether to intervene in a particular dispute may be quite complicated. Consider a female who witnesses a dispute between two of her offspring. Should she intervene? If so, which of her offspring should she support? When a male bonnet macaque is solicited by a higher-ranking male against a male who frequently supports him, how should he respond? In each case, the ally must balance the benefits to the victim, the costs to the opponent, and the costs to itself (**Figure 8.15**).

Given the complexity of even simple coalitions, knowledge of third-party relationships may be

FIGURE 8.15

Primates form coalitions that are more complicated than the coalitions of most other animals. (a) In a captive bonnet macaque group, members of opposing factions confront one another. (b) Three capuchin monkeys are in defense mode.

[a]

[b]

valuable because it enables individuals to predict how others will behave. Thus, animals who understand the nature of third-party relationships may have a good idea about who will support them and who will intervene against them in confrontations with particular opponents, and they may also be able to tell which of their potential allies are likely to be most effective in coalitions against their opponents.

Monkeys seem to rely on their knowledge of other monkeys' rank relationships when they make decisions about who to recruit as allies in agonistic interactions. Male bonnet macaques frequently solicit support from other males when they are involved in conflicts. A study of a large captive group found that males did not simply choose allies that outranked themselves. When they were faced with higher-ranking opponents, males chose allies that outranked both themselves and their opponents. In order to do this, males needed to know the relative ranks of other males and to track changes in male rank from month to month. Sooty mangabeys also seem to use third-party knowledge of rank relationships when they recruit allies.

We are beginning to gain insight into what monkeys and apes know about others' minds.

Monkeys and apes seem to be very good at predicting what other animals will do in particular situations. For example, we have seen that vervets groom monkeys who support them in coalitions, female langurs with newborn infants are fearful of new resident males, baboons express surprise when low-ranking animals elicit signs of submission from higher-ranking animals, and capuchins don't try to recruit support from monkeys who have closer bonds to their opponents than to themselves. These examples indicate that monkeys can predict what others will do and adjust their behavior accordingly.

Monkeys' ability to predict what other individuals will do may not seem very remarkable. After all, we know that many animals are very good at learning to make associations between one event and another. In the laboratory, rats, pigeons, monkeys, and many other animals can learn to pull a lever, push a button, or peck a key to obtain food. These are examples of associative learning, the ability to track contingencies between one event and another. Monkeys' ability to predict what others will do in particular situations might be based on sophisticated associative learning capacities, prodigious memory of past events, and perhaps some understanding of conceptual categories such as kinship and dominance. On the other hand, it is also possible that monkeys' ability to predict what others will do is based on their knowledge of the mental states of others—what psychologists call a **theory of mind**.

It may seem relatively unimportant whether monkeys and apes rely on associative learning to predict what others will do or whether they have a well-developed theory of mind. However, there may be some things that animals cannot do unless they understand what is going on in other animals' minds. For example, some researchers think that effective deception requires the ability to manipulate or take advantage of others' beliefs about the world. In the 1960s, the late Emil Menzel conducted a landmark set of experiments about chimpanzees' ability to find and communicate about the location of hidden objects (**Figure 8.16**). In a set of experiments, Menzel showed one chimpanzee where a food item was hidden and then released the knowledgeable chimpanzee and his companions into their enclosure. The group quickly learned to follow their knowledgeable companion; but he just as quickly learned that when he led others to the hidden food, he would not get a very big share of it. Menzel noticed that the knowledgeable chimpanzee sometimes led his companions in the wrong direction and then dashed off and grabbed the hidden treasure. How did the knowledgeable chimpanzee work out this tactic? He might have understood that his knowledge differed from the knowledge of other group members and then have come up with a way to take advantage of this discrepancy effectively. If this is what he did, then we would conclude that he had a well-developed theory of mind.

FIGURE 8.16

In Menzel's experiments, a researcher showed a young chimpanzee where food was hidden in the chimp's enclosure, as pictured here. Then the chimpanzee was reunited with the other members of the group, and the group was released inside the enclosure. The young chimpanzee often led the group back to the hidden food, but it also learned to divert the group so that it could get bits of food before others found the cache.

Although some researchers suggest that deception does not rely on a theory of mind, it seems clear that a theory of mind would allow for more complicated and successful deceptions. Similarly, the ability to pretend, empathize, take another's perspective, read minds, console, imitate, and teach relies on knowing what others know or how they feel. Humans do all of these things, but it is not clear whether other primates do.

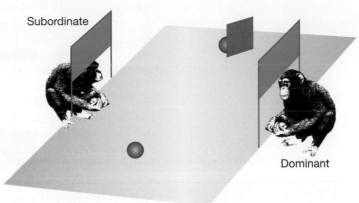

It is difficult to be sure what nonhuman primates know about the minds of other individuals. However, primatologists have begun to make some progress in this area.

Bryan Hare, now at Duke University, and Josep Call and Michael Tomasello of the Max Planck Institute for Evolutionary Anthropology in Leipzig, Germany, developed a protocol that evaluates individuals' ability to take advantage of discrepancies between their own knowledge and other individuals' knowledge in a competitive situation. In their experiments, they paired subordinate and dominant chimpanzees in the configuration illustrated in **Figure 8.17**. The experiments take advantage of the fact that subordinates cannot obtain food when dominants are present. Here, two pieces of food are visible to the subordinate, but the dominant can see only one; the other is hidden behind a barrier. Hare and his colleagues predicted that if the subordinate *knew* what the dominant could see, the subordinate would head for the piece of food that the dominant could not see, hoping to consume the hidden food item while the dominant was occupied with the other piece. This is exactly what the subordinate chimpanzees did. So the chimpanzees seemed to understand what other chimpanzees know.

Laurie Santos of Yale University and her colleagues applied the same reasoning in designing experiments with free-ranging rhesus macaques on Cayo Santiago. In these experiments, monkeys were given the chance to "steal" food from two human experimenters. In one experiment, one experimenter was facing toward the monkey, and the other was facing away. The monkeys were more likely to approach the experimenter who was facing away than the experimenter who was facing forward. In other experiments, the monkeys selectively approached experimenters whose faces were pointed away and experimenters whose eyes were averted (**Figure 8.18**). In another set of

FIGURE 8.17

In this experiment, one food item was hidden behind a barrier so that it could be seen by only the subordinate animal, and one food item was in plain sight of the dominant. Subordinates are unlikely to obtain food rewards when dominant animals are present. So if the subordinate knew what the dominant could see, the subordinate was expected to head for the item that was hidden from the dominant. This is what the chimpanzees did most of the time.

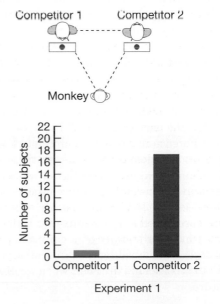

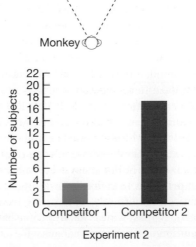

FIGURE 8.18

In this set of experiments, monkeys were presented with two human competitors who had food that the monkeys could potentially steal. The graphs show that monkeys were much more likely to approach the person who was looking away from them than the person who was facing them, and they were more likely to approach the person who was looking away from the food than the person who was looking toward the food. This suggests that monkeys know what others can and cannot see.

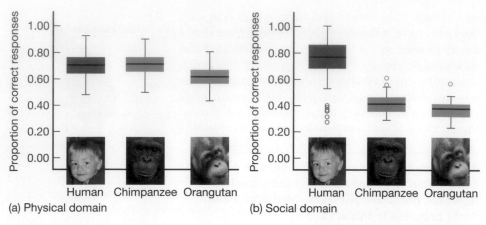

FIGURE 8.19

Children and apes were presented with an identical set of tasks that examined their physical and social cognition. (a) Children and apes showed no differences in tasks based on physical cognition. (b) But children were considerably more successful than apes on tasks that were based on social cognition. In this graph, the median value is indicated by the line that runs through each box and the box spans the interquartile range that includes 50% of the values. The lines outside the boxes (sometimes called "whiskers") represent maximum and minimum values, excluding outliers (*circles*), which are values that lie at least 1.5 times above or below the interquartile range.

experiments, Santos placed food rewards in two transparent boxes covered with bells. In one box, the ringers on the bells were removed, creating one noisy box and one quiet box. The experimenter baited the boxes, shook the boxes to display their auditory qualities, and then walked away and hid his face. The monkeys showed strong preferences for the quiet container, suggesting that they knew what the experimenter could hear. Santos and her colleagues argue that rhesus monkeys can accurately perceive what others see and hear in these competitive situations.

Human social cognition is more sophisticated than that of apes.

Although apes and monkeys can solve many complex cognitive tasks, there are still substantial differences between the cognitive skills of humans and other primates. These differences are most pronounced in tasks that involve social learning, communication, and knowledge of others' minds. In a comprehensive comparative study of ape cognition, Esther Herrmann and her colleagues from the Max Planck Institute for Evolutionary Anthropology evaluated the performance of 105 two-year-old humans on a battery of cognitive tasks and compared this with the performance of 106 chimpanzees and 32 orangutans of all ages on the same tasks. Some tasks focused on cognition about the physical world, such as tracking a reward after it has been moved or using a tool to retrieve a reward that is out of reach. Other tasks focused on social cognition, such as solving a novel problem after observing the demonstration of a solution or following a gaze to a target. There was little difference in the performance of children and apes on tasks that involved physical cognition (**Figure 8.19**). However, there were greater differences in the social domain. The human children were significantly more successful than the chimpanzees and orangutans on tasks that required social learning, communication, and knowledge of others' minds. Herrmann and her colleagues hypothesize that humans have "evolved some specialized socio-cognitive skills (beyond those of primates in general) for living and exchanging knowledge in cultural groups: communicating with others, learning from others, and 'reading the mind' of others in especially complex ways." We will come back to this idea in Part Four, when we discuss the evolution of the human capacity for culture.

The Value of Studying Primate Behavior

As we come to the end of Part Two, it may be useful to remind you why information about primate behavior and ecology plays an integral role in the story of human evolution. First, humans are primates, and the first members of the human species were probably more similar to living nonhuman primates than to any other animals on Earth. Thus, by studying living primates we can learn something about the lives of our ancestors. Second, humans are closely related to primates and similar to them in many ways. If we understand how evolution has shaped the behavior of animals that are so much like us, we may have greater insights about the way evolution has shaped our own behavior and the behavior of our ancestors. Both of these kinds of reasoning will be apparent in Part Three, which covers the history of our own human lineage.

CHAPTER REVIEW

Key Terms

social intelligence hypothesis (p. 196)

extracted foods (p. 197)

neocortex (p. 197)

neocortex ratio (p. 197)

third-party relationships (p. 199)

redirected aggression (p. 200)

theory of mind (p. 202)

Study Questions

1. There is a positive correlation between brain size and longevity in animal species. One interpretation of this correlation is that selection for longer life spans was the primary force driving the evolution of large brains. An alternative interpretation is that selection for larger brains was the primary force driving the evolution of longer life spans. Explain which of these interpretations is more likely to be correct and why this is the case.

2. We have argued that natural selection is a powerful engine for generating adaptations. If that is the case, then why do organisms grow old and die? Why can't natural selection design an organism that lives forever?

3. Life history traits tend to be bundled in particular ways. Explain how these traits are combined and why we see these kinds of combinations in nature.

4. Primates evolved from small-bodied insectivores that were arrayed somewhere along the fast/short end of the life history continuum. What ecological factors are thought to have favored the shifts toward slower/longer life histories in early primates, monkeys, and apes?

5. Primates take a relatively long time to grow up compared with other animals. Consider the costs and benefits of this life history pattern from the point of view of the growing primate and its mother.

6. What do comparative studies of the size and organization of primate brains tell us about the selective factors that shaped the evolution of primate brains? What are the shortcomings of these kinds of analyses?

7. Monkeys are quite skilled in navigating complicated social situations that they encounter in their everyday lives. They seem to know what others will do in particular situations and can respond appropriately. However, monkeys consistently fail theory-of-mind tests in the laboratory. How can we reconcile these two observations?

8. Monkeys seem to have some concept of kinship. What evidence supports this idea? What kind of variation might you expect to find in monkeys' concepts of kinship within and between species?

9. Suppose that you were studying a group of monkeys and you discovered convincing evidence of empathy or deception. How and why would these data surprise your colleagues?

10. Detailed studies of coalitionary behavior have provided an important source of information about primate cognitive abilities. Explain why coalitions are useful sources of information about social knowledge. What does the pattern of coalitionary support tell us about what monkeys know about other group members?

Further Reading

Byrne, R. W., and A. Whiten, eds. 1988. *Machiavellian Intelligence: Social Expertise and the Evolution of Intellect in Monkeys, Apes, and Humans.* New York: Oxford University Press.

Cheney, D. L., and R. M. Seyfarth. 2007. *Baboon Metaphysics: The Evolution of a Social Mind.* Chicago: University of Chicago Press.

Mitani, J., J. Call, P. Kappeler, R. Palombit, and J. B. Silk, eds. 2012. *The Evolution of Primate Societies.* Chicago: University of Chicago Press.

Platt, M. L., R. M. Seyfarth, and D. L. Cheney. 2016. "Adaptations for social cognition in the primate brain." *Philosophical Transactions of the Royal Society B: Biological Sciences*, 371(1687), 20150096.

Reader, S. M., and K. N. Laland. 2002. "Social Intelligence, Innovation, and Enhanced Brain Size in Primates." *Proceedings of the National Academy of Sciences U.S.A.* 99: 4436–4441.

Street, S. E., A. F. Navarette, S. M. Reader, and K. N Laland. 2017. "Coevolution of cultural intelligence, extended life history, sociality, and brain size in primates." *Proceedings of the National Academy of Sciences*, 114(30), 7908–7914.

Whiten, A. W., and R. W. Byrne, eds. 1997. *Machiavellian Intelligence II: Extensions and Evaluations.* New York: Cambridge University Press.

 Review this chapter with personalized, interactive questions through INQUIZITIVE.

 This chapter also features a Guided Learning Exploration on life history trade-offs and brain size.

PART
THREE

THE HISTORY
OF THE
HUMAN LINEAGE

FROM TREE SHREW TO APE

CHAPTER OBJECTIVES

By the end of this chapter you should be able to:

A Explain how the major changes in the position of the continents and world climates have influenced the course of primate evolution.

B Describe how paleontologists establish the age of fossils.

C Assess what we know about the earliest members of the primate lineage.

D Identify when and where apelike primates first appear in the fossil record.

During the Permian and early Triassic periods (**Table 9.1**), much of the world's fauna was dominated by therapsids, a diverse group of reptiles that possessed traits, such as being warm-blooded and covered with hair (**Figure 9.1**), that linked them to the mammals that evolved later. At the end of the Triassic, most therapsid groups disappeared, and dinosaurs radiated to fill all of the niches for large, terrestrial animals. One therapsid lineage, however, evolved and diversified to become the first true

TABLE 9.1 The Geologic Timescale

Era	Period	Epoch	Period Begins (Ma)	Notable Events
Cenozoic	Quaternary	Holocene	0.012	Origins of agriculture and complex societies
		Pleistocene	2.6	Appearance of *Homo sapiens*
	Tertiary	Pliocene	5	Dominance of land by angiosperms, mammals, birds, and insects
		Miocene	23	
		Oligocene	34	
		Eocene	54	
		Paleocene	66	
Mesozoic	Cretaceous		136	Rise of angiosperms, disappearance of dinosaurs, second great radiation of insects
	Jurassic		190	Abundance of dinosaurs, appearance of first birds
	Triassic		225	Appearance of first mammals and dinosaurs
Paleozoic	Permian		280	Great expansion of reptiles, decline of amphibians, last of trilobites
	Carboniferous		345	Age of Amphibians; first reptiles, first great insect radiation
	Devonian		395	Age of Fishes; first amphibians and insects
	Silurian		430	Land invaded by a few arthropods
	Ordovician		500	First vertebrates
	Cambrian		570	Abundance of marine invertebrates
Precambrian				Primitive marine life

(a)

(b)

FIGURE 9.1

Therapsids dominated Earth about 250 Ma, before dinosaurs became common. The therapsids were reptiles, but they may have been warm-blooded and they had hair instead of scales. The therapsid *Thrinaxodon*, whose (a) skeleton and (b) reconstruction are shown here, was about 30 cm (12 in.) long and had teeth suited for a broad carnivorous diet.

mammals. These early mammals were probably mouse-size, nocturnal creatures that fed mainly on seeds and insects. They had internal fertilization but still laid eggs. By the end of the Mesozoic era, 65 million years ago (Ma), placental and marsupial mammals that bore live young had evolved. With the extinction of the dinosaurs at the beginning of the next era (the Cenozoic) came the spectacular radiation of the mammals. All of the modern descendants of this radiation—including horses, bats, whales, elephants, lions, and primates—evolved from creatures that were something like a contemporary shrew (**Figure 9.2**).

To completely understand human evolution, we need to know how the transition from a shrewlike creature to modern humans took place. Remember that, according to Darwin's theory, complex adaptations are assembled gradually, in many small steps—each step favored by natural selection. Modern humans have many complex adaptations, such as grasping hands, **bipedal** locomotion (walking upright on two legs), toolmaking abilities, language, and large-scale cooperation. To understand human evolution fully, we have to consider each of the steps in the lengthy process that transformed a small, solitary, shrewlike insectivore scurrying through the leaf litter of a dark Cretaceous forest into someone more or less like you. Moreover, it is not enough to chronicle the steps in this transition. We also need to understand why each step was favored by natural selection. We want to know, for example, why claws were traded for flat nails, why quadrupedal locomotion gave way to upright bipedal locomotion, and why brains were so greatly enlarged.

In this part of the text, we trace the history of the human lineage. We begin in this chapter by describing the emergence of creatures that resemble modern lemurs and tarsiers, then we document the appearance of animals that look more like modern monkeys, and finally we investigate the origins of animals something like contemporary apes. In later chapters, we recount the transformation from hominoid to hominin. We will introduce the first members of the human tribe, Hominini; then the first members of our own genus, *Homo*; and finally the first known representatives of our own species, *Homo sapiens*. We know something about each step in this process, although far more is known about recent periods than about periods in the most distant past. We will see, however, that there is still a great deal left to be discovered and understood.

FIGURE 9.2

The first mammals probably resembled the modern-day Belanger's tree shrew.

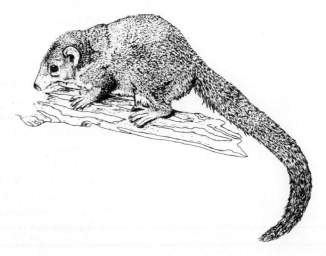

Continental Drift and Climate Change

FIGURE 9.3

Today, East African savannas look much like this. In the past, the scenery was probably quite different.

To understand the evolution of our species, it is important to understand the geologic, climatic, and biological conditions under which these evolutionary changes occurred.

When we think about the evolution of modern humans, we usually picture early humans wandering over open grasslands dotted with acacia trees—the same breathtaking scenery that we see in African wildlife documentaries. As we shift the time frame forward through millions of years, the creatures are altered but the backdrop is unchanged. This image is misleading, however, because the scenery has changed along with the cast of characters (**Figure 9.3**).

It is important to keep this fact in mind because it changes our interpretation of the fossil record. Remember that evolution produces adaptation, but what is adaptive in one environment may not be adaptive in another. If the environment remained the same over the course of human evolution, then the kinds of evolutionary changes observed in the hominin fossil record (such as increases in brain size, bipedalism, and prolonged juvenile dependence) would have to be seen as steady improvements in the perfection of human adaptations: Evolution would progress toward a fixed goal. But if the environment varied through time, then evolution would have to track a moving target. In this scenario, new characteristics seen in fossils would not have to represent progress in a single direction. Instead, these changes might have been adaptations to changing environmental conditions. As we will see, the world has become much colder and drier in the past 20 million years and has been extremely variable in the past 800,000 years, and these changes probably altered the course of human evolution. If the world had become warmer rather than colder during this period, then our human ancestors would probably have remained in the safety of the trees and would not have become terrestrial or bipedal. We would probably be stellar rock climbers but poor marathon runners.

The positions of the continents have changed in relation to one another and to the poles.

The world has changed a lot in the past 200 million years. One of the factors that has contributed to this change is the movement of the continents, or **continental drift**. The continents are not fixed in place; instead, the enormous, relatively light plates of rock that make up the continents slowly wander around the globe, floating on the denser rock that forms the floor of the deep ocean. About 300 Ma, all of the land making up the present-day continents was joined in a single, huge landmass called **Pangaea**. Pangaea broke apart into two pieces around 180 Ma (**Figure 9.4**). The northern half, called **Laurasia**, included what is now North America and Eurasia minus India; the southern half, **Gondwanaland**, was composed of the rest. By the time the dinosaurs became extinct (65 Ma), Gondwanaland had broken up into several pieces. Africa and India separated, and India headed north, eventually crashing into Eurasia, whereas the rest of Gondwanaland stayed in the south. Eventually, Gondwanaland separated into South America, Antarctica, and Australia, and these continents remained isolated from one another for many millions of years. South America did not drift north to join North America until about 5 Ma.

Continental drift is important to the history of the human lineage for two reasons. First, oceans serve as barriers that isolate certain species from others, so the position of the continents plays an important role in the evolution of species. As we will see, the long isolation of South America creates one of the biggest puzzles in our knowledge of primate evolution. Second, continental drift is one of the engines of climate change, and climate change fundamentally influences human evolution.

The climate has changed substantially during the past 65 million years—first becoming warmer and less variable, then cooling, and finally fluctuating widely in temperature.

The size and orientation of the continents have important effects on climate. Very large continents tend to have more extreme weather. This is why Chicago has much colder winters than London, even though London is much farther north. Pangaea was much larger than Asia and is likely to have had very cold weather in winter. When continents restrict the circulation of water from the tropics to the poles, world climates seem to become cooler. These changes, along with other poorly understood factors, have led to substantial climate change. **Figure 9.5** summarizes changes in global temperature during the Cenozoic era, which began about 66 Ma. **A Closer Look 9.1** explains how climatologists reconstruct ancient climates.

To give you some idea what these changes in temperature mean, consider that during the period of peak warmth in the early Miocene, palm trees grew as far north as what is now Alaska; rich temperate forests (such as those in the eastern United States today) extended as far north as Oslo, Norway; and only the tallest peaks in Antarctica were glaciated.

The Methods of Paleontology

Much of our knowledge of the history of life comes from the study of fossils, the mineralized bones of dead organisms.

In certain kinds of geologic settings, the bones of dead organisms may be preserved long enough for the organic material in the bones to be replaced by minerals (**mineralized**) from the surrounding rock. Such natural copies of bones are called **fossils**. Scientists who recover, describe, and interpret fossil remains are called **paleontologists**. **A Closer Look 9.2** provides more information about how fossils are formed.

A great deal of what we know about the history of the human lineage comes from the study of fossils. Careful study of the shapes of different bones tells us what early hominins were like—how big they were, what they ate, where they lived, how they moved, and even something about how they lived. When the methods of systematics described in Chapter 4 are applied to these materials, they also can tell us something about the phylogenetic history of long-extinct creatures. The kinds of plant and animal fossils found in association with the fossils of our ancestors tell us what the environment was like—whether it was forested or open, how much it rained, and whether rainfall was seasonal.

There are several radiometric methods for estimating the age of fossils.

To assign a fossil to a particular position in a phylogeny, we must know how old it is. As we will see in later chapters, the date that we assign to particular specimens can profoundly influence our understanding of the evolutionary history of certain lineages or traits.

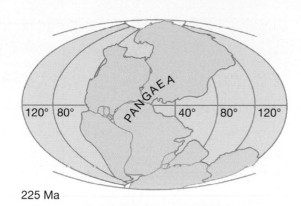

225 Ma

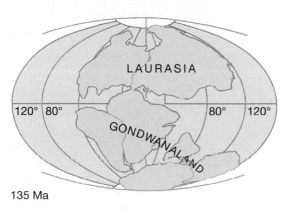

135 Ma

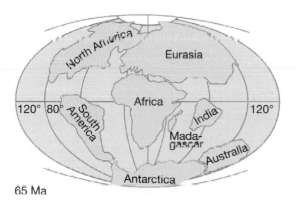

65 Ma

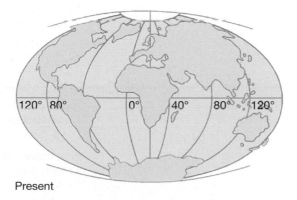

Present

FIGURE 9.4

The arrangement of the continents has changed considerably over the past 225 million years.

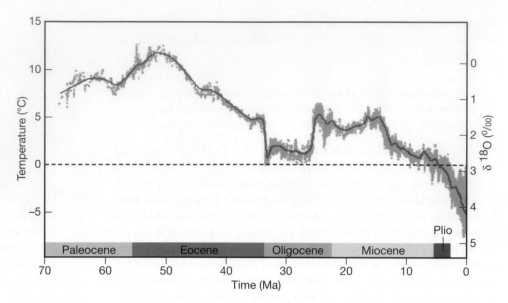

FIGURE 9.5

The gray points are estimates of average world temperature that are based on the ratio of ^{16}O to ^{18}O taken from deep-sea cores. The red line plots a statistically smoothed average value. The variability around the line represents both measurement error and rapid temperature fluctuations that last less than a million years. As we will see (Figure 12.1), the increase in variability in the past several million years has been the result of fluctuations in world temperature, some lasting only a few centuries.

Radiometric methods provide one of the most important ways to date fossils. To understand how radiometric techniques work, we need to review a little chemistry. All of the atoms of a particular element have the same number of protons in their nucleus. For example, all carbon atoms have six protons. However, different **isotopes** of a particular element have different numbers of neutrons in their nucleus. Carbon-12, the most common isotope of carbon, has six neutrons, and carbon-14 has eight. Radiometric methods are based on the fact that the isotopes of certain elements are unstable. This means they change spontaneously from one isotope to another of the same element or to an entirely different element. For example, carbon-14 changes to nitrogen-14, and potassium-40 changes spontaneously to argon-40. For any particular isotope, such changes (or **radioactive decay**) occur at a constant, clocklike rate that can be measured with precision in the laboratory. There are several radiometric methods:

1. **Potassium–argon dating** is used to date the age of volcanic rocks found in association with fossil material. Molten rock emerges from a volcano at a very high temperature. As a result, all of the argon gas is boiled out of the rock. After this, any argon present in the rock must be due to the decay of potassium. Because this decay occurs at a known and constant rate, the ratio of potassium to argon can be used to date volcanic rock. Then, if a fossil is discovered in a geologic **stratum** ("layer"; plural *strata*) lying under the stratum that contains the volcanic rock, paleontologists can be confident that the fossil is older than the rock. A new variant of this technique, which is called **argon–argon dating** because the potassium in the sample is converted to an isotope of argon before it is measured, allows more accurate dating of single rock crystals.

2. **Carbon-14 dating** (or **radiocarbon dating**) is based on an unstable isotope of carbon that living animals and plants incorporate into their cells. As long as the organism is alive, the ratio of the unstable isotope (carbon-14) to the stable isotope (carbon-12) is the same as the ratio of the two isotopes in the atmosphere. Once the animal dies, carbon-14 starts to decay into nitrogen-14 at a constant rate. By measuring the ratio of carbon-14 to carbon-12, paleontologists can estimate the amount of time that has passed since the organism died.

9.1 Using Deep-Sea Cores to Reconstruct Ancient Climates

Beginning about 50 years ago, oceanographers launched a program of extracting long cores from the sediments that lie on the floor of the deep sea (about 6,000 m, or 20,000 ft., below the surface). Data from these cores have allowed scientists to make much more detailed and accurate reconstructions of ancient climates. Figure 9.5 shows the ratio of two isotopes of oxygen, ^{16}O and ^{18}O, derived from different layers of deep-sea cores. Because different layers of the cores were deposited at different times over the past 65 million years and have remained nearly undisturbed ever since, they give us a snapshot of the relative amounts of ^{16}O and ^{18}O in the sea when the layers were deposited on the ocean floor.

The ratio of ^{16}O to ^{18}O allows us to estimate ocean temperatures in the past. Water molecules containing the lighter isotope of oxygen, ^{16}O, evaporate more readily than do molecules containing the heavier isotope, ^{18}O. Snow and rain have a higher concentration of ^{16}O than the sea does because the water in clouds evaporates from the sea. When the world is warm enough that few glaciers form at high latitudes, the precipitation that falls on the land returns to the sea, and the ratio of the two isotopes of oxygen

remains unchanged (**Figure 9.6a**). When the world is colder, however, much of the snow falling at high latitudes is stored in immense continental glaciers like those now covering Antarctica (**Figure 9.6b**). Because the water locked in glaciers contains more ^{16}O than the ocean does, the proportion of ^{18}O in the ocean increases. Therefore, the concentration of ^{18}O in seawater increases when the world is cold and decreases when it is warm. This means scientists can estimate the temperature of the oceans in the past by measuring the ratio of ^{16}O to ^{18}O in different layers of deep-sea cores.

FIGURE 9.6

Because water evaporating from the sea is enriched in ^{16}O, so is precipitation. (a) When the world is warm, this water returns rapidly to the sea, and the concentration of ^{16}O in seawater is unchanged. (b) When the world is cold, much precipitation remains on land as glacial ice, and so the sea becomes depleted in ^{16}O.

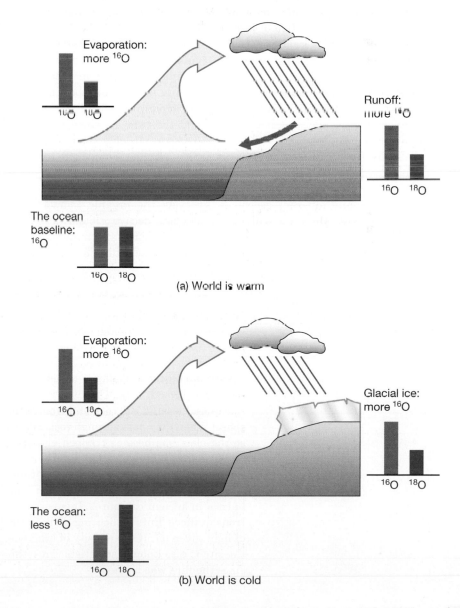

(a) World is warm

(b) World is cold

9.2 How Fossils Are Formed

A fossil is any preserved remains from a once living organism. Most of the fossils that we discuss in this book are the preserved bones of primates that lived in the past few tens of millions of years. But there are other kinds of fossils. For example, sometimes the tracks of long dead animals have been preserved. These are fossils, and you will see that one such fossil, a trackway made by hominins 3.5 Ma, plays an important role in our understanding of the evolution of human bipedality. Other kinds of fossils include insects and plants preserved in amber (chemically transformed tree resin); plant remains including leaves and seeds; preserved feces; and even molecules, especially DNA and proteins.

Given that fossils encompass such a variety of objects, you won't be surprised that they are formed through many different geochemical processes. Here we will focus on how hominin bones are fossilized because bones play such an important role in our understanding of human evolution.

The fossilization process begins when an organism dies and is covered with sediment. Bacteria, fungi, and other microorganisms in the soil rapidly strip away all of the soft tissue, leaving just the bones and teeth. Bones are made of living cells that contain organic materials such as DNA and collagen (a structural protein that also makes up your tendons and ligaments) and inorganic materials such as apatite (a crystalline material made up of calcium, phosphate, and fluorine). The organic component in bones slowly decays. Bacteria are thought to play an important role in this process, and the rate at which this occurs depends strongly on temperature; the organic component of bone decays more rapidly in hot places than it does in cold places.

As the organic component in the bone decays, groundwater seeps into the bone, and minerals in the water precipitate out in spaces, many very tiny, in the bone. Under the right conditions, these minerals make a very high-resolution stone "copy" of the original bone structure. A tiny fraction of these fossilized bones are then exposed by erosion or uncovered in excavations by a fortunate paleoanthropologist.

Ancient DNA is another kind of fossil that will play a big role in our story. When an organism dies in a cold, dry place, it may take a long time before all of the original organic material in the bone decays. If sufficient DNA remains when a fossil bone is found, geneticists can extract the DNA from the fossil and determine the genetic sequence. Some DNA has been recovered from hominin fossils that are more than half a million years old, and complete DNA sequences have been recovered from fossils that are about 100,000 years old. As we will see, these kinds of molecular fossils will play an important role in our understanding of hominin evolution.

3. **Thermoluminescence dating** is based on an effect of high-energy nuclear particles traveling through rock. These particles come from the decay of radioactive material in and around the rock and from cosmic rays that bombard Earth from outer space. When they pass through rock, these particles dislodge electrons from atoms, so the electrons become trapped elsewhere in the rock's crystal lattice. Heating a rock relaxes the bonds holding the atoms in the crystal lattice together. All of the trapped electrons are then recaptured by their respective atoms—a process that gives off light. Researchers often find flints at archaeological sites that were burned in ancient campfires. It is possible to estimate the number of trapped electrons in these flints by heating them in the laboratory and measuring the amount of light given off. If the density of high-energy particles currently flowing through the site is also known, scientists can estimate the length of time that has elapsed since the flint was burned.

4. **Electron-spin-resonance dating** is used to determine the age of **apatite crystals**, an inorganic component of tooth enamel, according to the presence of trapped electrons. Apatite crystals form as teeth grow, and initially they contain no trapped electrons. These crystals are preserved in fossil teeth and, like the burned flints, are bombarded by a flow of high-energy particles that generate trapped electrons in the crystal lattice. Scientists estimate the number of trapped electrons by subjecting the teeth to a variable magnetic field—a technique called *electron spin resonance*. To estimate the number of years since the tooth was formed, paleontologists must once again measure the flow of radiation at the site where the tooth was found.

5. **Uranium–lead dating** has long been used by geologists to date zirconium crystals found in igneous rocks. It is based on the fact that uranium decay produces a series of unstable elements but eventually yields a stable isotope of lead. By measuring the ratio of uranium to lead, the date of the formation of the crystal can be estimated. It is possible to use this method to date speleothems—stalactites, stalagmites, and flow stone formed by precipitation in limestone caves. This technology is used to date hominin sites in South Africa that lack the volcanic rocks necessary for potassium–argon methods.

Different radiometric techniques are used for different periods. Methods based on isotopes that decay very slowly, such as potassium-40, work well for fossils from the distant past. However, those methods are not useful for more recent fossils because their "clock" doesn't run fast enough. When slow clocks are used to date recent events, large errors can result. For this reason, potassium–argon dating usually cannot be used to date samples less than about 500,000 years old. Conversely, isotopes that decay quickly, such as carbon-14, are useful only for recent periods because all of the unstable isotopes decay in a relatively short time. Thus, carbon-14 can only be used to date sites that are less than about 40,000 years old. The development of thermoluminescence dating and electron-spin-resonance dating is important because these methods allow us to date sites that are too old for carbon-14 dating but too young for potassium–argon dating.

Absolute radiometric dating is supplemented by relative dating methods based on magnetic reversals and comparison with other fossil assemblages.

Radiometric dating methods are problematic for two reasons. First, a particular site may not always contain material that is appropriate for radiometric dating. Second, radiometric methods have relatively large margins for error. These drawbacks have led scientists to supplement such absolute methods with other relative methods for dating fossil sites.

One such relative method is based on the remarkable fact that, every once in a while, Earth's magnetic field reverses itself. This means, for example, that compasses now pointing north would at various times in the past have pointed south (if compasses had been around then, that is). The pattern of magnetic reversals is not the same throughout time, so for any given period the pattern is unique. But the pattern for a given time *is* the same throughout the world. We know what the pattern is because when certain rocks are formed, they record the direction of Earth's magnetic field at that time. Thus, by matching up the pattern of magnetic reversals at a particular site with the well-dated sequence of reversals from the rest of the world, scientists can date sites.

Another approach is to make use of the fact that sometimes fossils of interest are found in association with fossils of other organisms that existed for only a limited period. For example, during the past 20 million years or so there has been a sequence of distinct pig species in East Africa. Because each pig species lived for a known period (according to securely dated sites), some East African materials can be accurately dated from their association with fossilized pig teeth.

The Evolution of the Early Primates

The evolution of flowering plants created a new set of ecological niches. Primates were among the animals that evolved to fill these niches.

During the first two-thirds of the Mesozoic, the forests of the world were dominated by **gymnosperms**, trees like contemporary redwood, pine, and fir that produce seeds, but not flowers or fruit. With the breakup of Pangaea during the Cretaceous, a revolution in the plant world occurred. Flowering plants, called **angiosperms**, appeared and spread. The evolution of the angiosperms created a new set of ecological niches for animals. Many angiosperms depend on animals to pollinate them, and they produce colorful flowers with sugary nectar to attract pollinators. Some angiosperms also entice animals to disperse their seeds by enclosing them in nutritious and easily digestible fruits. Arboreal animals that could find, manipulate, chew, and digest these fruits could exploit these new niches. Primates were one of the taxonomic groups that evolved to take advantage of these opportunities. Tropical birds, bats, insects, and some small rodentlike animals probably competed with early primates for the bounty of the angiosperms.

The ancestors of modern primates were small-bodied nocturnal quadrupeds much like contemporary shrews.

To understand the evolutionary forces that shaped the early radiation of the primates, we need to consider two related questions. First, what kind of animal did natural selection have to work with? Second, what were the selective pressures that favored this suite of traits in ancient primates? Answers to the first question come from the fossil record. Answers to the second question come from comparative studies of living primates.

The **plesiadapiforms**, a group of fossil animals found in what is now Montana, Colorado, New Mexico, and Wyoming, give us some clue about what the immediate ancestors of the earliest primates may have been like. Plesiadapiforms are found at sites that date from the Paleocene epoch (66 Ma to 54 Ma), a time so warm and wet that broadleaf evergreen forests extended to 60°N (near present-day Anchorage, Alaska). The plesiadapiforms varied from tiny, shrew-size creatures to animals as big as marmots. It seems likely that they were solitary quadrupeds with a well-developed sense of smell. The teeth of these animals are quite variable, suggesting that their dietary specializations varied widely. Some members of this group were probably terrestrial, some were arboreal quadrupeds, and others may have been adapted for gliding. Most of the plesiadapiforms had claws on their hands and feet, and they did not have forward-facing eyes or stereoscopic vision.

One of the best-preserved plesiadapiform specimens comes from the Clarks Fork Basin in Wyoming. This specimen, which is dated to about 65 Ma, belongs to the species *Carpolestes simpsoni* (**Figure 9.7**). It had an opposable big toe with a flat nail but claws on its other digits. The claws on its feet and hands probably helped it climb large-diameter tree trunks, but it also could grasp small supports. *C. simpsoni* had low-crowned molars, which are suited for eating fruit. Its eyes were on the sides of the head, and the fields of vision did not overlap. These creatures probably used their hands and feet to grasp small branches as they climbed around in the terminal branches of fruiting trees and used their hands to handle fruit as they were feeding.

FIGURE 9.7

An artist's reconstruction of the plesiadapiform *Carpolestes simpsoni.* This creature, which lived about 56 Ma, had grasping hands and feet.

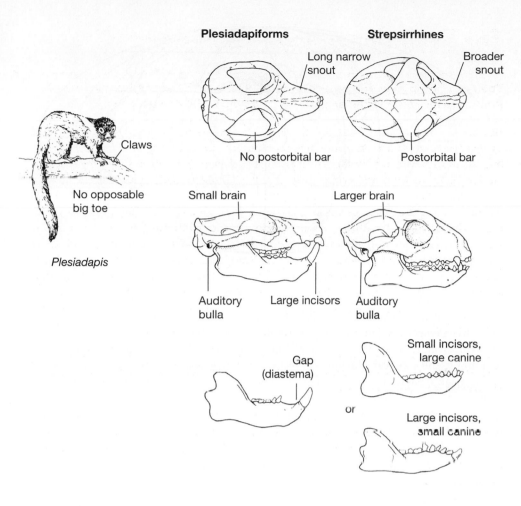

Plesiadapiforms

Long narrow snout

No postorbital bar

Small brain

Auditory bulla — Large incisors

Gap (diastema)

Claws

No opposable big toe

Plesiadapis

Strepsirrhines

Broader snout

Postorbital bar

Larger brain

Auditory bulla

Small incisors, large canine

or

Large incisors, small canine

FIGURE 9.8

Plesiadapiforms were once thought to be primates, but closer analysis indicates they lack many of the defining characteristics of early modern primates. For example, they had claws instead of nails; their eye orbits were not fully encased in bone; their eyes were placed on the sides of the head, so the fields of vision of the two eyes did not converge; and in some species, the big toe was not opposable.

Plesiadapiforms possess some but not all of the suite of traits characterizing modern primates, and many researchers don't think they should be classified as primates (**Figure 9.8**). The plesiadapiforms are important to know about, however, because they provide some information about the traits that characterized the common ancestor of modern primates.

There are several theories about why traits that are diagnostic of primates evolved in early members of the primate order.

In the 1970s, Matt Cartmill, an anthropologist now at Boston University, suggested that forward-facing eyes (orbital convergence) that provide binocular stereoscopic vision, grasping hands and feet, and nails on the toes and fingers all evolved together to enhance visually directed predation on insects in the terminal branches of trees. This idea is supported by the fact that many arboreal predators today, including owls and ocelots, have eyes in the front of the head. However, the discovery that grasping hands and feet evolved in a frugivorous plesiadapiform species before the eyes were shifted forward presents problems for this hypothesis.

Fred Szalay of Hunter College and Marian Dagosto of Northwestern University later suggested that grasping hands and feet and flat nails on the fingers and toes all co-evolved to facilitate a form of leaping locomotion. *C. simpsoni* also poses a problem for this hypothesis because it had grasping hands and feet but evidently didn't leap from branch to branch.

Robert Sussman of Washington University hypothesized that the suite of traits characterizing primates may have been favored because they enhanced the ability of early primates to exploit a new array of plant resources—including fruit, nectar,

FIGURE 9.9

Sites where Eocene prosimian fossils have been found. The continents are arranged as they were during the early Eocene.

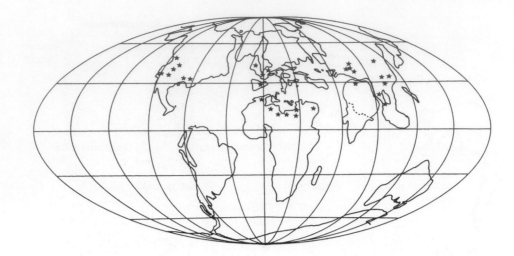

flowers, and gum—as well as insects. The early primates may have foraged and handled small food items in the dimness of the forest night, and this nocturnal behavior may have favored good vision, precise eye–hand coordination, and grasping hands and feet. However, *C. simpsoni* foraged on fruit before orbital convergence evolved.

Finally, the late Tab Rasmussen proposed that grasping hands and feet allowed early primates to forage on fruit, flowers, and nectar in the slender terminal branches of angiosperms. Later, the eyes were shifted forward to facilitate visually directed predation on insects. The idea that the evolution of grasping hands and feet preceded the movement of the eyes to the front of the face fits with the evidence from *C. simpsoni*.

Primates with modern features appeared in the Eocene epoch.

The Eocene epoch was even wetter and warmer than the preceding Paleocene, with great tropical forests covering much of the globe. At the beginning of the Eocene, North America and Europe were connected, but then the two continents separated and shifted farther apart. The animals within these continents evolved in isolation and became progressively more different. There was some contact between Europe and Asia and between India and Asia during this period, but South America was completely isolated. Primate fossils from this period have been found in North America, Europe, Asia, and Africa (**Figure 9.9**). More than 200 species of early primates have now been identified from fossil evidence. The Eocene primates were a highly successful and diverse group, occupying diverse ecological niches.

These Eocene primates exhibit many of the features that define modern primates (see Chapter 5). They had grasping hands and feet with nails instead of claws, hind limb–dominated posture, shorter snouts, eyes moved forward in the head and encased in a bony orbit, and relatively large brains for their body size.

The Eocene primates are classified into two families: Omomyidae and Adapidae (**Figure 9.10**). Although their phylogenetic affinities to modern primates are not known, most researchers compare the omomyids to galagos and tarsiers and the adapids to living lemurs (**Figure 9.11**). Some of the omomyids had large eye orbits, which may have helped them navigate under low light conditions. They had a dental formula of $\frac{2.1.4.3}{2.1.4.3}$, but the characteristics of their dentition were quite variable. Some seem to have been adapted for frugivory, and others for more insectivorous diets. Some omomyids have elongated **calcaneus** (heel) bones in their feet, much like those of modern dwarf lemurs, and they may have been able to leap from branch to branch.

The adapids had smaller eye orbits and were probably diurnal. They resemble living lemurs in many aspects of their teeth, skull, nasal, and auditory regions. However, the adapids do not display some of the unique derived traits that are characteristic of modern lemurs, such as the toothcomb, a specialized formation of

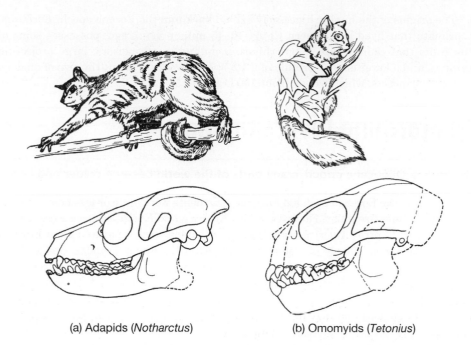

(a) Adapids (*Notharctus*)   (b) Omomyids (*Tetonius*)

FIGURE 9.10

Adapids were larger than omomyids, and they had longer snouts and smaller orbits than the omomyids had. (a) The size of the orbits suggests that the adapids were active during the day, and the shape of their teeth suggests that they fed on fruit or leaves. (b) Omomyids were small primates that fed mainly on insects, fruit, or gum. Their large eye orbits suggest that they may have been nocturnal.

incisors used for grooming. Their dietary adaptations varied, including insectivorous, folivorous, and frugivorous diets. They were generally larger than the omomyids, and their **postcranial** bones (the bones that make up the skeleton below the neck) indicate that some were active arboreal quadrupeds like modern lemurs, whereas others were slow quadrupeds similar to contemporary lorises. At least one species showed substantial sexual dimorphism, a feature that points to life in non-pair-bonded social groups.

One spectacularly well-preserved Eocene adapid specimen, *Darwinius masillae*, was discovered at a fossil site near Frankfurt, Germany, and dated to 47 Ma. As you can see in **Figure 9.12**, the specimen is so well-preserved that you can see the outline of its furry body and the contents of its last meal of fruit and leaves in its stomach cavity. It was about 24 cm (9.4 inches) long (without its tail) and probably weighed between 650 and 900 grams (23-32 ounces). Scientists think this specimen was female because it lacks a **baculum** (penis bone). She was probably still a juvenile and was recovering from a fracture of her right wrist at the time of her death.

FIGURE 9.11

The behavior of adapids and omomyids probably differed. (a) Adapids were probably diurnal. Here a group forages for leaves. (b) Omomyids were probably nocturnal. Several species are shown here.

(a)

(b)

The origins of the haplorrhines may extend back into the Eocene epoch. *Eosimias*, a primate that lived in southern China 40–45 million years ago, possesses some of the traits that characterize happlorhines, including small incisors, large projecting canines, and a lower dental formula of 2.1.3.3. *Eosimias* was about the size of modern pygmy marmosets, which weigh about 100 grams (3.5 oz).

Haplorrhine Diversification

During the Oligocene epoch, many parts of the world became colder and drier.

By the end of the Eocene epoch (34 Ma), the continents were more or less positioned on the globe as they are today. However, South America and North America were not yet connected by Central America, and Africa and Arabia were separated from Eurasia by the Tethys Sea, a body of water that connected the Mediterranean Sea and the Persian Gulf. South America and Australia had completed their separation from Antarctica, creating deep, cold currents around Antarctica (see Figure 9.4). Some climatologists believe that these cold currents reduced the transfer of heat from the equator to Antarctic regions and may have been responsible for the major drop in global temperatures that occurred during the Oligocene epoch (34 Ma to 23 Ma). These climatic changes had a major impact on plant and animal life. Throughout North America and Europe, tropical broadleaf evergreen forests were replaced by broadleaf deciduous forests. Africa and South America remained mainly warm and tropical. Many Eocene plant and animal species became extinct and were replaced by other taxa.

Primates vanished from North America as the Oligocene began.

Although primates were common in North America during the warm wet Eocene, they all disappeared from North America by about 35 Ma. The last fossil primate in North America was a lemur-like primate called *Ekgmowechashala*, which is known from sites in what are now Oregon and the Great Plains. It weighed about 2 kg (4.4 lbs) and probably fed on fruit. *Ekgmowechashala* is dated to about 29-28 Ma, about 6 million years after all the other North American primates disappeared. Some researchers believe that these animals evolved in Asia and migrated across Beringia, a landmass that once connected Siberia and Alaska.

The earliest unambiguous haplorrhine fossils are found at a site in the Fayum (also spelled "Faiyûm") Depression of Egypt.

The Fayum deposits straddle the Eocene–Oligocene boundary, 36 Ma to 33 Ma. The Fayum is now one of the driest places on Earth, but it was very different at the beginning of the Oligocene. Sediments of soil recovered from the Fayum tell us that it was a warm, wet, and somewhat seasonal habitat then. The plants were most like those now found in the tropical forests of Southeast Asia. The soil sediments contain the remnants of the roots of plants that grow in swampy areas, like mangroves, and the sediments suggest that there were periods of standing water at the site. There are also many fossils of waterbirds. All this suggests that the Fayum was a swamp during the Oligocene (**Figure 9.13**). Among the mammalian fauna at the Fayum are representatives of the suborder that includes porcupines and guinea pigs, opossums, insectivores, bats, primitive carnivores, and an archaic member of the hippopotamus family.

FIGURE 9.12

An Eocene primate specimen, *Darwinius masillae*, nicknamed Ida, was discovered in Germany in 2009 and dated to 47 Ma.

FIGURE 9.13

Although the Fayum Depression is now a desert, it was a swampy forest during the Oligocene.

The Fayum contains one of the most diverse primate communities ever documented. This community included at least five groups of strepsirrhines, one group of omomyids, and three groups of haplorrhine primates: the parapithecids, the propliopithecids, and the oligopithecids (**Figure 9.14**). To introduce a theme that will become familiar in the chapters that follow, the more that paleontologists learn about early primates from the Fayum, the more complicated the primate family tree becomes. Instead of a neat tree with a few heavy branches that connect ancient fossils with living species, we have a messy bush with many fine branches and only the most tenuous connections between most living and extinct forms. Nevertheless, we can detect certain trends in the primate fossil record at the Fayum that provide insight about the selective pressures that shaped adaptation within the primate lineage.

Among the earliest Fayum monkeys was a third group of Fayum haplorrhines, the oligopithecids. The oligopithecids, which may have ranged beyond the Fayum through North Africa and the Arabian Peninsula, weighed about 1.5 kg (3.3 lbs). The dental formula of some oligopithecids was the same as that of modern catarrhine monkeys and apes. It is not clear whether the reductions in the number of premolars in oligopithecids and propliopithecids represent independent evolutionary events or common ancestry.

The parapithecids were a very diverse group, currently they are divided into four genera and eight species. The largest parapithecids were the size of guenons (3 kg, or about 6.5 lb.) and the smallest were the size of marmosets (150 g, or about 5 oz.). These creatures have a dental formula of $\frac{0.1.0.0}{2.1.3.3}$, which has been retained in platyrrine monkeys but has been modified in catarrhine monkeys and apes, whose dental formula is $\frac{2.1.2.3}{2.1.2.3}$. (Dental formulas were discussed in Chapter 5.) Many aspects of parapithecid teeth and postcranial anatomy are also primitive, suggesting that they may have been the unspecialized ancestors both of more derived catarrhine monkey lineages and of the platyrrhine monkeys (see **A Closer Look 9.3**).

The propliopithecids are represented by at least three genera. These primates had the same dental formula as modern catarrhine monkeys and apes, but they lack other derived features associated with catarrhine monkeys. The largest and most famous of the propliopithecids is named *Aegyptopithecus zeuxis*, who is known from several skulls and several postcranial bones (**Figure 9.16**). *A. zeuxis* was a medium-size monkey, perhaps as big as a female howler monkey (6 kg, or 13.2 lb.). It was a diurnal, arboreal quadruped with a relatively small brain. The shape and size of the teeth suggest that it ate mainly fruit and leaves. Males were much larger than females, indicating that they probably did not live in pair-bonded groups. Other propliopithecids were smaller than *A. zeuxis*, but their teeth suggest that they also ate fruit, as well as seeds and perhaps gum. They too were probably arboreal quadrupeds with strong, grasping feet. Modern catarrhine monkeys and apes may have been derived from members of this family.

Among the earliest Fayum monkeys was a third group of Fayum haplorrhines, the oligopithecids. The oligopithecids, which may have ranged beyond the Fayum through North Africa and the Arabian Peninsula, weighed about 1.5 kg (3.3 lbs). The dental formula of some oligopithecids was the same as that of modern catarrhine monkeys and apes. It is not clear whether the reductions in the number of premolars in oligopithecids and propliopithecids represent independent evolutionary events or common ancestry.

FIGURE 9.14

The Fayum was home to a diverse group of primates, including propliopithecids such as *Aegyptopithecus zeuxis* (upper left) and *Propliopithecus chirobates* (upper right), as well as parapithecids such as *Apidium phiomense* (bottom).

9.3 Facts That Teeth Can Reveal

Much of what we know about the long-dead early primates comes from their teeth. Fortunately, teeth are more durable than other bones, and they are also the most common elements in the fossil record. If paleontologists had to choose only one part of the skeleton to study, most would choose teeth. There are several reasons for this choice. First, teeth are complex structures with many independent features, which makes them very useful for phylogenetic reconstruction. Second, tooth enamel is not remodeled during an animal's life, and it carries an indelible record of an individual's life history. Third, teeth show a precise developmental sequence that allows paleontologists to make inferences about the growth and development of long-dead organisms. Finally, as we saw in Chapter 5, each of the major dietary specializations (frugivory, folivory, insectivory) is associated with characteristic dental features.

Figure 9.15 shows one side of the upper jaw of three modern species of primates: an insectivore, a folivore, and a frugivore. The insectivorous tarsier (Figure 9.15a) has relatively large, sharp incisors and canines, which are used to bite through the tough external skeletons of insects. In contrast, the folivorous indri (Figure 9.15b) has relatively small incisors and large premolars with sharp crests that allow it to shred tough leaves. Finally, the frugivorous mangabey (Figure 9.15c) has large incisors that are used to peel the rinds from fruit. Its molars are small because the soft, nutritious parts of fruit require less grinding than leaves do.

Knowing what an animal eats enables researchers to make sensible guesses about other characteristics as well. For example, there is a good correlation between diet and body size in living primates: insectivores are generally smaller than frugivores, and frugivores are generally smaller than folivores.

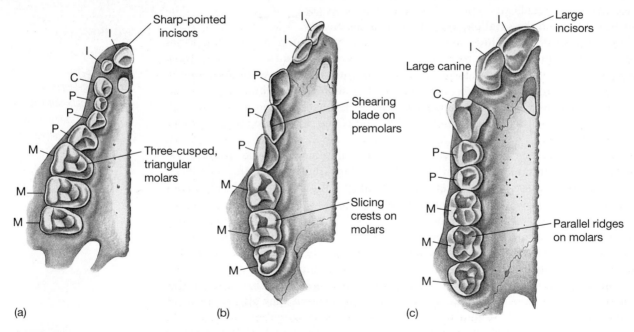

FIGURE 9.15

The right half of the upper jaws of (a) a tarsier, an insectivore; (b) an indri, a folivore; and (c) a mangabey, a frugivore. I = incisor, C = canine, P = premolar, M = molar.

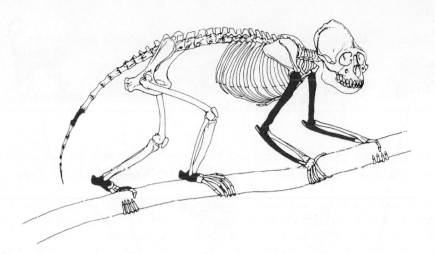

FIGURE 9.16

In this reconstruction of the skeleton of *Aegyptopithecus*, the postcranial bones that have been found are shown in red.

Primates appear in South America for the first time during the Oligocene, but where they came from or how they got there is unclear.

The earliest evidence of primates in South America comes from a few teeth found in late Eocene deposits of Peru and dated to about 36 Ma. These specimens, which are assigned to the genus *Perupithecus*, are very similar to early haplorrhine primates from North Africa. The next oldest South American primate fossils, assigned to the genus *Branisella*, come from a late Oligocene site in Bolivia dated to about 32-26 Ma. *Branisella* has three premolars, like modern platyrrhine monkeys, and were about the size of owl monkeys. The shape of their molars suggests they were frugivores. Sites in Argentina and Chile contain several monkey genera dating to the early and middle Miocene. They were part of a diverse animal population that included rodents, ungulates, sloths, and marsupial mammals. Most of these Patagonian primates were about the size of squirrel monkeys (800 g, or 1.8 lb.), though some may have been as large as sakis (3 kg, or 6.6 lb.). In Colombia, Miocene sites dated to 12 Ma to 10 Ma contain nearly a dozen species of fossil primates. Many of these species closely resemble modern platyrrhine monkeys (**Figure 9.17**). Pleistocene sites in Brazil and the Caribbean islands have yielded a mixture of extinct and extant species. Evidently, several species were considerably larger than any living platyrrhine primates. Although there are no indigenous primates in the Caribbean now, these islands once housed a diverse community of primates.

The origin of platyrrhine primates is a puzzle. The many similarities between the oldest platyrrhine monkeys and the Fayum primates suggest to many scientists that the ancestors of the platyrrhine monkeys came from Africa. But how did they get there? It is possible that monkeys, along with other reptiles that appeared in South America about the same time, rafted across the sea on massive mats of floating vegetation. Although South America and Africa have been separated for more than 100 million years, the distance between the continents has shifted over time. South America and Africa were separated by about 1,000 km (620 miles) 50 Ma and by about 1,500 km (930 miles) 40 Ma. During this period there were a series of large islands in the South Atlantic, which might have reduced migration distances, as well as westerly oceanic currents and winds that could have guided the vegetation mats and their passengers toward the South American coast.

Although there is no evidence of primates in South American until 36 Ma, this may not be a fatal liability for the rafting hypothesis. There is good reason to believe that the date of the earliest fossil we have discovered is likely to underestimate the actual age of a lineage. The method outlined in **A Closer Look 9.4** suggests that haplorrhines actually originated at least 52 Ma. Of course, there are also other possibilities (**Figure 9.18**).

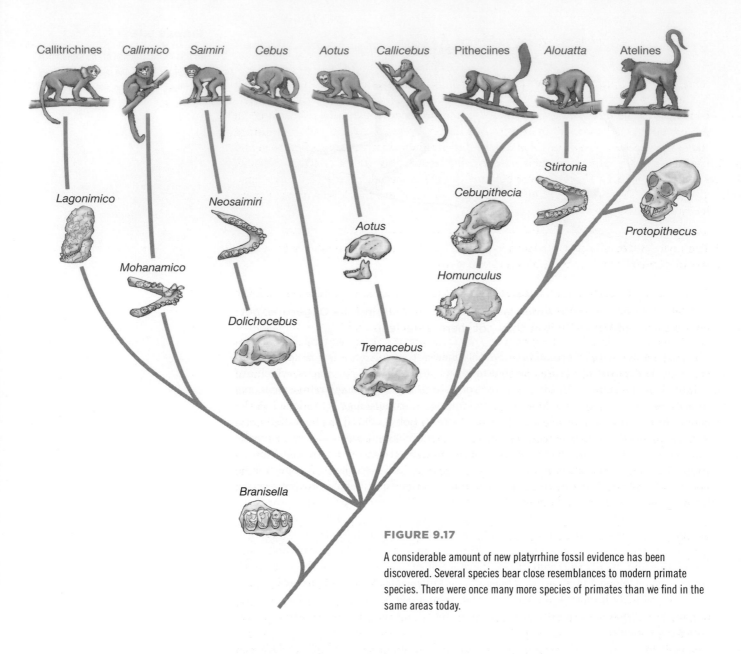

Callitrichines *Callimico* *Saimiri* *Cebus* *Aotus* *Callicebus* **Pitheciines** *Alouatta* **Atelines**

Stirtonia

Lagonimico

Cebupithecia

Neosaimiri

Aotus

Protopithecus

Mohanamico

Homunculus

Dolichocebus

Tremacebus

Branisella

FIGURE 9.17

A considerable amount of new platyrrhine fossil evidence has been discovered. Several species bear close resemblances to modern primate species. There were once many more species of primates than we find in the same areas today.

FIGURE 9.18

Perhaps this is how early monkeys *really* got from Africa to South America.

The Emergence of the Apes

Recall from Chapter 5 that apes are members of the superfamily Hominoidea, which includes the gibbons and siamangs, as well as orangutans, gorillas, chimpanzees, bonobos, and humans. Apes may have first evolved during the late Oligocene in Africa, but they flourished and diversified in the forests of Africa and Asia during the early Miocene. They lost their tails and evolved distinctive forms of posture and locomotion that influenced their skeletal morphology. By the end of the Miocene, as forests shrank and more open, seasonal habitats spread, relatively few ape species remained. Contemporary apes, including humans, are the descendants of these lucky survivors.

The early Miocene was warm and moist, but by the end of the epoch, the world had become much cooler and more arid.

At the end of the Miocene, as the world became considerably colder and more arid, the tropical forests of Eurasia retreated southward, and there was more open woodland habitat. India continued its slow slide into Asia, leading to the uplifting of the Himalayas. Some climatologists believe that the resulting change in atmospheric circulation was responsible for the late Miocene cooling. About 18 Ma, Africa joined Eurasia, splitting the Tethys Sea and creating the Mediterranean Sea. Because the Strait of Gibraltar had not yet opened, the Mediterranean Sea was isolated from the rest of the oceans. At one point, the Mediterranean Sea dried out completely, leaving a desiccated, searing hot valley thousands of feet below sea level. About the same time, the great north–south mountain ranges of the East African Rift began to appear. Because clouds decrease their moisture as they rise in elevation, there is an area of reduced rainfall, called a **rain shadow**, on the lee (downwind) side of mountain ranges. The newly elevated rift mountains caused the tropical forests of East Africa to be replaced by drier woodlands and savannas.

Contemporary apes differ from monkeys in posture and forms of locomotion, and this is reflected in their skeletal anatomy.

Some of the anatomic features that distinguish living apes from monkeys are related to their posture and locomotor behavior. Monkeys move along the tops of branches, using their hands and feet to grip branches and their tails for balance. They leap between gaps in the canopy. They sit on branches while they feed and have fleshy sitting pads to cushion their bottoms. In contrast, apes often hang below branches to feed (**Figure 9.21**), move underneath branches when they travel, and have no tails or fleshy pads on their bottoms. They use their long arms to "bridge" gaps in the canopy instead of leaping. By hanging below branches and using multiple branches to support their body weight, large-bodied apes can navigate the slender terminal branches of trees. Apes have little need of tails for balance, and the tail muscles have been restructured to strengthen the pelvic floor and provide support for the internal organs. This in turn allows apes to sustain upright postures. Suspensory feeding and locomotion have also favored relatively long arms and short legs, long fingers, more mobile limbs, and a short and stiff lumbar spine.

The earliest hominoid species possessed a combination of primitive and derived traits.

Several species of the early hominoids are known from sites in East Africa dated from 22.5 to 13.7 Ma. This group of species was originally categorized as a single genus, *Proconsul*, but some experts believe that they should be divided into two genera. We will refer to them collectively as **proconsulids**.

9.4 Missing Links

Anthropologist Robert Martin of the Field Museum of Natural History in Chicago has pointed out that most lineages are probably older than the oldest fossils we have discovered. The extent of the discrepancy between the dates of the fossils and the actual origin of the lineage depends on the fraction of fossils that have been discovered. If the primate fossil record were nearly complete, then the fact that no haplorrhines living more than 35 Ma have been found would mean that they didn't exist that long ago. However, we have reason to believe that the primate fossil record is quite incomplete (**Figure 9.19**).

Just as not all fossils of a given species are ever found, not all species are known to us either. How many species are missing from our data? Martin's method for answering this question is based on the assumption that the number of species has increased steadily from 66 Ma, when the first primates appeared, to the present. This means there were half as many species 32.5 Ma as there are now, three-fourths as many species

FIGURE 9.19

The sparseness of the fossil record virtually guarantees that the oldest fossil found of a particular species underestimates the age of the species. (a) The population size of a hypothetical primate species is plotted here against time. In this case, we imagine that the species becomes somewhat less populous as it approaches extinction. (b) The number of fossils left by this species is less than the population size. It is unlikely that the earliest and latest fossils date to the earliest and latest living individuals. (c) The number of fossils found by paleontologists is less than the total number of fossils.

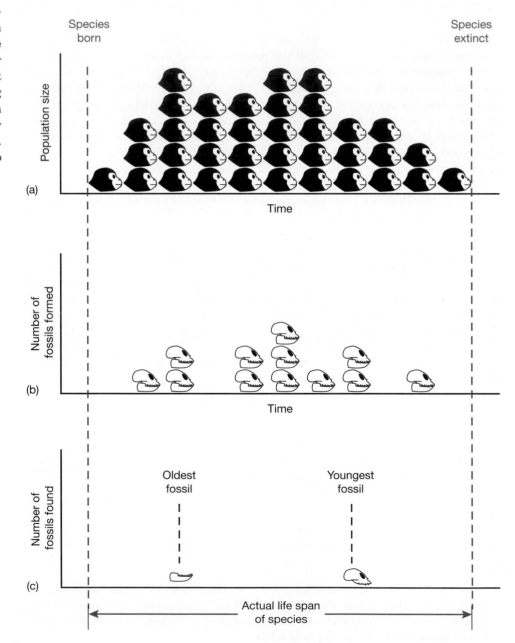

16.25 Ma as there are now, and so on. Assuming that each species has lived about 1 million years, the average life span of a mammalian species, Martin summed these figures to obtain the number of primate species that have ever lived. Then he took the number of fossil species discovered so far and divided that number by his estimate of the total number of species that have ever lived. According to these calculations, only 3% of all fossil primate species have been found so far.

Next, Martin constructed a phylogenetic tree with the same number of living species that we now find in the order Primates. One such tree is shown in **Figure 9.20a**. He then randomly "found" 3% of the fossil species. In **Figure 9.20b** the gray lines give the actual pattern of descent, and the red lines show the data that would be available if we knew the characteristics of all the living species but had recovered only 3% of the fossil species. The best phylogeny possible, given these data, is shown in **Figure 9.20c**.

Then Martin computed the difference between the age of the lineage from this estimated phylogeny and the actual age of the lineage from the original phylogeny. The discrepancy between these values represents the error in the age of the lineage that is due to the incompleteness of the fossil record. By repeating this procedure over and over on the computer, Martin produced an estimate of the average magnitude of error, which turned out to be about 40%. Thus if Martin is correct, living lineages are, on average, about 40% older than the age of the oldest fossil discovered.

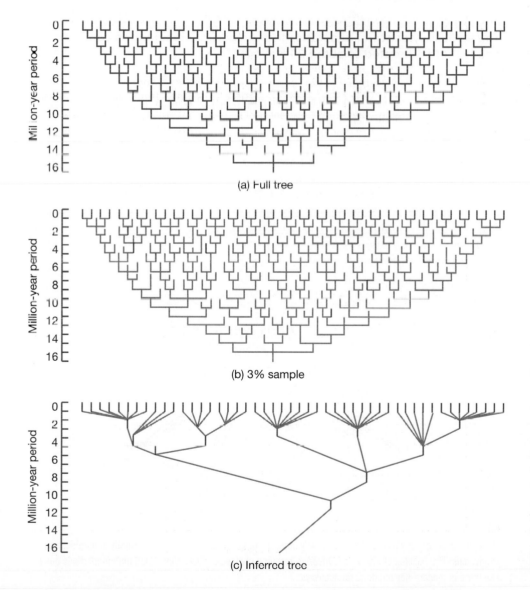

(a) Full tree

(b) 3% sample

(c) Inferred tree

FIGURE 9.20

(a) This tree shows the complete phylogeny of a hypothetical lineage. (b) This tree shows the same phylogeny when only a small percentage of the fossil species has been discovered. The red lines show the species that have been discovered, and the gray lines show the missing species. Note that all *living* species are known, but only 3% of the fossil species have been discovered. (c) The tree we would have to infer from the incomplete data at hand differs from the actual tree in two respects: (1) It links each species to its closest known ancestor, which often means assigning inappropriate ancestors to fossil and modern species, and (2) it often underestimates the age of the oldest member of a clade.

FIGURE 9.21

Apes sometimes hang below branches while they feed.

Proconsulids share several derived features with living apes and humans that we don't see in haplorrhine primates. For example, they didn't have tails and did not have the fleshy sitting pads that catarrhine monkeys and gibbons have. Proconsulids also had somewhat larger brains in relation to body size than similarly sized monkeys. Otherwise, proconsulids were similar to *Aegyptopithecus* and the other Oligocene monkeys. Their teeth had thin enamel, which is consistent with a frugivorous diet. Their postcranial anatomy, including the relative length of their arm and leg bones and narrow and deep shape of their thorax, was much like that of quadrupedal monkeys, but certain features of their feet and lower legs were more apelike. Proconsulids had a large grasping thumb, a feature that we see in humans but not in living apes or monkeys. On the basis of their postcranial anatomy, functional morphologists believe that proconsulids clambered through the trees, using their flexible, grasping limbs to reach branches on all sides and distribute its weight on multiple supports. All of the proconsulids show considerable sexual dimorphism, suggesting that they were not pair-bonded. (**Figure 9.22**). The smallest proconsulids were about the size of capuchin monkeys (3.5 kg, or about 7.5 lb.), and the largest were the size of female gorillas (50 kg, or a little over 100 lb.). The proconsulids seem to have occupied a variety of habitats, including the tropical rain forests in which we find apes today and the open woodlands where we now find only monkeys.

The earliest evidence of adaptations for suspensory locomotion comes from hominoid fossils dated to about 20 Ma.

Fossils that are now assigned to the species *Morotopithecus bishopi* were first collected at the site of Moroto, in Uganda, in the late 1950s and early 1960s. These finds were classified as hominoids, but they were not officially named until the mid-1990s, when Daniel Gebo of Northern Illinois University, Laura MacLatchy of the University of Michigan, and their colleagues resumed work in Moroto and collected additional material (**Figure 9.23**). According to MacLatchy and her colleagues, *M. bishopi* had several skeletal features that allowed it to move like an ape, not like a monkey. For example, several aspects of the femur suggest that *M. bishopi* might have climbed slowly and cautiously, it had a stiff lower back like apes, and the shape of the scapula indicates that it could have hung by its arms and brachiated (that is, swung) slowly through the trees. These features are shared with modern apes but not with the contemporary Miocene apes that we will meet next.

FIGURE 9.23

These fossil bones of *Morotopithecus* include parts of the right and left femurs, vertebrae, shoulder socket, and upper jaw. These remains suggest that these creatures moved like apes, not like monkeys.

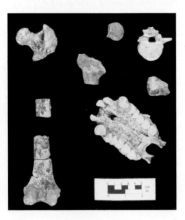

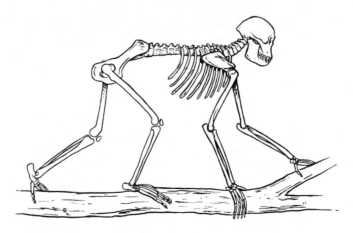

FIGURE 9.22

Members of the genus *Proconsul* were relatively large (15 to 50 kg, or 33 to 110 lb.), sexually dimorphic, and frugivorous. The skeleton of *Proconsul africanus*, reconstructed here, shows that it had limb proportions much like those of modern-day quadrupedal monkeys.

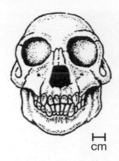

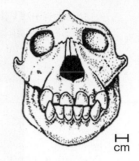

FIGURE 9.24

Miocene hominoids exhibit a diverse cranial morphology, but they share several derived characteristics with modern apes. From left to right: *Turkanapithecus*, *Micropithecus*, *Afropithecus*, and *Proconsul*.

The middle Miocene epoch saw a new radiation of hominoids and the expansion of hominoids throughout much of Eurasia.

Exploration of middle Miocene deposits (15 Ma to 10 Ma) has yielded an abundance of hominoid genera in Africa, Europe, and Asia. Examples include *Kenyapithecus* and *Nacholapithecus* from what is now East Africa; *Lufengpithecus* and *Sivapithecus* from Asia; and *Oreopithecus*, *Dryopithecus*, *Pierolapithecus*, *Danuvius*, and *Anoiapithecus* (aptly named, you might think) from Europe. The skulls and teeth of these hominoids typically differ from those of the proconsulids in several ways that indicate they ate harder or more fibrous foods than their predecessors did. Their molars had thick enamel for longer wear and rounded cusps, which are better suited to grinding. Their **zygomatic arches** (cheekbones) flared farther outward to make room for larger jaw muscles, and the lower jaw was more robust to carry the forces produced by those muscles. These features were probably a response to the climatic shift from a moist, tropical environment to a drier, more seasonal environment with tougher vegetation and harder seeds.

There was considerable diversity in locomotor and postural adaptations of the middle Miocene apes. The traits that distinguish contemporary apes from monkeys were combined in different ways in different Miocene ape lineages. For example, *Nacholapithecus* was adapted for more extensive forelimb dominated climbing and locomotion than were earlier hominoids but did not possess the full range of adaptations for below-branch feeding and travel seen in modern apes. It had long arms in relation to its legs as modern apes do, but the torso had not been restructured, and the shoulder was more monkeylike than apelike. *Danuvius* had relatively long forearms, flexible elbows, grasping thumbs, and curved fingers, which are associated with suspensory locomotion, but characteristics of the tibia suggest it may have also walked upright along tree branches.

Sivapithecus, known from sites in Asia, is thought to be closely related to modern orangutans on the basis of the morphology of the skull and facial structure. *Sivapithecus* had some traits that we find in modern apes that practice suspensory locomotion, including long fingers and toes, a strong big toe, and a flexible elbow. However, the orientation of the humerus, the upper arm bone, suggests that the shape of the thorax was more monkeylike. If this interpretation is correct, then some of the features that the modern great apes share may be the product of convergence rather than common descent. *Sivapithecus* had a large brain and a pattern of tooth development reflecting a long juvenile period, suggesting a prolonged life history strategy like that of modern great apes.

Oreopithecus, from Italy, provides another variation on suspensory adaptations (**Figure 9.25**). Its arms were considerably longer than its legs; it had a somewhat shortened lumbar spine and flexible hips. Although it seems to have been

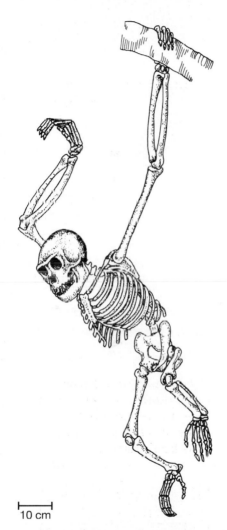

10 cm

FIGURE 9.25

Oreopithecus had several traits that are associated with suspensory locomotion, including a relatively short trunk, long arms, short legs, long and slender fingers, and great mobility in all joints. The phylogenetic affinities of this late Miocene ape from Italy are not well established.

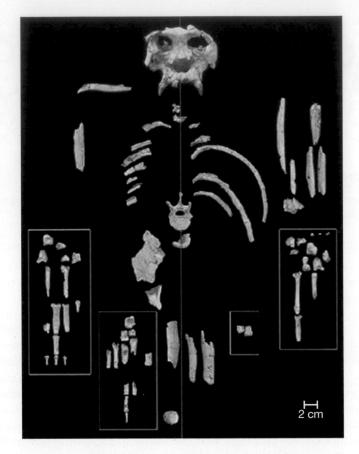

2 cm

FIGURE 9.26

This remarkably complete skeleton of the Miocene ape *Pierolapithecus catalaunicus* does not show morphological features associated with suspensory locomotion. This finding suggests that suspensory locomotion evolved independently in gibbons and the great apes after they diverged from a common ancestor.

adapted for below-branch movement, it does not have long, curved fingers as modern apes do.

The mosaic nature of ape adaptations is also illustrated by the middle Miocene ape *Pierolapithecus catalaunicus* (**Figure 9.26**). This specimen is important because it includes well-preserved cranial, dental, and postcranial material from a single individual that lived about 13 Ma. *Pierolapithecus* had a small, apelike face as well as several morphological features associated with upright posture and locomotion. The wrist is flexible, the rib cage is wide and shallow, and the lumbar region of the spine is somewhat short and stiff. However, these features are not as developed as they are in modern great apes, and the finger bones are not as long and curved as the fingers of orangutans. Again, this suggests that some features of modern ape morphology may have evolved independently after apes diverged from a common ancestor.

Climatic changes in the late middle Miocene reduced hominoid diversity in Asia and Europe.

During the middle Miocene there was a gradual cooling, and the climate became more seasonal in western and central Europe. Subtropical evergreen forests gave way to deciduous broadleaf woodlands. These climatic changes dramatically affected the mammalian fauna; many of the ape lineages that flourished in Africa, Asia, and Europe became extinct. A few species, including the ancestors of orangutans, survived in the remaining areas of evergreen forests. Others evolved adaptations that enabled them to survive in drier, more open, and more seasonal habitats.

Apes of the late Miocene are not well known, particularly in Africa. Recent finds fill in some of the gaps.

Crucial events in the history of the human lineage occurred during the late Miocene, 10 Ma to 5 Ma. Genetic data tell us that the last common ancestor of humans, gorillas, and chimpanzees lived between 9 Ma and 8 Ma and the last common ancestor of humans and chimpanzees lived about 8 Ma to 5 Ma. As we will see in Chapter 10, the oldest hominin fossils come from Africa, so the last common ancestor of humans and chimpanzees probably lived in Africa as well. Until recently, however, this logic did not fit the evidence because late Miocene apes were known only from sites in Europe. The presence of several species of late Miocene apes in Europe and Asia has led some researchers to suggest that the last common ancestor of the African apes diverged there and then migrated back to Africa. However, several discoveries of late Miocene apes in Africa have shifted the geographic focus of human origins back to Africa.

Chororapithecus abyssinicus was discovered at a site in Ethiopia dated 10.5 Ma to 10 Ma by a research team led by Gen Suwa of the University of Tokyo. The finds consist of teeth from at least three individuals. The molars have shearing crests for shredding foliage and thick enamel for feeding on hard and abrasive foods. Suwa and his colleagues have emphasized the similarity between the teeth of *Chororapithecus* and gorillas and suggest that *Chororapithecus* may be ancestral to modern gorillas. However, other researchers have pointed out that this thought is not consistent with divergence dates derived from genetic data and suggest that the dental similarities might not be evidence of shared ancestry.

Another large-toothed ape, *Nakalipithecus nakayamai*, dated 9.9 Ma to 9.8 Ma, comes from a site along the eastern edge of the Rift Valley in Kenya. From the size of its teeth and mandible, *Nakalipithecus* is thought to have been the size of a female gorilla. It had thick enamel on its molars, which suggests that its diet included some hard foods. *Nakalipithecus* bears several similarities to *Ouranopithecus*, a slightly more recent late Miocene ape known from sites in Greece and Turkey, but is more primitive. The third late Miocene ape from East Africa, *Samburupithecus kiptalami*, is known from a partial maxilla, premolars, and molars. This ape was found at a site dated 9.6 Ma and is thus slightly younger than *Nakalipithecus*. Its teeth retain more primitive traits than the teeth of any of the modern great apes.

There are no clear candidates for the ancestors of humans or any modern apes except perhaps orangutans.

The evolutionary history of the apes of the Miocene is still poorly understood. There were many species, and the phylogenetic relationships among them remain unclear. We have no strong candidates for the ancestors of any modern apes except for the orangutan, which shares several derived skull features with *Sivapithecus* of the middle Miocene. The teeth of *Chororapithecus* and *Nakalipithecus* bear similarities to those of modern gorillas, but it is not clear whether these similarities reflect common ancestry or convergence.

Carol Ward of the University of Missouri emphasizes that the suite of suspensory adaptations generally characterizing the hominoids developed piecemeal within lineages over time. There may have been multiple instances of convergence as various large-bodied apes evolved similar, but not identical, solutions to the challenges of terminal branch feeding and movement through the canopy. The last common ancestor of the great apes and humans may have been a fairly generalized ape, which lacked many of the highly specialized locomotor and postural adaptations that have evolved in extant apes. As we will see in Chapter 10, this view fits with new discoveries about what may be the oldest members of the human lineage.

During the early and middle Miocene, ape species were plentiful and monkey species were not. In the late Miocene and early Pliocene, many ape species became extinct and were replaced by monkeys.

Although apes flourished during the Miocene, all but a few genera and species eventually became extinct. Today there are only gibbons, orangutans, gorillas, and chimpanzees. We don't know why so many ape species disappeared, but many of them were probably poorly suited to the drier conditions of the late Miocene and early Pliocene. The fossil record of catarrhine monkeys is quite different from that of the apes. Monkeys were relatively rare and not particularly variable in the early and middle Miocene, but the number and variety of fossil monkeys increased in the late Miocene and early Pliocene.

Once again, the fossil record reminds us that evolution does not proceed on a steady and relentless path toward a particular goal. *Evolution* and *progress* are not synonymous. During the Miocene, there were dozens of ape species but relatively few species of catarrhine monkeys. Today, there are many monkey species and only a handful of ape species. Despite our tendency to think of ourselves as the pinnacle of evolution, the evidence suggests that, taken as a whole, our lineage was poorly suited to the changing conditions of the Pliocene and Pleistocene.

CHAPTER REVIEW

Key Terms

bipedal (p. 211)
continental drift (p. 212)
Pangaea (p. 212)
Laurasia (p. 212)
Gondwanaland (p. 212)
mineralized (p. 213)
fossils (p. 213)
paleontologists (p. 213)

radiometric methods (p. 214)
isotopes (p. 214)
radioactive decay (p. 214)
potassium–argon dating (p. 214)
stratum (p. 214)
argon–argon dating (p. 214)

carbon-14 dating, or radiocarbon dating (p. 214)
thermoluminescence dating (p. 216)
electron-spin-resonance dating (p. 217)
apatite crystals (p. 217)
uranium–lead dating (p. 217)
gymnosperms (p. 218)

angiosperms (p. 218)
plesiadapiforms (p. 218)
calcaneus (p. 220)
postcranial (p. 221)
baculum (p. 221)
rain shadow (p. 227)
proconsulids (p. 227)
zygomatic arches (p. 231)

Study Questions

1. Briefly describe the motions of the continents over the past 180 million years. Why are these movements important to the study of human evolution?

2. What has happened to the world's climate since the end of the Cretaceous period about 65 ma? Explain the relationship between climate change and the notion that evolution leads to steady progress.

3. What are angiosperms? What do they have to do with the evolution of the primates?

4. Why are teeth so important for reconstructing the evolution of past animals? Explain how to use teeth to distinguish among insectivores, folivores, and frugivores.

5. Which primate groups first appear during the Eocene? Give two explanations for the selective forces that shaped the morphologies of these groups.

6. Why is there a problem in explaining how primates arrived in the Americas?

7. Why does the oldest fossil in a particular lineage underestimate the true age of the lineage? Explain how this problem is affected by the quality and completeness of the fossil record.

8. Explain how potassium–argon dating works. Why can it be used to date only volcanic rocks older than about 500,000 years?

9. Some evidence suggests that the last common ancestor of humans and great apes evolved in Europe, whereas other evidence suggests the last common ancestor was an African species. Describe the logic and evidence underlying these two positions. What kind of evidence would help resolve this issue?

10. How does the evolutionary history of apes provide an example of the fact that evolution isn't always synonymous with progress?

Further Reading

Begun, D. R. 2007. "The Fossil Record of Miocene Apes." In W. Henke and I. Tattersall, eds., *Handbook of Paleontology*, Vol. 2, pp. 922–976. Berlin: Springer-Verlag.

Ciochon, R. and J. Fleagle. 2017. *Primate Evolution and Human Origins*. Abington, UK: Routledge.

Fleagle, J. G. 2013. *Primate Adaptation and Evolution*. 3rd ed. San Diego, CA: Academic Press.

Klein, R. G. 2009. *The Human Career: Human Biological and Cultural Origins*. 3rd ed. Chicago. University of Chicago Press.

MacLatchy, L. 2004. "The Oldest Ape." *Evolutionary Anthropology* 13: 90–103.

Ward, C. V. 2007. "The Locomotor and Postcranial Adaptations of Hominoids." In W. Henke and I. Tattersall, eds., *Handbook of Paleontology*, Vol. 2, pp. 1011–1030. Berlin: Springer-Verlag.

 Review this chapter with personalized, interactive questions through INQUIZITIVE.

10

THE EARLIEST HOMININS

CHAPTER OBJECTIVES

By the end of this chapter you should be able to:

A Describe why the earliest members of the human lineage were basically bipedal apes.

B Assess how the evolution of bipedal locomotion altered the postcranial skeleton in many important ways.

C Discuss why natural selection may have favored bipedal locomotion in early hominins.

D Summarize the key attributes of the hominin species that lived in Africa 5 Ma to 2 Ma.

E Understand why efforts to construct phylogenies of early hominins are unproductive.

During the Miocene, Earth's temperature began to fall. This global cooling caused two important changes in the climate of the African tropics. First, the total amount of rain that fell each year declined. Second, rainfall became more seasonal, so in many areas there were several months each year without rain. As a result, Africa became drier, moist tropical forests shrank, and woodlands and grasslands expanded. Like other animals, primates were affected by these ecological changes. Some species, including many of the Miocene apes, failed to adapt and became extinct. The ancestors of chimpanzees and gorillas remained in the shrinking forests and carried on their lives much as before. Changes brought about through generations of natural selection allowed a few species to move down from the trees, out of the rain forests, and into the woodlands and savannas. Our ancestors, the earliest **hominins**, were among these pioneering species.

The first hominins appear in the fossil record about 6 Ma. Between 4 Ma and 2 Ma, a diverse community of hominin species ranged through eastern and southern Africa.

These creatures were different from any of the Miocene apes in two ways. First, and most important, they walked upright. This shift to bipedal locomotion led to major morphological changes in their bodies. Second, some of the hominin species began to exploit new savanna and woodland habitats, and new kinds of food became available. As a result, the hominin chewing apparatus—including many features of the teeth, jaws, and skull—also changed. Otherwise, the behavior and life history of the earliest hominins were probably not much different from those of modern apes.

Several shared, derived characteristics distinguish modern humans from other living hominoids: bipedal locomotion, a larger brain, slower development, several features of dental morphology, and cultural adaptation.

To appreciate the evolutionary transitions that occurred in the human lineage, it is useful to think about how modern humans differ from other apes. Five categories of derived traits distinguish modern humans from contemporary apes:

1. We habitually walk bipedally.

2. Our dentition and jaw musculature are different from those of apes in several ways. For example, we have a wide parabolic dental arcade, thick molar enamel, reduced canine teeth, and larger molars in relation to the other teeth.

3. We have much larger brains in relation to our body size.

4. We develop slowly, with a long juvenile period.

5. We depend on an elaborate, highly variable material and symbolic culture, transmitted in part through spoken language.

In this chapter, we describe the species that constituted this early hominin community, and we discuss the selective forces that transformed an ancestral, arboreal ape into a diverse community of bipedal apes living in the forests, woodlands, and savannas of Pliocene Africa. In later chapters, we consider how one of these savanna apes became human.

Many of the morphological features that we focus on in this chapter may seem obscure, and you may wonder why we spend so much time describing them. These features help us identify hominin species and allow us to trace the origins of traits that we see in later species. We can also use some of these characteristics to reconstruct aspects of diet, social organization, and behavior.

At the Beginning

Genetic data indicate that the last common ancestor of humans and chimpanzees lived about 8 Ma. Since the turn of the millennium, fossil discoveries of several kinds of creatures have begun to shed light on this important period in the history of the human lineage. During this period, we see the first hints of some of the distinctive features that differentiate hominins from apes—evidence of bipedal locomotion (see **A Closer Look 10.1**, pp. 240–241), large posterior teeth, and canine reduction.

Sahelanthropus tchadensis

Sahelanthropus tchadensis, the earliest known hominin, has a mix of derived and primitive features.

In 2002, a team of researchers led by Michel Brunet of the University of Poitiers in France announced a find from a site in Chad. The fossil material consists of a nearly complete cranium (the skull minus the lower jawbone), four partial mandibles, and four teeth (**Figure 10.1**). Brunet and his colleagues named the fossil *Sahelanthropus tchadensis*. (The word *Sahelanthropus* comes from "Sahel," the vast dry region south of the Sahara desert, and *tchadensis* comes from "Chad," the country in which the fossil was found.) Although the site at which *S. tchadensis* was found is now a barren desert, millions of years ago lakes and rivers sustained forests and wooded grasslands in the region. The geology of the site does not allow radiometric or paleomagnetic dating. However, there is a close match between the other fossil animals found at the site and the fauna found at two sites in East Africa that are securely dated to between 7 Ma and 6 Ma.

Sahelanthropus possesses a mix of ancestral and derived anatomic features. For example, the **foramen magnum**, the hole in the skull through which the spinal cord passes, is located in the back of the skull in most quadrupedal primates. However, in *Sahelanthropus* the foramen magnum is located under the skull, as it is in modern humans, and this made Brunet and his colleagues think that these creatures walked upright. *Sahelanthropus* teeth are also different from the teeth of chimpanzees in several ways: The canines are smaller, the upper canine is not sharpened against the lower premolar as it is in chimpanzees, and the enamel is thicker. On the other hand, their brains were no bigger than the brains of chimpanzees. *Sahelanthropus tchadensis*'s brain measured about 370 cc (cubic centimeters); for comparison, the average size of chimpanzee brains is about 360 cc. Modern human brains average about 1,400 cc. The face is relatively flat, and there is a massive browridge over the eyes. These features are not characteristic of either chimpanzees or the australopiths, hominins that dominate the fossil record between 4 Ma and 2 Ma.

Orrorin tugenensis

Orrorin tugenensis is a second early fossil with similarities to humans.

In 2001, a team led by Brigitte Senut of the National Museum of Natural History in Paris and Martin Pickford of the Collège de France discovered 12 hominin fossil specimens in the highlands of Kenya. These fossils, which include parts of thigh and arm bones, a finger bone, two partial mandibles (lower jaws), and several teeth, are securely dated to 6 Ma (**Figure 10.2**). Senut and Pickford assigned their finds to a new

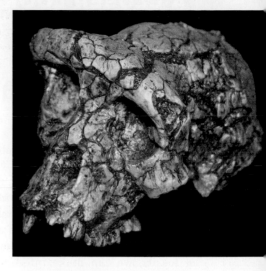

FIGURE 10.1

The cranium of *Sahelanthropus tchadensis* found in Chad dates to between 7 Ma and 6 Ma. It has a flat face and large browridge. Australopiths dated to between 4 Ma and 2 Ma have more apelike prognathic faces and lack browridges.

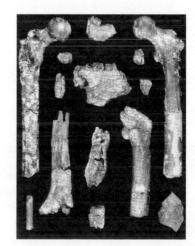

FIGURE 10.2

The fossils of *Orrorin tugenensis* include parts of the femur, an arm bone, lower jaw, a finger bone, and teeth.

10.1 What It Takes to Be a Biped

Bipedal locomotion distinguishes hominins from hominoids. The transition from a forest ape to a terrestrial biped involves many new adaptations. Several of these changes are reflected in the morphology of the skeleton. Thus, by studying the shape of fossils, we can make inferences about the animal's mode of locomotion. Changes in the pelvis provide a good example. In terms of shape and orientation, the human pelvis is very different from that of forest-dwelling apes such as the chimpanzee (**Figure 10.3**).

Other changes are more subtle but also diagnostic. When modern humans walk, a relatively large proportion of the time is spent balanced on one foot. Each time you take a step, your body swings over the foot that is on the ground, and all of your weight is balanced over that foot. At that moment, the weight of your body pulls down on the center of the pelvis, well inward from the hip joint (**Figure 10.4**). This weight creates a twisting force, or **torque**, that acts to rotate your torso down and away from the weighted leg. But your torso does not tip because the torque is opposed by **abductors**, muscles that run from the outer side of the pelvis to the femur. At the appropriate moment during each stride, these muscles tighten and keep you upright. (You can demonstrate this by walking around with your open hand on the side of your hip. You'll feel your abductors tighten as you walk and your torso tip if you relax these muscles. You might want to do this in private.) The abductors are attached to the **ilium** (plural, *ilia*), a flaring blade of bone on the upper end of the pelvis. The widening and thickening of the ilium and the lengthening of the neck of the **femur** (the thighbone) add to the leverage that the abductors can exert and make bipedal walking more efficient. In addition, the distribution of cortical bone in the femur is diagnostic of locomotor patterns. In humans the cortical bone is thickest along the lower edge of the femoral neck, whereas in chimpanzees the cortical bone is evenly distributed on the upper and lower edges.

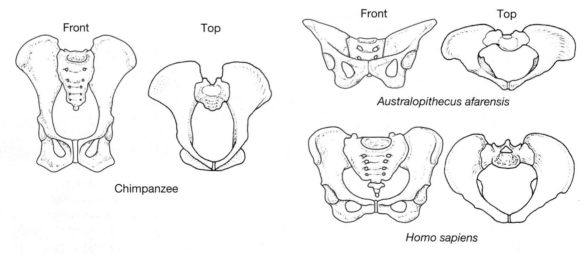

FIGURE 10.3

The pelvis of *Australopithecus afarensis,* an early hominin species, resembles the modern human pelvis more than it resembles the chimpanzee pelvis. Notice that the australopithecine pelvis is flattened and flared like that of the modern human. These features increase the efficiency of bipedal walking. However, the australopithecine pelvis is much wider from side to side and narrower from front to back than that of modern humans. Some anthropologists believe these differences indicate that australopithecines did not walk the same way that modern humans do.

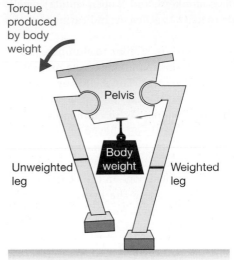

Torque produced by body weight

Pelvis

Unweighted leg

Body weight

Weighted leg

(a) Without abductors

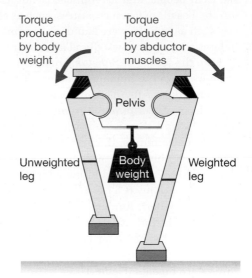

Torque produced by body weight

Torque produced by abductor muscles

Pelvis

Unweighted leg

Body weight

Weighted leg

(b) With abductors

FIGURE 10.4

The lower body at the point of the stride when all the weight is on one leg. Note that the body weight pulls down through the centerline of the pelvis, creating a torque, or twisting force, around the hip joint of the weighted leg. (a) If this torque were unopposed, the torso would twist down and to the left. (b) During each stride, the abductor muscles tighten to create a second torque that keeps the body erect.

The modern human knee joint is also quite different from the chimpanzee knee joint (**Figure 10.5**). Efficient bipedal locomotion requires the knees to lie close to the centerline of the body. As a result, the human femur slants down and inward, and its lower end is angled at the knee joint to make proper contact with the bones of the lower leg. In contrast, the chimpanzee femur descends vertically from the pelvis, and the end of the femur at the knee joint is not slanted. The feet also show several derived features associated with bipedal locomotion, including a longitudinal arch and a human-like ankle.

The presence or absence of these features in fossil skeletal material allows paleontologists to make strong inferences about the mode of locomotion that the animals used.

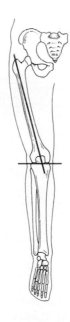

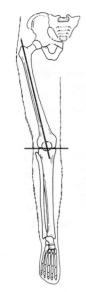

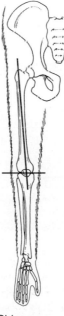

Human

Australopithecus afarensis

Chimpanzee

FIGURE 10.5

The knees of *Australopithecus afarensis* are more like the knees of modern humans than like the knees of chimpanzees. Consider the lower end of the femur, where it forms one side of the knee joint. In chimpanzees, this joint forms a right angle with the long axis of the femur. In humans and australopiths, the knee joint forms an oblique angle, causing the femur to slant inward toward the centerline of the body. This slant causes the knee to be carried closer to the body's centerline, which increases the efficiency of bipedal walking.

species, *Orrorin tugenensis*. The genus name means "original man" in the language of the local people, and it reminded Senut and Pickford of the French word for "dawn," *aurore*. Fossils of forest creatures, such as colobus monkeys, and of open-country dwellers, such as impala, were found in the same strata as *O. tugenensis*, indicating that the habitat was a mix of woodland and savanna.

Like the fossils of *Sahelanthropus*, these specimens are similar to chimpanzees in some ways and to humans in others. The incisors, canines, and one of the premolars are more like the teeth of chimpanzees than of later hominins. The molars are smaller than those of *Ardipithecus ramidus* (which you will meet next) and of later apelike hominins, and they have thick enamel like human molars. As in chimpanzees and some later hominins, the arm and finger bones have features believed to be adaptations for climbing. The morphology of the femur (thighbone) is intermediate between those of Miocene apes and later bipedal hominins. Senut and colleagues used computed tomography (CT) to assess the distribution of cortical bone in the *O. tugenensis* femur. In modern humans, the cortical bone is thicker along the lower edge of the femoral neck than it is along the upper edge, whereas in chimpanzees the cortical thickness is the same for both. Senut's team argues that the distribution of cortical bone in *O. tugenensis* is more humanlike than apelike, which suggests that these creatures were bipedal (**Figure 10.6**).

Ardipithecus

The genus *Ardipithecus* includes two species, *Ar. kadabba* and *Ar. ramidus*, both from the Middle Awash region of Ethiopia.

Some of the most important evidence of human origins comes from the northern end of the Rift Valley, in the Middle Awash region of Ethiopia. Although this area is now dry and desolate, it was once the site of woods and grasslands. As we will see, this is where several spectacular discoveries of hominin ancestors have been made.

About 5.8 Ma to 5.2 Ma, the Middle Awash region was occupied by an apelike hominin named *Ardipithecus kadabba*. Fossils from this species were first discovered in 1997 by Yohannes Haile-Selassie, now with the Cleveland Museum of Natural

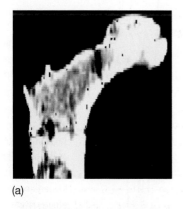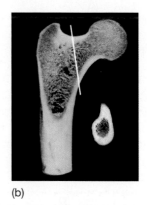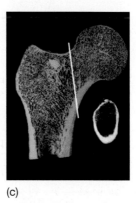

(a)   (b)   (c)

FIGURE 10.6

The morphology of the femur (thighbone) of *Orrorin tugenensis* is intermediate between those of apes and bipedal hominins. This figure compares a CT scan of part of the *O. tugenensis* femur (a) with X-ray images of the femurs of a chimpanzee (b) and a human (c). In modern humans, the cortical bone is thicker along the lower edge of the femoral neck than it is along the upper edge, whereas in chimpanzees the cortical thickness is similar in both areas. Senut's team argues that the distribution of cortical bone in *O. tugenensis* is more humanlike than apelike, which suggests that these creatures were bipedal. (Parts b and c have the orientation reversed from originally published versions for comparative purposes.)

History. The finds, which consist mainly of teeth and a single foot bone, were made at a site called Aramis, in the Middle Awash basin. Like *Sahelanthropus, Ardipithecus kadabba* possesses a mixture of primitive and derived dental traits. For example, the canine sharpens itself against the first premolar, a trait found in chimpanzees but not in modern humans. On the other hand, *Ar. kadabba* has thicker enamel on its molars than chimpanzees do, and the shape of its canines is like that of hominins that succeeded it in the area (**Figure 10.7**). The toe bone is similar to the toe bones of other bipedal hominins and may be diagnostic of bipedal locomotion.

Another member of the same genus, *Ar. ramidus*, appears in the fossil record about a million years after *Ar. kadabba*.

Fossils later assigned to *Ardipithecus ramidus* were originally discovered in 1992 by the members of an expedition led by Tim White of the University of California, Berkeley. During their first field season, the team found parts of the teeth and jaws, the lower part of the skull, and parts of the upper arms. The fossils were dated to 4.4 Ma. White and his colleagues Gen Suwa of the University of Tokyo and Berhane Asfaw of the Rift Valley Research Service in Addis Ababa named the species *Ardipithicus ramidus*, from the words *ardi* (meaning "ground" or "floor") and *ramis* (meaning "root") in the local Afar language. In hopes of finding additional material, the team returned to Aramis and combed the site for more fossils. Their efforts produced a wealth of skeletal material. Altogether, White's team collected more than 150,000 fossils of animals and plants, including 110 fossils from *Ardipithecus*.

The announcement of these finds created great anticipation in the paleontological community. No other hominins from this period were known, and there were tantalizing hints that *Ar. ramidus* might substantially change our picture of human origins. However, the fossils were in very poor condition—the bones crumbled when they were touched, and many had been trampled, crushed, and broken into many small pieces. It took painstaking care to excavate each specimen and reconstruct the bones. The process was finally completed, and a comprehensive description of *Ar. ramidus* was published in 2009.

Additional specimens of *Ar. ramidus* come from a second site near Aramis called Gona. At Gona, a team led by Sileshi Semaw of Consorcio CENIEH in Burgos, Spain, found fossils representing at least nine individuals.

ARA-VP-6/500, more commonly known as Ardi, represents most of the skeleton of a single individual.

White and his colleagues pieced together a nearly complete skeleton of one individual, which was assigned the accession number ARA-VP-6/500 (**Figure 10.8**). Paleontologists assign such numbers to help them keep track of the material in their collections. But because it's easier to remember names than accession numbers, particularly important fossils sometimes get nicknames. ARA-VP-6/500 is informally known as "Ardi."

Ardi weighed about 51 kg (112 lb.), stood 1.2 m (3.9 ft.) tall, and was probably female. This makes her somewhat larger than a wild chimpanzee male and smaller than a female gorilla. Her limb proportions were similar to those of quadrupedal catarrhine monkeys and the Miocene ape *Proconsul*, and she did not have the elongated arms and relatively short legs that we see in modern apes that are specialized for suspensory locomotion and below-branch feeding.

The thousands of fossils of plants and animals from Aramis provide a detailed picture of Ardi's habitat. Aramis was much wetter 4.4 Ma than it is today. Ardi lived in a woodland habitat dotted with patches of denser forest. Monkeys, including colobus monkeys and a small baboonlike monkey, were abundant. Kudu, large ungulates that now favor wooded areas, were also common.

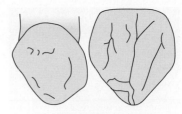

Australopithecus afarensis

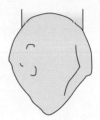

Ardipithecus ramidus

Ardipithecus kadabba

Chimpanzee   *Ar. kadabba*

FIGURE 10.7

The drawings show the canines of *Au. afarensis, Ar. ramidus*, and *Ar. kadabba*. The photograph shows chimpanzee canines on the left and *Ar. kadabba* canines on the right. In *Ar. kadabba* and chimpanzees, the canines are sharpened against the first premolar. This is not the case with later australopithecines (top). However, the shape of the canine of *Ar. kadabba* is similar to that of the canines of *Ar. ramidus* and *Au. afarensis*, which follow *Ar. kadabba*.

FIGURE 10.8

The *Ardipithecus ramidus* (ARA-VP-6/500) skeleton known as Ardi.

Ar. ramidus resembles *Sahelanthropus* and *Orrorin* in many features of its skull, face, and dentition.

Ardi had an ape-sized brain, with a cranial capacity of 300 to 350 cc. The upper part of her face was flatter than chimpanzee faces, whereas the middle part of her face was pushed out, much like the faces of chimpanzees. The skull was perched on top of the spine, indicating an upright posture. In all of these features, *Ar. ramidus* resembles *Sahelanthropus* and what is known of *Orrorin*.

Ar. ramidus has a distinctive suite of dental traits: thicker molar enamel, general reduction in the size and extent of sexual dimorphism in the canines, and no honing by the premolars.

The dental material includes 145 teeth, a wealth of material by paleontological standards. These teeth were analyzed by a team led by Suwa and provide important clues about the diet and social organization of *Ar. ramidus*. The dentition also reveals the first evidence of several distinctive features shared by hominin species, which we will meet later in this chapter.

The dental arcade, the shape made by the rows of teeth in the upper jaw, is U-shaped, as it is in chimpanzees. Overall, the teeth were similar in size to the teeth of chimpanzees, but *Ar. ramidus* had smaller incisors than those of chimpanzees. Chimpanzees use their incisors to process fruit before they chew it, so this suggests that *Ar. ramidus* was less frugivorous than modern chimpanzees.

The *Ar. ramidus* material includes 23 upper and lower canines from 21 individuals. The canines are about the size of the canines of female chimpanzees, but are not honed by the lower premolar as they are in other primates with large canines. As we discussed in Chapter 6, canines of a primate species provide important information about its social organization. In pair-bonded species, there is little sexual dimorphism in canine size, but in species that are not pair-bonded, male canines are generally much larger than female canines. Male chimpanzee canines are 19% to 47% larger than female canines, whereas in modern humans, males' canines are 4% to 9% larger than those of females (**Figure 10.9**).

Suwa and his colleagues cannot tell which teeth come from males and which teeth come from females, so they cannot assess the extent of sexual dimorphism directly. But they can assess the extent of variation within the full sample of canines to see whether there seemed to be some individuals with large canines, which could be males, and some individuals with small canines, which could be females. When they did this, they found surprisingly little variation in canine dimensions (**Figure 10.10**).

FIGURE 10.9

The teeth and jaws of (left) a modern human, (middle) *Ar. ramidus*, and (right) a modern chimpanzee. The top panels show that *Ar. ramidus* had larger molars than those of modern humans or chimpanzees. The lower panels show the molars of the three species, and the colors indicate enamel thickness ranging from thin enamel (*blue*) to thick enamel (*red*).

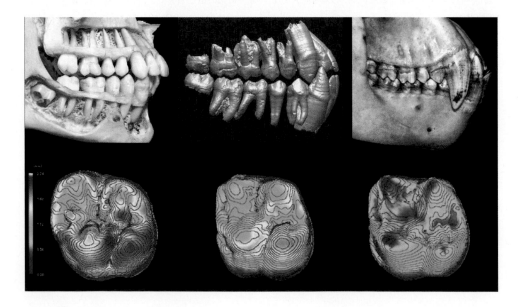

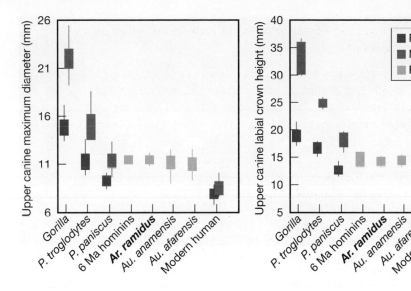

FIGURE 10.10

Sex differences in canine dimensions in great apes and hominins. The rectangles give the central 50% of each sample. *6 Ma hominins* refers to teeth from *Ar. kadabba* and *O. tugenensis*. The *labial crown height* is the height of the canine on the inside next to the tongue. Notice that the amount of variation in all *Ar. ramidus* canines is less than that in modern humans, a species in which there is little canine dimorphism. So even though we don't know which canines are from males and which are from females, we can say that the amount of sexual dimorphism was small.

They estimate that the canines of *Ar. ramidus* males might have been 10% to 15% larger than the canines of females. However, it is not yet clear whether a reduction in sexual dimorphism in canine size is linked to a reduction in sexual dimorphism in body size as well. As we will see later in this chapter, sexual dimorphism in body size and canine size seems to be decoupled in the early hominins. Canine size is reduced overall, but there still seems to be substantial variation in body size.

Ar. ramidus molar morphology also differs from the molar morphology of other great apes. Gorilla molars have thin enamel and tall shearing crests for shredding foliage; chimpanzee molars also have thin enamel, a broad basin in the middle of the tooth for crushing soft fruits, and moderate cusps for processing herbaceous plant material. *Ar. ramidus* had thicker enamel and more generalized low cusps on its molars (Figure 10.9).

Taken together, these features suggest that *Ar. ramidus* was a generalized omnivore and frugivore. It may have relied less on ripe fruit than modern chimpanzees do, less on tough fibrous material than gorillas do, and less on hard, tough foods than orangutans do.

Characteristics of the feet and pelvis indicate that *Ar. ramidus* walked upright.

Both *Sahelanthropus* and *Orrorin* have traits that are associated with bipedalism, but the limited postcranial material makes it hard to be certain about their locomotor patterns. For *Ardipithecus*, it is possible to reconstruct the postcranial skeleton with much greater accuracy, and features of the feet, pelvis, and hands provide important insights about how *Ardipithecus* moved around.

The *Ar. ramidus* foot combines characteristics that we see in modern apes and humans. Apes have flexible feet and an opposable (grasping) big toe, allowing them to climb trees and support their weight by grasping multiple small supports. By comparison, the human foot is fairly rigid, creating a better platform for transferring energy during walking and running. The big toe is not opposable, and we cannot grasp things with our feet. *Ar. ramidus* retained the opposable toe, but the other four toes were modified for bipedal walking.

The *Ardipithecus* pelvis has also been reconfigured for upright posture and bipedal locomotion. To understand the transformation of the pelvis, you need to know something about its structure. The pelvis is composed of three bones that form a circle of bone that supports the lower part of the vertebral column and protects the internal organs. The top part of the pelvis is called the ilium (or iliac crest), the central part is called the pubis, and the lower part is called the ischium. The pelvis rests on the leg bones and forms the hip joint. The neck of the femur is bent inward so the round head of the femur fits into a circular depression in the lower part of the pelvis. (Oddly, the bony protuberance below your waist is not your hip; it is your pelvis.)

FIGURE 10.11

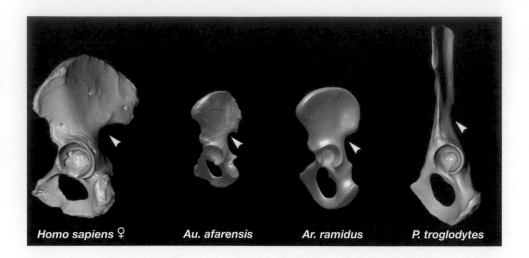

Homo sapiens ♀   **Au. afarensis**   **Ar. ramidus**   **P. troglodytes**

The pelvis of modern humans and the pelvis of chimpanzees (**Figure 10.11**) are strikingly different. In modern humans, the ilium is shorter and broader than in chimpanzees. This provides more room for the attachment of powerful muscles that keep the body upright during bipedal walking. The broadening of the top part of the ilium creates a distinctive curve (sciatic notch) in the lower part of the ilium, which is not seen in chimpanzees. The *Ar. ramidus* ilium is shorter and broader than the chimpanzee ilium and contains a sciatic notch. However, the lower part of the *Ar. ramidus* pelvis is more apelike and lacks some distinctive features that we see in modern humans. As in the foot, this combination of traits suggests that *Ar. ramidus* walked upright, but its gait may have been somewhat different from ours.

Characteristics of the hands and forelimbs suggest that *Ardipithecus* was not a knuckle walker like modern apes.

Ar. ramidus's hands are quite different from the hands of other African apes. Gorillas and chimpanzees bear their weight on their knuckles when they walk on the ground and often hang below branches when they feed or move through trees. They have long metacarpals (the bones in the palm of the hand), long phalanges (finger bones), and relatively short thumbs. These are derived traits linked to suspensory postures and below-branch locomotion. In *Ar. ramidus*, the palms and fingers are shorter, and the thumbs are longer and more robust. White and his colleagues hypothesize that *Ar. ramidus* did not move around the same way that other African apes do. Instead, *Ar. ramidus* walked along the tops of branches, bearing weight on its palms, and carefully bridged gaps in the canopy.

The Adaptive Advantages of Bipedalism

The shift from quadrupedal to bipedal locomotion is a defining feature of the hominins. However, why natural selection favored bipedalism is not entirely clear.

We take walking on two legs for granted, but it is actually a very unusual adaptation. Among mammals, the only habitual bipeds are macropods (kangaroos and wallabies), kangaroo rats, and springhares. Biomechanical analyses suggest that bipedalism and typical mammalian quadrupedalism are roughly equivalent in efficiency. If that is the case, then why did hominins adopt this odd form of locomotion? There are several possible explanations for the evolution of bipedalism in the hominins.

Bipedalism first evolved among arboreal Miocene apes as a feeding adaptation and was retained in hominins. Several of the Miocene apes seem to have had upright postures and may have been facultative arboreal bipeds, meaning they were capable of bipedalism but didn't primarily or exclusively utilize it. According to Robin Crompton of the University of Liverpool and his colleagues, these apes might have used their feet to grasp multiple small branches to support their body weight and might have used their hands for balance, to grasp branches to steady themselves, and to collect food items. Doing so may have allowed them to move in the slender terminal branches of trees where fruit is found and to make crossings from one tree to another. Today, orangutans and chimpanzees sometimes adopt bipedal postures when they are feeding in trees.

One problem with this explanation is that it does not fit neatly with the evidence that *Ar. ramidus* had monkeylike limb proportions and lacked the derived locomotor adaptations of living great apes.

Bipedal posture allows efficient harvesting of fruit from small trees. Kevin Hunt, an anthropologist at Indiana University, thinks that bipedal posture was favored because it allows efficient harvesting of fruit from the small trees that predominate in African woodlands. Hunt found that chimpanzees rarely walk bipedally but spend a lot of time standing bipedally as they harvest fruit from these small trees (**Figure 10.12**). Using their hands for balance, they pick the fruit and slowly shuffle from depleted patches to fresh ones. Standing upright allows the chimpanzees to use both hands to gather fruit, and the slow bipedal shuffling allows them to move from one fruit patch to another without lowering and raising their body weight.

Erect posture allowed hominins to keep cool. Heat stress is a more serious problem in open habitats than in the shade of the forest. If an animal is active in the open during the middle of the day, it must have a way to prevent its body temperature, particularly the temperature of its brain, from rising too high. Peter Wheeler of Liverpool John Moores University has pointed out that standing upright reduces heat stress in several ways (**Figure 10.13**).

This explanation does not seem to fit evidence that bipedalism evolved while hominins were living in wooded habitats, where heat stress would not be a major problem. However, there is some evidence that hominins occupied mosaic habitats, and it is possible that hominins sometimes ventured beyond the boundaries of the forest to forage.

Bipedal locomotion leaves the hands free to carry things. The ability to carry things is, in a word, handy. Quadrupeds can't carry things in their hands without interfering with their ability to walk and to climb. Consequently, they must carry things in their mouths. Some catarrhine monkeys pack great quantities of food into their cheeks to chew and swallow later. Other primates must eat their food where they find it—a necessity that can cause problems when food is located in a dangerous place or when there is a lot of competition over food. Bipedal hominins can carry more food in their hands and arms, and they can transport tools from one place to another.

Any or all of these hypotheses may be correct. Bipedalism might have been favored by selection because it was more efficient, because it allowed early hominins to keep

FIGURE 10.12

Chimpanzees sometimes stand bipedally as they harvest fruit from small trees. They use one hand for balance and feed with the other, shuffling slowly from one food patch to another.

FIGURE 10.13

Bipedal locomotion helps an animal living in warm climates to keep cool by reducing the amount of sunlight that falls on the body, by increasing the animal's exposure to air movements, and by immersing the animal in lower-temperature air.

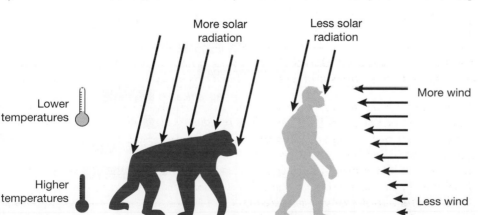

More solar radiation

Less solar radiation

Lower temperatures

Higher temperatures

More wind

Less wind

cool, because it enabled them to carry food or tools from place to place, or because it enabled them to feed more efficiently. And once bipedalism had evolved, it might have facilitated other forms of behavior, such as the use of tools.

The Hominin Community Diversifies

Several hominin species lived in Africa between 4 Ma and 2 Ma. They are divided into four genera: *Australopithecus*, *Paranthropus*, *Kenyanthropus*, and *Homo*.

Beginning about 4 Ma, the hominin lineage proliferated, and over the next 2 million years there were several species of relatively small-brained bipedal hominins living in Africa at any given time (**Figure 10.14**). Fossils can be classified in several ways, and there is little consensus about their phylogenetic relationships. We adopt the following taxonomic scheme:

1. *Australopithecus* includes six species: *Au. anamensis*, *Au. afarensis*, *Au. deyiremeda*, *Au. africanus*, *Au. garhi*, and *Au. sediba*. The genus name means "southern ape" and was first applied to a skull found in South Africa in the 1920s. These creatures were small bipeds with teeth, skull, and jaws adapted to a generalized diet. They had somewhat larger brains than those of chimpanzees but still developed rapidly like modern apes, reaching sexual maturity at 8 years old.

2. *Paranthropus* includes three species: *P. aethiopicus*, *P. robustus*, and *P. boisei*. The genus name means "parallel to man" and was coined by Robert Broom, who discovered the first specimens of *P. robustus*. These species were similar to the members of the genus *Australopithecus* from the neck down, but they had massive teeth and a skull modified to carry the enormous muscles necessary to power this chewing apparatus.

3. *Kenyanthropus* includes one species: *K. platyops*. The genus name means "Kenyan man" and was given to a specimen found in northern Kenya. It is distinguished from the contemporary *Au. afarensis* and *Au. deyiremeda* by a flattened face and small teeth.

4. Beginning around 2.7 Ma, members of our own genus *Homo* appear. By 2 Ma, there were three species: *H. habilis*, *H. rudolfensis*, and *H. erectus*. They had larger brains, smaller teeth, and developed more slowly than contemporary hominin species. Their appearance roughly coincides with the first flaked stone tools in the

FIGURE 10.14

Known hominin species that date from 8 Ma to 1 Ma. During most of that period, multiple hominin species were present.

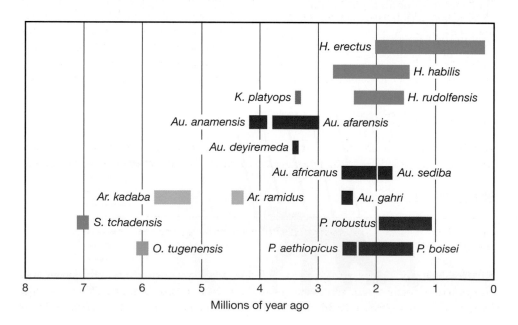

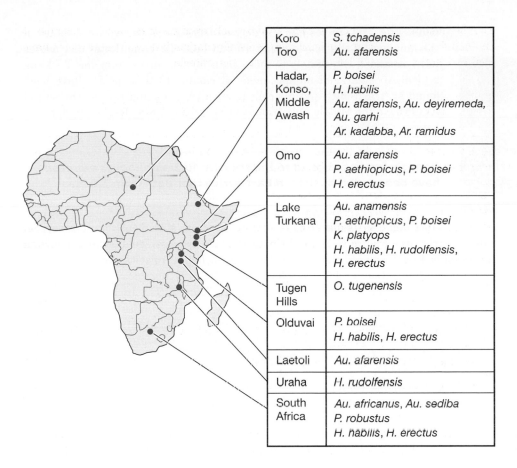

Koro Toro	S. tchadensis Au. afarensis
Hadar, Konso, Middle Awash	P. boisei H. habilis Au. afarensis, Au. deyiremeda, Au. garhi Ar. kadabba, Ar. ramidus
Omo	Au. afarensis P. aethiopicus, P. boisei H. erectus
Lake Turkana	Au. anamensis P. aethiopicus, P. boisei K. platyops H. habilis, H. rudolfensis, H. erectus
Tugen Hills	O. tugenensis
Olduvai	P. boisei H. habilis, H. erectus
Laetoli	Au. afarensis
Uraha	H. rudolfensis
South Africa	Au. africanus, Au. sediba P. robustus H. habilis, H. erectus

FIGURE 10.15

Sites in Africa where hominin fossils that date to the period from 8 Ma to 1 Ma have been found. *Homo erectus* fossils from this period have also been found at sites in the Republic of Georgia, Indonesia, and China.

archaeological record. Fossils from species in the genus *Homo* become much more common about 2 Ma, as we discuss in Chapter 11.

The sites at which early hominin fossils have been found in Africa are identified on the map in **Figure 10.15**.

You are excused if you feel as though you have mistakenly picked up a Russian novel with a long cast of characters with tongue-twisting, hard-to-remember names. Keep in mind, however, that all of these species shared important characteristics. They were bipedal on the ground but were probably also able to climb trees; their brains were somewhat larger than the brains of modern apes but still much smaller than those of modern humans; and they had smaller canines and incisors but bigger molars and premolars with thicker enamel than those of chimpanzees. They were smaller than most modern humans, and sexual dimorphism in body size was pronounced. These similarities have led many anthropologists to refer to all of these creatures, except *Homo*, collectively as "australopiths," a usage that we adopt here.

Australopithecus

In this section, we briefly describe the history and characteristics of each of the *Australopithecus* fossil species and then discuss the common features of the genus as a whole.

The cast of characters

The oldest australopith, *Australopithecus anamensis*, comes from sites in Kenya and Ethiopia.

In 1994, members of an expedition led by Meave Leakey of the National Museums of Kenya found fossils of a hominin species at Kanapoi and Allia Bay, two sites near Lake Turkana in Kenya (**Figure 10.16**). Leakey assigned them to the species *Australopithecus*

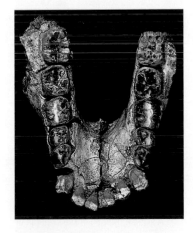

FIGURE 10.16

Australopithecus anamensis lived about 4 Ma, half a million years before *Au. afarensis*. Fragments of limb bones, the upper and lower jaws, and many teeth have been recovered.

anamensis. The species name is derived from *anam*, the word for "lake" in the language of the people living around Lake Turkana. Later discoveries have expanded the sample to more than 50 specimens from the Turkana region. An additional 30 specimens attributed to this species have been found by Tim White and his colleagues at Aramis and the nearby site of Asa Issie in the Middle Awash region of Ethiopia. These fossils are dated to between 4.2 Ma and 3.9 Ma.

FIGURE 10.17

The Awash River basin in Ethiopia is the site of several important paleontological discoveries, including many specimens of *Australopithecus afarensis*.

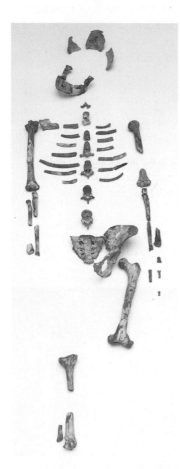

FIGURE 10.18

A sizable fraction of the skeleton of a single individual was recovered at Hadar. Because the skeleton is bilaterally symmetrical, most of the skeleton of this *Australopithecus afarensis* female, popularly known as Lucy, can be reconstructed.

Hominin fossils assigned to the species *Australopithecus afarensis* have been found at sites in East Africa that date from 3.6 Ma to 3.0 Ma.

Australopithecus afarensis is well known from specimens found at several sites in Africa (see Figure 10.15), but the most extensive fossil collections come from several sites in Ethiopia. In the early 1970s, a French and American team headed by Maurice Taieb and Donald Johanson began searching for hominin fossils at Hadar, in the Afar Depression of northeastern Ethiopia (**Figure 10.17**). In 1973, they found the bones of a 3-million-year-old knee that showed striking similarities to a modern human knee. The next year the team returned and found a sizable fraction of the skeleton of a single individual. They dubbed the skeleton "Lucy," after the Beatles' song "Lucy in the Sky with Diamonds" (**Figure 10.18**). Lucy, who lived about 3.2 Ma, was not the only remarkable find at Hadar. During the next field season, the team found the remains of 13 more individuals.

Just after the discovery of these fossils, the excavations were interrupted by civil war in Ethiopia, and paleontologists could not resume work at Hadar for more than a decade. Later a team led by Johanson returned to Hadar, while a separate team led by White searched for fossils in the nearby Middle Awash basin. Both groups discovered fossils from many *Au. afarensis* individuals, including a nearly complete skull (labeled AL 444-2). In 2001, the Ethiopian researcher Zeresenay Alemseged announced the discovery of a very well-preserved partial skeleton of a child at Dikika, a site in the Afar region of Ethiopia (**Figure 10.19**). The skeleton was embedded in sandstone, and it required painstaking efforts to extract the delicate fossils. The result was clearly worth the effort; the Dikika child's skeleton is even more complete than Lucy's and reveals elements of anatomy that are not known from any other early hominins. A partial skeleton, dated 3.6 Ma, has also been found at a site called Woranso-Mille. This find includes some delicate rib bones and one of the most complete *Au. afarensis* scapulas (shoulder blades) yet found.

Au. afarensis fossils have also been found at several sites elsewhere in Africa. During the 1970s, members of a team led by Mary Leakey discovered fossils of *Au. afarensis* at Laetoli in Tanzania that date to 3.5 Ma. In 1995, researchers announced interesting finds from Chad and South Africa. A French team published a description of a lower jaw from Bahr el Ghazal in Chad. This fossil, provisionally identified as *Au. afarensis*, is between 3.1 million and 3.4 million years old, judging by the age of associated fossils. In South Africa, Ronald Clarke of the University of the Witwatersrand in Johannesburg and the late Phillip Tobias found several foot bones that may also belong to *Au. afarensis*.

Au. anamensis and *Au. afarensis* may be part of a single lineage that evolved over time. Recall the *Au. anamensis* fossils that are dated to 4.2 Ma to 3.9 Ma. The similarities of *Au. anamensis* to *Au. afarensis* lead many researchers to think that *Au. afarensis* evolved from *Au. anamensis*, but the gap in the fossil record between 3.9 Ma and 3.6 Ma creates uncertainties about how and when this transformation occurred. In 2010, a team led by Yohannes Haile-Selassie announced the discovery of hominin fossil specimens at the site of Woranso-Mille in the Afar region of Ethiopia; these specimens are dated from 3.8 Ma to 3.6 Ma and share some characteristics with *Au. anamensis* and other characteristics with *Au. afarensis*, much as you would expect if *Au. anamensis* evolved into *Au. afarensis*.

Australopithecus deyiremeda lived about 3.5 Ma in Ethiopia at the same time as *Au. afarensis*.

In 2015, Yohannes Haile-Selassie described a new species—*Australopithecus deyiremeda*—known from two mandibles, two partial maxillae, and some teeth. Like *Au. afarensis*, these fossils have robust jaws and thickly enameled teeth, but they have smaller teeth and the shape of the face differs significantly from that of *Au. afarensis* (**Figure 10.20**).

Australopithecus garhi lived about 2.5 Ma in East Africa.

In 1999, a research team led by Asfaw and White announced the discovery of a new species, *Australopithecus garhi*, from the Awash valley of Ethiopia. *Garhi* means "surprise" in the Afar language. In 1996, Asfaw, White, and their colleagues had recovered hominin remains from a site called Bouri, the ancient site of a shallow freshwater lake, not far from where *Au. afarensis* was found. The findings included several postcranial bones and the partial skeleton of one individual. These specimens are well preserved and securely dated to 2.5 Ma, but they don't contain diagnostic features that can be used to assign them to a specific species. In 1997, however, the research team discovered several cranial remains nearby that came from the same stratigraphic level as the postcranial remains. The molars are large, similar to those found in *Paranthropus* (discussed later in this chapter).

FIGURE 10.19

Young, but old. The Dikika child's flat nose and projecting face look chimplike, but the Ethiopian fossil comes from a 3.3-million-year-old human ancestor that belongs to the same species as the famous Lucy skeleton.

FIGURE 10.20

A partial maxilla and a mandible from the fossils of *Australopithecus deyiremeda*.

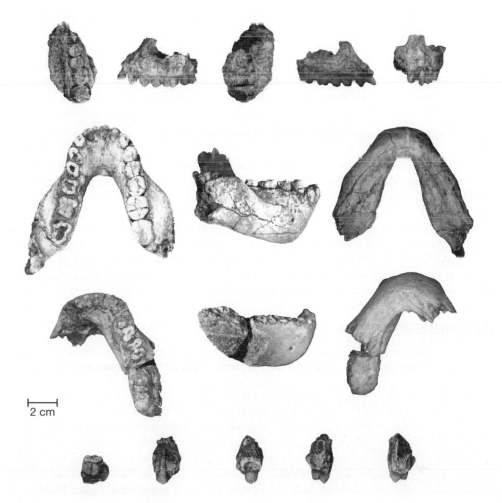

2 cm

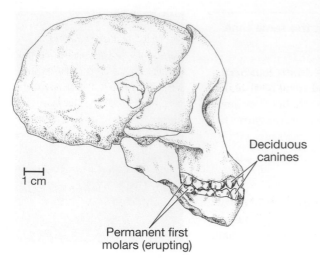

1 cm

Deciduous
canines

Permanent first
molars (erupting)

FIGURE 10.21

Raymond Dart identified in 1924 the first australopithecine specimen found in South Africa. Dart named it *Australopithecus africanus*, which means "the southern ape of Africa," but the specimen is often called the Taung child. Dart concluded that the Taung child was bipedal. Even though it had a very small brain, he believed it to be intermediate between humans and apes. Nearly 30 years passed before his conclusions were generally accepted.

Australopithecus africanus is known from several sites in South Africa that date from 3 Ma to 2.2 Ma.

At about the same time that *Au. afarensis* and then *Au. garhi* were living in East Africa, another australopith was living in South Africa. *Australopithecus africanus* was first identified in 1924 by Raymond Dart, an Australian anatomist living in South Africa. Miners brought him a piece of rock from which he painstakingly extracted the skull of an immature, small-brained creature (**Figure 10.21**). Dart formally named this fossil *Australopithecus africanus*, which means "the southern ape of Africa," and there is now evidence that members of the same species may have been distributed throughout southern Africa. Dart's fossil is popularly known as "the Taung child." Dart believed that the Taung child was bipedal because the position of the foramen magnum was more like that of modern humans than that of apes. At the same time, he observed many similarities between the Taung child and modern apes, including a relatively small brain. Thus he argued that this newly discovered species was intermediate between apes and humans. Members of the scientific community roundly rejected his conclusions because in those days most physical anthropologists believed that large brains had evolved in the human lineage before bipedal locomotion. Later, many adult skulls and postcranial bones were found at two other sites in South Africa: Makapansgat and Sterkfontein. The hip bone, pelvis, ribs, and vertebrae of *Au. africanus* are much like those of *Au. afarensis*. Dart's claim that the Taung child was bipedal has thus since been strongly supported.

Australopithecus sediba lived in South Africa about 2 Ma.

In 2008, a team led by Lee Berger of the University of the Witwatersrand in South Africa and Paul Dirks of John Cook University in Australia was searching for new hominin fossil sites in an area near Johannesburg that contains many caves. Near one cave, called Malapa, Berger's 9-year-old son spotted a hominin fossil embedded in a rock. Excavations in the Malapa cave eventually produced partial skeletons from two individuals, one an adult female and the other a juvenile, and the shinbone of a third individual. These finds are dated to 1.98 Ma (**Figure 10.22**). Berger and his colleagues assigned these fossils to a new species and chose the name *Australopithecus sediba*, which means "well-spring" in the local Sesotho language.

The shared characteristics of the genus

Australopithecus skull and dentition were intermediate between apes and humans.

The cranium (skull) of the australopiths has many apelike features (**Figure 10.23**). It is flared at the bottom, and the bone is full of air pockets (anatomists call this *pneumatized*); the face below the nose is pushed out—a condition known as **subnasal prognathism**; and the jaw joint is shallow. Many aspects of the dentition (teeth) are also apelike. For example, australopiths had relatively large and procumbent (forward-slanting) incisors, and in many specimens there is a space between the upper canine and incisor that accommodates the lower canine. Anatomists refer to this as a **diastema** (**Figure 10.24**).

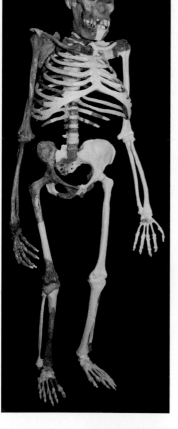

FIGURE 10.22

A reconstruction of the *Australopithecus sediba* skeleton.

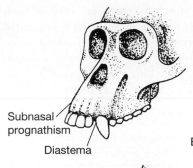
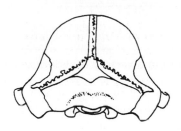

Temporal line

Nuchal line

Subnasal prognathism

Diastema

Broad (heavily pneumatized) cranial base

Chimpanzee, *Pan troglodytes*

Australopithecus afarensis

FIGURE 10.23

The cranium of *Australopithecus afarensis* has several primitive features, including a small brain, a shallow jaw joint, a pneumatized cranial base, and subnasal prognathism in the face.

Diastema

Diastema

(No diastema)

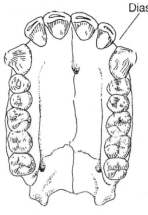
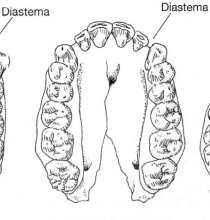
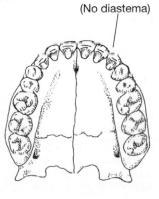

Chimpanzee

Australopithecus afarensis

Modern human

(a) Dental arcade

FIGURE 10.24

The teeth and jaws of *Australopithecus afarensis* have several features that are intermediate between those of apes and modern humans. (a) The dental arcade is less U-shaped than in chimpanzees and less parabolic than in modern humans. (b) Chimpanzees have larger and more sexually dimorphic canines than did *Au. afarensis*, which in turn had larger and more dimorphic canines than modern humans have. (c) The lower third premolars of chimpanzees have only one cusp, but those of modern humans have two cusps. In *Au. afarensis*, the second cusp is small but clearly present.

Male

Female

Chimpanzee

Male

Female

Australopithecus afarensis

Male   Female

Modern human

(b) Sexual dimorphism in canines

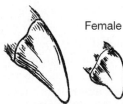

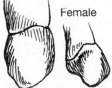

Chimpanzee

Australopithecus afarensis

Modern human

(c) Morphology of third premolar

There are also many derived features. Australopith front teeth are smaller than those of extant apes, but the premolars and molars are large and thickly enameled (Figure 10.24). Australopiths had a powerful chewing apparatus with robust jaws, large cheekbones, and robust and enlarged attachments for neck and chewing muscles. The upper part of the face is fairly vertical, although the lower face projects forward more than in humans (Figure 10.23). The dental arcade is also intermediate between the U-shaped dental arcade of apes and the parabolic dental arcade of humans.

The average **endocranial volume** (brain cavity capacity) of australopiths is 466 cc. This capacity is nearly 30% larger than the average endocranial volume for chimpanzees but much smaller than that of modern humans, which averages about 1,300 cc. The increase in brain size may have been associated with an increase in cognitive abilities, although we have no direct evidence to support this.

Australopiths lived in a variety of woodland and grassland habitats.

When paleontologists find the bones of hominins, they often find the bones of other animals as well. These bones provide valuable information about the kind of habitats that the hominins occupied. Asa Issie, one of the *Au. anamensis* sites, has many remains of colobine monkeys and bovids (the family including antelope and buffalo), animals that prefer closed or grassy woodlands. This habitat was quite similar to the habitat occupied by *Ar. ramidus* in the same area 200,000 years earlier. The faunal remains at Kanapoi and Allia Bay, two sites where *Au. anamensis* has been found, represent a more diverse range of habitats, including dry woodlands, river-filled gallery forests, and more open grasslands.

Australopithecus ate a wide variety of foods.

We have seen that between 7 Ma and 2 Ma, Africa became drier and more seasonal, with more woodland and grassland environments. Australopiths' large molars and robust faces and jaws suggest that they adapted to this environment by eating tough, hard-to-process foods such as nuts, seeds, roots, and tubers (enlarged underground stems, like potatoes).

Studies of the carbon isotopes found in fossilized teeth support this inference. Although the saying "You are what you eat" might not be correct in general, it is true of your tooth enamel. To see why, you have to know a little bit of botany and chemistry. Plants use one of three chemically distinct kinds of photosynthesis. Woody plants like trees, bushes, and shrubs use one type, which is called C_3 photosynthesis; grasses and sedges use a second type, called C_4 photosynthesis; and succulents and cacti use a third type, called CAM. C_4 plants have relatively high concentrations of the heavy isotope of carbon, ^{13}C, whereas C_3 or CAM plants have higher concentrations of the lighter isotope ^{12}C. Animals that eat plants (or eat things that eat plants) incorporate the carbon isotopes into the enamel of their teeth, and there is strong evidence that geologic processes after the death of an animal do not substantially affect the amount of these isotopes. Thus, the ratio of the two carbon isotopes, ^{13}C versus ^{12}C, denoted $\delta^{13}C$, in enamel of fossil australopith teeth tells an interesting story about what they ate. The range of $\delta^{13}C$ values observed in the teeth of *Australopithecus* species increased substantially over time, beginning with values indicating a mainly C_3 diet and then shifting to a mainly C_4 diet. *Ar. ramidus* and *Au. anamensis* have $\delta^{13}C$ values similar to that of chimpanzees living in open environments today, indicating that they concentrated on woodland resources. But beginning about 3.6 Ma, the mean $\delta^{13}C$ values increased, indicating that the australopiths ate more savanna foods such as seeds, roots, and tubers of grasses and sedges, as well as possibly animals that fed on C_4 foods, such as some antelope species. Specimens with higher $\delta^{13}C$ values tend to have larger molars and premolars. It is plausible that *Au. afarensis* and later members of the genus were dietary generalists who could exploit a wide variety of resources, as suggested

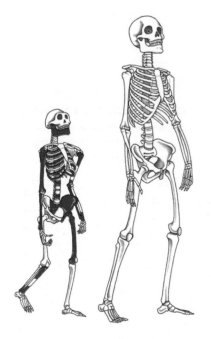

FIGURE 10.25

Lucy's skeleton (left) stands beside the skeleton of a modern human female. The parts of the skeleton that have been discovered are shaded. Lucy was shorter than modern females and had relatively long arms and a relatively small brain.

by the evidence that australopiths occupied a more diverse range of habitats than apes.

Anatomic evidence clearly indicates that australopiths were fully bipedal when on the ground.

We have several fossils of the postcrania (bones below the neck) for *Au. afarensis* and *Au. africanus* that clearly indicate the australopiths were committed bipeds (**Figure 10.25**). As in *Ar. ramidus*, the pelvis is short and wide, the neck of the femur is long, and the femur slants inward. However, australopiths had a suite of other morphological adaptations that made bipedal locomotion more efficient than it was for *Ardipithecus*. The most obvious differences are in the foot. The *Ar. ramidus* foot is a compromise—it is less flexible than the foot of other apes but allows use of the big toe to grasp tree branches. The australopith foot is much more like ours. It is relatively stiff with arches in both directions, and the big toe cannot grasp.

It is efficient for a biped to carry its weight balanced over the pelvis, which means that the spine must have an "S" shape that anatomists refer to as **lordosis**. The australopith pelvis and backbone had several specialized adaptations to facilitate this posture (**Figure 10.26**). In tree-climbing apes, the hamstring (the big muscle on the back of the thigh) is dominant over the quadriceps (the big muscle on the front of the thigh). *Ar. ramidus* had the same arrangement, probably to facilitate tree climbing. In australopiths and modern humans, the quadriceps are dominant to facilitate our striding gait.

Although there is no doubt that the australopiths were bipedal, some features of their pelvis and legs differ from those of modern humans. For example, their legs were shorter in relation to their body size than our legs, which may have reduced the efficiency and speed of their walking.

A trail of fossil footprints proves that a striding biped lived in East Africa at the same time as *Au. afarensis*.

The conclusion that australopiths had an efficient striding gait was dramatically strengthened by a remarkable discovery in 1978 by Mary Leakey and her co-workers at Laetoli in Tanzania. They uncovered a trail of footprints 30 m (about 100 ft.) long that was made by three bipedal individuals as they crossed a thick bed of wet volcanic ash about 3.5 Ma (**Figure 10.27**). Paleontologists estimate that the tallest of the individuals who made these prints was 1.45 m (4 ft. 8 in.) tall and the shortest was 1.24 m (4 ft. 1 in.) tall. The path of these three individuals was preserved because the wet ash solidified as it dried, leaving us a tantalizing glimpse of the past. Russell Tuttle, an anthropologist at the University of Chicago, has compared the Laetoli footprints with the footprints made by members of a group of people called the Machigenga, who live in the tropical forests of Peru. Tuttle chose the Machigenga because their average height is close to that of the makers of the Laetoli footprints, and they do not wear shoes. Tuttle found that the Laetoli footprints are functionally indistinguishable from those made by the Machigenga, and he concluded that the creatures who made the footprints walked with a fully modern striding gait.

Who made these footprints? The prime suspects are *Au. afarensis*, *Au. deyiremeda*, and *Kenyanthropus platyops* (a hominin you will meet later in this chapter) because they are the only hominins known to have lived in East Africa when the tracks were made. *Au. afarensis* is the most likely culprit because it is the only hominin whose fossil remains have been found at Laetoli.

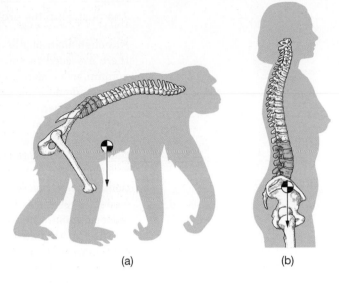

(a)　　　　　　　　(b)

FIGURE 10.26

The chimpanzee (a) backbone bows outward, whereas the human (b) backbone has an "S" shape, bowing inward at the bottom and then bowing outward.

FIGURE 10.27

Several bipedal creatures walked across a thick bed of wet volcanic ash about 3.5 Ma. Their footprints were preserved when the ash solidified as it dried.

The australopiths probably spent a good deal of time in trees.

Although australopiths lost the opposable big toe of *Ardipithecus* and the African apes, there are good reasons to think these creatures still did a lot of tree climbing. All nonhuman primates, except some gorillas, spend the night perched in trees, curled up in crude nests, or huddled on cliffs to protect themselves from nocturnal predators such as leopards. Several skeletal features suggest that the australopiths may have slept and foraged in trees, too. In chimpanzees, the **humerus**, radius, and ulna (the bones of the arm), as well as the femur, are about the same length. In modern humans, the bones of the forearm have become shorter than the humerus, and the femur has become longer than the humerus. Australopiths resemble chimpanzees in the proportions of these bones. In addition, the bones of their fingers are curved like those of modern apes. Their thumbs were short and the tips of their fingers tapered, whereas modern humans have longer thumbs and broader fingertips.

The fact that australopith morphology is not as well suited to climbing as that of chimpanzees does not mean that australopiths did not spend a lot of time in trees. Modern human foragers also frequently climb trees to forage, though human morphology is even less well adapted to tree climbing than was the morphology of the australopiths. Honey and bee larvae play an important role in the diets of many modern foraging groups. For example, during the 3-month honey season, Mbuti and Efe foragers get nearly half their calories from honey and bee larvae. To harvest honey and larvae, the Congo basin foragers climb as high as 50 m (150 ft.) into the canopy; consequently, falls are a significant source of mortality. Forest foragers climb more slowly than chimpanzees and tend to stay closer to the central tree trunk, perhaps because they cannot grasp smaller limbs with their feet.

The australopiths were sexually dimorphic in body size, suggesting that intersexual competition between males was important.

For most australopith species, we have a limited number of specimens and cannot reconstruct the whole skeleton. However, because the dimensions of certain bones are correlated with overall height and weight, paleontologists can estimate the size and weight of various australopiths without the entire skeleton. Identifying the sex of fossil specimens is usually difficult, so we cannot estimate the extent of sexual dimorphism directly. However, for sites in which bones from multiple individuals are present, we can get some idea of the range of variation within the population (as we saw earlier with canine dimorphism). This is informative because sexual dimorphism is linked to mating systems in other primates, which is reflected in the extent of variation in body size within groups. On the basis of comparative data from living primates, a pair-bonded species shows little variation in size between the sexes, whereas in species with high reproductive skew and considerable male–male competition, males tend to be considerably bigger than females.

At sites where bones from multiple individuals have been found, there seem to be big and small individuals. For example, at Hadar, bigger individuals were 1.51 m (5 ft.) tall and weighed about 45 kg (100 lb.). The smaller individuals were about 1.05 m (3.5 ft.) tall and weighed about 30 kg (65 lb.). Evidence from several sites with specimens from multiple individuals suggests that this variation represents sexual dimorphism, with the larger adults being male and the smaller ones female. The difference between the Hadar males and females approaches the magnitude of sexual dimorphism in modern orangutans and gorillas, and is considerably greater than that in modern humans, bonobos, and chimpanzees.

Recall from Chapter 6 that sexual dimorphism in body size is associated with sexual dimorphism in canine size in most primates. For example, male baboons are almost twice as big as females, and males also have much bigger canines than those of females. As we observed earlier in this chapter, in the australopiths, however, sexual dimorphism in body size has become decoupled from sexual dimorphism in canine size. Some

FIGURE 10.28

Kangaroos have polygynous mating systems and considerable body-size dimorphism but do not have canines. It is possible that in bipedal animals, such as kangaroos and humans, canines do not play an important role in combat.

researchers think that the lack of sexual dimorphism in canine size implies that australopiths were pair-bonded, whereas other researchers think that the presence of considerable sexual dimorphism in body size implies that australopiths were polygynous. It is possible that the absence of sexual dimorphism in canine size in australopiths reflects the fact that they stood upright and did not rely on their canines in fights with other males. Kangaroos, which are one of the only other bipedal mammals, have lost their canines altogether. Males grapple with their hands and arms and kick with their powerful legs (**Figure 10.28**). Most kangaroos have considerable sexual dimorphism in body size and polygynous mating systems.

The australopiths matured rapidly like chimpanzees, not slowly like humans.

We learned in Chapter 8 that life history varies among mammals. Some species develop rapidly, reproduce early, and die young. Others grow more slowly, delay reproduction, and have longer lives. These differences are correlated with brain size—species that develop more slowly have larger brains. Compared with other mammals, apes develop slowly and have large brains; modern humans develop even more slowly than our ape relatives.

Among living primates, the age at which molars, especially first molars (M1), erupt is a very good predictor of the age of sexual maturity, age at first reproduction, and overall life span. Species in which M1 erupt late also reach sexual maturity later, give birth to their first offspring later, and live longer (**Figure 10.29**). Thus, if we know the age of M1 eruption in extinct hominins, we can estimate how rapidly they developed overall and how long they lived. But how do we know how old these creatures were when their molars erupted? To age long-dead hominins, anthropologists rely on the fact that enamel is secreted as teeth grow. During each day the rate of secretion varies, and this variation creates minute parallel lines in the tooth enamel. Each line represents one day. These layers can be counted in microscopic images, allowing anthropologists to estimate the age at death of the individual whose remains were fossilized. Using a sample of six australopith fossil teeth from individuals who died around the age of M1 eruption, Jay Kelley and Gary Schwartz, colleagues at Arizona State University, estimated that the age of M1 eruption varied from 2.9 to 3.9 years, less than the age of M1 eruption for chimpanzees. Their results indicate that these creatures developed rapidly, not slowly like modern humans.

As we will discuss more fully later, these results are important because they suggest that australopith infants did not have as long a period of dependency as human children do. Many anthropologists believe that certain fundamental features of human foraging societies, such as home bases, pair bonding, sexual division of labor, and extensive food sharing, were necessitated by the fact that infants remained dependent on their parents for a long time. If australopith infants matured quickly, then it seems likely that these features were not yet part of the hominin adaptation and that australopiths lived lives like modern apes. Mothers raised one offspring at a time by themselves, and males invested little.

Paranthropus

Paranthropus aethiopicus is a hominin with teeth and skull structures specialized for heavy chewing.

Alan Walker of Pennsylvania State University discovered the skull of a very robust australopith at a site on the west side of Lake Turkana in northern Kenya. Walker called it the "Black Skull" because of its distinctive black color, but its official name

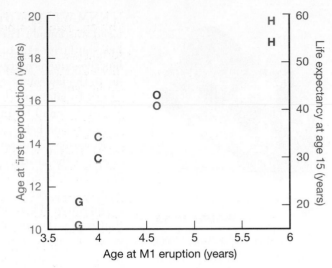

FIGURE 10.29

Age at reproduction (*blue*) and life expectancy at age 15 (*red*) as a function of the age at M1 eruption for four hominoid species, gorillas (G; M1 eruption at 3.38 years), chimpanzees (C; M1 eruption at 4.0 years), orangutans (O; M1 eruption at 4.6 years), and modern humans (H; M1 eruption at 5.8 years). The age of emergence of M1 is a good predictor of the age of first reproduction and longevity. The estimated age of M1 eruption for a sample of australopiths varies between 2.9 and 3.9 years, indicating that they developed more rapidly than modern humans.

FIGURE 10.30

The Black Skull (KNM-WT 17000) was discovered on the western side of Lake Turkana in Kenya.

is KNM-WT 17000 (**Figure 10.30**). (*WT* stands for "West Turkana," the site where the fossil was found.) This creature—*Paranthropus aethiopicus*—lived about 2.5 Ma and had similarities to *Au. afarensis*. For example, the hinge of the jaw in KNM-WT 17000 has the same primitive structure as the jaw in *Au. afarensis*, and both are very similar to the jaw joints of chimpanzees and gorillas. Later hominins had a modified jaw hinge. In addition, they had a similar postcranial anatomy and were equipped with relatively small brains in relation to body size. Like other australopiths, KNM-WT 17000 was probably quite sexually dimorphic in body size.

However, as shown in **Figure 10.31**, KNM-WT 17000 is very different from the skulls of the australopiths we have met so far. The molars are enormous, the lower jaw is very large, and the entire skull has been reorganized to support the massive chewing apparatus. For example, it has a pronounced **sagittal crest**, which enlarges the surface area of bone available for attaching the **temporalis muscle**, one of the muscles that works the jaw. You can easily demonstrate the function of the sagittal crest for yourself. Put your fingertips on your temple and clench your teeth; you will feel the temporalis muscle bunch up. As you continue clenching your teeth, slowly move your fingertips upward until you can't feel the muscle any more. This is the top point of attachment for your temporalis muscle, about an inch above your temple. All of the australopiths had bigger teeth than we do, so they needed larger temporalis muscles as well, which required more space for attachment to the skull. In australopiths, the muscles expanded so that they almost met at the top of the skull. *P. aethiopicus* had larger teeth and required even more space for muscle attachment, and the sagittal crest served this function (**Figure 10.32**). Other distinctive features of *P. aethiopicus* are also accommodations to the enlarged teeth and jaw musculature. For example, the cheekbones (zygomatic arches) flare outward to make room for the enlarged temporalis muscle, and the flaring causes the face to be flat or even pushed in.

Paranthropus robustus is a more recent species found in southern Africa.

In the late 1930s, a retired Scottish physician and avid paleontologist named Robert Broom discovered fossils at Kromdraai in South Africa that seemed very different from *Au. africanus* specimens previously found at the nearby site of Sterkfontein (**Figure 10.33**). The creatures at Kromdraai were much more robust (with more massive skulls and larger teeth) than their neighbors, so Broom called his find *Paranthropus robustus*. Many of his colleagues questioned this choice. They thought he had discovered another *Australopithecus*, and so for a time most paleontologists classified Broom's fossils as *Australopithecus robustus*. More recently, however, the consensus has been that the robust early australopiths are so different from the gracile (more lightly built) australopiths such as *Au. afarensis* that they

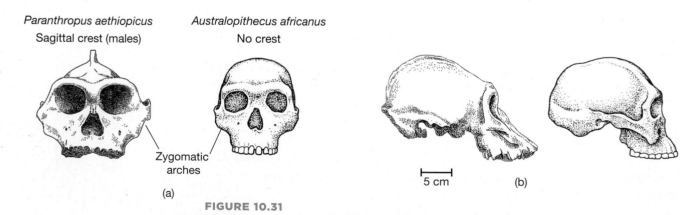

Paranthropus aethiopicus
Sagittal crest (males)

Australopithecus africanus
No crest

Zygomatic arches

(a)

5 cm

(b)

FIGURE 10.31

The skull of *Paranthropus aethiopicus* differed from that of *Australopithecus africanus* in having (a) a sagittal crest and (b) a flatter face and forehead.

should be placed in a separate genus, and Broom's original classification has been restored.

There is now a large sample of *P. robustus* fossils from Kromdraai and the nearby sites of Swartkrans and Drimolen. These creatures appeared about 1.8 Ma and disappeared about 1 Ma. Their brains averaged about 530 cc. Their postcranial anatomy shows that they were undoubtedly bipedal. They share many of the same derived features with humans that *Au. africanus* does. In addition, they share with *Au. africanus* several specialized features for heavy chewing, but these features are much more pronounced in *P. robustus*.

There was substantial sexual dimorphism in body size in *P. robustus*. The males stood about 1.3 m (4.5 ft.) tall and weighed about 40 kg (88 lb.); females were about 1.1 m (3.5 ft.) tall and weighed 32 kg (70 lb.). It looks as though males in this species continued to grow for a longer time than did the females, as is the case in other sexually dimorphic primate species. The late Charles Lockwood and his colleagues analyzed the size and age of 35 fossil specimens. They limited their sample to mature individuals, whose third molars had erupted, and then assessed the extent of tooth wear to rank them according to age. Although older individuals (with more worn teeth) are substantially larger than younger individuals, there is no relationship between size and age among the smallest specimens in the sample. If the smaller specimens are females and the larger specimens are males, then this pattern suggests that males continued to grow longer than females did.

What the robust australopiths were doing with their massive chewing apparatuses is unclear. Many anthropologists believe that *P. robustus* relied more on plant materials that required heavy chewing than *Au. africanus* did. In other animals, large grinding teeth are often associated with a diet of tough plant materials, and omnivorous animals typically have relatively large canines and incisors. Moreover, the wear patterns on *P. robustus* teeth suggest that it ate very hard foods such as seeds or nuts. However, the carbon isotope levels in the tooth enamel of *P. robustus* at Swartkrans are very similar to the values for *Au. africanus*, suggesting that the robust and gracile australopiths might have had considerable overlap in their diets. *P. robustus* obtained nearly a third of its diet from C_4 foods, such as grasses, roots, and tubers, or it consumed animals that fed on these kinds of foods.

Paranthropus boisei was a robust *P. robustus*.

Another robust australopith with large molars was discovered by Mary Leakey at Olduvai Gorge, Tanzania, in 1959. This specimen, officially labeled Olduvai Hominin 5 (OH 5), was first classified as *Zinjanthropus boisei*. *Zinj* derives from an Arabic word for "East Africa," and *boisei* comes from Charles Boise, who was funding Leakey's research at the time. Leakey's find was later reclassified *Paranthropus boisei* because of its affinities to the South African forms of *P. robustus*. The discovery of OH 5 was important partly because it ended nearly 30 frustrating years of work in Olduvai Gorge in which there had been no dramatic hominin finds. It was only the first of a very remarkable set of fossil discoveries at Olduvai.

Essentially, *P. boisei* is an even more robust *P. robustus*; that is, a hyper-robust australopith. Its body is somewhat larger than the body of *P. robustus*, and its molars are larger than those of *P. robustus*, even when the difference in body size is taken into account. The enamel is extremely thick, and the skull is even more specialized for heavy chewing.

In other primates, the proportion of leaves in the diet is inversely related to the ratio of the size of first and third molars. In *P. boisei*, this

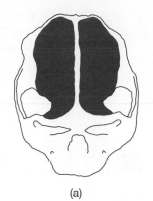
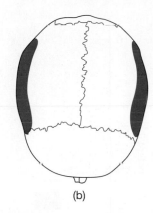

(a)   (b)

FIGURE 10.32

In these two skulls (a) *Paranthropus aethiopicus* and (b) modern human, the area of attachment of the temporalis muscle is shown in color.

FIGURE 10.33

A nearly complete skull of *Paranthropus robustus* was found at Sterkfontein in South Africa in 1999. This species has very large jaws and molars and relatively small incisors and canines.

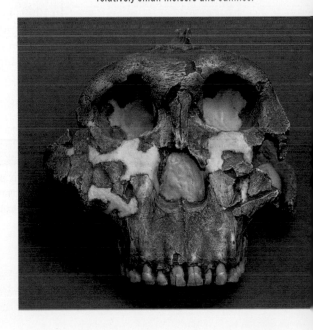

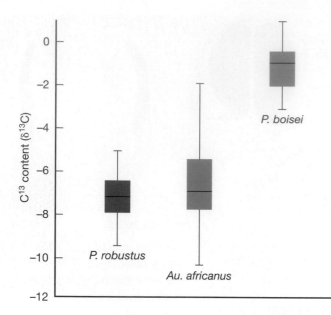

FIGURE 10.34

The range of δ¹³C values from *P. robustus*, *Au. africanus*, and *P. boisei* fossils. In each case, the horizontal line gives the median value, the box gives the range between the 25th and 75th percentiles, and the vertical lines give the full range of values. Larger δ¹³C values indicate a diet rich in grasses, whereas lower δ¹³C values indicate a diet of leaves and seeds. The diets of *P. robustus* and *P. boisei* seem quite different.

ratio is quite low, suggesting that its diet consisted of leaves or seeds. However, other elements of its teeth, including rounded molar cusps, suggest that it would not have processed leaves very efficiently. Instead, its diet might have consisted of seeds, tubers, bulbs, roots, and rhizomes.

Paranthropus boisei appears in the fossil record about 2.2 Ma in eastern Africa. It became extinct about 1.3 Ma, although the exact date of its disappearance is not well established.

P. boisei and *P. robustus* had different diets.

The teeth and facial structure of *P. boisei* and *P. robustus* are very similar, suggesting that they ate similar foods. However, two lines of evidence suggest that, in fact, their diets were very different. Carbon isotope data from *P. robustus* indicate that its diet was similar to that of *Au. africanus*, a variable mix of C$_3$ and C$_4$ resources. The teeth of *P. boisei* have very high ¹³C values consistent with a diet consisting almost completely of C$_4$ grasses and other savanna plants (**Figure 10.34**). Careful studies of microscopic wear patterns on the fossil teeth support this inference. A diet of hard objects such as seeds results in microscopic pits and other damage on the surface of teeth, and the teeth of *P. robustus* show exactly this pattern. However, the teeth of *P. boisei* show no sign of such damage and instead show wear patterns consistent with eating leaves or blades of grass (**Figure 10.35**). Thus, the teeth and jaws of *P. robustus* are consistent with the carbon isotope and dental microwear data, but those of *P. boisei* are not. It is possible that *P. boisei* ate grasses and leaves most of the time but occasionally, during droughts or some other crisis, had to rely on tough, hard-to-chew foods. Because surviving such crises was crucial, the argument runs, its teeth and jaws were adapted to life during these rare crises, but the isotope and microwear data reflect its everyday diets.

Kenyanthropus

Kenyanthropus platyops lived in East Africa between 3.5 Ma and 3.2 Ma.

In 1999, Justus Erus, a member of a research team led by Meave Leakey of the Kenya National Museum, found a nearly complete hominin cranium, labeled KNM-WT 40000, on the western side of Lake Turkana in Kenya (**Figure 10.36**). Argon dating methods indicate that KNM-WT 40000 is 3.5 million years old. This creature displays

FIGURE 10.35

The complexity of wear patterns on the teeth of fossils from several hominin species. More complex wear is associated with the use of teeth to crack open tough plant foods such as seeds by using vertical motions. Less complex wear is associated with the use of teeth to process softer plant foods such as grass stems by using sideways motions. The wear patterns on *P. robustus* and *P. boisei* are very different, suggesting that they had different diets.

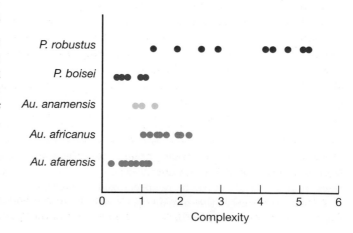

a distinctive mix of traits. As in chimpanzees, *Au. anamensis*, and *Au. ramidus*, the cranium has a small ear hole. Like most of the early hominins, the specimen has a chimpanzee-size braincase and thick enamel on its molars. However, its molars are substantially smaller than those of any other early hominin except *Au. ramidus*. As reconstructed, the face is broad and very flat, like the faces of *P. boisei* and early members of the genus *Homo*, which you will meet in Chapter 11. The fossils found in association with the cranium suggest that these creatures lived in a mix of woodland and savanna environments, and carbon isotope data indicate that their diet was a mix of C_3 and C_4 plants, similar to that of their contemporary, *Au. afarensis*.

This specimen combines features not found in other hominins, so Leakey and her co-authors placed it and a fragmentary upper jaw found nearby some years earlier in a new genus. They called the new genus *Kenyanthropus*, meaning "Kenyan man." The species name, *platyops*, is from the Greek for "flat face" and refers to the most notable anatomic feature of this specimen. Some researchers are skeptical about this attribution because of the distortion in the cranium that occurred when it was in the ground; this controversy will persist until more fossil material is found.

FIGURE 10.36

The fossil cranium of *Kenyanthropus platyops* was found near Lake Turkana in northern Kenya. It dates to about 3.5 Ma and has a unique combination of anatomic features.

Hominin Phylogenies

Inferring the phylogenetic relationships among the Plio-Pleistocene hominins is difficult.

One of the primary goals of our enterprise is to reconstruct the evolutionary history of the human lineage. To do so, we would like to understand the phylogenetic relationships among all of the early hominin species that we have described in this chapter and to identify the species that gave rise to the genus *Homo*, the genus to which modern humans belong. Unfortunately, as the hominin fossil record has become richer, this task has become harder.

The problem arises from the extensive convergence and parallelism in hominin evolution. Richard Klein of Stanford University points out that parallelisms are particularly likely to occur in a group of closely related species, such as the australopiths, because they share genes that increase the probability that they will respond in the same way to similar types of selection pressures. This means many alternative phylogenies are equally plausible, depending on what is assumed to be homologous and what is assumed to be convergent. This can lead to very different interpretations of the data.

Richard Klein proposes a "working phylogeny" of the hominins (**Figure 10.37**) that is based on the geographic locations in which fossils have been found, the times that various fossil species are believed to have lived, and the morphological characteristics of the fossil species. Klein places *Ardipithecus* at the base of the lineage and groups all of the robust australopiths together. As you can see, there are uncertainties about the relationships among *Au. afarensis*, *Au. africanus*, and *Au. garhi*. Klein suggests that *Au. garhi* may be most closely related to early members of the genus *Homo*, but he emphasizes that this is a very tentative hypothesis.

The absence of a secure phylogeny for the early hominins does not prevent us from understanding human evolution.

It is easy to be discouraged by the uncertainties in the hominin fossil record and to doubt that we know anything concrete about our earliest ancestors. Although there

FIGURE 10.37

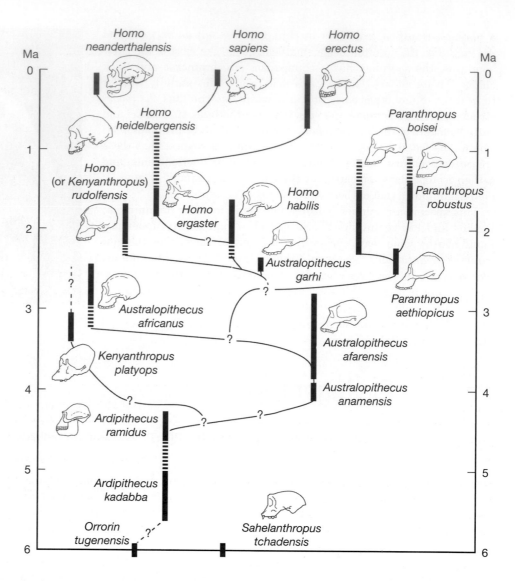

A possible phylogeny of the Plio-Pleistocene hominins. Note that there are many uncertainties in the tree.

are many gaps in the fossil record and much controversy about the relationship among the early hominin species, we do know several important things about them. In every plausible phylogeny, the human lineage is derived from a small biped who was adept in trees. Its teeth and jaws were suited for a generalized diet. The males were considerably taller and heavier than the females, their brains were the same size as those of modern apes, and their offspring developed faster than those of modern humans. This kind of creature links the apes of the Miocene to the earliest members of our own genus, *Homo*.

CHAPTER REVIEW

Key Terms

hominins (p. 238)
foramen magnum (p. 239)
torque (p. 240)
abductors (p. 240)

ilium (p. 240)
femur (p. 240)
subnasal prognathism
  (p. 252)

diastema (p. 252)
endocranial volume
  (p. 254)
lordosis (p. 255)

humerus (p. 256)
sagittal crest (p. 258)
temporalis muscle
  (p. 258)

Study Questions

1. What features distinguish modern humans from great apes?

2. Many researchers refer to the early hominins as bipedal apes. Is this description accurate? In what ways do you think that early hominins may have differed from other apes?

3. What was so distinctive about the fossil hominin material discovered in 2002 in Chad?

4. What circumstances might have favored the divergence and later diversification of hominin species in Africa 4 Ma to 2 Ma?

5. What do we mean when we label a trait "primitive"?

6. Outline three reasons natural selection may have favored bipedal locomotion in the hominin lineage.

In each case, explain why hominins became bipedal but other terrestrial primates, such as baboons, did not.

7. What evidence suggests that australopiths spent more time in trees than modern humans do?

8. What features do the *Australopithecus* species share? In what ways do they differ from *Paranthropus* and *Kenyanthropus*?

9. What features do the three earliest hominin species share? How are they different?

10. From the comparative and morphological evidence on hand, what can we say about the behavior and social organization of the early hominins?

Further Reading

Cartmill, M., F. H. Smith, and K. Brown. 2007. *The Human Lineage.* New York: Wiley-Blackwell.

Gibbons, A. 2007. *The First Human: The Race to Discover Our Earliest Ancestors.* New York: Doubleday.

Kelley, J., and G. T. Schwartz. 2012. "Life-History Inference in the Early Hominins." *International Journal of* Australopithecus *and* Paranthropus 33: 1332–1363.

Kimbel, W. 2007. "The Species and Diversity of Australopiths." In W. Henke and I. Tattersall, eds. *Handbook of Paleontology.* Vol. 3, pp. 1539–1574. Berlin: Springer-Verlag.

Klein, R. G. 2009. *The Human Career: Human Biological and Cultural Origins.* 3rd ed. Chicago: University of Chicago Press.

Reed, K. E., R. Leakey, and J. Fleagle. 2013. *The Paleobiology of Australopithecus.* Heidelberg: Springer Dordrecht.

Smith, T. 2013. "Teeth and Human Life History." *Annual Review of Anthropology* 42: 191–208, doi:10.1146/annurev-anthro-092412-155550.

Sponheimer, M., Z. Alemseged, T. E. Cerling, F. E. Grine, W. H. Kimbel, M. G. Loukoy, J. A. Loo Thorp, F. K. Manthi, K. E. Reed, B. A. Wood, and J. G. Wynn. 2013. "Isotopic Evidence of Early Hominin Diets." *Proceedings of the National Academy of Sciences USA* 110: 10513–10518, doi:10.1073/pnas.1222579110.

White, T. D., B. Asfaw, Y. Beyene, Y. Haile-Selassie, C. O. Lovejoy, G. Suwa, and G. WoldeGabriel. 2009. "*Ardipithecus ramidus* and the Paleobiology of Early Hominids." *Science* 326: 75–86.

White, T. D., C. O. Lovejoy, B. Asfaw, J. P. Carlson, and G. Suwa. 2015. "Neither Chimpanzee nor Human, *Ardipithecus* Reveals the Surprising Ancestry of Both." *Proceedings of the National Academy of Sciences USA* 112: 4877–4884, doi:10.1073/pnas.1403659111/-/DCSupplemental.

 Review this chapter with personalized, interactive questions through INQUIZITIVE.

11

EARLY *HOMO* AND *H. ERECTUS* (2.6–1 MA)

CHAPTER OBJECTIVES

By the end of this chapter you should be able to:

A Describe the morphology of the earliest members of the human genus, *Homo*.

B Describe the morphology, life history, and lifeways of *Homo erectus*.

C Explain how and when hominins left Africa.

D Describe the first stone tools made by hominins.

E Explain how the foraging techniques of modern foragers differ from those of other primates.

F Explain how reliance on complex foraging techniques has influenced human life history.

G Describe what we know about the foraging strategies of the creatures that made Oldowan tools.

H Explain why some experts think that Oldowan toolmakers hunted game, whereas other experts think they mostly scavenged meat.

Sometime before 2.8 Ma, a new kind of hominin appeared in Africa. These creatures were more like modern humans than the australopiths were, and so they are assigned to our own genus, *Homo*. They generally had larger brains than the australopiths, and their skulls and teeth have several derived features. They used flaked stone tools, may have controlled fire, and hunted large game. By 1.8 Ma some of these creatures left Africa and spread across southern Eurasia. Their skulls were quite variable, and it seems likely that they belonged to several species. There is considerable debate about which fossils should be assigned to which species, so we will collectively refer to the oldest specimens as early *Homo* and the later specimens as *Homo erectus*. Although we can see much of ourselves in these creatures, important differences remain. They had smaller brains than we have and probably developed much more rapidly than we do.

Early *Homo*

The Pleistocene epoch began 2.6 Ma and saw a cooling of the world's climate.

The Pleistocene is divided into three parts: the Lower, Middle, and Upper Pleistocene. The Lower Pleistocene began about 2.6 Ma, a date that coincides with a sharp cooling of the world's climate. The beginning of the Middle Pleistocene is marked by sharply increased fluctuations in temperature and the first appearance of immense continental glaciers that covered northern Europe about 800 thousand years ago (abbreviated ka), and its end is defined by the termination of the penultimate glacial period about 130 ka. The Upper Pleistocene ended about 12 ka when a warm, interglacial phase of the world climate began. This warm period has persisted into the present (**Figure 11.1**).

FIGURE 11.1

The pattern of world temperature over the past 6 million years. These estimates are based on the ratio of ^{16}O to ^{18}O in cores extracted from deep-sea sediments. During the Pliocene, world temperatures declined, and in the Pleistocene climate fluctuations increased, especially during the Middle and Upper Pleistocene.

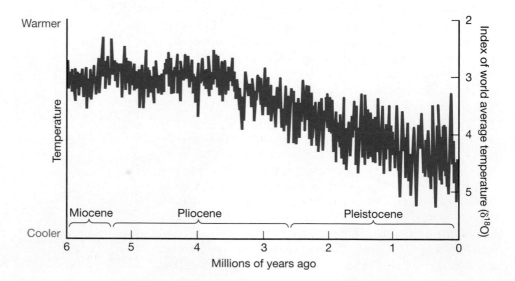

Fossils assigned to early *Homo* are found in East Africa at sites that date from 2.8 Ma to 1.4 Ma.

In 1960, while working at Olduvai Gorge with his parents, the renowned paleoanthropologists Louis and Mary Leakey, Jonathan Leakey found pieces of a hominin jaw, cranium, and hand. The Leakeys assigned this specimen, labeled Olduvai Hominin 7, or OH 7 (**Figure 11.2**), to the genus *Homo* because they believed the cranial bones indicated it had a much larger brain than that of the australopiths. Recent analyses by Fred Spoor of University College London and colleagues indicate that the Leakeys were right: The brain size of OH 7 was about 800 cc, substantially larger than the brains of australopiths.

Another member of the famous Leakey family found a second important early *Homo* fossil. Mary and Louis Leakey's son Richard set up his own research team and began work at a site called Koobi Fora on the eastern shore of Lake Turkana. In 1972, a member of Richard Leakey's team, Bernard Ngeneo, found a nearly complete skull of a hominin. This find is usually referred to by its official collection number, KNM-ER 1470 (**Figure 11.3**). (*KNM* stands for Kenya National Museum, and *ER* stands for East Rudolf—Lake Turkana was known as Lake Rudolf when Leakey began his work there.) Like OH 7, KNM-ER 1470 has a substantially larger endocranial volume than that of any australopith, about 775 cc.

Since the discovery of KNM-ER 1470, many other early *Homo* fossils have been found. These fossils have come from several sites in East Africa, including the Olduvai Gorge, the Omo River basin, and Ileret near Lake Turkana. Other fossils associated with early *Homo* come from Sterkfontein in South Africa and Uraha, a site near Lake Malawi. The oldest of these fossils, found by a team led by Kaye Reed of Arizona State University at Ledi-Geraru in Ethiopia, dates to 2.8 Ma; the youngest was found by a team led by Richard Leakey's daughter, Louise, and is dated to 1.4 Ma.

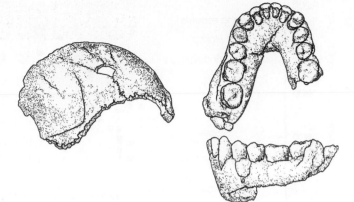

FIGURE 11.2

Computer-based reconstructions of OH 7. The fossil found by the Leakeys had been deformed during fossilization. Fred Spoor and colleagues digitally reconstructed the fossil by using three-dimensional images from computed tomography.

The skulls and jaws of fossils assigned to early *Homo* share several derived features but are also extremely variable. They probably belong to two species.

On average, fossils of early *Homo* had substantially larger brains than those of the australopiths. **Figure 11.4** plots the endocranial volume for australopiths, early *Homo*, and *Homo erectus*. You can see that endocranial volumes for australopiths range from a bit less than 400 cc to about 550 cc; for early *Homo* the range is from a little less than 550 cc to more than 800 cc with a mean of 658 cc, an increase of about 40%. The teeth of early *Homo* are smaller, on average, than those of the australopiths and have several other derived features.

The skulls, and especially the jaws, of early *Homo* are extremely variable. The dental arcade of some specimens, such as OH 7, is long and U-shaped—similar to that seen in the australopiths—whereas the dental arcade of other fossils, such as KNM-ER 62000, are more like those of later hominins, such as *H. erectus* and modern humans. There are also differences in the structure of the face. Some fossils, such as KNM-ER 1813, have prognathic faces that resemble those of australopiths, whereas others, such as KNM-ER 1470, have broad, flat faces unlike those of either australopiths or modern humans. These early *Homo* specimens differ more from one another than do living species such as chimpanzees and

FIGURE 11.3

The discovery of KNM-ER 1470 at a site on the eastern shore of Lake Turkana confirmed that there was at least one large-brained hominin in East Africa about 2 Ma.

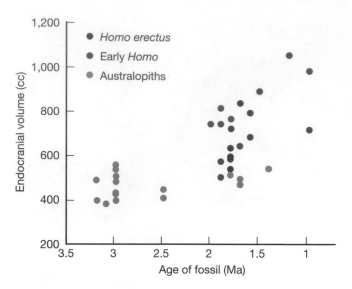

FIGURE 11.4

Endocranial volumes for australopiths (*green*), early *Homo* (*red*), and *H. erectus* (*blue*) plotted against the age of the fossil. The brains of the early *Homo* and the earliest *H. erectus* (2–1.5 Ma) were roughly the same size and substantially larger than the brains of australopiths. However, the range of brain sizes of both early *Homo* and *H. erectus* overlaps with that of the australopiths. The endocranial volumes of *H. erectus* increased over time. Notice that there are no fossils between 2.5 Ma and 2 Ma to allow the measurement of endocranial volume.

gorillas (**Figure 11.5**), suggesting that fossils within early *Homo* belonged to more than one species. However, there is much debate about exactly how to divide the fossils into species categories. Meave Leakey and Fred Spoor have suggested that fossils with more australopith-like faces, such as KNM-ER 1813 and OH 7, should be assigned to one species, *Homo habilis*, and those with more derived jaws and flat faces, such as KNM-ER 1470, should be assigned to another species, *Homo rudolfensis*.

There is much uncertainty about the postcrania of early *Homo*.

Paleoanthropologists have found a fair number of postcranial fossils from sites that have been dated to the same period as the early *Homo* fossils just discussed. The problem is that very little of this material is associated with fossil skulls, and since there were four species of hominins present in East Africa at this time (*P. boisei* and *H. erectus*, in addition to the early *Homo* species *H. habilis* and *H. rudolfensis*), it is not possible to say anything with certainty about the bodies of early *Homo*.

Homo erectus

Homo erectus appears in the African fossil record about 1.9 Ma.

Fossils of *Homo erectus* have been found at several sites in Kenya (Lake Turkana, Olorgesailie, and Ileret), as well as at Konso-Gardula (Ethiopia), Daka (Ethiopia), Olduvai Gorge (Tanzania), and Swartkrans (South Africa). **Figure 11.6** shows a very well-preserved skull (labeled KNM-ER 3733) from Lake Turkana that was found in 1976 by a team led by Richard Leakey. This fossil is now dated to between 1.8 Ma and 1.7 Ma. The oldest *H. erectus* fossil (labeled DH134) was recently found at Drimolen in South Africa. It consists of the top and back of a skull with features that are clearly diagnostic of *H. erectus*, and dates to a little more than 2 Ma. These findings suggest that early *Homo* and *H. erectus* may have coexisted in Africa for several hundred thousand years.

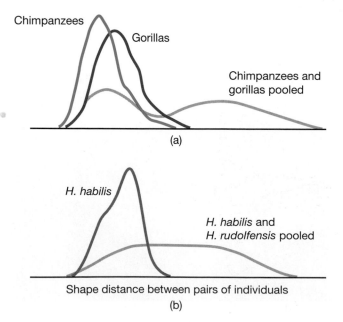

(a)

(b)

Shape distance between pairs of individuals

FIGURE 11.5

Part (a) plots the frequency of the difference in jaw shape between pairs of chimpanzees (*green*), pairs of gorillas (*red*), and pairs of both chimpanzees and gorillas (*gray*) pooled together. The horizontal axis gives the difference in jaw shape between pairs of specimens in a sample, and the vertical axis plots the frequency of pairs of specimens with the difference given on the horizontal axis. The pooled plot has two humps. The left hump comes from pairs of individuals of the same species that are similar, and therefore the differences between them are small. The right hump comes from pairs in which one member is a chimpanzee and the other is a gorilla, so the distance between the individuals is greater. Part (b) plots the frequencies of distances between pairs of fossils from *H. habilis* and *H. rudolfensis*. Notice that the pooled plot is broad like that of pooled chimpanzees and gorillas, whereas the plot of distances between individuals classified as *H. habilis* is similar to the plots of distances within living gorillas and chimpanzees. There are not enough *H. rudolfensis* fossils to plot this species alone. This supports the idea that *H. habilis* and *H. rudolfensis* were separate species.

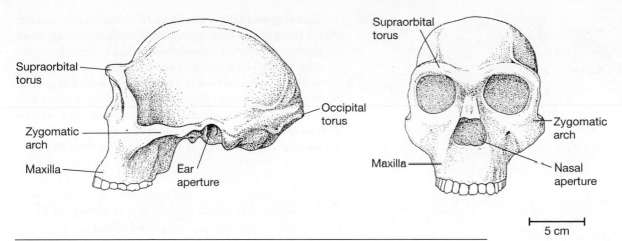

Supraorbital torus

Occipital torus

Zygomatic arch

Maxilla

Ear aperture

Supraorbital torus

Zygomatic arch

Maxilla

Nasal aperture

5 cm

FIGURE 11.6

Homo erectus skulls, like the skull of KNM-ER 3733 illustrated here, show a mix of primitive and derived features.

Skulls of *Homo erectus* differ from those of australopiths, early *Homo*, and modern humans.

Skulls of *Homo erectus* retain many of the characteristics of earlier hominins (**Figure 11.7**), including a marked narrowing behind the eyes, a receding forehead, and no chin. *H. erectus* also shows many derived features. Some of these are shared by modern humans, including a shorter and less prognathic face, a taller skull, smaller jaws and postcanine teeth, and fewer roots on the upper premolars. However, *H. erectus* also has some derived features that we don't see in earlier hominins or

FIGURE 11.7

Skulls of (a) *Australopithecus africanus*, (b) *Paranthropus robustus*, (c) *Homo erectus*, and (d) modern *Homo sapiens*.

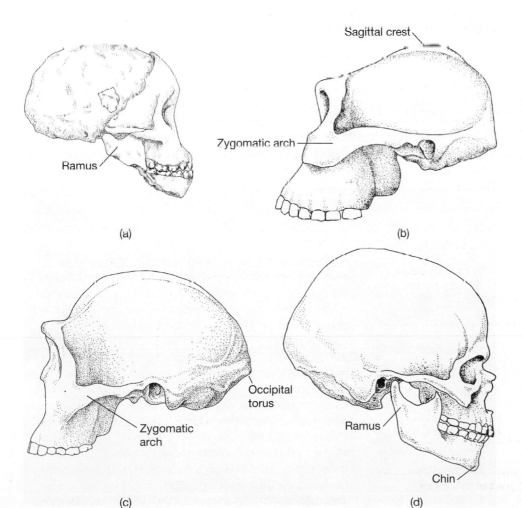

Sagittal crest

Zygomatic arch

Ramus

(a)

(b)

Occipital torus

Zygomatic arch

Ramus

Chin

(c)

(d)

Daka
Konso
Ledi-Geraru

Dmanisi

Yuanmou

L. Turkana
Olduvai
Sangiran
Uraha
Sterkfontein,
Swartkrans

● Early *Homo*

● Early *Homo* and *H. erectus*

● *H. erectus*

FIGURE 11.8

The locations of important fossil and archaeological sites. The sites in Eurasia date from 2.6 Ma to 0.8 Ma.

modern humans. For example, *H. erectus* has a horizontal ridge at the back of the skull (**occipital torus**), which gives it a pointed appearance when viewed from the side. It also has a shelflike browridge.

Many of the derived features of the skull of *H. erectus* may be related to its diet. These hominins were probably better adapted for tearing and biting with their canines and incisors and less suited to heavy chewing with their molars. Thus, all their teeth are smaller than the teeth of the australopiths, but their molars are reduced more in relation to their incisors. The large browridges and the point at the back of the skull may have been needed to buttress the skull against stresses created by an increased emphasis on tearing and biting.

The brain size of the earliest *H. erectus* found in Africa is about the same as that of the contemporary species *H. habilis* and *H. rudolfensis* (see Figure 11.4). The brains of *H. erectus* gradually increased so that by 1 Ma, specimens had brain volumes closer to 1,000 cc.

Homo erectus occupied almost all of Africa and extended its range into Eurasia.

Archaeological and paleontological evidence compiled by Richard Klein indicates that *H. erectus*'s range encompassed almost the entire African continent, except for areas of arid desert and the rain forests of the Congo River basin, suggesting that *H. erectus* could survive in a variety of environmental conditions. Around 1.5 Ma, *H. erectus* colonized the high-altitude plateaus of Ethiopia and made more intensive use of the dry edges of the Rift Valley than other hominins had. By 1 Ma, *H. erectus* had extended its range to the northernmost and southernmost parts of the continent (**Figure 11.8**).

We are not exactly sure how and when hominins first left Africa. But we do know that by around 1.8 Ma, hominins had reached the Caucasus Mountains of the Republic of Georgia. In 1991, archaeologists excavating beneath the ruins of a medieval town called Dmanisi (**Figure 11.9**) discovered the first of a remarkable series of

FIGURE 11.9

The excavation site at Dmanisi, at the foot of the Caucasus Mountains in the Republic of Georgia. Although archaeologists have worked there since 1936, only in 1983 did they begin to uncover remains from the Lower Pleistocene.

hominin fossils. In 1999, an international team of researchers, led by Leo Gabunia of the Republic of Georgia National Academy of Sciences, uncovered two nearly complete crania at Dmanisi that are very similar to African specimens of *H. erectus* (**Figure 11.10**). Later, the team discovered a very well-preserved cranium and associated mandible of a subadult with a very small brain (700 cc), which shows striking similarities to the specimen from Ileret, Kenya. In 2007, David Lordkipanidze and his colleagues published the first detailed descriptions of postcranial material from one adolescent and three adults found at Dmanisi. Radiometric dating of shards of volcanic glass found at the same levels as the hominin fossils indicates that they are about 1.8 million years old. The fauna associated with the hominin fossils are consistent with this date.

In 2013, Lordkipanidze and his colleagues published the first description of a remarkably well-preserved and intact cranium (**Figure 11.11**). The cranium matches a mandible found several years earlier in the same spot, and together they represent the most complete skull of *H. erectus* from any site. The skull presented paleontologists with several surprises. First, it had a very small brain. The cranial capacity was only 546 cc, making the brain smaller than that of any of the other specimens of *Homo* from Dmanisi and at the very low end of the distribution for *Homo* specimens from Africa. It also had a large and quite prognathic lower face with procumbent incisors. The skull was robust, with robust supraorbital tori (browridges), postorbital constriction, and a bulging glabellar region (the space between the eyebrows and above the nose). However, the skull also bears traits that link it to *Homo*, including the relatively vertical upper face and shape of the cranium.

Lordkipanidze and his colleagues emphasize that the Dmanisi hominins as a group display a diverse mix of primitive and derived features. They have relatively small brains (546 to 775 cc). They also retain primitive features in certain elements of their shoulder morphology. In modern humans, the elbow joint is rotated so that our palms face inward when our arms hung by our sides. In the Dmanisi hominins, the palms would be oriented more toward the front of the body. Conversely, the lower-limb morphology shows several derived features. The Dmanisi hominins have the same limb

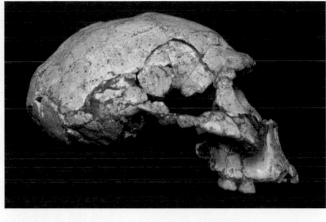

FIGURE 11.10

One of the crania found at Dmanisi. The Dmanisi fossils date to 1.8 Ma to 1.2 Ma and are very similar to *H. erectus* fossils of the same age from Africa. The Dmanisi site is at latitude 41° north, well out of the tropics, indicating that *H. erectus* could better adapt to a wider range of habitats than could previous hominins.

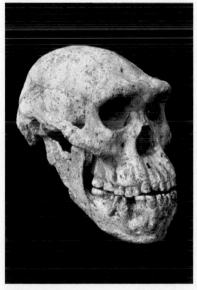

[a]

[b]

FIGURE 11.11

(a) A complete skull found at Dmanisi surprised paleontologists because it had a very small brain and large, robust jaws and teeth. (b) A reconstruction of the Dmanisi hominin.

proportions as modern humans, and their lower limbs and feet were well suited for long-distance walking and running. For example, they have relatively high femoral-to-tibia and humeral-to-femoral ratios, well-developed arches in their feet, and an adducted big toe (that is, it is in line with the rest of the toes).

Homo erectus arrived in eastern Asia between 1.8 Ma and 1.6 Ma.

Eugène Dubois (**Figure 11.12**) unearthed the first *Homo erectus* fossils near the Solo River in Java during the nineteenth century (**Figure 11.13**). Dubois named the species *Homo erectus*, or "erect man." Dubois's fossils were very poorly dated, and for many years they were believed to be about 500,000 years old. In the 1990s, however, Carl Swisher, now at Rutgers University, and Garniss Curtis used argon–argon dating techniques to determine the age of small crystals of rock from the sites at which Dubois's fossils had originally been found. Their analyses indicate that the two *H. erectus* sites in Java were actually 1.6 million to 1.8 million years old. The finding of two hominin incisors that are very similar to those associated with African *H. erectus* at Yuanmou in southern China that date to about 1.7 Ma supports the inference that *H. erectus* reached eastern Asia at this time.

Homo erectus may have developed more slowly than early hominins but more rapidly than modern humans.

Remember from Chapter 10 that the age of eruption of the first molar (M1) is a good predictor of important life history variables such as age of first reproduction and longevity, and that the age of M1 emergence can be estimated for fossils. Using the same methods that they used to estimate the age of australopiths, Kelley and Schwartz found that the M1 erupted at about 4.5 years in two *H. erectus* fossils. This finding suggests that *H. erectus* developed more slowly than australopiths but still faster than modern humans. If this interpretation is correct, then *H. erectus* did not have as long a childhood as modern humans have, suggesting that learning did not play as important a role in the lives of these creatures.

The postcranial skeleton of *Homo erectus* is more similar to the skeleton of modern humans than to that of earlier hominins, but it still differs from ours in interesting ways.

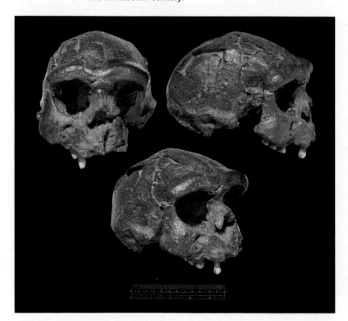

Fossils attributed to *H. erectus* differ considerably in size. The smallest ones, like KNM-ER 42700 from Ileret and BSN49/P27 from Gona, may have been the same size as early *Homo* and the australopiths, whereas others were considerably taller. The most complete information about the postcranial morphology of *H. erectus* comes from a spectacular specimen discovered by Kimoya Kimeu (**Figure 11.14**), the leader of the Koobi Fora paleontological team. The find was made on the western side of Lake Turkana, the same region where fossils of *Australopithecus anamensis*, *Paranthropus aethiopicus*, and *Kenyanthropus platyops* were found. The skeleton, formally known as KNM-WT 15000, belonged to a boy who was about 8 years old when he died. The skeleton provides us with a remarkably complete picture of the *H. erectus* body; even the delicate ribs and vertebrae are preserved (**Figure 11.15**).

Remember that earlier hominins were bipeds but still had long arms, short legs, and other features suggesting that they spent considerable time in trees. In contrast, KNM-WT 15000 had the same body proportions as people who live in tropical savannas today: long legs, narrow hips, narrow shoulders, and a barrel-shaped chest. KNM-WT 15000 also had shorter arms than those of earlier hominins. Taken together, these features suggest that *H. erectus* was fully committed to terrestrial life.

It is not clear how tall KNM-WT 15000 would have been as an adult. KNM-WT 15000 stood about 1.625 m (5.33 ft.) in height. If *H. erectus* growth patterns were comparable to those of modern humans, this boy would have been about 1.9 m (6 ft.) tall when he was fully grown. However, as we have seen, *H. erectus* may have developed more rapidly than modern humans, and if *H. erectus* had the same growth patterns as modern apes, he would have been close to his adult height when he died.

It is likewise not certain how dimorphic *H. erectus* was. Discoveries at Dmanisi and Gona indicate that *H. erectus* individuals came in a very wide range of sizes. But some of these differences are probably due to ecological variation, not differences between females and males. According to New York University anthropologist Susan Antón, taking the average over several fossils suggests that *H. erectus* males were 20% to 30% larger than females, making *H. erectus* less dimorphic than the australopiths but somewhat more dimorphic than modern humans. However, the small sample size of fossils and the difficulty of identifying the sex of fossils suggest that we should be cautious about firm conclusions.

Homo erectus may have been the first hominin that could run for long distances. Compared with most other mammals, modern humans are not good sprinters. However, we can outrun all but a few species over distances of several kilometers. Dennis Bramble, a biologist at the University of Utah, and Daniel Lieberman, an anthropologist at Harvard University, have argued that features of the *H. erectus* boy, such as long legs, narrow hips, and a barrel-shaped chest, are evidence that the capacity for long-distance running first appeared in this species. They believe that this talent may have been useful in long-distance scavenging and hunting in open country.

Stone Tools

Early hominins were probably tool users because tool use is common in apes.

Tool use almost certainly precedes the divergence from the lineage that led to modern great apes. Chimpanzees use long, thin branches, vines, stems, sticks, and twigs to poke into ant nests, termite mounds, and bees' nests to extract insects and honey. They sometimes scrape marrow and other tissue from the bones and braincases of mammalian prey with wooden twigs. They wad up leaves, dip them into pools of rainwater that have collected in hollow tree trunks, and then suck water from their leafy "sponges." In West Africa, chimpanzees pound stone hammers against heavy flat stones, exposed rocks, and roots to crack open hard-shelled nuts (**Figure 11.16**). Chimpanzees in West Africa use branches as spears to hunt galagos, and there is circumstantial evidence that they use digging sticks to extract underground storage organs too. There are examples of tool use in other apes as well. For example, orangutans use sticks to pry open fruits, and lowland gorillas sometimes use sticks to probe the depth of water as they cross swamps.

Chimpanzees sometimes modify natural objects for specific purposes. Caroline Tutin of the Centre International de Recherches Médicales de Franceville (CIRMF) in Gabon and William McGrew of Cambridge University observed that chimpanzees at Mount Assirik in Senegal use twigs to extract termites from their mounds. The twigs are first detached from the bush or shrub; then leaves growing from the stem are stripped off, the bark is peeled back from the stem, and the twig is clipped to an appropriate length. Jill Preutz of Texas State University and colleagues observed that chimpanzees sharpen the ends of the wooden branches that they use to hunt galagos.

FIGURE 11.14

Kimoya Kimeu is an accomplished field researcher. His many important finds include KNM-WT 15000 (see Figure 11.15).

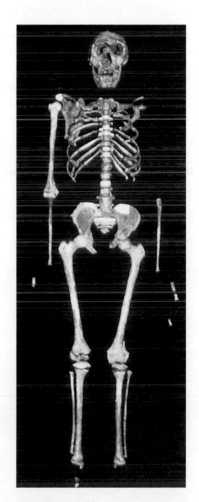

FIGURE 11.15

KNM-WT 15000, an *H. erectus* boy found on the western side of Lake Turkana, is amazingly complete.

It seems very likely that hominins used tools much as chimpanzees do today, but the only tools that have left traces in the archaeological record were those made from durable substances such as stone.

FIGURE 11.16

In West Africa, chimpanzees use stones to crack open hard-shelled nuts.

The earliest stone tools are probably very difficult to distinguish from naturally occurring rocks.

Early hominins may have first used naturally occurring stones as tools or made tools that are difficult to distinguish from naturally occurring stones. Simon Fraser University archaeologist Brian Hayden studied the manufacture and use of stone tools by aboriginal people in Australia's Western Desert and observed that they frequently use naturally occurring stones as tools both for butchering carcasses and for woodworking. He also found that most stone tools are manufactured using what archaeologists call the bipolar technique, in which the toolmaker places a small stone on a solid substrate and then smashes it with a larger stone. Some of the small flakes that result are useful tools, but these flakes are difficult to distinguish from stones broken as the result of ordinary geologic processes.

The oldest archaeological evidence for stone tool manufacture dates to 3.4 Ma.

For several years, the earliest known stone tools were dated to 2.6 Ma, having been found in Gona, a site in the Awash region of Ethiopia, in the 1990s. Then in 2010, at Dikika, paleoanthropologists found animal bones from 3.4 Ma that may have been marked by stone tools. Researchers thought it was possible that hominins used sharp-edged stones to scrape the flesh off bones of animal prey. But because no stone tools have been found in association with these bones, it is hard to exclude the possibility that the marks on the bones were made by other predators or by other natural processes. But in 2015, at Lomekwi, a site in West Turkana, researchers found an array of stone artifacts dating to 3.3 Ma, including **flakes** (small, sharp chips), **cores**, hammer stones, and debris from manufacturing (**Figure 11.17**). We think that these were tools because toolmaking (**knapping**) leaves telltale traces on the cores, so it's possible to distinguish natural breakage from deliberate modification. These tools are quite large, weighing up to 15 kg (33 lb.). The manufacturing techniques were very simple. Some were made using the bipolar technique, and others were made by smashing one stone onto a second stone sitting on the ground, a technique that archaeologists call the "passive hammer" technique. It's easy to see that it is only a small step from the chimpanzees' use of hammer stones to crack hard-shelled nuts to the bipolar or passive-hammer technique.

FIGURE 11.17

Lomekwi tools at the site where they were found.

More sophisticated Oldowan tools appear in the archaeological record at 2.6 Ma and become common around 2 Ma.

Beginning 2.6 Ma, somewhat more sophisticated tools appear in the archaeological record at Gona, a site in the Awash region of Ethiopia. The tool kit consists of rounded stones, like the cobbles once used to pave city streets. These cores were flaked (chipped) a few times to produce an edge (**Figure 11.18**). Ancient toolmakers who made these artifacts held the core in one hand and struck it with a second stone to produce the flakes. At a site on the western side of Lake Turkana dated to about 2.3 Ma, preservation conditions at the site were so good that workers were able to fit several flakes back onto the cores from which they had been struck. This reconstruction revealed that early hominins removed as many as 30 flakes from a single core, maintaining precise flaking angles throughout the toolmaking sequence.

These artifacts, collectively referred to as the **Oldowan tool industry** because they were identified at the Olduvai Gorge, are very simple. The Oldowan artifacts vary in their shape and size, probably because they were made of different raw materials. There is evidence that the flakes struck from these cobbles were at least as useful as the cores themselves (**A Closer Look 11.1**).

Up until about 2 Ma, Oldowan tools appear only sporadically in the archaeological record, but after this they become much more common. They continue to dominate the African archaeological record until about 1.6 Ma, when they were replaced by more sophisticated tools.

Chopper

Hammer stone

Flake scraper

Discoid

Polyhedron

Flake

├────────────┤
5 cm

Heavy-duty (core) scraper

FIGURE 11.18

Stone tools like these first appear in the archaeological record about 2.6 Ma. Researchers are not sure how these tools were used. Some think that the large cores shown here were used for a variety of tasks; others think that the small flakes removed from these cores were the real tools.

The Oldowan tool industry is an example of Mode 1 technology.

Tool industries such as the Oldowan refer to collections of tools that are found in a particular region and time. It is also useful to have a name for a particular method of manufacturing tools so that we can compare industries from different times and places. The late J. Desmond Clark devised a scheme for classifying modes of production that we will use here. According to this scheme, crude flaked pebble tools such as those associated with the Oldowan tool industry are classified as **Mode 1** technology. The distinction between industry and mode is not very meaningful for our discussion of early hominins, who used simple techniques to create a very limited set of tools, but it will become useful later when we examine the major regional and temporal variations in the composition of tool kits and modes of production (see Chapters 12 and 13).

We do not know which hominin species were responsible for making these early stone tools.

As you learned in Chapter 10, several hominin species were running around in East Africa between 3.3 Ma and 2 Ma, and it is not clear which of these species made the earliest stone tools. Many archaeologists have long assumed that early *Homo* was the first toolmaker. However, since the recent discovery of the stone tools in West Turkana dating to 3.3 Ma, this hypothesis has become problematic because the oldest *Homo* fossil dates to only 2.8 Ma—half a million years after those first stone tools appeared in the archaeological record. It is possible that one or all of the three hominin species in Africa 3.3 Ma (that is, *Au. afarensis*, *Au. deyiremeda*, and *K. platyops*) manufactured the first stone tools. However, because stone tools are more durable than bones, the archaeological record is usually more complete than the fossil record. This means the earliest tools typically appear in the fossil record before the first fossil of the creature that made them (see A Closer Look 9.4 for more discussion of this topic). Thus, the first stone toolmaker could have been an early member of the genus *Homo*. Until we find tools and fossils together at the same site, we will not be able to resolve this question.

After 2 Ma, stone tools become much more common in the archaeological record, and many paleoanthropologists believe that early *Homo* and *H. erectus* are the most likely makers of these tools. The only other possibilities are *P. boisei* in East Africa and *P. robustus* in South Africa.

11.1 Ancient Toolmaking and Tool Use

Kathy Schick and Nicholas Toth of Indiana University have done many experiments with simple stone tools. They have mastered the skills needed to manufacture the kinds of artifacts found at Oldowan sites and learned how to use them effectively. Their experiments have produced several interesting results. Schick and Toth believe the Oldowan artifacts that archaeologists have painstakingly collected and described are not really tools at all. They are cores left over after striking off small, sharp flakes, and the flakes are the real tools. Schick and Toth find that the flakes can be used for an impressive variety of tasks, even for butchering large animals such as elephants. In contrast, although the cores can be used for some tasks, such as for chopping down a tree to make a digging stick or a spear or for cracking bones to extract marrow, they are generally much less useful. Schick and Toth's conclusion is supported by microscopic analysis of the edges of a few Oldowan flakes, which indicates that they were used for both woodworking and butchery.

Schick and Toth have also been able to explain the function of enigmatic objects that archaeologists call spheroids. These are smooth, approximately spherical pieces of quartz about the size of a baseball. Several suggestions for their function have been put forth, including processing plants and bashing bones to extract marrow. Some researchers thought the spheroids were part of a bola, a hunting tool used on the grasslands of Argentina. In a bola, three stones are connected by leather thongs and thrown so that they tangle the prey's legs. Schick and Toth have shown there is a much more plausible explanation for these stones. If a piece of quartz is used as a hammer to produce flakes, bits of the hammer stone are inadvertently knocked off. The hammer surface is no longer flat, so the hammer stone is shifted in the toolmaker's hand. Thus the hammer gradually becomes more and more spherical. After a while, a quartz hammer becomes a spheroid.

Perhaps the most remarkable conclusion that Schick and Toth drew from their experiments is that early hominin toolmakers were right-handed. Schick and Toth found that right-handers usually hold the hammer stone in their right hand and hold the stone to be flaked in the left hand. After driving off the first flake, they rotate the stone clockwise and drive off the second flake. This sequence produces flakes in which the **cortex** (the rough, unknapped surface of the stone) is typically on the right side of the flakes (**Figure 11.19**). On flakes made by left-handers, the cortex is typically on the left side. With these data in mind, Schick and Toth studied flakes from sites at Koobi Fora, Kenya, dated to 1.9 Ma to 1.5 Ma. Their results suggest that most of the individuals who made the flakes were right-handed.

FIGURE 11.19

Demonstration of why right-handed flint knappers make distinctive flakes. When a right-handed person makes a stone tool, she typically holds the hammer stone in her right hand and the core to be flaked in the left hand. When the hammer stone strikes the core, a flake spalls off, leaving a characteristic pattern of rays and ripple marks centered around the point of impact (shown in *red*). The knapper then rotates the core and strikes it again. If she rotates the core clockwise, as shown, the second flake has the percussion marks from the first flake on the upper left and part of the original rough surface of the rock, or cortex (shown in *gray*), on the right. If she rotates the core counterclockwise, the cortex will be on the left and the percussion marks on the upper right. Modern right-handed flint knappers make about 56% right-handed flakes and 44% left-handed flakes, and left-handed flint knappers do just the opposite. A sample of flakes from Koobi Fora dated to between 1.9 Ma and 1.5 Ma contains 57% right-handed flakes and 43% left-handed flakes, suggesting that early hominin toolmakers were right-handed.

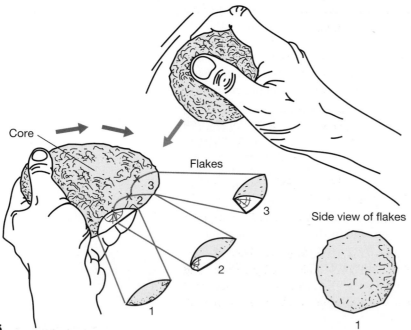

Core

Flakes

Side view of flakes

More sophisticated Mode 2 tools first appear about 1.75 Ma.

In Africa around 1.75 Ma, hominins added a new and more sophisticated type of tool to their kit. This totally new kind of stone tool, called a **biface**, appears almost simultaneously at Konso in Ethiopia and in West Turkana, and a bit later at Olduvai. To make a biface, the toolmaker strikes a large piece of rock from a boulder to make a core and then flakes this core on all sides to create a flattened form with a sharp edge along its entire circumference. The most common type of biface, called a **hand ax**, is shaped like a teardrop and has a sharp point at the narrow end (**Figure 11.20**). A **cleaver** is a lozenge-shaped biface with a flat, sharp edge on one end; a **pick** is a thicker, more triangular biface. Bifaces are larger than Oldowan tools, averaging about 15 cm (6 in.) in length and sometimes reaching 30 cm (12 in.). Bifaces are categorized as **Mode 2** technology. Paleoanthropologists call the Mode 2 industries of Africa and western Eurasia (that is, Europe and the Middle East) the **Acheulean industry** after the French town of Saint-Acheul, where hand axes were first discovered. However, Oldowan tools do not disappear when Acheulean tools make their debut. Hominins continued making simple Mode 1 tools, perhaps when they needed a serviceable tool in a hurry—as foraging people do today.

The standardized form of hand axes and other Mode 2 tools in the Acheulean industry suggests that toolmakers had a specific design in mind when they made each tool. The Mode 1 tools of the Oldowan industry have a haphazard appearance; no two are alike. This lack of standardization suggests that makers of Oldowan tools simply picked up a core and struck off flakes; they didn't try to create a tool with a particular shape that they had in mind beforehand. They may have done this because the flakes were the actual tools.

It is easy to see how a biface might have evolved from an Oldowan chopper by extending the fluking around the periphery of the tool. However, Acheulean tools are not just Oldowan tools with longer edges; they are designed according to a uniform plan and have regular proportions. The ratio of height to width to thickness is remarkably constant from one hand ax to another. *Homo erectus* must have started with an irregularly shaped piece of rock and whittled it down by striking flakes from both sides until it had the desired shape. Not all the hand axes would have come out the same if their makers hadn't shared an idea for the design.

There is no doubt that *H. erectus* made Acheulean tools because this was the only hominin species in Africa after 1.4 Ma. However, previously *H. erectus* had coexisted with early *Homo*, so it is possible that early *Homo* also manufactured Acheulean tools.

Hand axes were probably used to butcher large animals.

If hand axes were designed, what were they designed for? The answer to this question is not obvious because hand axes are not much like the tools made by later peoples. Several ideas about what hand axes were used for have been proposed:

1. *Butchering large animals.* *Homo erectus* acquired the carcasses of animals such as zebra or buffalo either by hunting or by scavenging and then used a hand ax as a modern butcher would use a cleaver—to dismember the carcass and cut it into useful pieces.

2. *Dispensing flake tools.* Hand axes weren't tools at all. Instead they were "flake dispensers" from which hominins struck flakes to be used for many everyday purposes.

3. *Woodworking.* Hand axes were used to shape spears, throwing sticks, and digging sticks.

Although we are not certain how hand axes were used, two kinds of evidence support the hypothesis that they were heavy-duty butchery tools. Kathy Schick and Nicholas Toth, whose investigations into the function of Oldowan tools we discussed in A Closer Look 11.1, have also done experiments using Acheulean hand axes for each

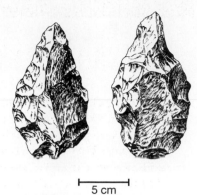

5 cm

FIGURE 11.20

Acheulean hand axes were teardrop-shaped tools created by removing flakes from a core. The smallest ones would fit in the palm of your hand, and the largest ones are more than 0.3 m (1 ft.) long.

of the tasks just listed. From these experiments, Schick and Toth conclude that hand axes are best suited to butchery. The sharp end of the hand ax easily cuts through meat and separates joints; the rounded end provides a secure grip. The large size is useful because it provides a long cutting edge, as well as the cutting weight necessary to be effective for a tool without a long handle. Schick and Toth's results are supported by the work of Lawrence Keeley, a paleoanthropologist at the University of Illinois at Chicago, who performed microscopic analysis of wear patterns on a few hand axes. He concluded that the pattern of wear is consistent with animal butchery.

Evidence from Olorgesailie, a site in Kenya at which enormous numbers of hand axes have been found, indicates that these axes may also have served as flake dispensers. A team led by Richard Potts of the Smithsonian Institution discovered most of the fossilized skeleton of an elephant, along with many small stone flakes dated to about 1 Ma. Chips on the edges of the flakes suggest that they were used to butcher the elephant, and the elephant bones show cut marks made by stone tools. Careful examination of the flakes reveals that they were struck from an already flaked core, such as a hand ax, not from an unflaked cobble. Moreover, when the flakes were removed from the hand axes, they sharpened the edge of the hand ax. Taken together, these data suggest that Acheulean toolmakers struck flakes from hand axes and used both hand axes and flakes as butchery tools.

An alternative view is that hand axes and other early stone tools were used mainly for woodworking. Recall that observations of contemporary aboriginal Australians show that they sometimes make stone tools. They use these tools to make spears and digging sticks much more often than they use them for animal butchery. This may have been true for the makers of hand axes as well.

The Acheulean industry remained remarkably unchanged for over 1 million years.

There is relatively little change over space and time in the Acheulean tool kit from its first appearance about 1.7 Ma until it was replaced around 300 ka. Amazingly enough, Acheulean tools that were made half a million years apart are just as similar as tools made at about the same time at sites located thousands of miles apart. Essentially, the same tools were made for more than 1 million years. Most anthropologists assume that the knowledge necessary to make a proper hand ax was passed from one generation to the next by teaching and imitation. However, some archaeologists find it hard to accept that such knowledge could have been faithfully transmitted and preserved for so long in a small population of hominins spread from Africa to eastern Eurasia. They hypothesize, instead, that early hominins had a genetically transmitted template analogous to those that regulate the production of artifacts by nonhuman species, such as birds' nests and beavers' dams, in which social learning, teaching, and imitation seem to play little role.

In East Asia, *Homo erectus* is associated with Mode 1 tools.

Acheulean tools appear by 1.75 Ma in Africa, but they are rarely found in eastern Asia. Instead, *H. erectus* in East Asia is usually associated with the simpler Mode 1 tools, similar to those of the Oldowan industry. Some paleoanthropologists think that the differences in tool technology between eastern Asia and Africa provide evidence of cognitive differences between eastern and western populations of *H. erectus*. Remember that *H. erectus* may have been present in eastern Asia by 1.7 Ma. If so, then *H. erectus* may have left Africa before Mode 2 tools first appeared there. It is possible that a cognitive change arose in the African *H. erectus* population after the first members of the genus had migrated north into eastern Asia and that this cognitive adaptation enabled later members of African *H. erectus* to manufacture bifaces. The absence of more symmetric Mode 2 tools in eastern Asia may mean that they lacked this cognitive ability.

Other researchers argue that differences in tool technology between eastern and western populations tell us more about their habitats than about their cognitive abilities. During the Middle Pleistocene, eastern populations of *H. erectus* lived in regions that were covered with dense bamboo forests. Bamboo is unique among woods because it can be used to make sharp, hard tools suitable for butchering game. In eastern Asia, *H. erectus* may not have made hand axes because they didn't need them. The discovery of many hand axes in the Bose basin in southern China supports this view (**Figure 11.21**). About 800 ka, a large meteor struck this area and set off fires that destroyed a wide area of the forest where bamboo grew. Grasslands replaced the forest. While grasslands predominated, and bamboo was presumably scarce, the residents of this area made Mode 2 tools like those seen in Africa around the same time. Before the forest was destroyed and after it regenerated, the inhabitants of this area made Oldowan-like Mode 1 tools. This evidence suggests that hand axes were an adaptation to open-country life.

FIGURE 11.21

Hand axes found in the Bose basin in southern China are the only Mode 2 tools discovered in East Asia. These tools date to about 800 ka. Mode 1 tools are found in the same area both before and after this date.

Complex Foraging Shapes Human Life History

Observations of how people subsist by hunting and gathering are useful for understanding human evolution.

Modern humans appear in the fossil record about 200 ka, and from at least that date until the origin of agriculture all people subsisted by hunting and gathering. During the twentieth century, anthropologists studied foraging societies in many parts of the world and learned that they all differ from the societies of other primates in two important ways. First, modern foragers depend on complex, hard-to-learn foraging techniques; second, food sharing and division of labor are common. To understand the transition from early apelike hominins to modern humans, it is useful to consider these features of modern foraging societies in more detail.

Anthropologists divide the foods acquired by foragers into three types according to the amount of knowledge and skill required to obtain them. These are, in order of increasing difficulty of acquisition, collected foods, extracted foods, and hunted foods.

Hillard Kaplan and Jane Lancaster of the University of New Mexico, along with Kim Hill and A. Magdalena Hurtado of Arizona State University, have argued that the evolution of modern human life history was driven by a shift to valuable but hard-to-acquire food resources. They rank food resources into three categories according to how difficult it is to acquire them:

1. **Collected foods** can be simply gathered from the environment and eaten. Examples include ripe fruit and leaves.

2. **Extracted foods** come from things that don't move but are protected in some way. These things must be processed before the food can be eaten. Examples include fruits in hard shells, tubers or termites that are buried deep underground, honey hidden in hives high in trees, and plants containing toxins that must be extracted before the plants can be eaten.

FIGURE 11.22

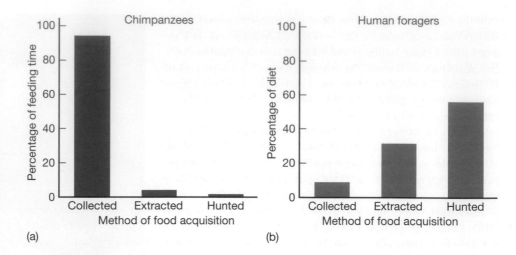

3. **Hunted foods** come from things that run away and must be caught or trapped. They may also need to be extracted and processed before consumption. Vertebrate prey are the prime example of hunted foods for both humans and chimpanzees.

Apes are the brainiacs of the nonhuman primate world, at least when it comes to foraging. Gorillas and orangutans use elaborate routines to process some plant foods. Both orangutans and chimpanzees use tools to process some kinds of foods. Moreover, as we will see in Chapter 16, different ape populations use different techniques to acquire extracted foods such as hard-shelled nuts. Chimpanzees have a broader diet than that of other apes, including both collected and extracted foods. They also hunt small game, including colobus monkeys. However, even these clever apes do not come close to human expertise in foraging.

Humans depend on hard-to-learn skills to acquire food.

Kaplan and his colleagues emphasize the importance of the fact that contemporary foraging peoples depend on extracted and hunted foods to a much greater extent than chimpanzees do. **Figure 11.22** compares the average dependence of chimpanzees and humans on collected, extracted, and hunted foods. The general pattern is clear: Chimpanzees are overwhelmingly dependent on collected resources, but human foragers get almost all their calories from extracted or hunted resources.

Unlike other predators, humans must learn a very diverse set of hunting skills. Most large mammalian predators capture a relatively small selection of prey species by using one of two methods: They wait in ambush or they combine a stealthy approach with fast pursuit. Once the prey is captured, they process it with tooth and claw. In contrast, human hunters use a vast number of methods to capture and process a huge variety of prey species. For example, the Aché, a group of foragers who live in Paraguay, hunt 78 species of mammals, 21 species of reptiles, 14 species of fish, and more than 150 species of birds by using a vast array of techniques that depend on the prey type, the season, the weather, and many other factors (**Figure 11.23**). The Aché track some animals, a difficult skill that entails a great deal of ecological knowledge. They call other animals by imitating the prey's mating or distress sounds. Still other animals they capture with snares or traps or by smoking them out of burrows. They capture and kill animals by using their hands, arrows, clubs, or spears. And this is just the one group of foragers in one habitat; if we included all human habitats, the list would be immeasurably longer.

It takes a long time to learn this range of skills. Jeremy Koster of the University of Cincinnati led a project to assess changes in men's hunting efficiency over the life course. He and his colleagues assembled a massive data set, which included data on about 23,000 hunts by 1,800 individuals in 40 societies. They used these data to

FIGURE 11.23

Meat makes up about 70% of the diet of the Aché, a group of foragers from Paraguay. Here an Aché man takes aim at a monkey.

evaluate men's hunting returns as a function of age. They found that, on average, men's hunting returns are highest when they are between 30 and 35 years of age (**Figure 11.24**). But the high point is more like a gentle hill than a sharp peak: by 18 years of age men are about 89% as productive as they are at their peak, and they remain within 89% of peak productivity until their mid-fifties.

Efficient extraction of resources also requires considerable skill. Nicholas Blurton Jones, an anthropologist at the University of California, Los Angeles, who has studied two African foraging groups, the Hadza and the !Kung, describes digging up deeply buried tubers from rocky soil as a complex mining operation involving much clever engineering of braces and levers. Among the Hiwi, a group of foragers living in the tropical savanna of Venezuela, women do not achieve maximum efficiency in gathering roots until they are between 35 and 45 years old. Ten-year-old girls get only 10% as much as older, highly skilled women. Among the Aché, rates of starch extraction from palms and honey extraction also peak when people are in their twenties.

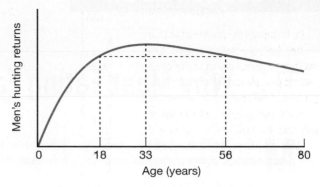

FIGURE 11.24

Men's productivity peaks when they are in the early thirties. By the time they are 18 years old, men are about 89% as successful as they are at their peak, and they drop below 89% of peak proficiency when they are in their middle fifties.

A reliance on hunting and extractive foraging favors food sharing and division of labor in contemporary foraging groups.

In all contemporary foraging groups, hunting and extractive foraging are associated with extensive food sharing and sexual division of labor. In nearly all foraging groups, men take primary responsibility for hunting large game, and women take primary responsibility for extractive foraging (**Figure 11.25**). This division of labor makes sense for two reasons: First, hard-to-learn techniques reward specialization. It takes a long time to learn how to be a good hunter, and it takes a long time to learn how to dig tubers. This means that everyone is better off if some individuals specialize in hunting and others specialize in extractive foraging, as long as people share the food they obtain. Second, because child care is more compatible with gathering than with hunting, and lactation commits women to child care for a substantial portion of their adult lives, it makes sense that men specialize in hunting and women specialize in extractive foraging. Again, this works only if members of the group regularly share food.

FIGURE 11.25

Data on foraging behavior for three well-studied modern foraging groups: the Aché, the Hadza, and the Hiwi. Men and women in these groups specialize in different foraging tasks. Men hunt, and women concentrate on extractive foraging.

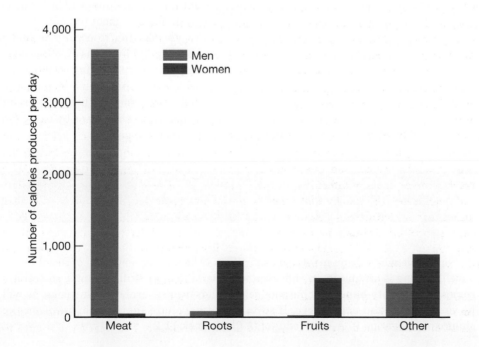

11.2 Why Meat Eating Favors Food Sharing

Many anthropologists believe that a heavy dependence on meat makes food sharing necessary. Let's examine how food sharing provides insurance against the risks inherent in hunting. Hunting, especially for hunters who concentrate on large game, is a boom-or-bust activity. When a hunter makes a kill, a lot of high-quality food becomes available. Hunters are often unlucky, however, and each time one sets out to hunt there is a fairly high probability of returning empty-handed and hungry. Food sharing greatly reduces the risks associated with hunting by averaging returns over several hunters.

To see why this argument has such force, consider the following simple hypothetical example. Suppose there are five hunters in a group that subsists entirely on meat. Hunters can hunt every day, and each hunter has a 1-in-5 (0.2) chance of making a kill and a 4-in-5 (0.8) chance of bringing back nothing. Further, suppose that people starve after 10 days without food. We can calculate the probability of starvation for each hunter over a 10-day period by multiplying the probability of failing on the first day (0.8) by the probability of failing on the second day (0.8), and so on to get

$$0.8 \times 0.8 \times 0.8 \times 0.8 \times 0.8 \times 0.8 \\ \times 0.8 \times 0.8 \times 0.8 \times 0.8 \approx 0.1$$

Thus there is about a 10% chance that a hunter will starve over any 10-day period. With these odds, it is impossible for people to sustain themselves by hunting alone.

A comparison with chimpanzee hunting provides good reason to think that these probabilities are realistic for early hominins. Craig Stanford of the University of Southern California and his colleagues have analyzed records of hunting by chimpanzees at Gombe Stream National Park in Tanzania. In about half of the hunts, the chimpanzees killed at least one monkey, and sometimes they made more than one kill. The average number of monkeys killed per hunt was 0.84. However, the hunting groups contained seven males on average. Dividing 0.84 by 7 gives the average number of monkeys killed per male per hunt, which comes out to 0.12. Thus, on any given day, each male chimpanzee had only a 12% chance of making a kill, which is less than the 20% chance we posited for the five human hunters at the outset of our example.

Now let's consider how sharing food alters the probability of starvation for our human hunters. If each hunter has a 0.8 chance of coming back empty-handed, then the chance that all five hunters will come back on a given evening without food is

$$0.8 \times 0.8 \times 0.8 \times 0.8 \times 0.8 \approx 0.33$$

Thus on each day there is a 1-in-3 chance that no one will make a kill. If the kill is large enough to feed all members of the group, then no one will go hungry as long as someone succeeds. The chance that all five hunters will face starvation during any 10-day period is

$$0.33 \times 0.33 \times 0.33 \times 0.33 \times 0.33 \\ \times 0.33 \times 0.33 \times 0.33 \times 0.33 \\ \times 0.33 \approx 0.000015$$

Sharing reduces the chance of starvation from 1 chance in 10 to roughly 1 chance in 60,000. Clearly, the risks associated with hunting could be reduced even further if there were alternative sources of food that unsuccessful hunters might share. For example, suppose one member of the group hunted while the others foraged, and they all contributed food to a communal pot.

The fact that food sharing is mutually beneficial is not enough to make it happen, however; as we pointed out in Chapter 7, food sharing is an altruistic act. Each individual will be better off if he or she gets meat but does not share it. For sharing to occur among unrelated individuals, as it often does in contemporary foraging societies, those who do not share must be punished in some way, such as by being excluded from future sharing or by being forced to leave the group.

For people who rely on meat, food sharing may be a necessary form of social insurance. Hunting is an uncertain endeavor, and even the most skilled hunter sometimes comes home empty-handed. If his bad luck extends over several days or weeks, he will be very hungry and may starve. If several hunters share their catch, however, the chance of starvation is greatly reduced (**A Closer Look 11.2**).

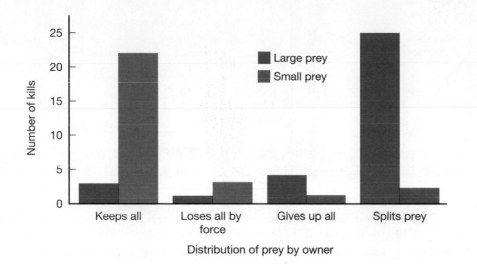

FIGURE 11.26

In the Taï Forest, chimpanzees sometimes share their kills. Infant and juvenile monkeys (small prey) are generally consumed by the captor, but adult monkey carcasses are generally divided by the captor and shared with other chimpanzees.

Food sharing occurs in chimpanzees but plays a much less important nutritional role than it does in humans. Once they are weaned, chimpanzees obtain virtually all their own food themselves.

Unlike most other primates, chimpanzees sometimes share food. Mothers share plant foods with their infants, and adults sometimes share meat. The patterns of food sharing between mothers and their infants have been most carefully analyzed at Gombe, in northeastern Nigeria. There, mothers are most likely to share foods that are diffi cult for the infants to obtain or to process independently (see Chapter 7). For example, infants have a hard time opening hard-shelled fruits and extracting seeds from sticky pods. A mother will often allow an infant to take bits of these items from her own hand or sometimes she will spontaneously offer them to her infant. When chimpanzees capture vertebrate prey, they dismember, divide, and sometimes redistribute the meat among members of the foraging party. In the Taï Forest, small prey are generally retained by the captor, and larger prey are typically divided among several individuals (**Figure 11.26**). The distribution of meat spans a continuum from outright coercion to apparently voluntary donations. High-ranking males sometimes take kills away from lower-ranking males, and adult males sometimes take kills away from females. At Gombe, about one-third of all kills are appropriated by higher-ranking individuals. More often, however, kills are retained by the captor and shared with others who cluster closely around. Males, generally the ones who control the kills, share food with other males, adult females, juveniles, and infants (**Figure 11.27**). However, in all these communities, the amounts of food obtained through various forms of food sharing constitute just a small fraction of all calories consumed.

FIGURE 11.27

Male chimpanzees sometimes share their kills with other males, sexually receptive females, and immature individuals.

In foraging societies, food sharing and division of labor lead to extensive flows of food between people of different ages and sexes.

The self-sufficiency after weaning that we see in chimpanzees is probably the ancestral state in the hominin lineage. The economy of human foragers, on the other hand, is strikingly different: Some people produce much more food than they consume, and others consume much more than they produce. Anthropologists have quantitatively studied the subsistence economies of several foraging groups. In these societies, anthropologists observed people's daily behavior, measuring how much food they produced and how much they consumed. For three of these groups—the Aché, the Hiwi, and the Hadza—researchers have

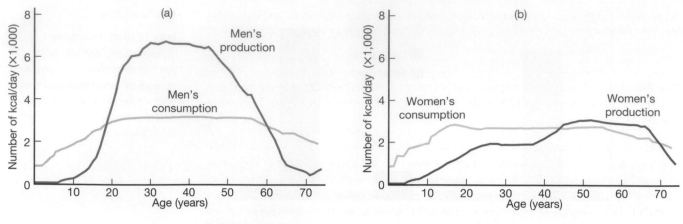

FIGURE 11.28

Data from three contemporary foraging groups show that both (a) men and (b) women do not become self-sufficient in terms of food production until they are adults. Adult men produce many more calories than they consume; adult women are approximately in balance.

---

computed average food production and consumption for men and women of different ages. Kaplan and his colleagues compiled these data to compare patterns of food production across societies. Their analysis reveals striking differences in the foraging economy of humans and chimpanzees and important changes in lifetime productivity.

Figure 11.28 shows that human children continue to depend on others for food long after they are weaned. Men become self-sufficient around age 17, and women do not produce enough to feed themselves until they are in their late forties. Older men also depend on others for their daily needs. These deficits are made up by the production of young and middle-aged men and, to a lesser extent, by the efforts of postmenopausal women. In contrast, chimpanzees obtain very little of their food from others after they are weaned.

Less detailed data from other foraging groups are consistent with this pattern. **Figure 11.29** shows that men contribute more than half of the total calories consumed in seven of the nine foraging groups for which the necessary data are available. Notice that all these groups live in tropical habitats. It seems likely from historical and

FIGURE 11.29

Males contribute significantly more calories than do women in nine contemporary foraging groups from different parts of the world.

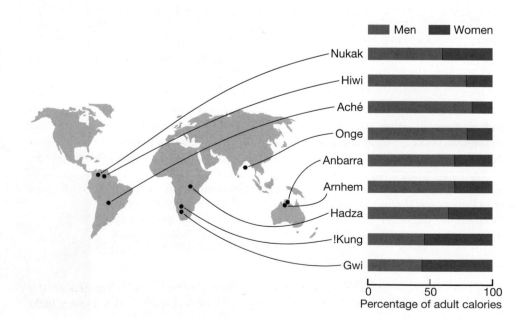

ethnographic accounts that temperate and arctic foragers depend even more on meat than do tropical foragers, and thus in these societies men may contribute even more calories.

Selection may have favored larger brains, a prolonged juvenile period, and a longer life span because these traits make it easier to learn complex foraging methods.

Complex, learned foraging techniques allow humans to acquire highly valuable or otherwise inaccessible food resources. Meat is a much better source of most of the nutrients that animals need than are leaves and ripe fruit, the foods on which most other primates rely. Meat is rich in energy, essential lipids, and protein. It is also dense enough to be economically transported from the kill site to home base. Some extracted resources such as honey, insect larvae, and termites are also concentrated sources of important nutrients. Extractive foraging can unlock vast new food supplies. Tubers are a prime example. Several tropical savanna plants store their reserve supplies of energy underground as various kinds of tubers, protected from the teeming herds of grazers and browsers by as much as a meter of rocky soil. By learning to recognize which plant species have tubers, how to use tools to dig them up, and when such work is likely to be profitable, humans gained access to a large supply of food for which there was relatively little competition from other organisms.

If learning is valuable, natural selection will favor adaptations that make a better learner. Thus, a shift to hunting and extractive foraging may have favored larger brains and greater intelligence. Reliance on complex, learned foraging skills would also favor the evolution of a prolonged juvenile period. As we all know, learning takes time. You can't become a proficient skier, baker, or computer programmer in a day; practice and experience are needed. Similarly, it takes years to learn the habits of animals and the lore of plants and to acquire the knowledge for tracking animals, and it takes years of practice to be able to hit a moving animal with an arrow or blowgun and to become adept at extracting starch from baobab pulp. Thus, it is plausible that selection favored a longer juvenile period to allow human children the time to acquire the skills they needed to become skilled foragers.

A prolonged juvenile period generates selection for a long life span. It is often said that time is money. But in evolution, time is fitness. To see why, suppose that two genotypes, A and B, have the same number of children on average, but type A completes reproduction in 30 years and type B completes reproduction in 60 years. If you do a bit of math, you will see that type A will have twice the population growth rate as type B, and it will quickly replace type B in the population. However, what if the offspring of type B individuals were twice as likely to survive as offspring of type A individuals? Although a prolonged juvenile period can be costly, it may be favored by natural selection because it causes people to have more surviving children over their life span. Human childhood is like a costly investment; it costs time, but the added time allows learning that produces more capable adults. Like any expensive investment, it will pay off more if it is amortized over a longer period. (The same logic explains why you are willing to spend more on something that you will use for a long time, such as a watch or a new pair of shoes, than on something you will use for a short time and then discard, such as a notepad.) Selection favors a longer life because it allows people to get more benefit from the productive foraging techniques they learned during the necessary, but costly, juvenile period.

Food sharing and division of labor lead to reduced competition between males and reduced sexual dimorphism.

We saw in Chapter 6 that the intensity of competition between males depends on the amount of male investment in offspring. In most primate species, males do very little for their offspring, and selection consequently favors traits in males that enhance their ability to compete with other males for matings. This increased competition leads to the pronounced sexual dimorphism seen in most primate species. When males do invest in offspring, there is less male–male competition and reduced sexual dimorphism.

Food-sharing patterns seen in contemporary foraging societies mean that males are making substantial investments in offspring. This is clear from the data shown in Figure 11.29: Males produce the bulk of the surplus calories that sustain children and teenagers. Thus, we would expect selection to favor behavioral and morphological traits that make men good providers, and we would also expect selective pressures favoring traits linked to male–male competition to be reduced. The reduction in male–male competition, in turn, should lead to reduced sexual dimorphism such as that seen in modern humans.

Evidence for Complex Foraging by Early Toolmakers

Let's stop and review for a second. So far, we have made two points. First, the early toolmakers (whichever species may have been the culprit) are plausible candidates for the species that links early apelike hominins to later hominins who have more humanlike life history patterns. Second, contemporary foragers rely on complex, hard-to-learn foraging techniques much more than other primates do, and this shift can explain the evolution of the main features of human life history. To link these points, we need to consider the evidence that toolmaking hominins had begun to rely on extractive foraging and hunting to make a living.

Wear patterns on bone tools from South Africa suggest that they were used to excavate termite mounds.

Most of the discussion of the subsistence patterns of early hominins focuses on meat eating, but researchers have recently begun to investigate the importance of extractive foraging as well. Modern peoples often use wooden digging sticks to extract tubers, a laborious business. There are reports of chimpanzees using sticks to dig for tubers as well. This kind of foraging activity is likely to leave few traces in the archaeological record because wood does not preserve well. However, one interesting piece of evidence suggests that Plio-Pleistocene hominins were extractive foragers. During their excavations at Swartkrans, Robert Brain and his co-workers identified a sizable number of broken bones that had wear patterns suggesting that they had been used as tools. Lucinda Backwell of the University of the Witwatersrand and Francesco d'Errico of the Institut de Préhistoire et de Géologie du Quaternaire in France analyzed these bones to find out how they were used. First the researchers used freshly broken bones to do several foraging tasks, including digging in hard soil for tubers and digging in a termite mound. Each activity creates a distinctive wear pattern that can be detected under microscopic analysis. Then they compared these wear patterns with those on the fossil bones found at Swartkrans.

This analysis indicated that the fossil tools were used to dig in termite mounds. **Figure 11.30** shows the wear patterns on the fossil tool and two experimental tools, one used to dig in soil and the other used to excavate a termite mound. All the tools have a smooth, rounded point. However, the tool experimentally used to dig in soil has deep marks of different depths going in all directions. In contrast, the tool used to dig in termite mounds has fine, parallel grooves. The fossil-bone tools resemble the experimental tool used to dig in termite mounds, so these tools were probably used for that purpose. If this is correct, then Oldowan hominins were using tools to do extractive foraging.

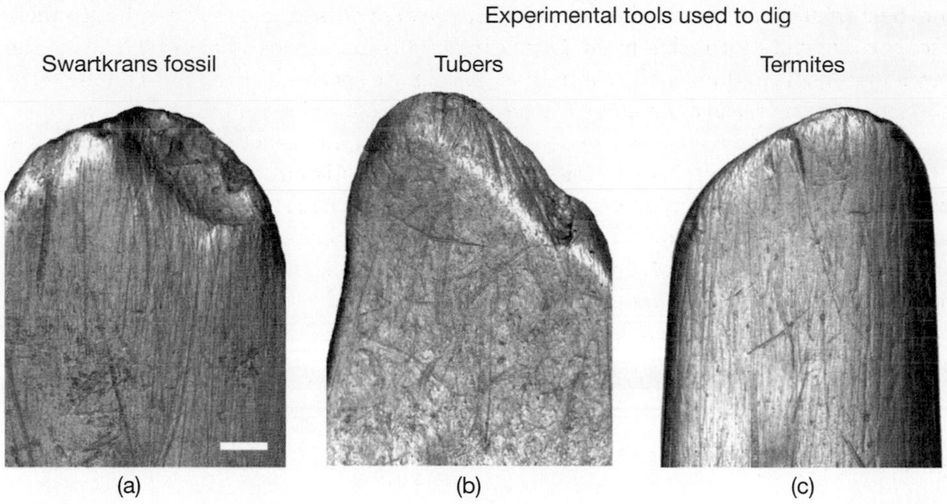

Experimental tools used to dig

Swartkrans fossil

Tubers

Termites

(a)

(b)

(c)

FIGURE 11.30

Experiments indicate that bone tools found at Swartkrans were used to excavate termite mounds. The tool shown in panel (a) is a cast of the original tool found at Swartkrans. The experimental tools in panels (b) and (c) were used for subsistence tasks: digging for tubers and digging for termites. The wear pattern on the Swartkrans fossils was most similar to that on the experimental tool used to dig for termites.

Archaeological Evidence for Meat Eating

At several archaeological sites in East Africa, tools have been found along with dense concentrations of animal bones.

Archaeological sites with early stone tools occur at the Olduvai Gorge of Tanzania, Koobi Fora in Kenya, and several sites in Ethiopia. Several sites in Bed I of the Olduvai Gorge excavated by Mary Leakey have been analyzed the most extensively. These sites, which are dated 2 Ma to 1.5 Ma, measure only 10 to 20 m (33 to 66 ft.) in diameter, but they are littered with fossilized animal bones (Figure 11.31). The densities of animal bones in the archaeological sites are hundreds of times higher than those in the surrounding areas or in modern savannas. The bones belong to a variety

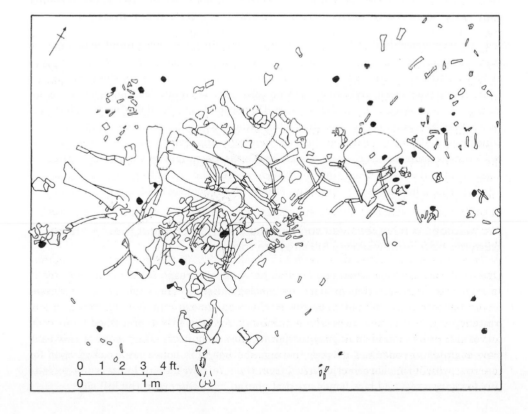

FIGURE 11.31

In this site map for one level of Bed I at Olduvai Gorge, most of the bones are from an elephant. Tools are shown in black.

0   1   2   3   4 ft.

0   1 m

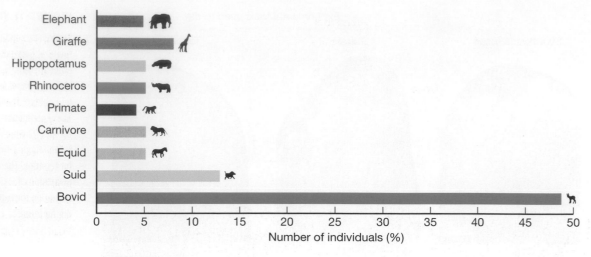

FIGURE 11.32

The bones of many mammals were found at one archaeological site at Olduvai. Bovids (including antelope, gazelles, sheep, goats, and cattle) clearly outnumber all other taxa.

of animal species, including bovids (such as present-day antelope and wildebeests), suids (pigs), equids (horses), elephants, hippopotamuses, rhinoceroses, and assorted carnivores (**Figure 11.32**).

At some of these sites, Mary Leakey also found several kinds of stone artifacts: cores, flakes, battered rocks that may have been used as hammers or anvils, and some stones that show no signs of human modification or use. The artifacts were manufactured from rocks that came from various spots in the local area; some were made from rocks found several kilometers away.

The association of hominin tools and animal bones does not necessarily mean that early hominins were responsible for these bone accumulations.

It is easy to conclude that the association of hominin tools and animal bones means that the hominin toolmakers hunted and processed the prey whose bones we find at these sites. If that is our assumption, these may have been sites where hominins lived, like modern foragers' camps, or they may have been butchery sites where hominins processed carcasses but did not live. However, there are also other possibilities. Bones may have accumulated at these sites without any help from early hominins. The bones may have been deposited there by moving water (which has now disappeared) or by other carnivores, such as hyenas. These sites also might be where many animals died of natural causes. Hominins might have visited the sites after the bones accumulated, perhaps hundreds of years later, and left their tools behind.

Archaeologists have resolved some of the uncertainty about these sites by studying how contemporary kill sites are formed.

One way to determine whether hominin hunting was responsible for these sites is to study the processes that produce archaeological sites. The study of the processes of site formation are called **taphonomy**. Archaeologists examine the characteristics of contemporary kill sites—spots where animals have been killed, processed, and eaten by various predators, including contemporary human hunters. They monitor how each type of predator consumes its prey, noting whether limb bones are cracked open for marrow, which bones are carried away from the site, how human hunters use tools to process carcasses, and how bones are distributed by predators at the kill site.

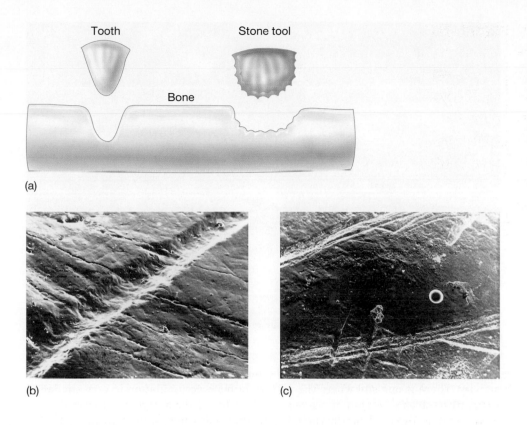

(a)

(b)

(c)

FIGURE 11.33

The marks made on bone by teeth differ from the marks made by stone tools. (a) The smooth surfaces of teeth leave broad, smooth grooves on bones; the edges of stone tools have many tiny, sharp points that leave fine parallel grooves. Cut marks made by (b) carnivore teeth and (c) stone tools can be distinguished when they are examined with a scanning electron microscope. These are scanning electron micrographs of 1.8-million-year-old fossil bones from Olduvai Gorge.

They also examine the marks that are left on bones when they have been chewed on by predators, processed with stone tools, or left out in the open for long periods. When carnivores gnaw on meat, their teeth leave distinctive marks on the bones. Similarly, when humans use stone tools to butcher prey, their tools leave characteristic marks. Flaked-stone tools have microscopic serrations on the edges and make very fine parallel grooves when they are used to scrape meat away from bones (**Figure 11.33**).

These data enable archaeologists to develop a profile of the characteristics of kill sites created by different types of predators. They can also assess many of the same characteristics in archaeological sites. By comparing the features of archaeological and contemporary sites, they can sometimes determine what happened at an archaeological site in the past.

Analyses at Olduvai Gorge suggest that the bones at most of these sites did not accumulate by natural processes.

Studies of the Olduvai sites where both animal bones and stone tools have been found tell us, first of all, that the bones were not deposited by moving water. Animals sometimes drown as they try to cross a swollen river or when they are swept away in flash floods (**Figure 11.34**). The bodies are carried downstream and accumulate in a sinkhole or on a sandbar. As the bodies decompose, the bones are exposed to the elements. The study of modern sites shows that sediments deposited by rapidly moving water have several distinctive characteristics. For example, such sediments tend to be graded by size because particles of different sizes and weights sink at different spots. Sediments surrounding Olduvai sites show none of these distinctive features.

Study of these sites also tells us that the dense concentrations of bones were not due to the deaths of many animals at one spot. Sometimes many animals die in the same place. In severe droughts, for example, many animals may die near water holes. Mass deaths usually involve members of a single species, and there is typically little mixing of bones from different carcasses. By contrast, the bones at the Olduvai sites come from several species, and bones from different carcasses are jumbled together.

FIGURE 11.34

Sometimes many wildebeest drown when trying to cross swollen rivers.

At some sites, however, bone accumulations *do* seem to be the product of carnivore activity. There is one site where the pattern of bone accumulation is very similar to the pattern of bone accumulations near modern hyena dens. Hyenas often carry and drag carcasses from kill sites to their dens so that they can feed their young and avoid competition with other carnivores.

Study of cut marks on fossil bones of prey species suggests that hominins were active at two of the Olduvai sites and used tools at these sites to process carcasses.

At two sites, FLK Zinjanthropus and Bell's Karongo, there is good evidence that hominins processed carcasses. There are large quantities of bones from multiple animal species as well as stone tools. Moreover, many of the bones bear cut marks and percussion marks (**Figure 11.35**), clear signs of hominin activity. Some of these bones also show tooth marks, suggesting that carnivores were also present at these sites.

Three other lines of circumstantial evidence suggest that *Homo erectus* ate meat.

FIGURE 11.35

This bone shows linear cut marks (*A*) and round percussion marks (*B*).

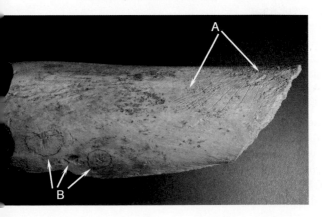

One line of evidence that these hominins ate meat comes from the skeleton of an *H. erectus* woman (KNM-ER 1808) that was discovered at Koobi Fora by Alan Walker and his colleagues. The long bones of this woman, who died about 1.6 Ma, are covered with a thick layer of abnormal bone tissue. This kind of bone growth is symptomatic of vitamin A poisoning (**Figure 11.36**). How could a hunter-gatherer get enough vitamin A to poison herself? The most likely way would be to eat the liver of a large predator, such as a lion or a leopard. The same symptoms have been reported for Arctic explorers who ate the livers of polar bears and seals. If this woman's bones were deformed because she ate a large predator's liver, then we can assume that *H. erectus* ate meat. Of course, we don't know how this woman obtained the liver that poisoned her. She might have scavenged the liver from a predator's carcass or she might have killed the predator in a contest over a kill.

The ability of *H. erectus* to survive in temperate latitudes also supports the idea that these hominins ate meat. *H. erectus* moved from the tropics to more temperate habitats outside Africa. Recall from Chapter 5 that other primates rely heavily on fruit and tender parts of plants for survival. To survive a temperate winter when such things are not available, *H. erectus* might have relied on meat. Very few primates except humans live in places with severe winters, and these are only small, peripheral populations of species that are usually more tropical.

Additional evidence for meat eating comes from an unexpected source—our intestinal parasites. Humans are the terminal hosts of three species of tapeworm in the genus *Taenia*. African carnivores are the terminal hosts of closely related members of this genus. Tapeworms have a complex life cycle, diagrammed in **Figure 11.37**. Herbivores ingest tapeworm eggs while they are feeding, and the eggs develop into larvae, which are encased in cysts in the flesh of the host. When a carnivore consumes the cysts in meat, the larvae develop into adults and produce eggs, which are shed in the host's feces. Domestic cattle and pigs are now the intermediate hosts of human tapeworms, and it had been assumed that we acquired our tapeworms from domesticated livestock. However, studies of the molecular phylogeny of several *Taenia* species now suggest otherwise. Two of the human tapeworm species, *T. saginata* and *T. asiatica*, diverged from a common ancestor 1.7 Ma to 0.8 Ma. This suggests that humans had become the terminal host for the ancestral species (and consumed meat regularly) before this point. This is well before humans domesticated animals and coincides with the emergence of *H. erectus* in Africa.

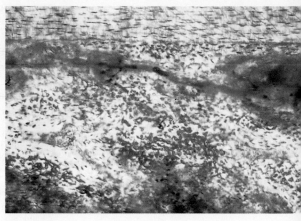

FIGURE 11.36

A microscopic view of the bone structure of KNM-ER 1808. Note that the small band of normal bone at the top looks very different from the puffy, irregular, diseased bone in the rest of the picture.

FIGURE 11.37

The life cycle of two tapeworm species. Humans are the terminal hosts of sister species *Taenia saginata* and *T. asiatica*, and a third species, *T. solium*.

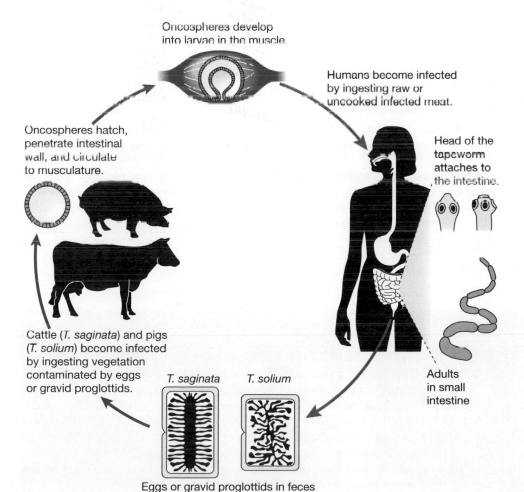

Oncospheres develop into larvae in the muscle.

Humans become infected by ingesting raw or uncooked infected meat.

Oncospheres hatch, penetrate intestinal wall, and circulate to musculature.

Head of the tapeworm attaches to the intestine.

Cattle (*T. saginata*) and pigs (*T. solium*) become infected by ingesting vegetation contaminated by eggs or gravid proglottids.

T. saginata   *T. solium*

Adults in small intestine

Eggs or gravid proglottids in feces are passed into the environment.

Hunters or Scavengers?

There has been controversy about whether the hominins were hunters or scavengers.

The archaeological evidence indicates that Oldowan hominins processed the carcasses of large animals, and we assume that they ate the meat they cut from the bones. But eating meat does not necessarily imply hunting. Many carnivores rely at least partly on scavenging; they steal kills from other predators or rely on opportunistic discoveries of carcasses.

There has been considerable dispute about how early hominins acquired the meat they ate. Some researchers have argued that early hominins killed the prey found at the archaeological sites; others have argued that early hominins could not have captured large mammals because they were too small, too poorly armed, and not smart enough. They contend that the early hominins were scavengers who occasionally appropriated kills from other predators or collected carcasses they found.

For most contemporary carnivores, scavenging is as difficult and dangerous as hunting.

To resolve the debate about whether early hominins were hunters or scavengers, we must first rethink popular conceptions of scavengers. Although scavengers have an unsavory reputation, scavenging is not an occupation for the cowardly or lazy. Scavengers must be brave enough to snatch kills from the jaws of hungry competitors, shrewd enough to hang back in the shadows until the kill is momentarily left unguarded, or patient enough to follow herds and take advantage of natural mortality. Studies of contemporary carnivores show that almost all scavenged meat is acquired by taking a kill away from another predator. Most predators respond aggressively to competition from scavengers. For example, lions jealously guard their prey from persistent scavengers that try to steal bits of meat or to drag away parts of the carcass. These contests can be quite dangerous (**Figure 11.38**).

Most large mammalian carnivores practice both hunting and scavenging.

We also tend to think that some carnivores, such as lions and leopards, only hunt, and others, such as hyenas and jackals, only scavenge. But the simple dichotomy between scavengers and hunters collapses when we review the data on the behavior of the five largest African mammalian carnivores: lion, hyena, cheetah, leopard, and wild dog. The fractions of meat obtained by scavenging vary from none for the cheetah to 33% for hyenas, with the others ranging somewhere in between. Contrary to the

FIGURE 11.38

Competition among predators at kill sites is often intense. Here (a) lions have scavenged prey from a pack of hyenas, and (b) the hyenas fight to get it back.

[a]

[b]

usual stereotype, the noble lion is not above taking prey from smaller competitors, including female members of his own pride (**Figure 11.39**), and hyenas are in fact accomplished hunters. For most large carnivores in eastern Africa, hunting and scavenging are complementary activities.

No mammalian carnivores subsist entirely by scavenging. It would be difficult for any large mammal to do so. For one thing, many prey species create movable feasts, migrating in large herds over long distances. Although natural mortality in these herds might make scavenging feasible, their migratory habits eliminate this option. Mammalian carnivores cannot follow these migrating herds very far because they have dependent young that cannot travel long distances (**Figure 11.40**). Only avian scavengers that can soar over great distances, such as vultures, rely entirely on scavenging (**Figure 11.41**). When migratory herds are absent, mammalian carnivores rely on other prey, such as waterbuck and impalas. Natural mortality among resident species is not high enough to satisfy the caloric demands of carnivores, so they must hunt and kill much of their own prey, though they still scavenge when the opportunity arises.

FIGURE 11.39

Male lions sometimes take kills from smaller carnivores and from female lions.

Scavenging might be more practical if carnivores switched from big game to other forms of food when migratory herds were not present. This is a plausible option for early hominins. Most groups of contemporary foraging people rely heavily on gathered foods—including tubers, seeds, fruit, eggs, and various invertebrates—in addition to meat. A few, such as the Hadza, obtain meat from scavenging as well as from hunting. It is possible that early hominins relied mainly on gathered foods and scavenged meat opportunistically.

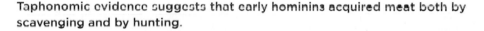

Taphonomic evidence suggests that early hominins acquired meat both by scavenging and by hunting.

As we saw earlier, predators often face stiff competition for their kills. An animal that tries to defend its kill risks losing it to scavengers. For this reason, leopards drag their kills into trees and eat the meat in safety. Other predators, such as hyenas, sometimes rip off the meaty parts of the carcass, such as the hindquarters, and drag their booty away to eat in peace. This means that limb bones usually disappear from a kill site first, and less meaty bones, such as the vertebrae and skull, disappear later or remain at the kill site (**Figure 11.42**). If hominins obtained most of their meat from scavenging,

FIGURE 11.40

As dusk falls, three cheetah cubs wait for their mother to return from hunting.

FIGURE 11.41

Only avian scavengers with very large ranges, such as the vultures shown here, are able to rely entirely on scavenging.

FIGURE 11.42

After other predators have left, vultures consume what remains at the kill site.

we would expect to find cut marks made by tools mainly on bones typically left at kill sites by predators, such as vertebrae. If hominins obtained most of their meat from their own kills, we would expect to find tool marks mainly on large bones, such as limb bones.

There is a vigorous debate about the taphonomic evidence for hunting versus scavenging. Robert Blumenschine of Rutgers University has argued that carnivores killed prey and partially consumed the flesh; then hominins acquired pieces of the carcass, which they used to process marrow; finally, other carnivores gnawed on the bones. He argues that opportunistic scavenging would not require novel technological skills or behavioral adaptations and is therefore an evolutionarily conservative hypothesis. However, scavenging would mark a departure from chimpanzees, which hunt often but very rarely take advantage of scavenging opportunities.

Manuel Domínguez-Rodrigo from the Universidad Complutense de Madrid takes a different view. He argues that hominins had first access to carcasses. He bases his conclusions on two sources of evidence. First, he believes that the carnivore–hominin–carnivore model is based on a misreading of the taphonomic evidence from one of the two sites at which hominins were active. He and his colleagues argue that some natural biochemical marks on bones may have been mistaken for carnivore tooth marks. Genuine tooth marks, Domínguez-Rodrigo observes, are mainly limited to the ends of long bones, which indicates that they were processed by carnivores after they were processed by hominins.

Second, many of the bones excavated at another site at Olduvai, Bell's Karongo, bear cut marks and percussion marks (see Figure 11.35) but don't tend to display tooth marks. If carnivores had had first access to these carcasses, it is unlikely there would be as many hominin-made cut marks on the limb bones because the carnivores would have already defleshed them, negating the need for the hominin processing. Blumenschine rejects these interpretations based on analyses of the marks on the bones, and the debate is not resolved.

Control of fire probably played an important role in human evolution, but we do not know when this event occurred.

The ability to control fire may have played an important role in human evolution. Fire provides warmth, which may have enabled a tropical primate to expand its range into temperate habitats. Fire may also have played an important role in food preparation. Food is digested as it passes through the intestine, and the length of the intestine influences the amount of nutrients that can be absorbed. Humans have much shorter intestines than other apes, and this means they cannot extract as many nutrients from their food. Richard Wrangham of Harvard University argues that hominids solved this problem by cooking their food. Cooking increases the available caloric content in food and makes both meat and vegetables easier to chew and digest. As hominins moved into more open habitats, they may have shifted to a diet rich in meat and tubers and begun to use fire for cooking. Wrangham thinks that cooking was a critical adaptation that allowed *H. erectus* to support the higher energetic costs of their larger brains. Since the anatomic evidence suggests that *H. erectus* had a shorter intestine than other mammals of similar body size, Wrangham believes that they must have learned how to control fire and cook their food.

The problem with this hypothesis is that there is no strong evidence that *H. erectus* controlled fire or cooked their food (**Figure 11.43**). At three Kenyan sites dated to about 1.5 Ma, archaeologists have found areas of baked earth associated with Oldowan tools. But it is not clear whether these areas are the remains of a

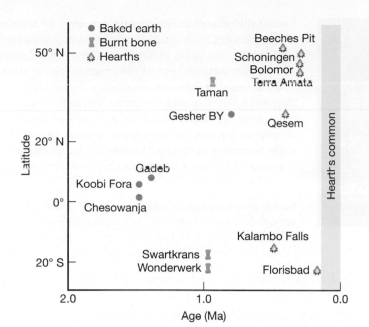

FIGURE 11.43

A summary of archaeological evidence for use of fire by hominins. The horizontal axis is the date of the site, and the vertical axis is its latitude. These data suggest that regular use of fire did not occur until about 300 ka.

campfire or a slow-burning natural fire. Excavations at Swartkrans Cave in South Africa have yielded many fragments of burned bones of antelope, zebras, warthogs, baboons, and *P. robustus* that date to about 1 Ma. Burned bones have also been found at two other sites in South Africa that date to about the same time. To determine whether the Swartkrans bones were burned in campfires, C. K. Brain of the Transvaal Museum in Pretoria and Andrew Sillen of the Synergos Institute burned modern antelope bones at a range of temperatures. They found that the very high temperatures characteristic of long-burning campfires produce changes in the microscopic structure of bone—changes that can also be seen in the burned fossils at Swartkrans. This finding suggests that hominins may have used fire, but the absence of burned bones at many other sites means that it may not have been a regular occurrence. It is important to remember that hominins may have used fire before they could control or produce fire. That is, they may have exploited naturally occurring fires that ignited after lightning strikes but not have been able to sustain fires or produce fire on their own.

Beginning about 350 ka, there is evidence of stone hearths at several sites in Africa and Europe. Especially clear evidence comes from Tabun Cave in Israel, where a group led by Ron Shimelmitz of the University of Haifa has studied fire use in a long sequence of human occupation beginning before 415 ka and ending about 215 ka. There is no evidence for use of fire at this site before about 350 ka. Between 350 ka and 300 ka, burnt flints become relatively common, and by 200 ka they are very common. This evidence suggests that hominins may have occasionally used fire beginning around 1 Ma, but regular control of fire did not occur until about 300 ka. If this scenario is correct, then cooking may not have played as important a role in the lives of *H. erectus* as Wrangham has suggested.

Home Bases of Early Toolmakers

As discussed previously, we have reason to believe that early hominin toolmakers used their tools for extractive foraging and processing prey carcasses. Early hominins may have obtained these carcasses through a mix of hunting and scavenging. Thus, they had probably come to rely on complex foraging skills that were difficult to master. In modern foraging societies, reliance on complex foraging skills is also linked to food sharing, sexual division of labor, and the establishment of **home bases**.

Nearly all contemporary foraging peoples establish a temporary gathering spot, where food is shared, processed, cooked, and eaten. The camp is also the place where people weave nets, manufacture arrows, sharpen digging sticks, string bows, make plans, resolve disputes, tell stories, and sing songs (**Figure 11.44**). Because foragers often move from one location to another, their home bases are simple, generally consisting of modest huts or shelters built around several hearths. If early hominins established home bases, we might be able to detect traces of their occupation in the archaeological record.

Some archaeologists have interpreted the dense accumulations of stones and bones as home bases, much like those of modern foragers, but this view is not consistent with some of the evidence.

Some archaeologists, particularly Glyn Isaac, have suggested that the dense accumulations of animal bones and stone tools found at some sites mark the location of hominin home bases. They have speculated that early hominins acquired meat by hunting or scavenging and then brought pieces of the carcass home, where it could be shared. The dense collections of bones and artifacts were thought to be the result of prolonged occupation of the home base (and sloppy housekeeping). At one Olduvai Bed I site, there is even a circle of stones (**Figure 11.45**), dated to 1.9 Ma, similar to the circles of stones anchoring the walls of simple huts constructed by some foraging peoples in dry environments today. Many Oldowan tools and bone fragments from a variety of prey species are found at the same site.

However, several other observations are inconsistent with the idea that these sites were home bases:

- Both hominins and nonhominin carnivores were active at many of the Olduvai sites. Many of the bones at the Olduvai sites were gnawed by nonhominin carnivores. Sometimes the same bones show both tooth marks and cut marks, but some show only the marks of nonhominin carnivores.

- Hominins and nonhominin carnivores apparently competed over kills. The bones of nonhominin carnivores occur more often than would be expected on the basis of their occurrence in other fossil assemblages or modern carnivore densities. Perhaps the carnivores were killed (and eaten) when attempting to scavenge hominin kills or when hominins attempted to scavenge their kills. Hominins may not have always won such contests; some fossilized hominin bones show the tooth marks of other carnivores.

- Modern kill sites are often the scene of violent conflict among carnivores. This conflict occurs among members of different species as well as the same species. It is especially common when a small predator, such as a cheetah, makes a kill. The kill attracts many other animals, most of which can displace the cheetah.

- The bones accumulated at the Olduvai sites are weathered. Bones laying on the surface of the ground crack and peel in various ways. The longer they remain exposed on the surface, the greater the extent of weathering. Taphonomists can calibrate the weathering process and use these data to determine how long fossil bones were on the ground before being buried. Some of the bones at the Olduvai sites were exposed to the elements for at least 4 to 6 years (**Figure 11.46**).

- The Olduvai sites do not show evidence of intensive bone processing. The bones at these sites show cut marks and tooth marks, and many bones were apparently smashed with stone hammers to remove marrow. However, the bones were not processed intensively, as they are by modern foragers.

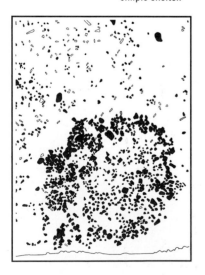

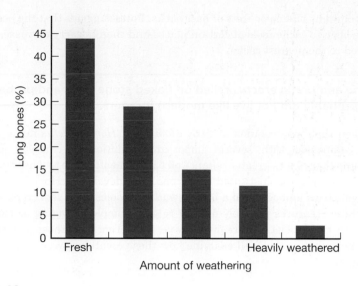

FIGURE 11.46

Many of the bones at Bed I sites at Olduvai are heavily weathered, suggesting that they were deposited and exposed to the elements over a fairly long period.

These observations are difficult to reconcile with the idea that the Olduvai sites were home bases—places where people eat, sleep, tell stories, and care for children. First, foragers today do everything they can to prevent carnivores from entering their camps. They often pile up thorny branches to fence their camps and keep dogs that are meant to chase predators away. It is hard to imagine that early hominins could have occupied these sites if lions, hyenas, and saber-toothed cats were regular visitors. Second, bones at the Olduvai sites appear to have accumulated over a period of years. Contemporary foragers usually abandon their home bases permanently after a few months because the accumulating garbage attracts insects and other vermin. Even though they revisit the same areas regularly, they don't often reoccupy their old sites. Finally, the fossilized bones found at Olduvai were not processed as thoroughly as modern foragers process their kills.

Hominins may have brought carcasses to these sites and processed the carcasses with flakes made from previously cached stones.

Richard Potts, an anthropologist at the Smithsonian Institution, suggests that these sites were not home bases but butchery sites—places where hominins worked but did not live. He believes that hominins brought their kills to these sites and dismembered their carcasses there. Some of the carcasses were scavenged by hominins from other carnivores, and some of the hominins' kills were lost to scavengers. Hominins may have carried bones and meat away to other sites for more intensive processing. This would explain why bones accumulated over such a long time, why bones of nonhominin carnivores were present, and why bones were not completely processed.

At first glance, it might seem inconvenient for hominins to schlep their kills to butchery sites. Why not process the carcass at the kill site? We have seen that hominins used tools to process the meat. However, they couldn't be sure of finding appropriate rocks for toolmaking at the sites of their kills. And they couldn't leave their kills unguarded while they went off to fetch their tools, lest a hungry scavenger steal their supper. So they must have had to carry the meat to where their tools were kept or to keep their tools with them all the time. Remember that these tools were fairly heavy, and early hominins had no pockets or backpacks. Potts suggests that the best strategy would have been to cache tools at certain places and then to carry carcasses that they had acquired to the nearest cache.

Early *Homo* and *Homo erectus* relied on flaked stone tools and probably ate meat but probably did not live like modern foragers.

We have seen that the evolution of early *Homo* and *Homo erectus* during the early Pleistocene coincided with several important evolutionary changes. Brain size increased, and there was a greater reliance on eating meat. Stone tool use became ubiquitous, and new, more sophisticated tool types were developed. Hominin populations spread out of Africa and occupied a broad swath of tropical and semitropical Eurasia. However, these creatures probably did not rely on complex foraging or food sharing and as a result their societies were probably very different from those of modern foragers. In the next chapters we will examine how these creatures were transformed into modern humans.

CHAPTER REVIEW

Key Terms

occipital torus (p. 270)
flakes (p. 274)
cores (p. 274)
knapping (p. 274)

Oldowan tool industry (p. 275)
Mode 1 (p. 275)
cortex (p. 276)
biface (p. 277)

hand ax (p. 277)
cleaver (p. 277)
pick (p. 277)
Mode 2 (p. 277)
Acheulean industry (p. 277)

collected foods (p. 279)
extracted foods (p. 279)
hunted foods (p. 280)
taphonomy (p. 288)
home base (p. 295)

Study Questions

1. When and where do the first members of the genus *Homo* appear? What features distinguish them from earlier hominins?

2. There is some disagreement about whether material assigned to early *Homo* represents one species or two species. Explain the basis of this dispute.

3. What are the main differences between early *Homo* and *H. erectus*? How would you explain the evolution of these differences?

4. Which derived features are shared by modern humans and *H. erectus*? Which derived features are unique to *H. erectus*?

5. What is a hand ax, and what did early hominins use it for? How do we know?

6. What evidence suggests that early hominins controlled fire?

7. What evidence suggests that early hominins may have eaten both extracted and hunted foods?

8. *Homo erectus* is the first hominin to leave Africa. How do *H. erectus* fossils from Dmanisi, Java, and Africa differ?

9. *Homo erectus* in Africa is associated with Mode 2 tools, including hand axes, whereas *H. erectus* in China is mainly associated with Mode 1 tools. Discuss possible explanations for this difference in tool technology.

10. Why is it difficult for us to determine precisely who the first toolmakers were? Which hominin species are possible candidates for this?

Further Reading

Antón, S. C. 2012. "Early Homo Who, When, and Where." *Current Anthropology* 53: S278–S298.

Antón, S. C., R. C. Potts, and L. C. Aiello. 2014. "Evolution of Early *Homo*: An Integrated Biological Perspective." *Science* 345, doi:10.1126/science.1236828.

Conroy, G. 2005. *Reconstructing Human Origins: A Modern Synthesis*. 2nd ed. New York: W. W. Norton.

De la Torre, I. 2016. "The Origins of the Acheulean: Past and Present Perspectives on a Major Transition in Human Evolution." *Philosophical Transactions of the Royal Society*, B, 371, doi:10.1098/rstb.2015.0245.

Domínguez-Rodrigo, M. 2009. "Are Oldowan Sites Palimpsests? If So, What Can They Tell Us about Hominid Carnivory?" In E. Hovers and D. R. Braun, eds., *Interdisciplinary Approaches to the Oldowan*, pp. 129–147. Dordrecht, The Netherlands: Springer.

Kaplan, H., K. Hill, J. Lancaster, and A. M. Hurtado. 2000. "A Theory of Human Life History Evolution: Diet, Intelligence, and Longevity." *Evolutionary Anthropology* 9: 156–185.

Kimbel, W. H., and B. Vilmoare. 2016. "From Australopithecus to *Homo*: The Transition That Wasn't." *Philosophical Transactions of the Royal Society*, B, 371, doi:10.1098/rstb.2015.0248.

Klein, R. G. 2008. *The Human Career: Human Biological and Cultural Origins*. 3rd ed. Chicago: University of Chicago Press.

Potts, R. 2012. "Environmental and Behavioral Evidence Pertaining to the Evolution of Early *Homo*." *Current Anthropology* 53, doi:10.1086/667704.

Schick, K. D., and N. Toth, eds. 2009. *The Cutting Edge: New Approaches to the Archaeology of Human Origins*. Bloomington, Ind.: Stone Age Institute Press.

Walker, A., and P. Shipman. 1996. *The Wisdom of the Bones: In Search of Human Origins*. New York: Knopf.

 Review this chapter with personalized, interactive questions through INQUIZITIVE.

 This chapter also features a Guided Learning Exploration on the evolution of foraging complexity.

THE NEANDERTHALS AND THEIR CONTEMPORARIES

CHAPTER OBJECTIVES

By the end of this chapter, you should be able to:

A Explain the role of changing climates on evolutionary events of the Early and Middle Pleistocene.

B Describe the morphology of the first successors to *Homo erectus.*

C Describe the Middle Pleistocene hominins, including the Neanderthals.

D Discuss the difficulty in classifying Middle Pleistocene hominins.

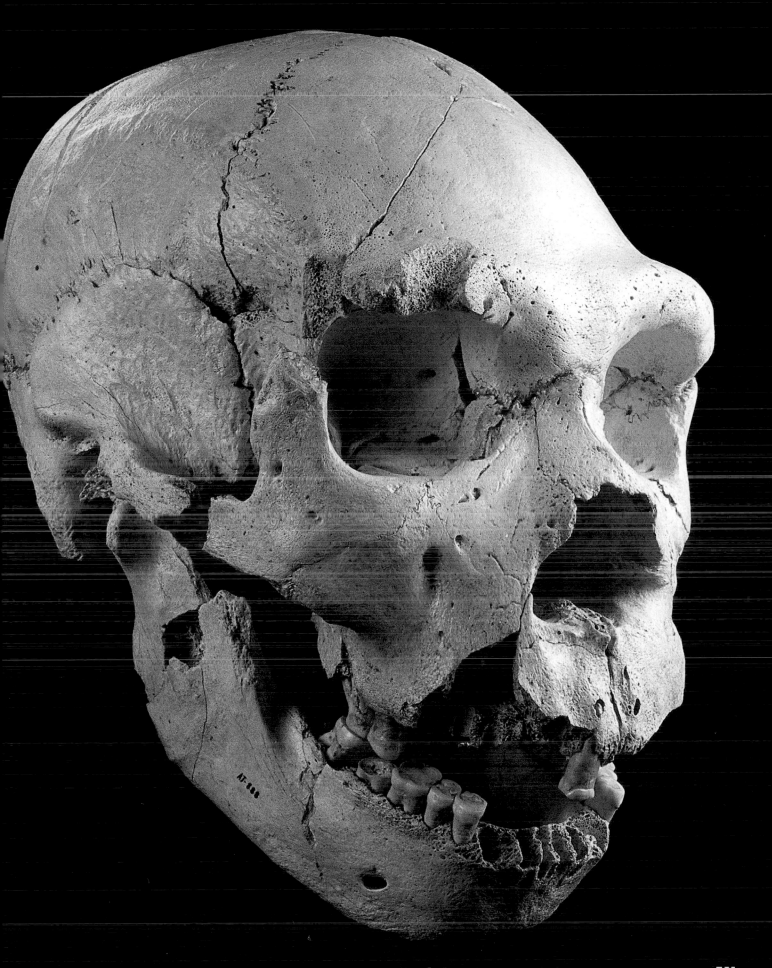

For about a million years, *Homo erectus* roamed around much of Africa and Eurasia with hand axes (or serviceable wooden alternatives), hunting or scavenging game and butchering their kills at central work sites. Then, about 900 ka, Earth began to experience a series of major climate fluctuations that completely transformed its habitats. In this chaotic and rapidly changing world, natural selection reshaped the hominin lineage once again. *H. erectus* was transformed into a smarter and more versatile creature. In eastern Asia, meanwhile, *H. erectus* persisted with little change until perhaps 30 ka, though other hominins appeared in the region around 200 ka. Specifically, hominins in Africa and western Eurasia began to develop more sophisticated technology and behavior about 300 ka. This process of change continued, particularly in Africa, until the behavior and technology of hominins became indistinguishable from those of modern humans.

Although many of these events can be established relatively well in the fossil and archaeological records, there is considerable controversy about the phylogeny of the human lineage during this period. There are fierce debates about how many species of hominins there were and how different forms were related. We will turn to these controversies later in the chapter. But first you need to learn something about the climatic events that transformed the world during the Middle and Upper Pleistocene.

Climate Change during the Middle Pleistocene (770 ka to 126 ka)

The world's climate became colder and much more variable during the Middle Pleistocene.

As you already know, modern humans are not the first primate species to be affected by climate change. Hominins of the Middle and Upper Pleistocene contended with major climatic changes over relatively short timescales. During the Middle and Upper Pleistocene, there were many long, cold glacial periods punctuated by short, warmer interglacial periods. **Figure 12.1** shows estimates of global temperatures for this period. Global temperatures have fluctuated over the past million years, but around 700 ka the difference between the peaks and troughs in temperatures increased. From geologic evidence, we know that during the cold periods, glaciers covered North America and Europe and arctic conditions prevailed. These cold periods were intermittently interrupted by shorter warm periods during which the glaciers receded and the forests returned.

FIGURE 12.1

The climate record for the past million years. The *y* axis is a measure of the ratio of ^{18}O and ^{16}O. Smaller values (up on the graph) represent warmer temperatures. This record has been filtered so that only changes that take place slowly are represented. Notice that the present climate is relatively warm, but that the world was much colder just 25 ka.

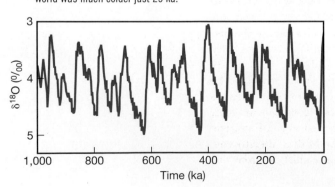

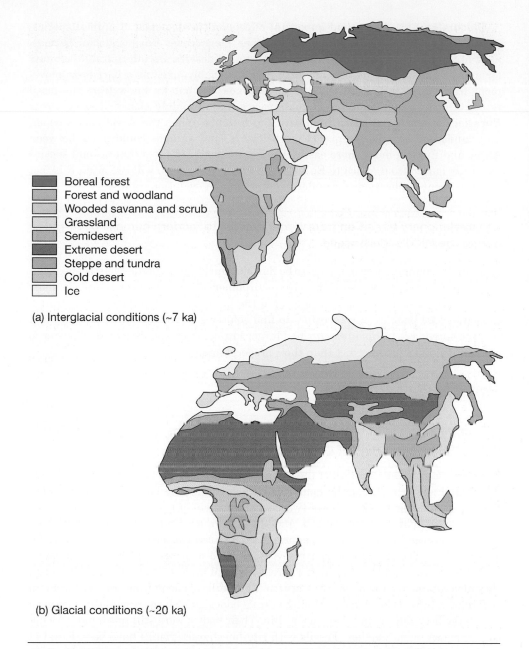

FIGURE 12.2

(a) A reconstruction of biological habitats about 7 ka during the warmest, wettest part of the present interglacial period. Much of Eurasia and Africa were covered with forest and were connected by a broad swath of grassland across northern Africa and southwestern Eurasia. (b) The habitats during the coldest, driest part of the last glacial period, about 20 ka. There was very little forest cover. Grassland and scrub predominated in central Africa and Southeast Asia. Northern Eurasia was covered with cold, dry steppe and desert. Central and southern Africa were separated from Eurasia by a band of extreme desert across northern Africa, the Arabian Peninsula, and central Asia.

Boreal forest
Forest and woodland
Wooded savanna and scrub
Grassland
Semidesert
Extreme desert
Steppe and tundra
Cold desert
Ice

(a) Interglacial conditions (~7 ka)

(b) Glacial conditions (~20 ka)

During glacial periods, the world was dry, and Africa and Eurasia were isolated from each other by a massive desert. During interglacial periods, the world was much wetter.

Temperature fluctuations have massive effects on the world's biological habitats, which in turn influence the distribution of plants and animals. **Figure 12.2** shows how different the world looked about 7 ka during the warmest and wettest part of the current interglacial period and about 20 ka during the coldest and driest part of the last glacial period. During the warm interglacial period, much of Eurasia and Africa were covered by forests, East and North Africa and Arabia were grasslands, and deserts were limited to small bits of southwestern Africa and central Asia. At the depth of the last glacial period, the middle latitudes were dominated by vast expanses of extreme desert, as dry as the Sahara Desert is today. North and south of this desert, grassland, scrub, and open woodland predominated, and forests were restricted to small regions of central Africa and Southeast Asia. Since the beginning of the Middle Pleistocene, the planet has oscillated between these two extremes, sometimes shifting from one extreme to the other in just a few hundred years.

Fluctuations in climate had important effects on the dispersal of animal species, including hominins, in the Middle and Upper Pleistocene. During glacial periods, most of the northern part of Africa was desert, and deserts are inhospitable habitats for many animals. The deserts of North Africa probably acted as a barrier to movement from Africa into Eurasia. This corresponds to what we know from the fossil record. During glacial periods, animal species moved mainly east and west across Eurasia, not north and south from Africa to Eurasia. When the world was warmer, grasslands and savannas replaced most deserts, and animals could move between Africa and Eurasia much more easily. The fossil record indicates that animal species generally moved from Africa to Eurasia, not vice versa. As we will see later, this fact has important implications for understanding human evolutionary history.

The evolutionary transition from *Homo erectus* to modern humans occurred during the Middle Pleistocene.

In a rapidly changing world, it pays to be flexible. During the Middle Pleistocene, hominins were forced to cope with massive environmental changes over relatively short timescales. These events may have favored the evolution of more powerful cognitive capacities that enabled our ancestors to find solutions to novel problems. As we will see, the successors to *Homo erectus* had larger brains, more sophisticated subsistence technology, and a broader diet than their predecessors.

Homo heidelbergensis

During the first half of the Middle Pleistocene, hominins with larger brains and more modern skulls appear in the fossil record.

Sometime during the first half of the Middle Pleistocene, *H. erectus* was succeeded in Africa and western Eurasia by hominins with substantially larger brains and more modern skulls. **Figure 12.3** shows two nearly complete crania from this period: the Petralona cranium found in Greece and the Kabwe (or Broken Hill) cranium from Zambia. These individuals had larger brains than *H. erectus*, measuring between 1,200 and 1,300 cc. The skulls also share several derived features with modern humans, including more vertical sides, higher foreheads, and a more rounded back. However, they also retained many primitive features. The skulls are long from front to back and still relatively low. They have very thick cranial bones and a large prognathic face with very large browridges, and they lack a chin. Their bodies were still much more robust than modern human bodies. Fossils with similar characteristics have been found at a handful of other sites in Africa (**Figure 12.4**). During this period, the first hominin fossils also appear in western Europe and bear similarities to the fossils from Africa. A mandible from Mauer near Heidelberg, Germany, combines primitive features, such as a wide ascending ramus, and derived features, including small canines and incisors. A massive tibia from this period has also been found in southern England at a site called Boxgrove. There has been no sign of these kinds of fossils in eastern Asia during this period, although they do appear later.

The fossils that make up this group are quite variable and are widely distributed in space and time. Some of the most complete specimens are not well dated, creating controversy about whether they represent one evolving lineage or multiple species. Here we adopt what is perhaps the most common opinion and treat them as members of a single species. Because the Mauer mandible was the first of this group to be found and named, the species is known as ***Homo heidelbergensis***.

Scientists are uncertain about exactly when or where *Homo heidelbergensis* first appeared. The oldest candidate for inclusion in this species is a cranium found at Buia, in Eritrea, dated to 995 ka. This skull shares some characteristics with *H. erectus*, including a brain of only 1,000 cc, but it also has several derived features

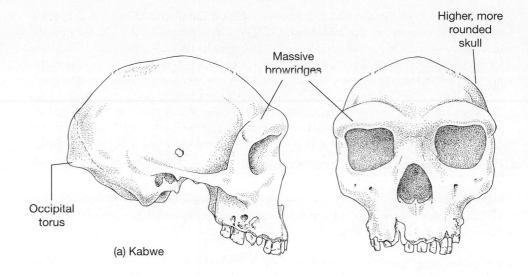

Massive browridges

Higher, more rounded skull

Occipital torus

(a) Kabwe

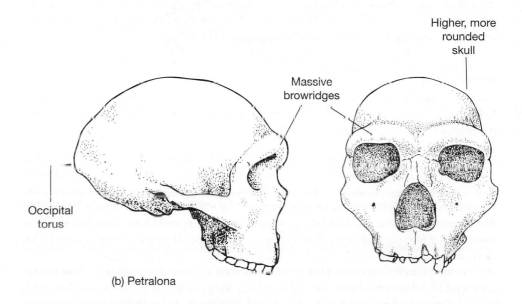

Higher, more rounded skull

Massive browridges

Occipital torus

(b) Petralona

FIGURE 12.3

Sometime between 800 ka and 500 ka, hominins with higher, more rounded crania and larger brains first appear in the fossil record; by 400 ka, *Homo heidelbergensis* was common in Africa and western Eurasia. These *H. heidelbergensis* fossils, from (a) Kabwe (sometimes called Broken Hill) in Zambia and (b) Petralona in Greece, are 400,000 years old.

that are associated with other fossils included in *H. heidelbergensis*. Another candidate comes from Trinchera Dolina (also called Gran Dolina) in the Sierra de Atapuerca in northern Spain. This site has yielded several hominin fossils, including part of an adolescent's lower jaw and most of an adult's face. Although the Trinchera Dolina fossils are too fragmentary to provide an estimate of endocranial volume, they exhibit several facial features that are seen in more modern hominins. These fossils have been dated to about 800 ka by means of paleomagnetic methods, but rodent fossils found at the same site suggest a more recent date, perhaps 500 ka. Given uncertainty about the ages of the fossils assigned to *II. heidelbergensis*, the best that we can do is to bracket the first appearance of these creatures between 1 Ma and 500 ka.

There is evidence that *Homo heidelbergensis* hunted big game.

The tools used by early *H. heidelbergensis* are similar to those used by *H. erectus*. Tool kits are dominated by Acheulean hand axes and other core tools at most sites, but in some cases the hand axes are more finely worked. There is also some evidence

FIGURE 12.4

Locations of fossil and archaeological sites mentioned in the text (some in the next chapter). *H. erectus* and *H. heidelbergensis* are found in East Asia, Neanderthals in Europe, and both *H. heidelbergensis* and *H. sapiens* in Africa.

that wooden tools were used in hunting. Three wooden spears were found in an open-pit coal mine in Schoningen, Germany, dated to about 300 ka. It seems likely that the spears were used for hunting because they were found along with the bones of hundreds of horses, and many of the bones show signs of having been processed with stone tools.

It seems likely that these spears were thrown at prey during hunts. The spears look a lot like modern javelins and modern throwing spears. They are about 2 m (6 ft.) long and are thickest and heaviest near the pointed end, gradually tapering to the other end. Some modern people use similar spears when they are hunting, and it seems plausible that *H. heidelbergensis* used them the same way. In experiments using replicas, track and field athletes who participate in javelin throwing were able to accurately cast the spears 20 m (66 ft.).

Taphonomic evidence from several Middle Pleistocene sites also indicates that *H. heidelbergensis* was a competent hunter. For example, the Boxgrove site, which is dated to about 500 ka, has produced many stone tools—mainly hand axes—and the remains of many large mammals, including rhinoceros, horse, and red deer (elk). At one location within the site, GTP17, fossilized horse bones have been found. These are confined to a narrow range of sediments and do not exhibit very much weathering, suggesting that this assemblage represents a single hunt. This inference is consistent with the fact that the pieces of many broken stone tools at the site can be refitted. This makes it likelier that these tools were all used at the same time. There are several additional reasons to think that these horses were killed by hominins. First, most of the long bones show stone tool marks, but not carnivore tooth marks. Second, all of the bones of a single horse are present, which is not typical of fossil remains left by scavengers. Third, the shoulder blade of one of the horses appears to have been fractured by a spear.

Homo heidelbergensis used a variety of plant and animal resources.

We saw in Chapter 11 that modern human foragers typically dine on many kinds of foods and that they use complex processing techniques to acquire and process these foods. The site of Gesher Benot Ya'aqov in northern Israel provides the first evidence of similar levels of dietary diversity. This site, now in the dry Dead Sea valley, was on the shore of a lake when hominins lived there 790 ka. Nira Alperson-Afil and colleagues from Bar Ilan University in Israel have found evidence that the hominins who lived there (probably _H. erectus_ or _H. heidelbergensis_) used a variety of plant foods, including oak acorns, water lily seeds, and water chestnuts. Modern foragers roast these starchy nuts to make it easier to peel away the inedible shell and reduce the tannin content, and it is plausible that the hominins of Gesher Benot Ya'aqov did the same. Archaeologists have also found the remains of several different species of freshwater fish, a species of crab, and turtles, indicating that aquatic resources were also important for subsistence at this site. There are also remains of larger mammals such as elephants and fallow deer.

Hominins of the Later Pleistocene (300 ka to 50 ka)

About 300 ka, and then slightly later in western Eurasia, hominins in Africa shifted to a new stone tool kit.

Beginning about 300 ka, hand axes become much less common and are replaced by carefully crafted flake tools. Whereas earlier Oldowan flakes were struck from cobble cores and were irregular in shape and size, the hominins of the later Pleistocene produced large, symmetric, regular flakes by using more complicated techniques. One method, called the **Levallois technique** (after the Parisian suburb where such tools were first identified), involves three steps. First, the knapper prepares a core with one precisely shaped convex surface. Then the knapper makes a striking platform at one end of the core. Finally, the knapper hits the striking platform, knocking off a flake, the shape of which is determined by the original shape of the core (**Figure 12.5**). A skilled knapper can produce a variety of tools by modifying the shape of the original core. Such prepared core tools are classified as **Mode 3** technology. (Mode 1 and Mode 2 technologies are discussed in Chapter 11.)

Microscopic analyses of the wear patterns on tools made during this period suggest that some tools were **hafted** (attached to a handle). Hafting is an important innovation because it greatly increases the efficiency with which users can apply force to stone tools. (Try using a hammer without a handle.) _Homo heidelbergensis_ probably hafted pointed flakes onto wooden handles to make stone-tipped spears, a major breakthrough for a big-game hunter.

During the second half of the Middle Pleistocene, _H. heidelbergensis_ appeared in eastern Asia, where it may have coexisted with _H. erectus._

Up until the middle of the Middle Pleistocene, _H. erectus_ was the only hominin in eastern Asia. Hominins with larger brains and more rounded skulls begin to appear about 200 ka. The

FIGURE 12.5

The process of making a Levallois tool. (a) The knapper chooses an appropriate stone to use as a core. The side and top views of the unflaked core are shown. (b) Flakes are removed from the periphery of the core. (c) Flakes are removed radially from the surface of the core, with the flake scars on the periphery being used as striking platforms. Each of the red arrows represents one blow of the hammer stone. (d) The knapper continues to remove radial flakes until the entire surface of the core has been flaked. (e) Finally, a blow is struck (_red arrow_) to free one large flake (outlined in _red_). This flake will be used as a tool. (f) At the end, the knapper is left with the remains of the core (left) and the tool (right).

Flake the margin of the core.

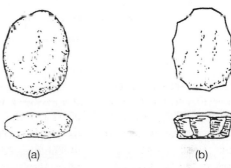

(a)   (b)

Prepare the surface of the core.

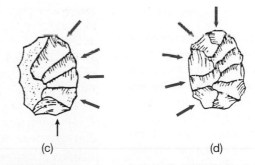

(c)   (d)

Remove Levallois flake.

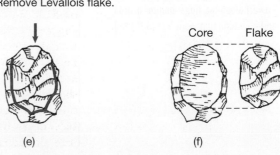

Core   Flake

(e)   (f)

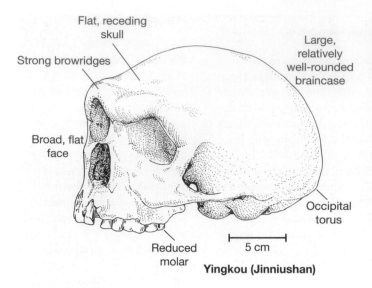

Flat, receding
skull

Strong browridges

Large,
relatively
well-rounded
braincase

Broad, flat
face

Occipital
torus

Reduced
molar

5 cm

Yingkou (Jinniushan)

FIGURE 12.6

Hominins with the characteristics of *H. heidelbergensis* appeared in eastern Asia later than in western Eurasia. The specimen illustrated here from Yingkou in northern China is approximately 200,000 years old.

most complete and most securely dated fossil is from the Jinniushan site, Yingkou, in northern China (see Figure 12.4). This specimen (**Figure 12.6**), which consists of a cranium and associated postcranial bones, is similar to early *H. heidelbergensis* fossils from Africa and Europe. Like the crania found at Kabwe and Petralona, it has a larger braincase (about 1,300 cc) and a more rounded skull, but it also has massive browridges and other primitive features. Similar fossils have been found in Dali in northern China and Maba in southern China, and they are probably somewhat younger than the Yingkou specimen. These fossils are associated with Oldowan-type tools. It is unclear whether these hominins were immigrants from the west or the result of convergent evolution in eastern Asia.

Such large-brained hominins may have coexisted with *H. erectus* in eastern Asia during this period. Fossils of *H. erectus* that are between 200,000 and 300,000 years old have been found at Hexian (also called He Xian), in southern China. The fossils of other animals found in association with *H. erectus* fossils at two sites in Java (Ngandong and Sambungmachan) are consistent with an age of 250,000 to 300,000 years old.

A tiny, small-brained hominin called *Homo floresiensis* lived on the Indonesian island of Flores during the Upper Pleistocene.

In the fall of 2004, a team of Indonesian and Australian researchers reported the discovery of what has been called "the most surprising fossil hominin found in the last 50 years" from the Indonesian island of Flores. Flores is one of many small islands located in a region known as Wallacea, which lies between the continental landmass of Asia and what is now Australia and Papua New Guinea (**Figure 12.7**). Excavations at a cave site called Liang Bua revealed the remains of between 9 and 14 individuals. The fossil specimens included a complete skull (**Figure 12.8**) and much of the rest of the skeleton (**Figure 12.9**).

The first surprise about these creatures, named *Homo floresiensis*, was that they were very, very small. They were slightly more than 1 m tall (about 3 ft.), much

FIGURE 12.7

Fossils of *Homo floresiensis* were discovered on the island of Flores in eastern Indonesia. The shaded green areas are now underwater but were then dry land, showing that Flores was isolated from both Asia and Australia even when sea levels were at their lowest levels. As a result, Flores was home to an odd mix of creatures, including a dwarf elephant, huge monitor lizards, and hominins that were 1 m (3 ft.) tall.

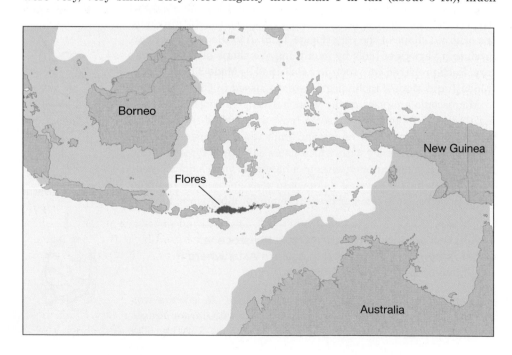

smaller than any other member of the genus *Homo*. Their brains were also very small (385 to 417 cc)—so small, in fact, that they may have been less encephalized than *H. erectus*. The second surprise about these creatures was their age. Most of the *H. floresiensis* fossils date to between 100 ka and 60 ka, when the environment surrounding the cave was a dry grassland. The discovery of these tiny creatures stimulated a flurry of public interest, speculation about their origins, and controversy about their place in the human lineage. Some researchers were convinced that they were the descendants of an early occupation by either *H. habilis* or *H. erectus* that became isolated on the island of Flores. Their small size was thought to be an example of evolutionary dwarfism, which occurs when animal populations are confined on islands. Biologists think that natural selection favors smaller body size on small islands because islands typically have fewer predators and more limited food supplies. If islands are isolated, there will be little gene flow from continental populations and animals will adapt to local conditions. This interpretation was challenged by other experts who were firmly convinced that the occupants of Liang Bua were members of a modern human population afflicted with a pathology that led to small bodies and very small brains. For example, Laron syndrome results from a mutation in a gene that reduces sensitivity to growth hormone and is associated with very small body size. However, new evidence supports the original island dwarfism hypothesis. In 2014, at the site of Mata Menge located about 74 km (46 mi.) from Liang Bua, Gerrit van den Bergh of the University of Wollongong in Australia and his colleagues discovered fragments of a hominin mandible and six teeth from at least three individuals. These specimens are even smaller than the ones from Liang Bua and are dated to about 700 ka. These finds demonstrate that a small-bodied hominin was present on Flores long before modern humans evolved.

Archaeological evidence also supports the idea that hominins and an odd collection of other animals have been present on Flores for a very long time. The Mata Menge site has yielded hundreds of artifacts, including tools, dated to 840 ka to 700 ka, along with the bones of Komodo dragons, giant rats, freshwater crocodiles, and a tiny elephant species, another example of island dwarfism. At this time, the site was probably savanna grassland. These tools are relatively crude with flakes struck off cores (**Figure 12.10**). They are more similar to Oldowan Mode 1 tools than to Mode 2 tools associated with *H. erectus* elsewhere. The assemblage of artifacts at Mata Menge is very similar to the assemblage of artifacts at Liang Bua, suggesting that there was little change in technology over this long period. Stone tools have also been found at another site on Flores dated to about 1 Ma.

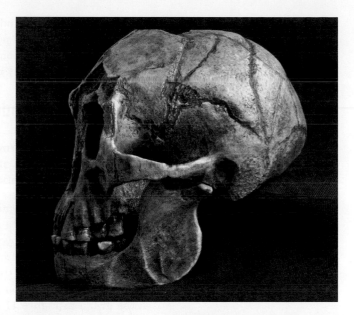

Side view of the skull of one of the *H. floresiensis* specimens (LB1). The skull is very small but shares several derived features with early *Homo*. The creature's brain was about the same size as the brain of australopiths.

A second hominin called *Homo luzonensis* lived on Luzon, the largest island in the Philippines, between 60 ka and 50 ka.

More recently, a number of fossils were found in a large limestone cave in northern Luzon. Like Flores, the islands of the Philippines are part of Wallacea and were never connected to mainland Southeast Asia. These fossils include part of a femur, finger and toe bones, and a number of teeth. These fossil remains show a strange mix of ancestral and derived features. The teeth are small and have simplified cusps, like those of modern humans, but their overall shapes are more like *H. erectus*;

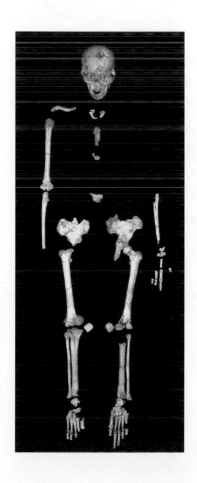

FIGURE 12.9

The skeleton of the *H. floresiensis* specimen (LB1). This individual would have been about 1 m (3 ft.) tall. The skeleton shows several features similar to early *Homo*, including the morphology of the wrists, pelvis, and feet.

FIGURE 12.10

Archaeological evidence suggests that hominins were present on Flores for about a million years. The Mata Menge tools are similar to Oldowan Mode 1 tools.

the finger and toe bones are similar to those seen in australopiths. The descriptions published in 2019 by the scientists examining the fossils suggest these hominins were smaller than modern people, but it is not clear whether they were as small as *H. floresiensis*.

The Neanderthals and the Denisovans

The last warm interglacial period before the current interglacial period lasted from about 130 ka to about 105 ka. Then the world cooled, and beginning about 75ka the global climate became much colder.

The next chapter in hominin history took place during a time when world temperatures began to cool. The data in **Figure 12.11** give a detailed picture of global temperatures over the past 123,000 years. These data are based on the ^{18}O to ^{16}O ratios of different layers of cores taken from deep inside the Greenland ice cap. Because snow accumulates at a higher rate than sediments on the ocean bottom do, ice cores provide more detailed information on past climates than do ocean cores like those used to construct the graph in Figure 11.1. You can see from Figure 12.11 that during the last warm interglacial period, about 105 ka, the world was substantially warmer than it is today, and plants and animals were distributed quite differently from how they are now. Plankton species currently living in subtropical waters (such as those off the coast of Florida) extended their range as far as the North Sea during the last interglacial period. Animals now restricted to the tropics had much wider ranges. For example, the remains of a hippopotamus have been found under Trafalgar Square in the center of London. In Africa, rain forests extended far beyond their current boundaries, and in temperate areas, broadleaf deciduous forests extended much farther north than they do today.

As the last glacial period began, the world slowly cooled. In Europe, temperate forests shrank and grasslands expanded. The glaciers grew, and the world became colder and colder—not steadily, but with wide fluctuations from cold to warm. When the glaciation was at its greatest extent (about 20 ka), huge continental glaciers covered most of Canada and much of northern Europe. Sea levels dropped so low that the outlines of the continents were altered substantially: Asia and North America were connected by a land bridge that spanned the Bering Sea; the islands of Indonesia joined Southeast Asia in a landmass called Sundaland; and Tasmania, New Guinea, and Australia formed a single continent called Sahul. Eurasia south of the glaciers was a vast, frigid grassland, punctuated by dunes of loess (fine dust produced by glaciers) and teeming with animals—woolly mammoths, woolly rhinoceroses, reindeer, aurochs (wild oxen that are ancestral to modern cattle), musk ox, and horses.

During the Middle Pleistocene, the morphology of *H. heidelbergensis* in western Eurasia diverged from the morphology of its contemporaries in Africa and Asia.

A large sample of fossils from a site in Spain provides evidence that *H. heidelbergensis* in Europe had begun to diverge from other hominin populations during the Middle Pleistocene. This site, called Sima de los Huesos ("Pit of Bones"), is located in the Sierra de Atapuerca, only a few kilometers from Trinchera Dolina. There, paleoanthropologists

Increasing amounts of ^{18}O

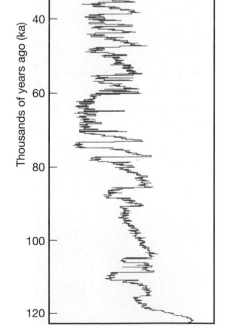

Thousands of years ago (ka)

Colder ← → Warmer

FIGURE 12.11

Fluctuations in the ratios of ^{18}O to ^{16}O over the past 123,000 years taken from an ice core drilled in northern Greenland. These data indicate that between 120 ka and 80 ka, the world's climate got colder and less stable. In ice cores like this one, a higher ratio of ^{18}O to ^{16}O indicates higher temperatures. (In deep-sea cores, more ^{18}O in relation to ^{16}O indicates lower temperatures.)

excavated a small cave 13 m (43 ft.) below the surface and found 2,000 bones from at least 24 individuals. These bones, which date to about 434 ka, include several nearly complete crania as well as many bones from other parts of the body. **Figure 12.12** shows one of the crania, labeled SH 5. Like other fossils of *H. heidelbergensis*, these skulls mix derived features of modern humans and primitive features associated with *H. erectus*. However, the crania from Sima de los Huesos also share several derived characteristics not seen in hominins living at the same time in Africa. Their faces bulge out in the middle and have double-arched browridges, and the backs of some of the skulls are rounded. The fossils from Sima de los Huesos also have relatively large cranial capacities, in one specimen reaching 1,390 cc, close to the average value for modern humans. These characteristics are significant because they are shared by **Neanderthals**, the hominins who dominate the European fossil record from 127 ka to 30 ka.

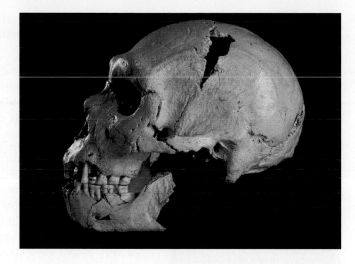

FIGURE 12.12

The many hominin fossils found at Sima de los Huesos in Spain provide evidence that hominins in Europe began to evolve a distinctive cranial morphology at least 434 ka. The features—which include a rounded browridge, a large pushed-out face, a skull with a rounded back, and a large brain—are important because they are shared with the Neanderthals, the hominins that dominate the European fossil record during the Upper Pleistocene.

The Neanderthals were an enigmatic group of hominins that lived in Europe and western Asia from about 127 ka to 30 ka.

In 1856, workers at a quarry in the Neander Valley in western Germany found some unusual fossil bones. The bones made their way to noted German anatomist Hermann Schaafhausen, who declared them to be the remains of a race of humans who had lived in Europe before the Celts. Many experts later examined these curious finds and drew different conclusions. Thomas Henry Huxley, one of Darwin's staunchest supporters, suggested that they belonged to a primitive, extinct kind of human. The Prussian pathologist Rudolf Virchow, on the other hand, proclaimed them to be the bones of a modern person suffering from a serious disease that had distorted the skeleton. Initially, Virchow's view held sway, but as more fossils were discovered with the same features, researchers became convinced that the remains belonged to a distinctive, extinct kind of human. The Germans called this extinct group of people *Neanderthaler*, meaning "people of the Neander Valley" (*thal*, now spelled *tal*, is German for "valley"). Today, we call them the Neanderthals.

The Neanderthals are for the most part a dead-end branch of the human family tree, but we know more about them than we do about any other extinct hominin species. There are two main reasons for this: First, Neanderthals lived in Europe, and paleontologists have studied Europe much more thoroughly than they have studied Africa or Asia. Second, geneticists have managed to extract DNA from several Neanderthal fossils, providing important insights about their origins and their relationship to modern humans.

FIGURE 12.13

The skulls of Neanderthals, like this one from Shanidar Cave, Iraq, are large and long, with large browridges and massive faces.

Neanderthals were characterized by several distinctive morphological features.

Neanderthals have become the archetypical image of "early man" for the general public. This is largely based on their distinctive morphology.

- *Large brains.* The Neanderthal braincase is much larger than that of *H. heidelbergensis*, ranging from 1,245 to 1,740 cc, with an average size of about 1,520 cc. In fact, Neanderthals had larger brains than those of modern humans, whose brains average about 1,400 cc. It is unclear why the brains of Neanderthals were so large. Some anthropologists point out that the Neanderthals' bodies were much more robust and heavily muscled than those of modern humans, and they suggest that the large brains of Neanderthals reflect the fact that larger animals usually have bigger brains than smaller animals.

- *More rounded crania* (**Figure 12.13**). The Neanderthal skull is long and low, much like the skulls of *H. heidelbergensis*, but relatively thin-walled. The back of the

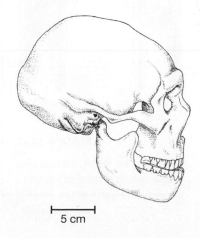

5 cm

FIGURE 12.14

Artist's conception of a Neanderthal man.

skull has a characteristic rounded bulge or bun and does not come to a point at the back like an *H. erectus* skull does. There are also detailed differences in the back of the cranium.

- *Big faces.* Like *H. erectus* and *H. heidelbergensis*, the skulls of Neanderthals have large browridges, but the Neanderthal's browridges are larger and rounder and they stick out less to the sides. Moreover, the browridges of *H. erectus* are mainly solid bone, whereas those of Neanderthals are lightened with many air spaces. The function of these massive browridges is not clear. The face, particularly the nose, is enormous: Every Neanderthal was a Cyrano, or perhaps a Jimmy Durante (**Figure 12.14**).

- *Small back teeth and large, heavily worn front teeth.* Neanderthal molars are smaller than those of *H. erectus*. They had distinctive **taurodont roots** in which the pulp cavity expanded so that the roots merged, partially or completely, to form a single broad root (**Figure 12.15**). Neanderthal incisors are relatively large and show very heavy wear. Study of these wear patterns indicates that Neanderthals may have pulled meat or hides through their clenched front teeth. There are also microscopic, unidirectional scratches on the front of the incisors, suggesting that these hominins held meat in their teeth while cutting it with a stone tool. It is interesting that the direction of the scratches suggests that most Neanderthals were right-handed, just as the Oldowan toolmakers were.

- *Robust, heavily muscled bodies.* Like *H. erectus* and *H. heidelbergensis*, Neanderthals were extremely robust, heavily muscled people (**Figure 12.16**). Their leg bones were much thicker than ours, the load-bearing joints (knees and hips) were larger, the **scapulae** (shoulder blades; singular *scapula*) had more extensive muscular attachments, and the rib cage was larger and more barrel shaped. All these skeletal features indicate that Neanderthals were very sturdy and strong, weighing about 30% more than contemporary humans of the same height. A comparison with data on Olympic athletes suggests that Neanderthals most closely resembled hammer, javelin, and discus throwers and shot-putters. They were a few inches shorter, on average, than modern Europeans, and had larger torsos and shorter arms and legs.

This distinctive Neanderthal body shape may have been an adaptation to conserve heat in a very cold environment. In cold climates, animals tend to be larger and have shorter and thicker limbs than do members of the same species in warmer environments. This is because the rate of heat loss for a body is proportional to its surface area, so any change that reduces the amount of surface area for a given volume will conserve heat. The ratio of surface area to volume in animals can be reduced by increasing overall body size or by reducing the size of the limbs. In contemporary human populations there is a consistent relationship between climate and body proportions. One way to compare body proportions is to calculate the **crural index**, which is the ratio of the length of the shinbone (tibia) to the length of the thighbone (femur). As **Figure 12.17** shows, people in warm climates tend to have relatively long limbs in proportion to their height. Neanderthals resemble modern peoples living above the Arctic Circle.

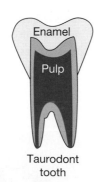

Nontaurodont tooth   Taurodont tooth

FIGURE 12.15

In Neanderthal molars, the roots often fuse partially or completely to form a single massive taurodont root. The third root is not shown.

Neanderthals made Mode 3 tools and hunted large game.

Although Neanderthals are popularly pictured as brutish dimwits, this characterization is unfair. As we noted earlier, their brains were larger than ours. The archaeological evidence suggests that they were skilled toolmakers and proficient big-game hunters. The Neanderthal's stone tool kit is dominated by Mode 3 tools, which are characterized by flakes struck from prepared cores. Their stone tool industry is called the **Mousterian industry** by archaeologists. There is also evidence that Neanderthals made compound tools by hafting stone points onto wooden shafts.

Neanderthals were proficient hunters who regularly killed large animals. Neanderthal sites are littered with stone tools and the bones of red deer, fallow deer, bison, aurochs, wild sheep, wild goats, gazelle, horses, and rhinoceros (**Figure 12.18**). The evidence indicates that these prey were acquired by hunting not scavenging. Animal remains at sites from this period are often dominated by the bones of only one or two prey species. At Mauran, a site in the French Pyrenees, for example, more than 90% of the assemblage is from bison and aurochs. The same pattern occurs at other sites scattered across Europe. It is hard to see how an opportunistic scavenger would acquire such a nonrandom sample of the local fauna. Moreover, the age distribution of prey animals does not fit the pattern for modern scavengers such as hyenas, which prey mainly on the most vulnerable members of prey populations: sick or wounded animals, the old, and the very young (**Figure 12.19**). At these European Neanderthal sites, the bones of apparently healthy, prime-age adults are well represented. The distribution of animal bones is what we would expect to see at sites of catastrophic events in which whole herds of animals are killed. At several sites, such as Combe Grenal in France, the bones from the meatiest parts of prey animals are overrepresented, and the cut marks on these bones suggest that Neanderthals stripped off the fresh flesh. (Remember that hunters often haul away the meatiest bones from kill sites to eat in peace.)

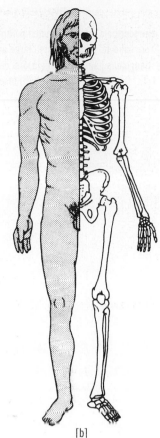
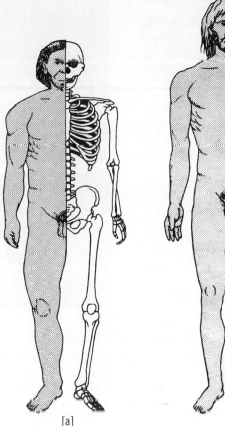

[a] [b]

FIGURE 12.16

The Neanderthals were very robust people. Compare the bones of (a) a Neanderthal with those of (b) a modern human.

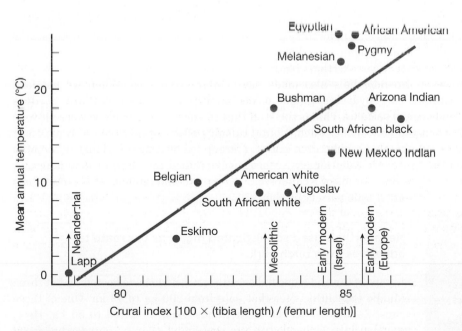

FIGURE 12.17

People have proportionally longer arms and legs in warm climates than in cold climates. Local temperature is plotted on the vertical axis; crural index is plotted on the horizontal axis. Smaller values of the crural index are associated with shorter limbs in relation to body size. Populations in warm climates tend to have high crural index values, and vice versa. Neanderthals had a crural index similar to those of present-day Lapps, who live above the Arctic Circle.

FIGURE 12.18

Neanderthals and their contemporaries are believed to have hunted large and dangerous game, such as (a) red deer and (b) bison.

[a]   [b]

The Neanderthal diet included both animals and plants.

Studies of nitrogen isotopes found in Neanderthal fossils indicate that meat played a significant role in their diets. The ratio of ^{15}N, the heavier stable isotope of nitrogen, to ^{14}N, the lighter stable isotope, in plant and animal tissues increases as one ascends the tropic pyramid. The ratio in the tissues of herbivores is higher than that of the plants they eat, and the ratio in carnivores is higher than that in herbivores. The nitrogen ratio measured in fossil Neanderthal bones from all over western Eurasia is uniformly high and is similar to that found in top carnivores such as lions and hyenas. This suggests that Neanderthals regularly ate large herbivores.

However, Neanderthals also made use of plant resources. When we eat food that contains carbohydrates, plaque forms on our teeth. This eventually hardens to become **dental calculus**. This is the stuff that the dental hygienist scrapes off your teeth when you get them cleaned. Dental calculus often captures microscopic plant parts called phytoliths and starch grains. A group led by Amanda Henry of the Max Planck Institute for Evolutionary Anthropology studied the dental calculus from a wide sample of Neanderthal fossil teeth and found that Neanderthals ate a wide variety of plants.

FIGURE 12.19

Hyenas often scavenge kills, and they prey mainly on very young and very old individuals.

Neanderthals used sophisticated methods to create tar adhesives from birch bark.

Lumps of tar and stone flakes with tar residue have been found at three Middle Paleolithic sites that date from 200 ka to 43 ka. One of these sites, Zandmotor on the Netherlands coast, is dated to 50 ka. Here, a stone flake embedded in tar was found near a Neanderthal fossil (**Figure 12.20**). (Tar here means a sticky material extracted from the bark of trees, most often birch trees, not the petroleum residue used today to make asphalt streets and seal flat roofs.) The tar may have been used as an adhesive to bind stone flakes to wooden handles or to make a handle for the flakes.

Extracting tar from birch bark is a tricky business. Today the bark is heated to a high temperature (200°C to 400°C) in air-tight metal or ceramic containers so that the tar does not ignite. Modern experiments suggest that there are a number of ways to produce tar without such containers. The simplest is to burn tightly rolled birch bark near a vertical stone. Smoke that condenses on the stone contains small amounts of tar. More complex methods, which involve burying birch bark in an earthen oven, produce much higher yields. The amount of tar in the Zandmotor sample, its chemistry, and the nature of the impurities it contains all suggest that earthen ovens were used to extract the tar; this implies that Neanderthals were able to solve complex technological problems like their contemporaries, modern humans in southern Africa.

FIGURE 12.20

The stone tool with birch tar residues found at Zandmotor in the Netherlands. The tar adheres to the bottom section of the tool and probably served as an adhesive to bind the stone flake to a handle.

Neanderthals probably wore clothes.

Neanderthals lived in Europe and western Asia during glacial periods when the climate was very cold—winter temperatures may have averaged −25°C—so it seems plausible that they may have needed clothes to stay warm. Clever research by Nathan Wales of the University of York suggests that this indeed was the case. Wales collected data on the percentage of the body that is covered by clothing in 245 contemporary hunting and gathering groups. Using information about winter and summer where these groups live, he worked out the relationship between climate and clothing coverage. Then, he adjusted this relationship to account for the fact that Neanderthals were better able to cope with the cold because they had thicker torsos and shorter limbs than modern humans do. Finally, using data on the climates in western Eurasia when Neanderthals were present, he generated predictions of how much clothing Neanderthals would have needed to wear during the winter in different parts of their range. His findings suggest that during the winter, Neanderthals would need to be fully covered by clothing across much of their range (**Figure 12.21**). We do not know whether Neanderthals made tailored clothes, like pants and parkas, or simply wrapped themselves in loose cloaks.

FIGURE 12.21

How much clothing Neanderthals needed to wear in different parts of Europe during the periods 74 ka to 66 ka and 59 ka to 37 ka. The triangles mark sites where Mousterian tools have been found.

There is little evidence for the existence of shelters or even organized camps at Neanderthal sites.

There are many Neanderthal sites, some very well preserved, where archaeologists have found concentrations of tools, abundant evidence of toolmaking, many animal remains, and concentrations of ash. Most of these sites are in caves or **rock shelters**, places protected by overhanging cliffs. This doesn't necessarily mean that Neanderthals preferred these kinds of sites. Cave sites are more likely to be found by researchers because they are protected from erosion, and they are relatively easy to locate because the openings of many caves from this period are still visible. Most archaeologists believe that cave sites represent home bases, semipermanent encampments from which Neanderthals sallied out to hunt and to forage.

The archaeological record suggests that Neanderthals did not build shelters. Most Neanderthal

0-10 10-20 20-30 30-40 40-50 50-60 60-70 70-80 80-90 90-100

Maximum percentage of body covered by clothing

FIGURE 12.22

Two perforated mollusk shells from Cueva de los Aviones in Spain. This site dates to about 50 ka and is associated with Middle Paleolithic tools. The perforations may have allowed the shells to be strung on a cord for use as ornamentation.

sites lack evidence of postholes and hearths, two features generally associated with simple shelters. The few exceptions to this rule occur near the end of the period. For example, hearths were built at Vilas Ruivas, a site in Portugal dated to about 60 ka.

Neanderthals probably buried their dead.

The abundance of complete Neanderthal skeletons suggests that, unlike their hominin predecessors, Neanderthals often buried their dead. Burial protects the corpse from dismemberment by scavengers and preserves the skeleton intact. Study of the geologic context at sites such as La Chapelle-aux-Saints, Le Moustier, and La Ferrassie in southern France also supports the conclusion that Neanderthal burials were common.

It is not clear whether these burials had a religious nature or if Neanderthals buried their dead just to dispose of the decaying bodies. Anthropologists used to interpret some sites as ceremonial burials in which Neanderthals were interred along with symbolic materials. In recent years, however, skeptics have cast serious doubt on such interpretations. For example, anthropologists used to think that the presence of fossilized pollen in a Neanderthal grave in Shanidar Cave, Iraq, was evidence that the individual had been buried with a garland of flowers. More recent analyses, however, revealed that the grave had been disturbed by burrowing rodents, and it is quite possible that they brought the pollen into the grave.

Neanderthals may have used painted seashells as personal ornaments.

Humans devote substantial creativity and lots of resources to personal adornment. In the modern world, huge amounts of money are spent on clothes, jewelry, and makeup, and both history and anthropology suggest that personal adornment is a universal human trait. As we will see in the next chapter, there is good evidence that the earliest modern humans also decorated themselves with pigments and wore jewelry. Thus, it is clearly of interest to know whether earlier hominins shared this aspect of our psychology, but until recently the evidence was equivocal. Beautiful personal ornaments were found in association with Neanderthal fossils at a site in France called Arcy sur Cure, but the dates for this site vary wildly, and it seems possible that ornaments made by later, modern human occupants got mixed with Neanderthal layers, perhaps when the moderns dug postholes.

FIGURE 12.23

The perforated scallop shell found at Cueva Antón. The left image shows the naturally red inside of the shell. The right shows the naturally white outside of the shell that was stained with the orange mineral pigments goethite and hematite.

However, a recent discovery suggests that Neanderthals used seashells as adornments. Several pierced shells have been found at Cueva Antón and Cueva de los Aviones, two sites in southeastern Spain. These shells have holes that could have been used to string them on a leather thong **(Figure 12.22)**. They also have been treated with orange mineral pigments **(Figure 12.23)**. It seems likely that this use was decorative because these pigments come from at least 5 km (3.1 mi.) away and appear only on the shells, not on other artifacts found at the site. Similar shell ornaments have been found at contemporaneous African sites occupied by modern humans.

Neanderthals seem to have lived short, difficult lives.

Study of Neanderthal skeletons indicates that Neanderthals didn't live very long. The human skeleton changes throughout the life cycle in characteristic ways, and these changes can be used to estimate the age at which fossil hominins died. For example, human skulls are made up of separate bones that fit together in a three-dimensional jigsaw puzzle. When children are first born, these bones are still separate, but later they fuse, forming tight, wavy joints called **sutures**. As people age, these sutures are slowly obliterated by bone growth. By assessing the degree to which the sutures of fossil hominins have been obliterated, anthropologists can estimate how old the individual was at death. Several other skeletal features can be used in similar ways. All these features tell the same story: Neanderthals died young. Few lived beyond the age of 40 to 45 years.

Many of the older Neanderthals suffered disabling disease or injury. For example, the skeleton of a Neanderthal man from La Chapelle-aux-Saints shows symptoms of severe arthritis that probably affected his jaw, back, and hip. By the time this fellow died, around the age of 45, he had also lost most of his teeth to gum disease. Another individual, Shanidar 1 (from the Shanidar site in Iraq), suffered a blow to his left temple that crushed the orbit (**Figure 12.24**). Anthropologists believe that his head injury probably caused partial paralysis of the right side of his body, and this in turn caused his right arm to wither and his right ankle to become arthritic. Other Neanderthal specimens display bone fractures, stab wounds, gum disease, withered limbs, lesions, and deformities.

In some cases, Neanderthals survived for extended periods after injury or sickness. For example, Shanidar 1 lived long enough for the bone surrounding his injury to heal. Some anthropologists have proposed that these Neanderthals would have been unable to survive their physical impairments—to provide themselves with food or to keep up with the group—had they not received care from others. Some researchers further argue that these fossils are evidence of the origins of caretaking and compassion in our lineage.

There are reasons to be cautious of such claims. In some contemporary societies, disabled individuals can support themselves, and they do not necessarily receive compassionate treatment from others. In addition, nonhuman primates sometimes survive despite permanent disabilities. At Gombe Stream National Park, a male chimpanzee named Faben contracted polio, which left him completely paralyzed in one arm. Despite this impairment, he managed to feed himself, climb the steep slopes of the community's home range, keep up with his companions, and even climb trees (**Figure 12.25**).

Analyses of ancient DNA provide a new source of information about the origins of the Neanderthals and their relationships to other hominin taxa.

In the film *Jurassic Park*, scientists extract dinosaur DNA from the bodies of bloodsucking insects trapped in amber and use the DNA to clone living dinosaurs. The gap between science fiction and science has narrowed in the laboratory of Svante Pääbo at the Max Planck Institute for Evolutionary Anthropology in Leipzig, Germany. Over the past two decades, Pääbo and his colleagues have made a series of extractions and sequenced the DNA from several Neanderthal fossils.

It is extraordinarily difficult to obtain genetic material from fossils that have been buried in the earth for thousands of years. Most of the genetic material has simply disappeared. Much of what remains is broken up into small pieces, making it difficult to reconstruct the original DNA sequence. There is often chemical damage to DNA that alters the original identity of some of the bases within the DNA molecule. Finally, samples may be contaminated with DNA from other organisms at the original site, from researchers who have handled them in the field, and from researchers in the labs where they have been stored. Molecular geneticists have made great strides in dealing with these technical challenges in recent years.

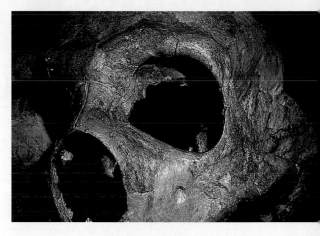

FIGURE 12.24

Many Neanderthal skeletons show signs of injury or illness. The orbit of the left eye of this individual was crushed, the right arm was withered, and the right ankle was arthritic.

FIGURE 12.25

At Gombe Stream National Park, a male chimpanzee named Faben contracted polio, which left one arm completely paralyzed. After the paralysis, Faben received no day-to-day help from other group members, but he was able to survive for many years. Here, Faben climbs a tree one-handed.

12.1 The Origins of Language

So far, we have said almost nothing about the origins of language. This may seem like an odd omission because language is such a fundamental feature of our lives. Without it, how could we explain complex ideas, reminisce about the past, make plans for the future, or gossip? There is a reason for our reticence: We don't really know when or how language evolved. Like behavior, language leaves little trace in the fossil record. Most of the anatomic features that enable us to produce spoken language, such as our long and flexible tongue, are made of soft tissue that does not fossilize. Similarly, most of the cognitive adaptations that allow us to create meaning from sound leave no traces. This means theories about the origins of language are difficult to test, and there is considerable room for controversy about how and when language first emerged. Here we briefly outline some of the evidence about key questions in the evolution of language.

Did Human Language Evolve from a Vocal or Gestural Communication System?

Vocal communication is not unique to humans. Chimpanzee infants whine pitifully when their mothers reject their attempts to nurse, rhesus macaques scream to recruit allies in agonistic disputes, and female baboons grunt to announce their intentions to behave peacefully. In a classic study, Dorothy Cheney and Robert Seyfarth of the University of Pennsylvania showed that vervet monkeys give acoustically distinct alarm calls when they spot different types of predators; when listeners hear these calls they know what kind of predator has been spotted even though they haven't seen it themselves. These calls are said to have *referential meaning* because the calls are arbitrary acoustic referents to something in the real world.

It seems logical that vocal communication is the foundation for spoken language, and you might be able to imagine how natural selection might go from calls with referential meaning to words. But there is a big gulf between the vocalizations that other animals make and human language. In spoken language, acoustically distinct sounds (phonemes) are combined in meaningful units (morphemes), and these units can be combined into a nearly infinite set of unique messages. In contrast, primate vocal repertoires are limited to a relatively small number of species-specific calls. There is little flexibility in how calls are produced or used, and learning seems to play a relatively limited role in vocal behavior.

An alternative hypothesis is that language evolved from gestural communication. Primates use a variety of facial expressions and manual gestures to communicate with conspecifics. For example, chimpanzee infants use manual gestures to initiate play, beg for food, and get the attention of others. Gestures are used in flexible ways across different contexts, and new gestures are sometimes acquired and incorporated into a group's repertoire. For example, leaf clipping, in which an individual grasps a leaf and pulls pieces of the leaf off with its teeth to produce a ripping sound, occurs in different contexts in different chimpanzee communities. In the Mahale Mountains of Tanzania, males use leaf clipping to attract the attention of sexually receptive females, and in Boussou in West Africa leaf clipping is used to attract play partners. Of course, proponents of the gestural theory of language have to explain how a nonvocal system of communication was transformed to a vocal one.

When Did the Capacity for Spoken Language Arise?

Although most of the anatomic features that are linked to speech and hearing are made of soft tissues that do not fossilize, a few of the structures related to speech production and audition (hearing) are made of bone. For example, the tiny hyoid bone lies in the throat and anchors the tongue. These anatomic clues give paleontologists some purchase on tracing the emergence of language.

- The production of spoken language depends on finely coordinated movements of the tongue, mouth, vocal cords, and muscles in the diaphragm and thorax (upper chest) that regulate breathing. The nerves that regulate the actions of the muscles of the diaphragm and thorax are connected to the spinal cord, which is protected by the bony vertebral canal. The vertebral canal in the thoracic region of the back is much larger in modern humans than it is in apes, and it contains a proportionally thicker spinal cord. All the extra nerves that enlarge the spinal cord are connected to the muscles of the rib cage and diaphragm. Ann MacLarnon of Roehampton University reconstructed the thoracic vertebrae of the *H. erectus* skeleton WT-15000 and found that it had a narrow vertebral canal. The vertebral canals of Neanderthals are like those of modern humans.

- Chimpanzees and other great apes have air sacs that are attached to their vocal apparatus and fill with air as it flows into the lungs. Models of sound moving through the vocal tract suggest that air sacs make it harder to distinguish vowel sounds. Air sacs don't fossilize, but all of the apes with air sacs also have hyoid bones with a distinctive bulge, or bulla, in the middle. Modern humans do not have air sacs, and their hyoids lack a bulla. Although hyoids are very small bones, and few have been preserved in the fossil record, we know from the Dakika child that australopiths had hyoids like those of chimpanzees. The hyoids of *H. heidelbergensis* and Neanderthals are very human-like (**Figure 12.26**).

- In primates and other mammals, the larynx (Adam's apple) sits above the trachea (windpipe) and houses the vocal cords. Air passes over the vocal cords, which vibrate and produce an audible sound. In infants and other primates, the larynx is positioned

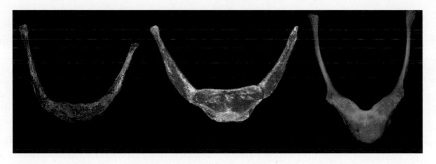

FIGURE 12.26

The hyoid bones of a modern human (*left*), Neanderthal (*middle*), and chimpanzee (*right*) are shown here.

at the back of the mouth. As human infants mature, the larynx moves down into the throat. Lowering the larynx makes the horizontal and vertical portions of the vocal tract about the same length, and Philip Lieberman of Brown University has argued that a 1:1 ratio of the horizontal and vertical portions is necessary to articulate the full range of vowel sounds. Several researchers have attempted to estimate the horizontal and vertical portions of the vocal tract in several *H. heidelbergensis* and Neanderthal specimens. They conclude that the ratios for these fossils is about 0.8, lower than the value for adult humans but approximately the same as that of 10-year-old human children.

Molecular genetics also gives us clues about the origins of language. One gene, FOXP2, produces a protein that has myriad effects on other proteins and genes during development. FOXP2 also plays some role in language. Studies of the members of one family who have inherited a mutation in FOXP2 genes show that they have trouble with some aspects of grammar and difficulty controlling the muscles of the face and mouth that are used in pronouncing words. Geneticists have sequenced the FOXP2 gene in chimpanzees and modern humans. There have been only two changes in the amino acid sequence of the FOXP2 allele since humans diverged from chimpanzees, and three since humans diverged from mice. Neanderthals have the human version of the FOXP2 allele.

Taken together, these data suggest that Neanderthals might have been able to produce sounds that are like those that we produce. But language relies on more than the ability to articulate vowels and consonants, and we do not yet know whether Neanderthals had the capacity for language.

The first analyses of Neanderthal DNA were based on mitochondrial genes. The mitochondria are organelles that are responsible for the basic energy processing that goes on inside cells. They contain small amounts of DNA (about 0.05% of the DNA contained in chromosomes) called **mitochondrial DNA (mtDNA)** that codes for a few proteins and a few transfer RNAs (tRNAs). Comparatively, **nuclear DNA** codes for most of the genome in multi-celled organisms. Mitochondrial DNA is transmitted without recombination from mothers to both sons and daughters, but only daughters pass mtDNA on to their own offspring. mtDNA thus behaves like a single genetic locus, and provides information about a small subset of an individual's ancestors (your mother, your mother's mother, your mother's mother's mother, and so on). An organism's autosomal nuclear DNA is much more informative because it contains many thousands of independent genes inherited from both male and female ancestors from both your maternal and paternal lines. However, it is easier to get mtDNA out of fossils than nuclear DNA because there are hundreds or even thousands of mitochondria in each cell, but only two copies of each nuclear gene.

These early analyses of Neanderthal mitochondrial DNA showed that it was outside the range of variation observed in living humans. While this suggested that Neanderthals had not interbred with anatomically modern humans, it is also possible that interbreeding had occurred but Neanderthal mtDNA was lost in modern humans due to chance.

As technology improved, it became possible to reconstruct sequences of ancient nuclear DNA. Pääbo and his colleagues reported sequencing the complete nuclear genome from three Neanderthal fossils found in Vindija Cave in Croatia that date from 44 ka to 38 ka. Despite limits on the existing technology and the condition of the genetic material, the Leipzig team produced a low-resolution sequence of the genome. You can think of this as a "rough draft" of the genome. A few years later, technical advances coupled with exceptional preservation of DNA allowed Pääbo and his colleagues to create a very high-resolution sequence from a Neanderthal toe bone found at Denisova Cave in the Altai Mountains of southern Siberia. This fossil is named the Altai Neanderthal. Remarkably, the resolution of the Alta Neanderthal sequence is as high as the resolution of sequences derived from modern humans. The Leipzig researchers also reported a lower-resolution sequence from a fifth Neanderthal found at Mezmaiskaya in the foothills of the Caucasus Mountains in southern Russia. Comparisons between the genomes of modern humans and these Neanderthals indicate that the last common ancestor of Neanderthals and modern humans lived between 765 ka and 550 ka. Consistent with the morphological affinities, ancient DNA research has also shown that the 434 ka *Homo heidelbergensis* fossils from Sima de Los Huesos (see p. 310) were early Neanderthals or closely related to the ancestors of Neanderthals.

Genetic data from fossils found at Denisova Cave indicate that another genetically distinct hominin was living in Eurasia at the same time as the Neanderthals.

Excavations at Denisova Cave turned up one very odd specimen: a huge molar tooth. The molar is as big as those of early *Homo* and much larger than the molars of other Asian fossils. It also lacks several derived features found in the molars of other Asian fossils. A toe bone was also found near the molar. The dates for these fossils are uncertain, but the stratigraphy suggests that the Altai Neanderthal lived at an earlier time than the creatures that left these bones. Exceptional preservation of the DNA allowed Pääbo and his colleagues to construct a very high-quality genome sequence from these fossils. They found that the tooth and toe bone came from different individuals who were members of a single population. Later, the Leipzig group published low-resolution sequences of the mtDNA and nuclear DNA from two more fossils found at the same site. Their analyses show that these individuals were genetically distinct from Neanderthals. Denisovans occupied the cave at least as early as about 300 ka and as recently as about 55 ka, while Neanderthal presence begins later (200 ka) and ends earlier (100 ka).

In 1980, a robust, chinless jawbone with huge molars was found in a cave high on the Tibetan plateau, and in 2019, scientists working on the fossil announced that it appeared to be Denisovan. They were unable to extract DNA from this 160 ka specimen but were able to analyze ancient collagen proteins from one of the teeth and found that the individual resembled Denisovans more closely than it did Neanderthals or modern humans. This evidence for Denisovan presence in caves in both Siberia and in Tibet suggests that Denisovans were probably fairly widespread across eastern Eurasia.

Neanderthal and Denisova genomes are more similar to each other than to modern humans, but they had different evolutionary histories.

Pääbo and his colleagues compared the genomes of modern humans, Neanderthals, and the Denisova fossil. The results of this analysis are summarized by the genetic distance tree shown in **Figure 12.27**. The Neanderthal genome and the Denisova genome are a bit more similar to each other than either is to the genome of modern humans. This suggests that while the last common ancestor of Neanderthals, Denisovans, and modern humans lived from 765 ka to 550 ka, Neanderthals and Denisovans were also descended from a more recent common ancestor who lived about 450 ka. This 450 ka split time is consistent with the finding that the Sima de los Huesos hominins of 434 ka were more closely related to

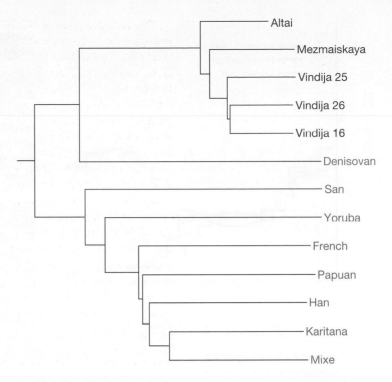

FIGURE 12.27

This tree gives the overall genetic distance between the genomes of several Neanderthal fossils (*red*), the Denisovan fossil (*green*), and several living people from around the globe (*blue*). The length of the lines is proportional to the genetic distance. The simplest interpretation of the data is that Neanderthals and Denisovans share a more recent common ancestor with each other than with modern humans, but Denisovans were genetically distinct from Neanderthals.

Neanderthals than to Denisovans, as this indicates that Neanderthals and Denisovans must have begun diverging before this time. Furthermore, the Neanderthal fossils are much more similar to each other than any are to the Denisova fossil, even though the Neanderthals come from widely dispersed sites in Germany, Spain, Croatia, and Russia. This suggests that all Neanderthals shared an even more recent common ancestor who lived about 140 ka, well after the Neanderthal lineage split from the lineage leading to the Denisovans. Recall from Chapter 3 that genetic drift removes variation from a population, and this occurs more quickly when populations are smaller. The most likely explanation for the low levels of genetic variation across the Neanderthal range is that Neanderthal populations suffered a severe reduction in size, often called a *population bottleneck*, after they diverged from the Denisovans.

In addition, there are unusual fragments in the Denisovan genome suggesting that after they diverged from the Neanderthals, Denisovans may have received gene flow from an older hominin that diverged from the common ancestor of Denisovans, Neanderthals, and modern humans about 2 Ma. This "super-archaic" lineage, so-named because it diverged from the Neanderthal/Denisovan/modern human lineage much earlier than the "archaic" Neanderthal/Denisovan lineage diverged from modern humans, is not associated with any fossil material or direct ancient DNA data. Super-archaics are an example of what geneticists call a **ghost lineage**: a population or species from whom we do not have a direct ancient DNA sequence in unadmixed form, but whose existence can be statistically inferred on the basis of patterns of genetic variation in populations or species with which they interbred. Some authors have recently proposed that these superarchaics may have also interbred with the ancestors of Neanderthals and Denisovans before they began diverging from one another.

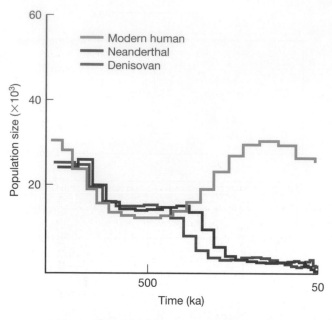

FIGURE 12.28

Estimates of population sizes for Neanderthals and Denisovans over time compared with a modern human population, the San of southern Africa. See Chapter 13 for an explanation of how geneticists derive such estimates from complete genomes.

FIGURE 12.29

This cranium found at Florisbad in South Africa shows a mixture of features of *H. heidelbergensis* and modern *H. sapiens*. Although still quite robust, it has reduced browridges and a more rounded shape. It dates to between 300 ka and 200 ka.

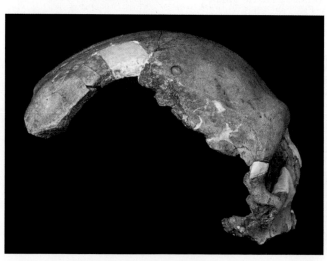

Genetic data indicate that the populations of Neanderthals and Denisovans were very small.

Geneticists have derived methods that allow them to use data from the genome of a single individual to estimate the sizes of ancient populations through time. Applying this method to the high-resolution Neanderthal and Denisovan genomes indicates that their populations were very small, perhaps 10 times smaller than the populations of modern *Homo sapiens* that lived at the same time (**Figure 12.28**). The genetic data also indicate that the parents of the Altai Neanderthal were closely related: half siblings, uncle and niece, aunt and nephew, or grandparent and grandchild. Furthermore, a 90-ka bone fragment in Denisova Cave was shown to belong to an individual with a Neanderthal mother and a Denisovan father. Neanderthals and Denisovans may have had to mate with close relatives or with individuals outside of their own "species" because they lived in small, isolated groups where more ideal mates were hard to come by. (Note that "species" is in quotes here because many biologists do not classify Denisovans and Neanderthals as distinct species.)

Modern research reveals functional genetic differences between modern humans and Neanderthals/Denisovans.

Although modern humans are very genetically similar to Neanderthals and Denisovans, analyses of the functions of the genes that differ offer insight into the evolution of modern human phenotypic differentiation from these two hominins. Specifically, there are 571 protein-altering genes with variants that are present in more than 90% of humans but for which Neanderthals and Denisovans have the same gene as chimpanzees. There are also notable differences in gene regulation between modern humans and Neanderthals/Denisovans. Collectively, these differences point to changes in the molecular mechanisms in cell division, as well as changes to networks affecting cellular features of neurons, that occurred as modern humans evolved apart from Neanderthals/Denisovans. Although these changes likely resulted in some differences in brain growth trajectory and cognitive traits, much more work is required to really understand the consequences of these genetic differences on phenotypic differences between modern humans and their relatives.

Africa: The Road to *Homo sapiens*?

Hominins living in Africa during the later Middle Pleistocene were more similar to modern humans than were Neanderthals.

What was happening in Africa while the Neanderthals were living in Europe and Asia is not entirely clear. The fossil record for this time in Africa is poor. Several fossils dated to later in the Middle Pleistocene (300 ka to 200 ka) have robust features similar to those of *H. heidelbergensis*, though recent discoveries suggest the continued presence in Africa of hominins whose morphology more closely resembles that of early *Homo* (**A Closer Look 12.2**). Examples of fossils with more robust features include the Florisbad cranium found in South Africa (**Figure 12.29**)

12.2 *Homo naledi*: Mysteries of the Rising Star Cave

Although professional paleontologists are quick to dismiss Indiana Jones's spectacular finds as fiction, sometimes life imitates art. After Lee Berger and his team discovered the *Australopithecus sediba* remains in the Malapa Cave, he issued an alert for cavers to be on the lookout for fossils in other South African caves. With this in mind, Steven Tucker and Rick Hunter set off to explore the Rising Star Cave, about 10 miles from Malapa. The cave was well known to cavers, and many of its passages and chambers had already been explored, named, and mapped. Tucker and Hunter made their way deep into the cave, squeezing through impossibly small passageways and down narrow shafts, eventually discovering a small chamber deep in the cave. The floor of the chamber was littered with bones, some of which looked very humanlike. They photographed what they found and showed them to Berger, who realized it was a very intriguing find.

Berger mounted a full-scale scientific expedition. The first challenge was to get trained excavators into the cave to the fossil site. Berger and the other paleontologists he knew were too big to get through the narrow passageways that led to the bones. So he recruited a team of researchers with slight builds who had the appropriate technical skills. Then he and his team threaded 2 miles of cable down into the cave so that they could monitor the excavations, communicate with the excavators from a command post at the surface, and broadcast the proceedings on social media.

The cave eventually yielded 1,550 fossil specimens, the largest collection of a single species found anywhere in Africa (**Figure 12.30**). At least 15 individuals are represented in the sample. Berger put together a large team of researchers to analyze the material, and the first comprehensive descriptions of the fossils were published in 2015, just 2 years after they were discovered.

These creatures display a distinctive mixture of primitive and derived traits, and Berger assigned them to a new species, *Homo naledi* (*Naledi* means "star" in the Sotho language.) Their brains are 465 to 560 cc, at the low end of values for early *Homo*. Their teeth and jaws are generally small, unlike those of the australopiths, but they retain several primitive features. They have humanlike hands and feet but primitive shoulders, rib cage, and pelvis. Surprisingly, they may have been well adapted for both tool use and climbing trees. Like later hominins and modern humans, they have long and robust thumbs with attachment sites for muscles that enhance the strength and precision of the grip. This might have allowed them to grasp and manipulate tools. However, their fingers and toes were curved, a trait that is linked to tree climbing.

When the *Homo naledi* fossils were first published, they had not been dated, but, based on the morphology, Berger and his colleagues suspected that they were about 2 Ma. However, in 2017 the fossils were dated and turned out

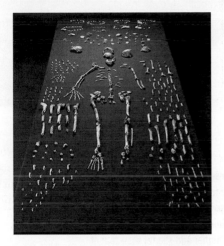

FIGURE 12.30

A sample of fossils from the Rising Star Cave. The "skeleton" in the center is a composite of elements from various individuals.

to be only 250,000 years old. This is interesting because it implies there were several species of *Homo* living in Africa around this time.

It is a mystery how the fossils got into the cave. The skeletons were unusually complete, with even the small bones of the hand and foot preserved. The bones of *H. naledi* are not commingled with the bones of other animals. Berger argues that bodies of the dead were deliberately deposited in the cave, not accumulated by predators or other natural processes. As we saw earlier, burials do not become common in the archaeological record until about 100 ka and are thought to be associated with hominins with much larger brains and probably more advanced cognitive abilities. So it would be surprising if the small-brained *H. naledi* followed this practice.

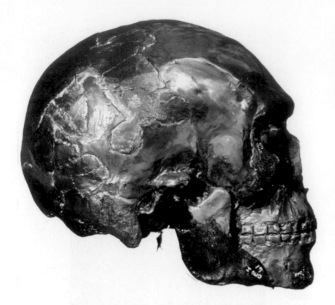

FIGURE 12.31

Called Omo Kibish 1, this fossil skull found in southern Ethiopia in 1963 was dated to 190 ka. It lacks the distinctive features of Neanderthals: The face does not protrude, and the braincase is higher and shorter than that of Neanderthals.

and the Ngaloba cranium (LH 18) from Laetoli, Tanzania. Like the Neanderthals, these African hominins show large cranial volumes ranging from 1,370 to 1,510 cc. However, none of the African fossils shows the complex of specialized features that are diagnostic of the European Neanderthals. Although the African fossils from this period are variable and some are quite robust, many researchers believe that they are more like modern humans than are the Neanderthals or earlier African hominins.

Hominins belonging to our own species began to appear in Africa sometime between 300 ka and 100 ka.

There is debate about the age of the oldest fossil classified as *Homo sapiens*. In 1960, a number of fossils, including a nearly complete skull, were found at Jebel Irhoud in Morocco, but these were not well dated. More recent excavations have yielded more fossils and stone tools, all securely dated in 2017 to 315 ka. The Jebel Irhoud fossils have small faces like those of modern humans but retain the long, low skulls characteristic of *H. heidelbergensis*. Other early fossils were excavated at a site in southern Ethiopia called Omo Kibish, where paleoanthropologists found most of two fossil skulls and several other bone fragments. These skulls are quite robust, with prominent brow-ridges and large faces. However, one of them has several modern features—most notably a high, rounded braincase (**Figure 12.31**). These specimens have been dated to about 190 ka by radiometric methods. Several similar skulls were uncovered at Herto, another site in Ethiopia, and are dated to 160 ka. Other more fragmentary fossil materials from other sites in Africa also suggest that more modern hominins were present in Africa between 200 ka and 100 ka. The archaeological record indicates that Africans during this period developed more sophisticated technology and social behavior than did their contemporaries in Europe or Asia—a pattern that is consistent with the idea that this period saw the gradual accumulation of the cognitive and behavioral characteristics that make modern humans so different from other hominins. We turn to this evidence in the next chapter.

The Sources of Change

Europe may have been populated repeatedly by hominins from Africa during the Middle Pleistocene.

The changes in hominin morphology and technology that we have described in this chapter may be the product of several kinds of processes. It is possible that the changes we see reflect adaptive modifications in tool technology in particular regions of the world. This is probably what has happened in Africa, which has been occupied continually by members of the genus *Homo* since *Homo* first appeared around 2.8 Ma. This means that the makers of Mode 1 tools evolved slowly into the makers of Mode 2 tools, and the makers of Mode 2 tools evolved into the makers of Mode 3 tools. It may be that the same thing happened in Europe.

However, it is also possible that some of the changes in the fossil record are the product of the replacement of one population of hominins by another. Marta Lahr and Robert Foley, anthropologists at Cambridge University, have suggested that Europe and western Eurasia were subjected to repeated invasions by hominins from Africa. They argue that the technological shifts seen in Eurasia were associated with the migration of hominins from Africa during interglacial periods. Remember that during glacial periods, Africa and Eurasia were separated by a formidable desert barrier (see Figure 12.2), and Eurasia was a cold, dry, and inhospitable habitat for primates. During these glacial periods, hominin populations in western Eurasia may have shrunk or disappeared altogether. When the glacial periods ended and the world became warmer, there was substantial movement of animal species from Africa into Eurasia. It is possible that hominins staged repeated invasions of Eurasia during each of these warm interglacial periods, bringing new technologies along with them.

Lahr and Foley's argument is consistent with some archaeological evidence. Remember that the earliest evidence for hominins in Europe dates to about 800 ka. Up until about 500 ka, only Mode 1 tools are found in Europe. Then Mode 2 technologies appear in Europe and persist until about 250 ka, when they are replaced by Mode 3 technologies. In each case, Foley and Lahr argue that new technology appears first in Africa and then in Europe and (sometimes) in Asia (**Figure 12.32**). The appearance of new tool technologies at 500 ka and 250 ka coincides with the timing of interglacial periods, consistent with the repeated replacement of Eurasian populations by African populations. Other archaeologists dispute this interpretation, arguing that the data do not support the idea that Mode 3 tools first appeared in Africa.

The Muddle in the Middle

Anthropologists strongly disagree about how to classify Middle Pleistocene hominins.

You may have noticed that we avoided assigning species names to the Neanderthals or Denisovans. This is partly because much disagreement exists about how to classify Middle Pleistocene hominins. The disagreement stems from different ideas about the processes that shaped human evolution during this period. According to G. Philip Rightmire of Harvard University, all the early Middle Pleistocene hominins represent one species, *H. heidelbergensis*, and all the hominins of the later Middle Pleistocene in Europe and the Neanderthals represent another species, *H. neanderthalensis* (**Figure 12.33a**). Although it may seem peculiar to include specimens without the distinctive Neanderthal features in the species *H. neanderthalensis*, this name is used because it has historical

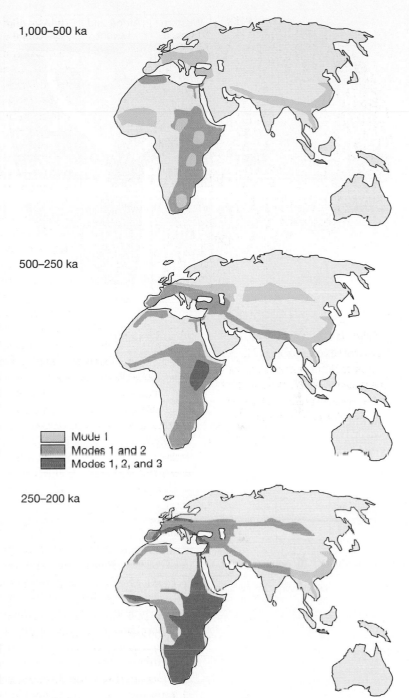

FIGURE 12.32

The geography of tool technologies through time suggests that Eurasia was subjected to repeated invasions of hominins from Africa. Between 1 Ma and 500 ka, Mode 2 technologies were confined to Africa, and hominins in Eurasia were restricted to Mode 1 tools. Beginning about 500 ka, Mode 2 technologies appeared in Eurasia. The introduction of Mode 2 coincided with a relatively warm, moist period, which could have facilitated the movement of hominins from Africa to Eurasia. About 300 ka, elements of Mode 3 technology appeared in East Africa; by about 250 ka, Mode 3 technology had spread throughout Africa and southern Europe. Once again, this spread coincided with a period of warmer climate.

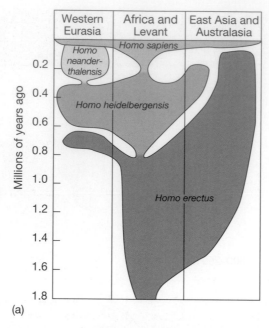

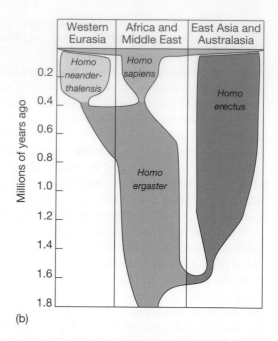

(a)

(b)

FIGURE 12.33

Many phylogenies have been proposed to account for the temporal and geographic patterns in hominin evolution during the Pleistocene. Two such proposals are presented here. (a) G. Philip Rightmire believes that both African and Asian specimens of *H. erectus* should be classified as a single species. Approximately 800 ka, a larger-brained species, *H. heidelbergensis*, evolved in Africa and eventually spread to Europe and perhaps East Asia. In Europe, *H. heidelbergensis* gave rise to the Neanderthals; in Africa, it gave rise to *H. sapiens*. (b) Richard Klein argues that *H. ergaster* evolved in Africa about 1.8 Ma and soon spread to Asia, where it differentiated to become a second species, *H. erectus*. About 500 ka, *H. ergaster* spread to western Eurasia, where it evolved into *H. neanderthalensis*. In Africa, *H. ergaster* evolved into *H. sapiens* at about the same time.

priority. Meanwhile, Richard Klein believes that hominin populations in Africa, western Eurasia, and eastern Eurasia were genetically isolated from each other during most of the Pleistocene and represent three species (**Figure 12.33b**). In Africa, *H. erectus* (which Klein believes was a separate species, *H. ergaster*) gradually evolved into *H. sapiens* about 500 ka. *H. erectus* was isolated in Asia and persisted through the Early and Middle Pleistocene. Klein considers the development of larger brains and more modern-looking skulls in the eastern Eurasia fossils, such as those found at Yingkou, to be the result of convergent evolution and includes them in *H. erectus*. Once hominins reached Europe about 500 ka, they became isolated from African and east Asian populations and diverged to become *H. neanderthalensis*.

This kind of disagreement is to be expected because the timescales of change become progressively shorter as we move closer to the present. Instead of being interested in events that took place over millions of years, we are now interested in events that took place in just a hundred thousand years. This may be roughly how long it took for two species to diverge during allopatric speciation (see Chapter 4). Thus, as hominins spread out across the globe and encountered new habitats, regional populations may have become isolated and responded to different selective pressures—conditions that would eventually lead to speciation. The rapidly fluctuating climates of the Pleistocene, however, caused the ranges of hominins and other creatures to shift as well. As their ranges expanded and contracted, some populations may have become extinct, some may have become fully isolated and even more specialized, and others may have merged. New fossils and new genetic data have allowed real progress on these questions over the past two decades, but, at present, there are still important questions that need to be answered.

CHAPTER REVIEW

Key Terms

Homo heidelbergensis (p. 304)

Levallois technique (p. 307)

Mode 3 (p. 307)

hafted (p. 307)

Neanderthals (p. 311)

taurodont roots (p. 312)

scapulae (p. 312)

crural index (p. 312)

Mousterian industry (p. 312)

dental calculus (p. 314)

rock shelters (p. 315)

sutures (p. 317)

mitochondrial DNA (mtDNA) (p. 320)

nuclear DNA (p. 320)

ghost lineage (p. 321)

Study Questions

1. Using present-day examples, describe the variation in climate during the Middle Pleistocene. Why is this variation important for understanding human evolution?

2. Why would it be more challenging for animals to live during a time when temperatures are fluctuating than in a time when temperatures are more stable?

3. How is *H. heidelbergensis* different from *H. erectus*?

4. Why do archaeologists think that *H. heidelbergensis* hunted big game?

5. Describe how knappers make Levallois tools.

6. Why was the discovery of *Homo floresiensis* so startling? Describe their characteristics and the controversy about their origins.

7. What important technological transition occurred about 300 ka? Why was it important?

8. What is the crural index and what does it measure? How does the crural index of the Neanderthals differ from that of modern tropical peoples?

9. What are the distinctive derived characteristics of the Neanderthals?

10. Neanderthals are popularly pictured as brutish dimwits. Summarize archaeological evidence suggesting that this characterization is not accurate.

Further Reading

Hublin, J. J. 2009. "The Origin of the Neanderthals." *Proceedings of the National Academy of Sciences U.S.A.* 106: 16022–16027.

Humphrey, L., and C. Stringer. 2018. *Our Human Story.* London: Natural History Museum.

Klein, R. G. 2008. *The Human Career: Human Biological and Cultural Origins.* 3rd ed. Chicago: University of Chicago Press.

McBrearty, S., and A. S. Brooks. 2000. "The Revolution That Wasn't: A New Interpretation of the Origin of Modern Human Behavior." *Journal of Human Evolution* 39: 453–563.

Papagianni, D., and M. Morse. 2015. *Neanderthals Rediscovered: How Modern Science Is Rewriting Their Story*, 2nd ed. London: Thames and Hudson.

Rightmire, G. P. 2008. "*Homo* in the Middle Pleistocene: Hypodigms, Variation, and Species Recognition." *Evolutionary Anthropology* 17: 8–21.

 Review this chapter with personalized, interactive questions through INQUIZITIVE.

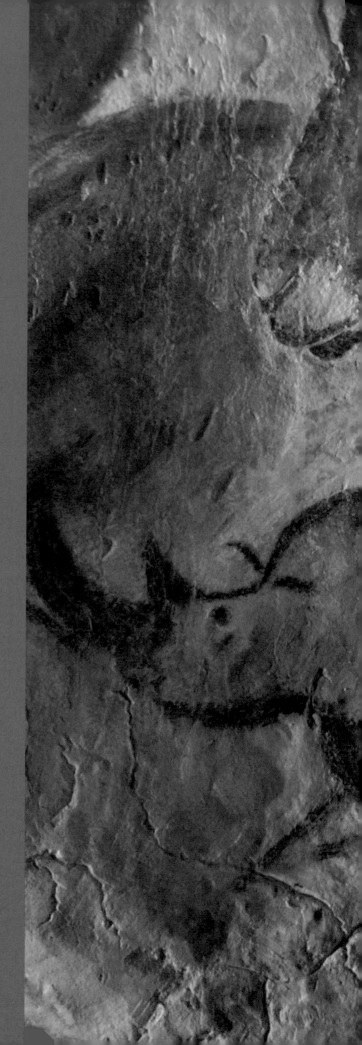

13

HOMO SAPIENS AND THE EVOLUTION OF MODERN HUMAN BEHAVIOR

CHAPTER OBJECTIVES

By the end of this chapter you should be able to:

A Describe how modern humans differ morphologically from earlier hominins.

B Explain how genetic data allow us to reconstruct the expansion of modern humans out of Africa.

C Assess how genetic data indicate that modern humans interbred with earlier hominins.

D Discuss how we know that modern human behavior emerged in Africa by 100 ka.

E Explain how humans spread across the globe beginning 60 ka.

F Describe the lifeways of early modern humans in Europe.

Beginning about 60 ka, the fossil record outside Africa documents a striking change: Neanderthals and other robust hominins disappeared and were replaced by people who looked much like people in the world today, with high foreheads, well-defined chins, and less robust physiques. The archaeological record suggests that they also behaved much like modern people, using sophisticated tools, trading over long distances, and making jewelry and art. Anatomically and behaviorally similar people appeared in Australia around the same time and reached North and South America about 14 ka. Neither area had been occupied by previous hominins.

In this chapter, we describe what the fossil and archaeological records and the genetic data tell us about these early modern people. You will see that evidence from the fossil record and molecular genetic studies tell us that anatomically modern humans evolved in Africa between 200 ka and 100 ka. The archaeological evidence also indicates that the components of modern human behavior and technology evolved in Africa along with modern human morphology over about 200,000 years and that fully modern people were living on the southern coast of Africa by 70 ka. People from one or more populations left Africa and spread across the world, replacing other hominin populations, and there was a modest amount of gene flow between modern humans and other hominins. We are not completely sure when the first exodus of modern humans took place. There is no doubt that there was a major migration out of Africa about 80 ka to 60 ka, but there may have been a previous migration out of Africa around 120 ka that left few descendants among people living today. By about 40 ka, archaic hominins such as the Neanderthals had disappeared, and modern humans occupied Africa and Eurasia from the tropics to the edge of the Arctic Ocean.

Modern Humans in Africa before 60 ka

Fossils classified as anatomically modern *Homo sapiens* share several important derived features with contemporary humans (**Figure 13.1**):

FIGURE 13.1

The skulls of modern humans have higher, rounder crania and smaller faces than earlier hominins had, as illustrated by this skull from a man who lived about 25 ka near the Don River in Russia.

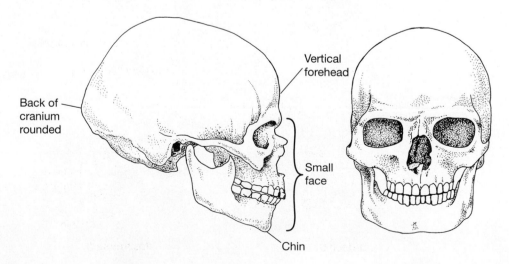

Vertical forehead

Back of cranium rounded

Small face

Chin

- *Small, flat face with protruding chin.* These individuals had smaller faces and smaller teeth than earlier hominins had. The face is flat and is tucked under the braincase. The lower jaw had a jutting chin for the first time (**Figure 13.2**). Some anthropologists believe that the smaller face and teeth were favored by natural selection because these individuals did not use their teeth as tools as much as earlier people had. There is no agreement about the functional significance of the chin.

- *Rounded skull.* Like modern people, these individuals had high foreheads, a distinctive rounded back of the cranium, and greatly reduced browridges (see Figure 13.2).

- *Cranial capacity of at least 1,350 cc.* Cranial capacity varies to some extent across populations but is generally at least 1,350 cc. This value is smaller than the value for Neanderthals but greater than the value for other hominins in the late Middle Pleistocene.

- *Less robust postcranial skeleton.* The skeleton of these individuals was much less robust than Neanderthal skeletons. These individuals had longer limbs with thinner-walled bones; longer, more lightly built hands; shorter, thicker pubic bones; and distinctive shoulder blades. Although these individuals were less robust than Neanderthals, they were still more heavily built than any contemporary human population. Erik Trinkaus of Washington University in St. Louis suggests that these individuals relied less on body strength and more on elaborate tools and other technological innovations to do their work.

- *Relatively long limbs and short trunks.* The body proportions of these creatures were similar to those of peoples who live in warm climates and may reflect their African origins.

The earliest anatomically modern human fossils come from Africa and are dated to between 315 and 160 ka

Two early fossils suggest that the transition to modern human morphology occurred in Africa during this time period. In 1960, a nearly complete skull was unearthed during

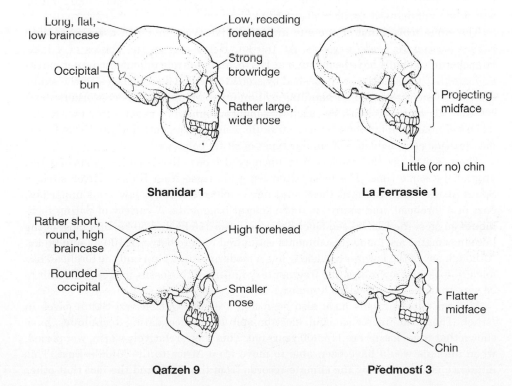

Shanidar 1

La Ferrassie 1

Qafzeh 9

Předmostí 3

FIGURE 13.2

Neanderthals, represented here by Shanidar 1 from Iraq and La Ferrassie 1 from France, differ from modern *Homo sapiens*, like Qafzeh 9 from Israel and Předmostí 3 from the Czech Republic. Modern humans had higher foreheads, smaller browridges, smaller noses, more rounded skulls, and more prominent chins than Neanderthals.

FIGURE 13.3

Side and front views of one of the hominin crania found at Herto, Ethiopia. This specimen (BOU-VP-16/1) is intermediate between *H. heidelbergensis* and modern *H. sapiens*, displaying prominent browridges but a high, rounded braincase.

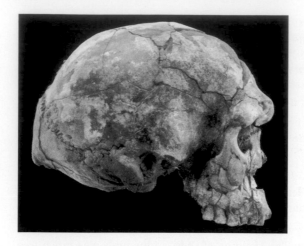
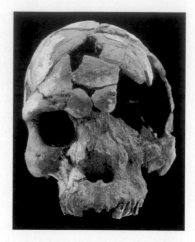

a mining operation at Jebel Irhoud in Morocco. Subsequent excavations yielded a partial braincase and many other fossils. However, the site was not well dated. Recent excavations have produced several new fossils and many stone tools, all securely dated to 315 ka. The Jebel Irhoud fossils have similar faces to those of modern humans, but retain the long, low skulls characteristic of *Homo heidelbergensis*. The second early fossil comes from Omo Kibish in southern Ethiopia and was discovered in 1963. Though incomplete, this fossil skull seems more modern than the Jebel Irhoud fossils and dates to 190 ka.

A team led by Tim White, who has made many important discoveries in the Middle Awash region of Ethiopia, found the fossilized crania of two adults and one immature individual along with several other fragments at a site called Herto. These skulls are intermediate between those of modern humans and older African hominins classified as *H. heidelbergensis*. The skulls are longer and more robust than those of most living people. They have prominent browridges, pointed occipital bones, and other features that link them to earlier African hominins. However, they also have some very modern features—most notably, a high, rounded braincase (**Figure 13.3**). Argon–argon dating was used to date these fossils to about 160 ka.

The stone tools found at the site are much like those found in association with earlier African *H. heidelbergensis*. At 160 ka, Herto was on the shores of a lake. Hippopotamus skulls have been found at Herto, and multiple unambiguous cut marks indicate that the skulls were defleshed by stone tools. Polished surfaces on the skulls are consistent with repeated handling by hominins. White and his co-workers point out that the mortuary practices used by some contemporary peoples in New Guinea leave a similar combination of marks on skulls, and they suggest that the Herto fossils may provide early evidence of similar types of ritual behavior.

There is evidence that modern humans may also have lived in other parts of Africa around the same time. The most extensive finds come from Klasies River sites in South Africa. Excavations at these sites have yielded five lower jaws, one upper jaw, part of a forehead, and many smaller skeletal fragments. A variety of dating techniques suggest the Klasies sequence dates between 110 ka and 50 ka, and most of the human remains are found in sediments estimated to date between 100 ka and 90 ka. Although one of the lower jaws clearly has a modern jutting chin and the forehead has modern-looking browridges, the fragmentary nature of the fossils makes it difficult to be sure that these are modern humans.

Modern human fossils have also been found at the Qafzeh and Skhūl caves in Israel. Thermoluminescence and electron-spin-resonance dating techniques have shown that these fossils are 115,000 years old. This was a relatively warm, wet period, when animals would have been able to move from Africa to the Middle East. This inference is supported by the climate records from the caves and the fact that other fossilized animals found at these sites are primarily African species. Thus, although these sites are outside Africa geographically, they were ecologically similar to Africa during this period.

Modern humans and Neanderthals may have coexisted in Israel for some time. Neanderthal fossils have been found at three sites located quite close to Qafzeh and Skhūl: Kebara, Tabun, and Amud. Two of these sites, Kebara and Amud, are dated by thermoluminescence and electron-spin-resonance methods to 60 ka to 55 ka. We do not know what kinds of interactions modern humans and Neanderthals had, but the data suggest that they lived in the same area at about the same time.

The African Archaeological Record for the Later Pleistocene

Modern human behavior is more complex and more variable than the behavior of earlier hominins.

Remember that *Homo erectus* and *H. heidelbergensis* used Acheulean tools throughout Africa and western Eurasia for more than a million years. Thus, the same tools and presumably similar adaptive strategies were used over a variety of habitats for a very long time. In contrast, present-day foragers use a vast array of highly specialized tools and techniques to adapt to diverse environments. They also engage in elaborate and varied symbolic, artistic, and religious behavior that is unparalleled among other creatures. The extraordinary geographic range and sophistication of modern humans is due partly to our cognitive abilities. Humans can solve problems that would completely stump other creatures (see Chapter 8). However, being smarter than the average bear (or primate) is only part of our secret. Our other trick is having the ability to accumulate and transmit complex adaptive and symbolic behavior over successive generations. People rarely solve difficult problems entirely on their own. Even the most brilliant of us couldn't construct a seaworthy kayak, invent perspective drawing from scratch, or figure out how to bake bread on our own. Instead, we gain skills, knowledge, and techniques from being instructed by and watching others. The variety and sophistication of modern human behavior are a product of our ability to acquire information in this way (see Chapter 16).

Archaeological evidence suggests that early modern humans living in Africa were able to accumulate complex adaptive and symbolic behavior.

From about 250 ka to 40 ka, the African archaeological record is dominated by a variety of stone tool kits that, for the most part, emphasize Mode 3 tools. These industries are collectively labeled the **Middle Stone Age (MSA)**. Until recently, most archaeologists thought that the MSA in Africa was qualitatively similar to the much better-known Mousterian tool industries associated with Neanderthals in Europe. Signatures of modern human behavior, such as more complex tools, long-distance exchange networks, and art and ritual practices, which are well documented in Europe by 30 ka, were thought to be largely absent in Africa. However, this view has been challenged by several twenty-first-century discoveries.

In 2000, archaeologists Sally McBrearty at the University of Connecticut and Alison Brooks at George Washington University published a paper in which they argued that the MSA is not qualitatively similar to the Mousterian in Europe and that most of the signatures of modern human behavior developed in Africa between 250 ka and 60 ka, well before they appeared in Europe. Their point of view was later strengthened by several remarkable archaeological discoveries mainly on the coast of South Africa, and it now seems likely that people living in Africa achieved this kind of complex and varied behavior by at least 70 ka. We'll explore the archaeological evidence for this behavior now.

FIGURE 13.4

Bifacial Still Bay points found at Blombos Cave that date to about 70 ka. Similar points have been found at several other sites on the coast of South Africa that date to about the same age. These points were probably produced using a sophisticated technique called pressure flaking, which allows great control over the shape of the edge. Previously, the earliest examples of this technique were tools found in southern France that date to about 20 ka.

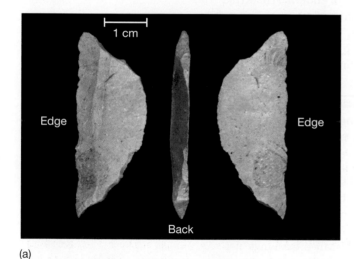

(a)

(b)

FIGURE 13.5

(a) Microlith found at Pinnacle Point on the coast of South Africa. Many such microliths have been found in layers that date from 72 ka to 60 ka. (b) A complete arrow found at a site in Europe that dates to about 8 ka. Note how the microlith was hafted to make a composite tool.

Complex adaptive technologies appear in association with MSA archaeological sites.

Blades are long, thin stone tools that are commonly found at modern human sites in Europe after 40 ka. And for a long time, archaeologists thought that blades were indeed first made by modern humans. However, at the Kapthurin Formation in Kenya, which has been dated to between 280 ka and 240 ka by the argon–argon technique, roughly 25% of the tools are blades. Archaeologists can show that the knappers at this site were highly skilled at blade production, making few mistakes and wasting little raw material. Blades, which are designated as **Mode 4** tools, have also been found at several other MSA sites that date between 250 ka and 60 ka.

Around 70 ka, two sophisticated types of stone tools appear in southern Africa. Finely made, symmetric, leaf-shaped points were manufactured at a site called Still Bay near the coast of South Africa (**Figure 13.4**). Modern experiments suggest that the final phase of shaping the points involved a technique called **pressure flaking**. Instead of using a hammer stone to shape points, the knapper uses a hard, pointed tool and applies pressure to remove small flakes. This technique allows great control over the flaking process and was used to make the beautiful Solutrean points found in southern France (dated to 20 ka) and Clovis points made by early Native Americans (dated to 12 ka). The second new tool type is made up of very small stone tools called **microliths**, which have a sharp edge on one side and a carefully flattened surface on the other side (**Figure 13.5a**). Archaeologists classify these as **Mode 5** tools. Similar tools were made throughout the world over the next 50,000 years and were used either as arrow points (**Figure 13.5b**) or to tip light spears thrown with an atlatl. An atlatl is a notched throwing stick (**Figure 13.6**) that increases the length of the arm and greatly increases the distance a spear can be thrown. The use of bows or atlatls represents a significant increase in technological complexity compared with that of earlier peoples.

To make these tools, people had to use complex techniques. Tools at Still Bay were manufactured from silcrete, a kind of stone that is usually very difficult to flake. Kyle Brown of the University of Cape Town and his colleagues have shown that these early humans were able to make the refined Still Bay points and microliths because they heat-treated the silcrete, gradually raising its temperature to around 350°C, probably by burying it in sand under a campfire, transforming it into a much harder, easily flaked material. Heat treating first appears around 164 ka and was regularly used in southern Africa at 72 ka.

The use of bone is another signature of modern human sites, and bone tools also occur at several MSA sites. Alison Brooks and her co-workers recovered several exquisite bone points, some of them elaborately barbed, from Katanda, a site on the northeastern border of the Democratic Republic

FIGURE 13.6

An atlatl is a tool that lengthens the arm. This allows a light spear to be thrown with much greater velocity than a spear thrown without an atlatl.

of the Congo (**Figure 13.7**). The MSA layer containing the points has been dated to between 90 ka and 60 ka. At Blombos Cave in South Africa, several polished bone points have been found that date to about 72 ka.

There is also evidence for shelters and hearths in the MSA. At the Mumbwa Caves in Zambia, there are three arcs of stone blocks that probably served as windbreaks. Elaborate stone hearths were constructed inside these structures. There are many more MSA sites at which people seem to have constructed hearths and built huts, but for each of these it is impossible to rule out the possibility that natural processes produced the features that archaeologists have documented.

Later populations in southern Africa were probably more adaptable than earlier peoples. About 190 ka the world experienced a sharp drop in temperature that lasted for about 60,000 years. Archaeological sites dating to this period are rare, suggesting that the human population in Africa contracted sharply. Then, during a warmer period that began about 130 ka, the population expanded. When the world climate cooled between 105 ka and 75 ka, there was no population contraction; instead there was a period of cultural flowering with new tool types and, as we shall see, increased symbolic behavior.

MSA peoples probably had large social networks.

MSA peoples sometimes transported raw material great distances. Most of the stone used to make tools at MSA sites comes from a short distance away. At several MSA sites, however, small amounts of raw materials were transported much farther. At several sites in East Africa, for example, tools were made from obsidian carried 140 to 240 km (about 90 to 150 mi.). The long-distance movements of these resources may mean that Upper Paleolithic peoples ranged over long distances or that they formed long-distance trade networks.

Modern people in Africa made decorative carvings and beads and used pigment.

In 2007, a team led by Abdeljalil Bouzouggar of the Institut National des Sciences de l'Archéologie et du Patrimoine of Morocco announced the discovery of 41 perforated shell beads from the site of Grotte de Pigeons, which is dated to 82 ka (**Figure 13.8**). The shells are all punctured in a similar way, and some have been painted with ocher. The surfaces of the shells are worn in a way that suggests they may have been strung on a cord or sewn onto clothing. The shells were brought to the site from the coast, which was at least 40 km away. Similar perforated shells have been found at three other sites in northern Africa.

FIGURE 13.7

Beautiful bone points like this one found at Katanda in the Democratic Republic of the Congo have been dated to between 90 ka and 60 ka. If these dates are correct, then Middle Stone Age peoples could produce bone tools that rival the best of the European Upper Paleolithic tools.

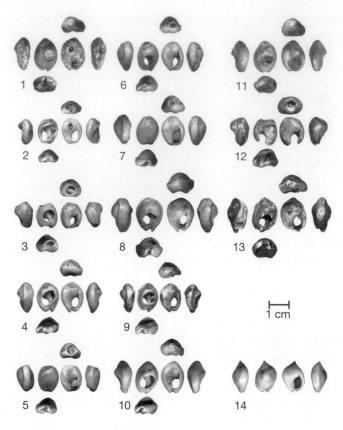

FIGURE 13.8

These shells come from a site in Morocco dated to 82 ka. The shells were gathered and brought to the site, perforated, and covered in red ocher. They may have been strung on a cord.

FIGURE 13.9

These are small pieces of ostrich eggshell found at Diepkloof Rock Shelter in South Africa. They are about 2.5 cm across and date to about 60 ka. They are decorated with a distinctive cross-hatched pattern also found on many such fragments at this site. Modern southern African foragers use ostrich eggs as containers for liquids, usually water.

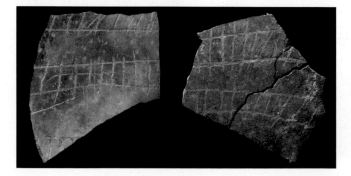

There is also evidence of ornaments and decorative carving at the other end of Africa. A team led by Christopher Henshilwood of the University of Bergen discovered a cache of shell beads at Blombos Cave that are dated to 76 ka. The shells come from a site 20 km away.

Even more striking, Pierre-Jean Texier of the Université Bordeaux and colleagues have found many engraved fragments of ostrich eggshell at Diepkloof Rock Shelter in South Africa (**Figure 13.9**). Modern foragers in southern Africa often use ostrich eggshells as containers—they hold about a liter of water. At Diepkloof, the shells have been engraved with complex geometric patterns, which Texier and colleagues believe symbolized group identity in the same way that pottery and basket decoration do today. These artifacts date to about 60 ka and are associated with the more advanced Howieson's Port stone tools that are found at some South African sites.

There is strong evidence for the use of red ocher at several sites. The earliest evidence comes from the Kapthurin Formation, which dates to between 280 ka and 240 ka. Two elaborately engraved pieces of red ocher have been found at Blombos Cave, the same site that produced shell beads (**Figure 13.10**). Many present-day African peoples use red ocher to decorate themselves and for symbolic purposes, but some archaeologists have argued that red ocher might also be used for utilitarian purposes, such as tanning hides. Few images are associated with MSA sites, but the lack of artwork is probably due to the fact that rock surfaces there are constantly peeling away.

Modern human behavior may have been caused by either increased cognitive ability or cultural innovations.

Remember that the first anatomically modern humans appear in the fossil record by 160 ka. This means the first evidence of complex behaviors associated with modern people occurred well after the first appearance of people who *look* fully modern. However, we need to remember that morphology and behavior can be decoupled. People who look fully modern could have evolved new cognitive abilities that were not reflected in their skeletal anatomy. For example, Richard Klein has suggested that the human revolution may have been caused by a mutation that allowed fully modern speech. It could be that linguistic ability evolved late in the human lineage and gave rise to the technological sophistication and symbolic behavior of the Upper Paleolithic peoples. It seems more likely, however, that cognitive innovations like those that support language evolved gradually through the accumulation of small changes. Such a process requires no special macromutations or unlikely chance events. As we saw in Chapters 7 and 8, behavior is subject to the same kinds of evolutionary forces that shape morphology and physiology.

Alternatively, the striking changes in human behavior may have resulted from cultural, not genetic, changes. We know that something like this happened later in human history. About 10 ka, an equally profound transformation was associated with the adoption of agriculture. Agriculture led to sedentary villages, social inequality, large-scale societies, monumental architecture, writing, and many other innovations that we see in the archaeological record. The transition we see in the archaeological record in the MSA could have been caused by a similar kind of technological innovation that allowed for much more efficient food acquisition, which in turn led to a greater economic surplus, more economic specialization, and greater symbolic and ritual activity. There is no clear evidence for such an innovation, but many kinds of technology leave no trace in the archaeological record.

Out of Africa

We have seen that modern human behavior gradually evolved in Africa beginning about 250 ka, and by 70 ka fully modern people were living on the southern coast of the continent. Sometime in the past 200,000 years, some of these individuals may have left Africa and spread across the world. Although there is some evidence of an early expansion out of Africa around 200 ka, if this early migration did occur, the emigrants were not particularly successful. Few of their genes survive in contemporary populations, and they did not replace archaic hominins. However, there is no doubt that a major expansion out of Africa did occur about 60 ka, and from there modern humans eventually spread throughout the world. By about 30 ka, archaic hominins like the Neanderthals had disappeared, and modern humans occupied Africa and Eurasia from the tropics to the edge of the Arctic Ocean. In this section we present the fossil, genetic, and archaeological evidence of the origins and movements of modern humans.

Fossil Evidence

There are many anatomically modern human fossils outside of Africa that date to after 63 ka. There are two sites with earlier dates, but the dates of both are problematic.

In 1980, two partial fossil skulls were found in a limestone cave at Apidima, in southern Greece. Both were distorted and originally identified as Neanderthals. Recently, a team led by Katarina Harvati of the University of Tübingen used computer techniques to compensate for the distortion of the fossil. Their reconstructions show that the back of one of the skulls is rounded like the skull of modern humans and lacks the bulge characteristic of Neanderthals. Limestone material encrusting this skull has produced a range of dates (40 ka to 300 ka) with an average of 210 ka, much earlier than other modern fossils outside of Africa. In southeastern China, researchers discovered 47 teeth at a cave site in Daoxian (**Figure 13.11**). These teeth are well within the size range and look much like the teeth of modern humans. They lack the taurodont roots of Neanderthals or other features that distinguish them from modern human teeth. The fossils were found between flow-stone layers that have been dated from 120 ka to

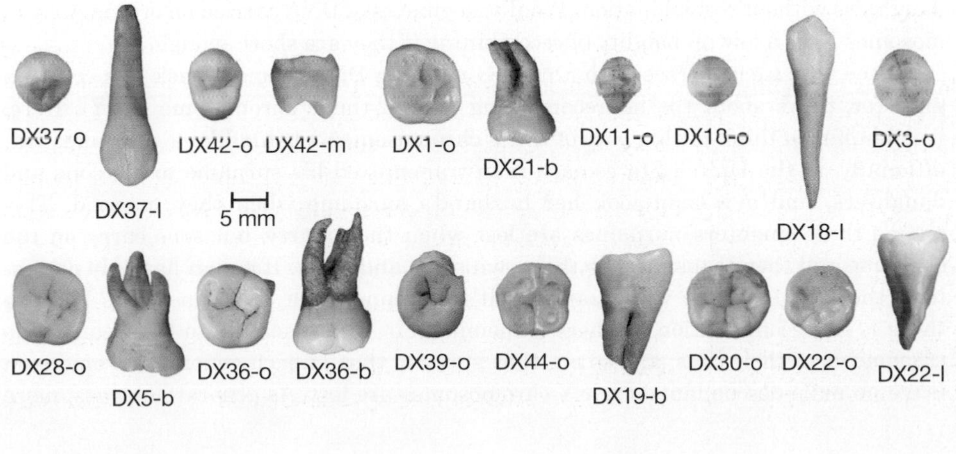

FIGURE 13.10

One of two engraved pieces of red ocher found at Blombos Cave in South Africa. The artifact dated to 77 ka is associated with MSA tools. The engravings are similar to cave art found in other parts of the world.

DX37 o   DX42-o DX42-m   DX1-o   DX11-o DX10-o   DX3-o
                                   DX21-b

DX37-l   5 mm   DX18-l

DX28-o   DX36-o DX36-b DX39-o DX44-o   DX30-o DX22-o DX22-l
    DX5-b   DX19-b

FIGURE 13.11

Examples of modern human teeth found at Daoxian Cave in southeastern China. The size and morphology of the teeth are within the human range and are different from teeth associated with archaic humans.

80 ka by using uranium–thorium dating methods. However, because of doubts about the geology of the site, not all researchers are convinced that these dates are correct.

Other modern fossils found outside Africa date to after 63 ka. In Asia, a partial cranium and mandible from Tam Pa Ling Cave in Laos are dated from 60 ka to 46 ka; a partial cranium and some postcranial material from Niah Cave in Borneo date from 46 ka to 34 ka. At Tianyuan Cave near Beijing, paleontologists found a modern human mandible and femur dating from 42 ka to 39 ka. In Australia, a cranium and partial skeleton dating from 42 ka to 38 ka were found at Lake Mungo. Notably, this fossil shows evidence of cremation—the oldest evidence of this behavior. The oldest modern human fossils in Europe were found in Romania and date to about 40 ka. Beginning about 30 ka, there is a wealth of evidence of the presence of modern humans across Europe.

Genetic Evidence

Patterns of genetic variation in living people and in DNA recovered from fossils provide information about the origin and expansion of modern humans.

The patterns of genetic variation within living people and genetic material extracted from fossils tell us a lot about the history of human populations. From genetic data we have learned the following:

1. Modern humans evolved in Africa between 300 ka and 90 ka.

2. Modern humans outside Africa are descended from one or more populations that left Africa between 120 ka and 40 ka.

3. A small amount of interbreeding occurred between expanding modern human populations and the hominins already living in Eurasia, including the Neanderthals and the Denisovans.

To understand how geneticists use data from fossils and living people to trace the history of modern populations, you need to learn a little bit more about genetics. Although the algorithms that geneticists use to make these calculations are very complicated, we will explain their basic logic.

All the copies of any extant DNA sequence transmitted without recombination can be traced back to a single copy in an individual who lived in the past.

We saw in Chapter 2 that genes on most chromosomes frequently recombine. However, some genetic material is transmitted without recombination. For example, 95% of the material on the Y chromosome does not recombine with material on the X chromosome. This DNA is faithfully transmitted from fathers to sons. We've seen earlier that a few genes carried on mitochondria are transmitted from mothers to both sons and daughters without recombination. Finally, segments of DNA carried on ordinary chromosomes have a low probability of recombining if they are short enough.

To see why we can trace such a nonrecombining DNA segment back to a common ancestor, think about the nonrecombining part of the Y chromosome (NRY). Here, it is helpful to think of this part of the Y chromosome as being like a surname. Traditionally, in the United States each man transmitted his surname to his sons and daughters, and a woman took her husband's surname when they married. This means that daughters' surnames are lost when they marry, but sons carry on the surname and then transmit it to their own sons, and so on. If a man has only daughters, then his surname will disappear. It's the same with Y chromosomes because there is no recombination. Fathers transmit their Y chromosome to their sons, who transmit it to their sons, and so on. Now suppose that in each generation some men leave no male descendants; their Y chromosomes are lost. As generations pass, more and more Y chromosomes are lost until eventually every man carries the name and the Y chromosome of a single man—the **most recent common ancestor (MRCA)** for the Y chromosome. The same process applies to the mitochondria and to any segment of DNA carried on ordinary chromosomes that is short enough to avoid recombination—all are copies of the DNA segment carried by a single individual sometime in the past.

The number of generations that have passed since the MRCA for a nonrecombining bit of DNA lived depends on the size of the population. The larger the population, the more generations, on average, you have to go back to reach the MRCA. To see why, pick a Y chromosome from the current generation. If mating takes place at random, the probability that a second Y chromosome derives from the same father is equal to 1 divided by the population size (number of fathers). If we assume that population size is constant from one generation to the next, it follows that if one man has two offspring, then another man must have none. Therefore, the probability that a man's lineage becomes extinct each generation is proportional to 1 divided by the population size. Thus, the rate at which a population loses Y-chromosome lineages declines as population size increases.

Each bit of the genome has its own ancestry. The MRCA of our mitochondria is sometimes given the colorful but misleading name "Eve." But she is not the mother of us all, only the mother of our mitochondria. The MRCA of our NRY was not Eve's mate; in fact, we now know that he lived tens of thousands of years after Eve. The MRCAs of each of our genes were different people, who lived at very different times as well.

The accumulation of mutations allows us to derive the phylogenetic history of nonrecombining sections of DNA.

Even though the Y chromosomes of all living people are replicas of a single ancestral Y chromosome, they are not identical because mutation occasionally introduces new genetic variants. There is no recombination, so when a mutation occurs in a particular man, it is carried by all of his descendants (unless they have a mutation themselves). The same goes for mitochondrial DNA and nonrecombining bits of autosomal genes. Mutation has two important consequences. First, we can use the methods of phylogenetic reconstruction discussed in Chapter 4 to reconstruct a tree of descent based on derived similarities. To make such reconstructions (sometimes called **gene trees**) feasible, mutations must occur frequently enough to allow the branches of the gene tree to be distinguished.

Mutation also makes it possible to estimate the time that has passed since the MRCA of that gene lived, using the genetic-distance measures that we discussed in Chapter 4. Because there is a known relationship between population size and the age of the MRCA, if we know the age of the MRCA we can estimate population sizes in the past. We repeat this for each of the many genes we carry and use these data to reconstruct the size of human populations in the past.

Gene trees support an African origin for modern humans.

Geneticists have constructed gene trees for mitochondria and the nonrecombining portion of the Y chromosome. **Figure 13.12** shows the gene tree based on the entire DNA sequences of the mitochondria of several hundred people. The colored blob at the end of each branch represents the mtDNA of a single individual. To find the last common mtDNA ancestor of two individuals, trace backward through the tree from each individual until you reach the node where the two paths meet. This node represents the common ancestor of those individuals. The length of any branch in the tree is proportional to the number of mutations accumulated along that branch. This means that the age of the common ancestor of the mitochondrial DNA (mtDNA) of two individuals is the average distance that must be traversed through the tree from these two individuals to their common ancestor. The shorter this distance, the more recent the common ancestor.

This gene tree for the complete mitochondrial genomes of 277 people from all over the world indicates that all modern humans are descended from an African population. Each branch represents the mtDNA sequence of a single individual, and the length of the branch represents the number of mutations along that branch. Thus the longer branches for Africans reflect the greater genetic variation on that continent. Almost all non-Africans are descended from a single node in the tree and from a more recent common ancestor than that of many Africans.

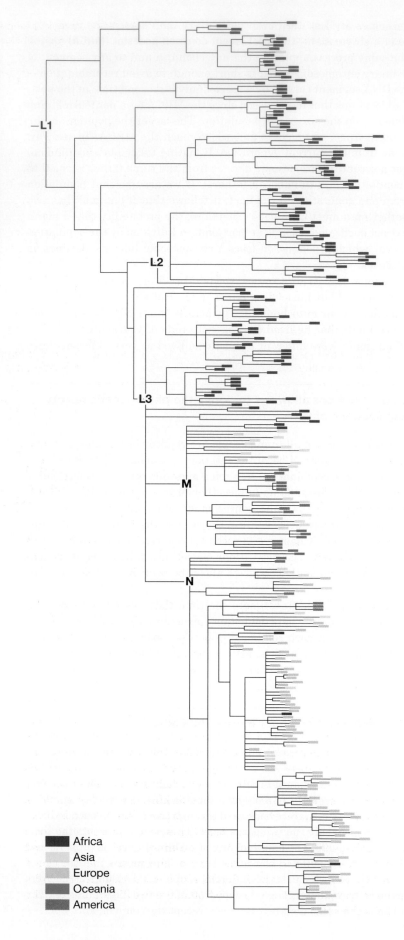

- Africa
- Asia
- Europe
- Oceania
- America

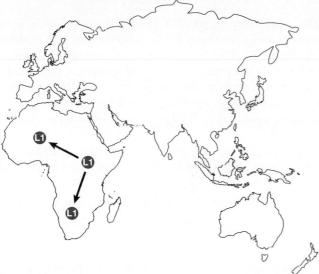

150 ka: Modern humans evolve somewhere
in Africa and L1 lineage spreads

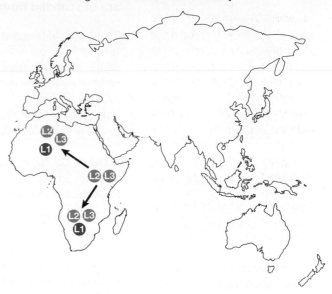

80–60 ka: New expansion spreads L2 and L3
lineages. L1 reduced to a minority

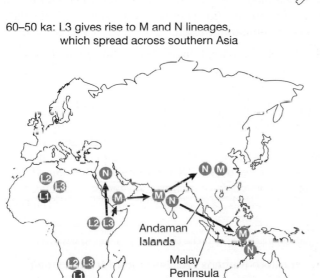

60–50 ka: L3 gives rise to M and N lineages,
which spread across southern Asia

Andaman
Islands

Malay
Peninsula

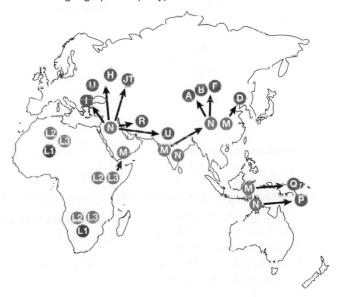

50–30 ka: M and N lineages give rise to main
geographic haplotypes in Eurasia

FIGURE 13.13

The current geographic distribution
of mitochondrial DNA haplotypes is
consistent with an initial spread of
anatomically modern peoples out
of Africa and along the southern
coasts of Asia and then north into
Asia and Europe.

This tree supports the idea that modern humans originated in Africa and then migrated out of Africa and spread across the rest of the world. Notice that all Africans in the sample are descended from nodes L1, L2, and L3 and are connected by the longest branches in the tree. All mtDNA carried by non-Africans descends from two nodes, M and N, that are connected to L3 by very short branches. The branches linking non-Africans are much shorter than branches linking many Africans. This suggests that the population containing the last common ancestor, L1, came from Africa. The genetic distance along branches indicates that this individual lived about 130 ka. Later, perhaps 90 ka to 60 ka, L2 and L3 arose in Africa and spread through the continent. mtDNA that is descended from a particular node is called a **haplogroup**. A population containing M and N haplogroups left Africa around 60 ka and spread across the rest of the world over the next 50,000 years (**Figure 13.13**). The Y-chromosome tree tells the same story.

Complete DNA sequences also indicate that most modern humans are descended from people who emigrated from Africa about 80 ka to 60 ka.

The patterns suggested by the trees based on mtDNA and the Y chromosome have largely been confirmed by high-resolution analyses of the complete genomes of thousands of people from all over the world. These new studies are important because complete genomes include hundreds of thousands of coding and noncoding genes. Like mtDNA and the Y chromosome, each of these genes can be traced back to a common ancestor, and that ancestor can be dated. Because each gene will have its own history, the date of the MRCA will vary from gene to gene. To determine when two populations diverged (that is, became relatively isolated from one another due to a lack of gene flow between them), researchers calculate the median MRCA of hundreds of thousands of genes in the two populations. These high-resolution analyses tell us a number of important things about the history of our genes:

1. The deepest population split in living humans—between the Khoe-San (hunter-gatherers who now live in southern Africa but who ancient DNA shows were once much more widespread across the continent) and all other humans—occurred about 350 ka to 260 ka. This split time provides a lower bound for the timing of the origin of modern humans, which must have occurred before the Khoe-San began diverging from other human populations. This is broadly consistent with the date of the oldest anatomically modern humanlike fossil.

2. The second-deepest population split in living humans is also fairly old, at around 200 ka. This split divided central African hunter-gatherers from all other humans (excluding the Khoe-San).

3. A third split, which divided the population that were not Khoe-San or central African hunter-gatherers, occurred around 80 ka to 60 ka. A subset of humans left Africa about this time and carried with them a subset of the large amount of genetic diversity that is seen in African people today. Genetic studies show that this dispersal gave rise to the vast majority of the DNA of contemporary non-African people. However, there is not yet enough resolution in the genetic data to exclude the possibility of earlier dispersal from Africa that contributed up to 2% of the DNA of contemporary non-Africans.

Genetic data indicate that all modern humans are descended from a small population.

Remember that mutation introduces new genes at a very low rate and that genetic drift eliminates genetic variation at a rate that depends on the size of the population. Drift removes genetic variants more slowly in large populations than in small populations. The effects of mutation and genetic drift will eventually balance as the amount of variation reaches an equilibrium (**Figure 13.14**). The amount of variation at equilibrium depends on the size of the population. For a given mutation rate, bigger populations will have larger amounts of genetic variation at equilibrium. You can combine the observed amount of genetic variation and estimates of the mutation rate to calculate population size.

Calculations based on whole genomes allow us to generate more detailed estimates of past population sizes. The genome of a single individual contains hundreds of thousands of genetic loci. At each polymorphic locus, there are two alleles;

FIGURE 13.14

A physical analogy for the mutation–drift equilibrium can help explain why large populations have more genetic variation at equilibrium than small ones do. Mutation introduces new genetic variants at a constant rate (u), much like a stream of water entering a tank. Genetic drift removes variants, like a drain at the bottom of the tank, at a rate that depends on the population size, N. The amount of variation (m) is analogous to the volume of water in the tank: As the level of water in the tank rises, the pressure increases and the water drains out more rapidly. Eventually, the rate of outflow equals the rate of inflow, and the depth of the water remains constant. In the same way, as the amount of variation in a population increases, drift removes variation faster; eventually, a steady state is reached at which the amount of variation is constant. To calculate the equilibrium amount of variation, set the rate of inflow (u) equal to the rate of outflow ($m/2N$), and solve for m.

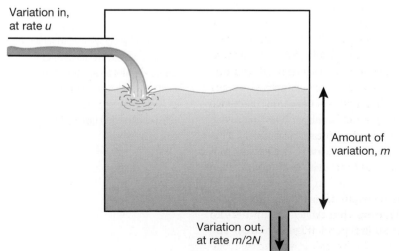

Variation in, at rate u

Amount of variation, m

Variation out, at rate $m/2N$

the age of the MRCA of each pair of alleles can be calculated using the methods described previously. Geneticists use a computer algorithm to find the history of population sizes that is most consistent with the ages of the MRCA calculated across all loci. **Figure 13.15** shows the results of this procedure for high-resolution human genomes drawn from six contemporary non-African populations around the world. Estimates of the population size (approximately the number of breeding adults) derived from people in each of the six populations are very similar for genes with MRCA dated from 500 ka to about 100 ka and then begin to diverge slightly. After 50 ka, there are major differences in the estimates of human population sizes derived from different populations. The data also show that human populations reached a maximum of around 30,000 individuals about 300 ka and then dropped precipitously until about 50 ka. After that date, population sizes diverge, probably because these human populations had different demographic histories after they left Africa.

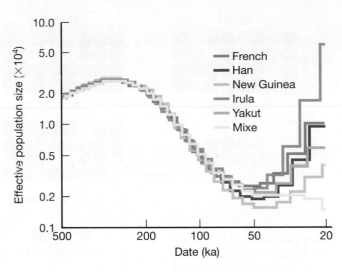

FIGURE 13.15

The history of human population sizes calculated from the genomes of people drawn from six populations. Along with the French and New Guinean populations, the Han are Chinese, the Irula are from southern India, the Yakut are a Siberian population, and the Mixe live in southern Mexico.

Genetic variation decreases with the distance from Africa, consistent with expansion of human populations out of Africa.

When human populations expanded out of Africa, they probably didn't know where they were going. They just moved next door, and next door happened to be a nearby continent. After that population expanded, things got crowded, and another group of emigrants moved again—not back to Africa, because it was already occupied, but farther into Eurasia. In this way, modern human populations may have spread across Eurasia bit by bit. This kind of process could have led to a very rapid geographic expansion, even if no single person moved very far in his or her own lifetime. For example, suppose individuals moved only 25 km (about 15 mi.) in their lifetime. If we assume 25-year generations and no movement backward, this would mean that populations would spread 100 km (about 60 mi.) in a century, fast enough to move from Africa to Australia in about 10,000 years.

This kind of expansion means that the amount of genetic variation within populations will decrease as distance from Africa increases. Each time people leave their natal population, they carry with them a subset of the genes present in that population. **Figure 13.16** illustrates how this works. Initially, there are four sites, and only site 1 is inhabited. There are 32 individuals with four genotypes represented by the colors blue, red, light blue, and green. Initially, there are eight of each genotype. Then eight individuals from site 1 move to neighboring site 2. By chance, the blue and green genotypes are overrepresented among the migrants. This population grows to fill the site and, because the genes are assumed to be neutral, their frequencies are the same as among the founders. Then the process is repeated. Eight migrants colonize site 3. By chance no red genotypes made the trip. Then that population grows, and site 4 is colonized. At each step, the migrants carry only a fraction of the genetic variation present in their natal home. If they move only short distances, the amount of variation will decrease smoothly as they move farther and farther from their site of origin. We would see much the same pattern if people made a smaller number of long-distance migrations. Again, each migration would lead to decreased variation as humans moved out of Africa. Initially, the variation would decrease in a steplike pattern, but local migration over the past 50,000 years would smooth out the variation between adjacent locations.

This is exactly what we see in modern human populations. Two groups of geneticists, one led by Sohini Ramachandran of Brown University and a second by Franck Prugnolle of the University of Montpellier, independently computed the genetic variation in 51 populations from around the globe by using data on a large

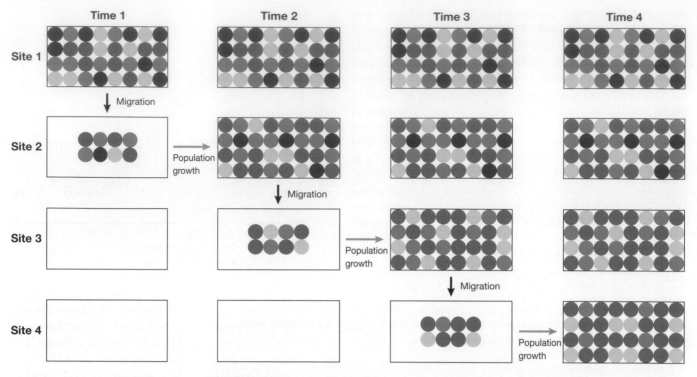

FIGURE 13.16

Spreading across Eurasia and the Americas through a series of expansions is expected to lead to reduced genetic variation. At time 1, a population of 32 individuals occupies site 1. There are four genotypes: red, blue, light blue, and green. Sites 2 through 4 are unoccupied. A subset of the population colonizes site 2, and because the number of emigrants is small, the blue and green genotypes are overrepresented by chance. By time 2, these colonists have reproduced, and site 2 has become saturated. Because the genes are not subject to natural selection, they exist in the same frequency as the colonists—blue and green are more common, and red and light blue are less common than in the original population. A subset drawn from site 2 then colonizes site 3. Again, chance affects the genotypes present among the colonists and the red genotype is lost, and blue is even more common. Once again, the population grows and sends off new emigrants to site 4. As the spreading proceeds, variation is lost because at each stage, chance causes some genotypes to be overrepresented and others to be underrepresented. Eventually, this leads to the loss of genotypes.

sample of microsatellite loci collected as part of the Human Genome Diversity Project. **Microsatellite loci** are noncoding, repetitive, and highly variable DNA sequences (often used in criminal investigations for individual identification). The researchers also computed the distance of each population from East Africa along the most likely migration route (avoiding migrations across large bodies of water, and so on). As you can see in **Figure 13.17**, human genetic variation decreases with distance from East Africa.

Notably, the same seems to be true for phenotypic variation. The Cambridge group, led by Frank Prugnolle, collaborated with Tsunehiko Hanihara, now at the Kitasato University Medical School in Japan, who had compiled 37 measurements on each of 4,666 human skulls drawn from 105 populations around the world. For each measurement, they calculated the amount of phenotypic variation in each population. The amount of phenotypic variation decreases with geographic distance from Africa, but the relationship is not as strong as the relationship between genetic variation and geographic distance from Africa. This makes sense even though the genes that affect variation in skull morphology should be subject to the same sampling processes that affect the microsatellite loci. But because the genes that affect morphology are expressed in the phenotype, they are subject to natural selection. Local conditions favor different skull shapes, partially masking the effects of migration.

Ancient genomes indicate that the ancestors of modern humans sometimes interbred with Neanderthals and Denisovans in Eurasia, but not in Africa.

As they left Africa and spread into Eurasia, modern humans encountered Neanderthal and Denisovan populations. As far as we know, Neanderthals and Denisovans never lived in Africa. If there was interbreeding between modern humans and Neanderthals or Denisovans, then contemporary Europeans and Asians should be more similar to Neanderthals and Denisovans than are Africans.

To test this idea, geneticists have compared the Neanderthal and Denisovan genomes with the genomes of people around the world. They have found that Africans have only trace amounts of Neanderthal DNA, but the genomes in all Eurasian and Pacific populations contain about 2% Neanderthal DNA. African and European populations don't carry any Denisovan genes. East Asian populations carry a small fraction (less than 0.5%) of Denisovan genes, whereas Australians, Papuans (people from New Guinea), and some other Pacific peoples carry between 3% and 6% Denisovan DNA (**Figure 13.18**). The simplest explanation for these facts is that Neanderthals mated with early modern humans in Southwest Asia before the modern human population split to colonize

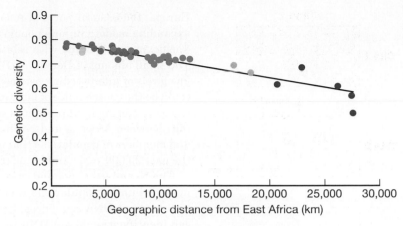

- American populations
- Asian populations
- European populations
- Middle Eastern populations
- Oceanian populations
- African populations

FIGURE 13.17

This figure plots the amount of genetic variation of several human populations from a large sample of microsatellite loci against distance from East Africa along the likely path of human expansion through the world. The fact that the amount of genetic variation declines supports the hypothesis that humans emerged from Africa and spread along the hypothesized path.

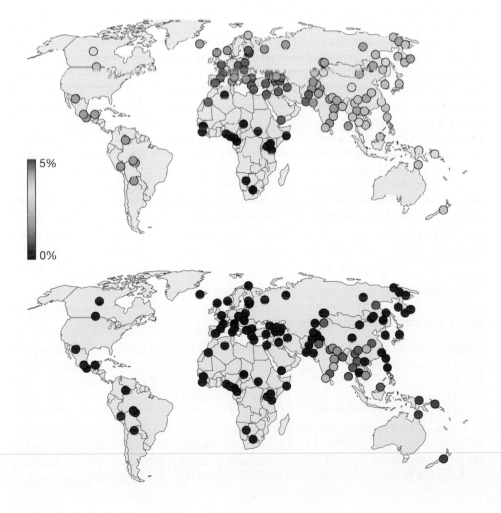

FIGURE 13.18

Map of the percentages of Neanderthal (*top*) and Denisovan (*bottom*) ancestry in modern human populations. Each circle represents a different modern population. Colors give the percentage of Neanderthal and Denisovan ancestry in a population. Neanderthal ancestry ranges from 0% (*black*) through blue and green to a maximum of 3% (*yellow*). Denisovan ancestry ranges from 0% (*black*) to 0.5% (*blue*) on mainland Asia (darker blue means less than 0.5%) and 5% (*red*) in Oceania.

Europe, the eastern parts of Asia, and the Pacific. After this split, members of the expanding modern human population, who eventually settled in New Guinea and the Pacific, met and mated with relatives of the Denisovans, perhaps in Southeast Asia. The small amount of Denisovan DNA found in East Asians could be the result of very low levels of interbreeding between Denisovans and the ancestors of East Asians, or it could be the consequence of later matings between East Asian and Pacific peoples. The trace amounts of DNA present in some African populations were not produced by matings between Africans and Neanderthals, but come from matings between Africans and members of populations that migrated from West Eurasia back into Africa within the past 20,000 years and brought Neanderthal genes along.

The Neanderthal population that contributed genes to contemporary non-African modern human populations was closely related to the Neanderthals for whom we have ancient DNA sequences, but this was not the case for Denisovans. Scientists can identify stretches of DNA in modern humans that look very similar to the corresponding stretches of DNA observed in Neanderthals or Denisovans. Stretches of DNA are called **haplotypes**, and the term **introgression** refers to the incorporation of one haplotype into the genome of another population. Introgressed Neanderthal haplotypes in humans match those observed in Neanderthal ancient genomes very closely, suggesting that non-Africans received gene flow from a Neanderthal population that was very similar to the one from which we actually have archaic sequences. However, the introgressed Denisovan haplotypes are not as closely matched to the Denisovan genome from the Siberian cave. The amount of sequence divergence between the introgressed Denisovan haplotypes observed in humans and the haplotypes observed in the Denisovan genome from the Siberian cave suggests that the population of Denisovans that actually mated with modern humans had diverged from the population of Denisovans represented in the Siberian cave about 340 ka.

The presence of highly divergent haplotypes in some populations of modern humans in Africa suggests that Africans may have mated with another archaic hominin population for whom we do not have direct ancient DNA data (that is, a *ghost lineage*; see p. 321). There are some highly divergent haplotypes in some contemporary African hunter-gatherer populations, which suggests that their ancestors mated about 40 ka with one or more archaic hominin populations. This archaic population would have diverged from the line leading to modern humans, Neanderthals, and Denisovans 700 ka to 1 Ma. Evidence of interbreeding between Africans and archaic hominins is less conclusive than evidence of interbreeding between modern humans, Neanderthals, and Denisovans because we can't compare the modern African haplotypes that are thought to be introgressed directly with a DNA sequence from the haplotypes' archaic source. However, the analytic method that was used to identify deeply diverged African haplotypes successfully identified introgressed haplotypes in non-African populations that came from Neanderthals, suggesting that the method is reliable.

The genomes of several early modern humans indicate that interbreeding with Neanderthals occurred early in the spread of modern humans around 60 ka and continued for a long time.

Geneticists have extracted DNA from three early modern human fossils: one in central Siberia at a site near Denisova Cave, a second from a site in western Russia, and a third from a site in Romania. These fossils have been dated using radiocarbon methods. The two European fossils date to between 38 ka and 36 ka, and the Siberian fossil dates to about 45 ka.

The length of the Neanderthal DNA segments allows geneticists to estimate how much time has passed since the ancestors of the fossil interbred with Neanderthals. When Neanderthals and modern humans mated, their children would have had long segments of Neanderthal DNA. When these people then mated mainly with modern humans, recombination gradually broke the Neanderthal DNA segments into shorter and shorter segments. The Neanderthal DNA segments in the Siberian fossil are short, indicating that the Neanderthal DNA came from matings occurring between 7,000 and 13,000 years before the age of the fossil. This fact suggests that interbreeding occurred about 50 ka. The Neanderthal DNA segments in the Romanian fossil are long, indicating that the Neanderthal–modern human matings occurred only four to six generations before the age of the fossil, which suggests that interbreeding between the expanding modern human populations and Neanderthals continued until at least 36 ka. If there were such a long period of gene flow, then only about one mating between the two populations every 100 years would be necessary to account for the amount of Neanderthal DNA in modern human populations.

Modern human DNA found in the Altai Neanderthal genome suggests that modern humans and Neanderthals interbred about 100 ka.

We have seen that dates for the MRCA of genes in the modern human genome vary widely. Some of these genes are so old that we share them with Denisovans and Neanderthals. These very old segments are short because there has been a long time for recombination to chop them into successively shorter pieces. However, a recent analysis of the Altai Neanderthal genome shows that between 0.1% and 2% of this genome was shared with modern Africans, but the MRCA for the modern African segments is dated to between 230 ka and 120 ka, well after the split between modern humans and Neanderthals. Moreover, these segments are much longer than other segments shared between modern humans and Neanderthals. These facts suggest that Neanderthals and modern humans interbred well before 60 ka. Notably, there is no evidence for such early gene flow between modern humans and Denisovans.

Although evidence shows that selection acted against most Neanderthal and Denisovan genes after they entered the modern human population, a few genes were beneficial.

The fitness of a particular gene usually depends on how its products fit with other genes in the genome. Neanderthals had lived in Eurasia for several hundred thousand years, and it seems likely that on average genes that were common in Neanderthals were favored by selection. The same is true for Denisovan genes. However, there is evidence that most of the Neanderthal genes that entered modern human populations were selected against, probably because they did not fit so well with the modern human genome.

Evidence supporting this conjecture comes from mapping the location of Neanderthal and Denisovan genes on the chromosomes of modern non-African people. Remember that the genome has regions in which functional genes are common as well as regions in which they are rare. It turns out that Neanderthal and Denisovan genes are especially rare or depleted in regions with many functional genes. Moreover, there is even more depletion of Neanderthal and Denisovan ancestry in sequences that regulate gene expression than in sequences that code for proteins. In particular, genes expressed in areas of the brain that are responsible for cognitive functions such as language are depleted of introgressed archaic genetic material. There are four large regions in the modern genome that have very few Neanderthal genes. The biggest of these is the X chromosome. On average a bit more than 1% of genes in non-Africans come from Neanderthals, but on the X chromosome only about 0.2% of the genes have Neanderthal ancestry. Also notable is that there are no Neanderthal genes expressed in the testes. When genes from one species are expressed in the testes of a second species (as a result of mating), it can very commonly lead to male sterility.

However, there are some areas of the genome of modern humans where a Neanderthal/Denisovan sequence is enriched rather than depleted, suggesting that the sequence was favored by natural selection. Most of the segments in modern humans that are enriched in Neanderthal/Denisovan ancestry harbor genes that relate to skin biology and immune function. Beyond these very general functions, however, the effect of archaic introgression on the modern human phenotype is poorly understood. It will take many years of painstaking, controlled laboratory cell culture research to determine how having the Neanderthal variant actually affects physiologic function, and how these differences in function may have helped modern humans adapt to the new environments they encountered when they left Africa.

The best understood case of the benefits of archaic ancestry in modern humans involves people who live on the Tibetan plateau, which stands 4,500 m above sea level.

The levels of oxygen on the Tibetan plateau are 40% lower than the levels of oxygen at sea level. Native Tibetan women have lower infant mortality and higher fertility than do women who have resided for long periods on the plateau but lack Tibetan ancestry, partly because their physiologic responses to low oxygen levels differ. For women whose ancestors come from low-altitude sites, acclimatization to low oxygen involves an increase in blood hemoglobin levels. Although this has the benefit of increasing the amount of available oxygen, it comes at the cost of increased blood viscosity and cardiac problems. In contrast, native Tibetans have only a very limited hemoglobin increase at high altitude. Although several genes are involved in the low hemoglobin response and other high-altitude adaptations of Tibetans, a particularly important gene is EPAS1. A variant of the EPAS1 gene that is associated with low hemoglobin levels is very common in native Tibetans and absent in most other modern human populations. The Tibetan EPAS1 variant is also present in the Denisovan genome. As we learned in Chapter 12, there is evidence of Denisovan presence on the Tibetan plateau 160 ka. The length of the haplotype surrounding the Tibetan/Denisovan EPAS1 variant suggests that it was introduced into the ancestors of Tibetans via interbreeding with Denisovans or a population closely related to Denisovans about 40 ka and then spread rapidly among Tibetans because it increased their ability to cope with high altitudes, and was therefore favored by natural selection.

Archaeological Evidence

Archaeological data suggest modern humans first entered southern Asia 45 ka and brought along microliths similar to those used in southern Africa.

At Mehtakheri, a site in the Narmada Valley of northwestern India, archaeologists have recovered microliths in sediments that date from 50 ka to 45 ka. These backed, crescent-shaped tools are strikingly similar to those found on the coast of southern Africa. Although India is rather poorly explored archaeologically during this period, there are several other sites that date from 40 ka to 35 ka and have produced tools like those found in Africa (**Figure 13.19**). Beads and decorated ostrich eggshells that are very similar to those found in Africa have also been found at these sites.

Modern humans first entered Australia at least 40 ka, bringing sophisticated technology.

The fossil and archaeological record in Australia is quite good. Many well-dated archaeological sites indicate that modern *Homo sapiens* entered Australia at least 40 ka, perhaps as early as 65 ka, and occupied the entire continent by 30 ka. The most complete early site is in southeastern Australia at Lake Mungo. The finds here include three hominin skulls, hearths, and ovens that have been dated to 32 ka by using

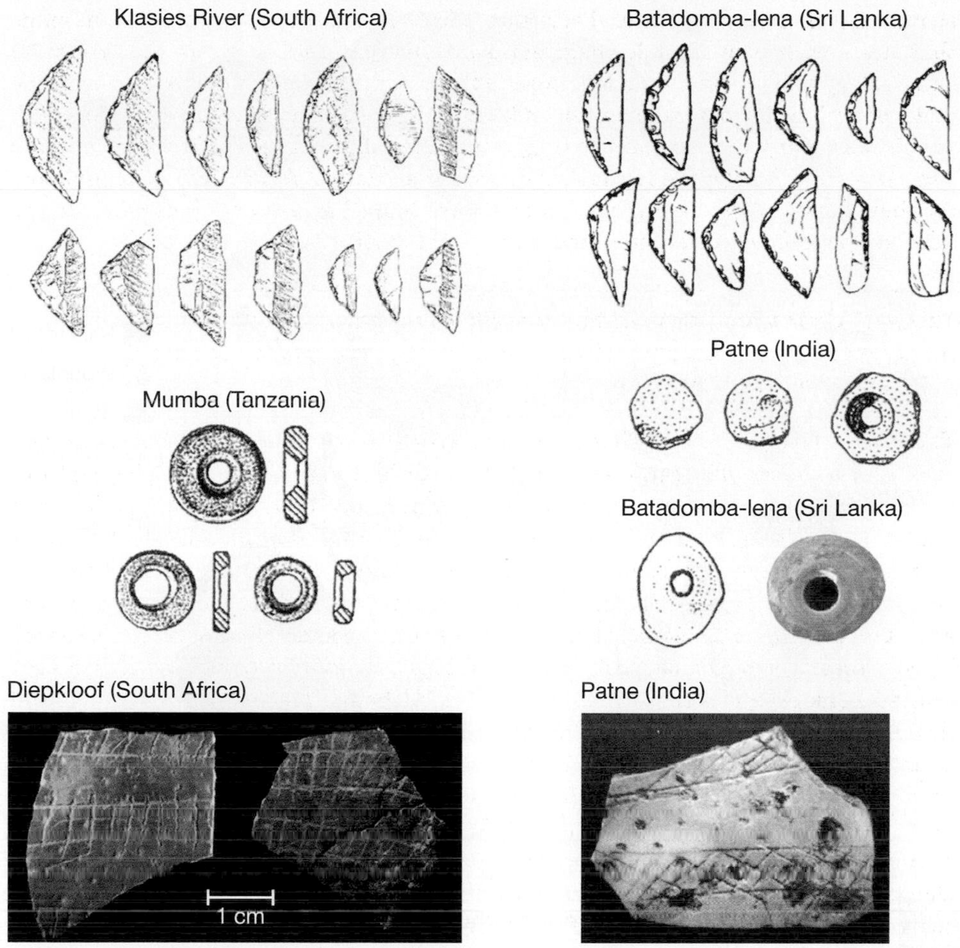

Klasies River (South Africa)

Batadomba-lena (Sri Lanka)

Mumba (Tanzania)

Patne (India)

Batadomba-lena (Sri Lanka)

Diepkloof (South Africa)

1 cm

Patne (India)

FIGURE 13.19

Tools, beads, and decorated ostrich eggshells found at African (left) and Indian (right) sites that date from 40 ka to 35 ka.

carbon-14 dating. The skull specimens are fully modern and well within the range of variation of contemporary Australian aborigines.

The earliest Australians made sophisticated tools, and there is evidence of symbolic behavior. Some of the tools found at these early sites were made from bone, there are cave paintings dated to 17 ka, and there is evidence of both ceremonial burials and cremation. About 15 ka, Australians seem to have been the first people to have used polished stone tools, which are made by grinding rather than flaking. In other parts of the world, polished stone tools do not appear in the record until agriculture is introduced about 10 ka.

The fact that people got to Australia at all is evidence of their technological sophistication. Much of the world's water 40 ka was tied up in continental glaciers, so New Guinea, Australia, and Tasmania formed a single continent named Sahul (**Figure 13.20**). However, at least 100 km (62 mi.) of open ocean separated Asia and Sahul. This gap was wide enough to prevent nearly all movement of terrestrial mammals across it and thus to preserve the unique, largely marsupial fauna of Australia and New Guinea. The colonization of Sahul cannot be dismissed as a single lucky event because people also crossed another 100 km of ocean to reach the nearby islands of New Britain and New Ireland at about the same time.

Clearly, then, the people who first settled these islands could build seaworthy boats. There is a cave site at Jerimalai on the eastern tip of the island of Timor that dates to 42 ka. At this site, archaeologists have found many stone tools and the bones of many species of fish. About half of the fish bones are from pelagic species, mainly tuna, which are found well away from shore. This suggests that the people who lived there had watercraft that allowed them to engage in offshore fishing.

FIGURE 13.20

A reconstruction of the distribution of habitats when humans left Africa (72 ka to 60 ka). The world was colder and drier than it is now, so sea levels were lower. Sri Lanka was connected to the Eurasian continent, and what are now the islands of southeastern Asia were part of a large peninsula called Sunda. Australia and New Guinea were linked in an isolated continent called Sahul. Forests were much less widespread and mainly limited to central Africa, Sunda, and Sahul. Much of Asia was covered with desert or steppe tundra.

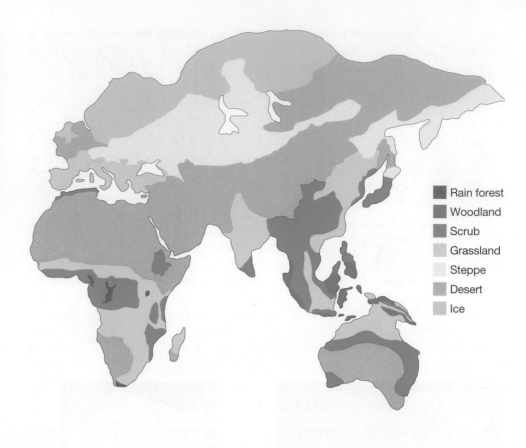

Rain forest
Woodland
Scrub
Grassland
Steppe
Desert
Ice

There is evidence that modern humans arrived in eastern Asia around the same time.

Modern humans must have come to Australia by way of southern Asia, but signs of modern humans in southern Asia are scarce. Only one site in southern Asia contains the kind of sophisticated artifacts seen in Upper Paleolithic Europe. The Batadomba-lena Cave in Sri Lanka is dated to about 28 ka, and it has yielded modern human remains, elaborate stone tools, and tools made from bone.

Evidence for modern human occupation has been found at several sites in northern China, Mongolia, and Siberia. Especially notable are two sites north of the Arctic Circle. Archaeologists found a mammoth carcass that shows clear evidence that the animal was killed and butchered at a site on the Yenisei River at latitude 72° north. This site is securely dated to 45 ka. At Berelekh, a site 500 km (about 300 mi.) north of the Arctic Circle where the Yana River empties into the Arctic Ocean, archaeologists have recovered sophisticated stone tools and several artifacts made from bone, ivory, and horn (**Figure 13.21**). Radiocarbon methods indicate that these tools were made about 30 ka. From pollen data we know that this area then had a cool, dry climate in which grasslands were mixed with stands of larch and birch trees. Many processed bones of horse, musk ox, bison, and mammoth indicate that big game was plentiful. Nonetheless, life at this site must

FIGURE 13.21

Spear foreshafts found at Berelekh, a site 500 km (about 300 mi.) north of the Arctic Circle in eastern Siberia. These rhino horn foreshafts were fitted to the front end of the spear. If hunters struck an animal without killing it, they could remove the spear, fit a second foreshaft to the spear, and reuse it.

have been challenging for a recent African emigrant. Today, the winters are long and dark, and January temperatures average −37°C (about −34°F)—a bracing thought, given that the world was much colder 30 ka than it is today.

One Expansion out of Africa or Two?

There was a major expansion out of Africa around 60 ka.

The fossil, genetic, and archaeological evidence all points to a major expansion about 60 ka. Although modern human fossils appear in Africa almost 200 ka, with one important exception the earliest modern human fossils elsewhere date to after 60 ka. Evidence from mtDNA, Y chromosomes, and an extensive collection of complete DNA sequences of living people indicates that the MRCA of genes carried by contemporary Africans dates to more than 100 ka, whereas the MRCA of most genes carried by non-Africans dates to after 60 ka. This evidence suggests that most contemporary non-Africans are descended from people who left Africa about 60 ka. The negative correlation between distance from Africa and levels of genetic variation is also consistent with an out of Africa expansion. Archaeological evidence for modern humans appears in southern Africa well before 60 ka, whereas elsewhere most evidence is dated after 60 ka. Moreover, the pattern of dates is consistent with a spread from Africa moving east and north into Eurasia and then Australia.

There may have been an earlier expansion more than 100 ka.

Along with the preceding evidence of a major expansion around 60 ka, there *is* tentative evidence that modern humans also left Africa more than 100 ka. A partial skull with modern human features found in Greece dates to between 300 ka and 40 ka with a mean of 210 ka. However, there are doubts about whether the skull belonged to a modern human, and the wide range of dates is troubling. Modern human teeth found in China have been dated from 120 ka to 80 ka by using uranium thorium dating. Here too, though, there is controversy, as debate about the geology within this cave site has led some investigators to think that the dates may be inaccurate. But the DNA sequence of the Altai Neanderthal suggests that modern human genes entered Neanderthal populations more than 100 ka, and since there were no Neanderthals in Africa at this time, modern humans must have entered Eurasia before that date. Some paleoanthropologists think that this interbreeding could have occurred in the Levant, where modern humans and Neanderthals may have overlapped at an early date. One of the three recent studies reporting DNA sequences from a large, worldwide sample of modern humans indicates that about 2% of the genome of modern Papuans represents genes from an earlier human expansion.

Even if modern humans did expand before 100 ka, this expansion was unsuccessful. Virtually all the genes carried by modern non-Africans are descended from people who left Africa after 60 ka. Either populations in the early expansion were mainly extinct before 60 ka or they were replaced by the people in the latter expansion without much interbreeding. There are precedents for such extinctions, which we cover in the next chapter.

Life in the European Upper Paleolithic

There is a very rich record of modern human settlement in western Eurasia.

The archaeological record for western Eurasia is exceptionally rich and detailed, and we have much more archaeological information about the expansion of modern humans in this part of the world than we do in other areas. We also have much more information from ancient DNA, as the cold and dry climate of Europe aids DNA preservation. Ancient DNA is valuable for answering a question that has long vexed archaeologists: When we see cultural change in a given area over time (for example, a distinctive tool kit is replaced by another very distinctive tool kit), did this change occur because a new population bearing that tool kit moved into the area, perhaps even totally replacing the former residents, or did this change occur because the resident population changed their culture themselves (that is, a local genius innovated the new tool kit herself or copied the tool kit from a neighboring culture)? In the rest of this chapter, we take advantage of this rich archaeological and genetic record to paint a detailed picture of the lifeways of the earliest modern humans.

The first modern humans in western Eurasia created several tool industries in various areas and times. This diversity stands in striking contrast to the Acheulean tool kit, which remained much unchanged throughout more than half of Africa and Eurasia for more than a million years. Archaeologists refer to these tool industries collectively as **Upper Paleolithic** industries to distinguish them from the earlier tool industries associated with Neanderthals and other earlier hominins. The earliest Upper Paleolithic tools appear in the Near East about 50 ka, and they disappear from the archaeological record about 10 ka. Archaeologists refer to this period in western Eurasia as the Upper Paleolithic period and refer to the people who made the tools as Upper Paleolithic peoples.

Upper Paleolithic industries varied in time and space.

The oldest Upper Paleolithic industry, labeled the **Initial Upper Paleolithic** (IUP), has been found at sites that date to around 50 ka in the Levant (Israel and southern Turkey) and a bit later at sites in central Europe and western Asia (**Figure 13.22a**). In the Levant, the IUP develops gradually from the existing Middle Paleolithic industry, but in central Europe the transition from Mousterian industries is more abrupt, suggesting that the IUP was brought into central Europe by an immigrant population. Some archaeologists believe the immigrants were modern humans who spread into central Europe. Unfortunately, the fossil evidence is ambiguous. A modern human femur dated to 45 ka has been found at Ust'-Ishim, a site in Russia, but it is not associated with any tools. There are several hominin fossils associated with IUP industries, but none can be clearly identified as modern human or Neanderthal. Ancient DNA analyses show that the Ust'-Ishim individual, along with a 40-ka modern human from Romania, were not closely related to hunter-gatherers that lived in Europe only a few thousand years later. This lack of genetic continuity suggests that the IUP toolmakers may have been pioneering populations that went extinct and were replaced by later waves of hunter-gatherer migration.

FIGURE 13.22

(a) The earliest Upper Paleolithic industry, the Initial Upper Paleolithic (IUP), is found in the Levant and central and eastern Europe with dates from 50 ka to 45 ka. (b) By 42 ka, a second Upper Paleolithic industry, the Aurignacian, had spread across much of Europe. At this time there were two industries intermediate between the Mousterian and the Aurignacian: the Châtelperronian in France and the Uluzzian in Italy. The white map shows the current European coastlines, and the light gray shows the coastline 45 ka.

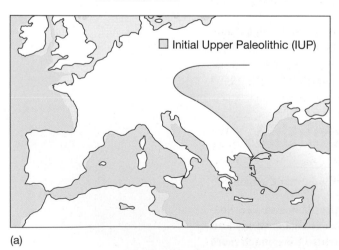

(a)

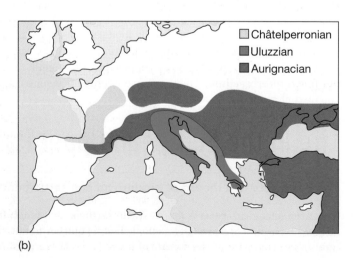

(b)

By 42 ka, a second industry, the **Aurignacian**, was widespread in Europe (**Figure 13.22b**). It is characterized by certain types of large blades, burins, and bone points (**Figure 13.23**). Modern human fossils have been found associated with very early Aurignacian tools at two sites in northern Italy. The enamel structure of fossil teeth from one site clearly indicates that they were modern humans, and at the second site mtDNA extracted from a fossil tooth unambiguously identifies it as modern human. Modern human fossils have also been found at later Aurignacian sites, so most archaeologists believe that the Aurignacian is an indicator of modern human occupation. It thus appears that anatomically modern humans and Neanderthals overlapped for several thousand years in some regions. The oldest ancient autosomal DNA from a modern human associated with Aurignacian tools comes from a 35,000-year-old individual found in Belgium. Unlike the modern humans that preceded them in Europe, members of this individual's population contributed ancestry to later hunter-gatherers in Europe, as well as to Europeans living today. Indeed, ancient DNA analyses indicate that almost all hunter-gatherers living in Europe between 37 ka and 14 ka were descended from a single population that was isolated from non-European populations.

Several enigmatic stone tool industries found in Europe are intermediate between the Mousterian and the Aurignacian and are sometimes termed "transitional" industries (see Figure 13.22b). For example, in southern France, during the transition between the Mousterian and the Aurignacian, a third industry called **Châtelperronian** is observed at some sites. Châtelperronian tools are associated with Neanderthal fossils at Saint-Césaire, Arcy-sur-Cure, and Grotte du Renne in France. Many anthropologists believe that the Châtelperronian is the result of Neanderthals borrowing ideas and technology from modern humans, whereas others believe that it represents the independent development of new tool types by Neanderthals. Other transitional industries, such as the **Uluzzian** from Italy and the Szeletian from central Europe, are not associated with fossils that are easily identified as either modern human or Neanderthal.

About 30 ka, the Aurignacian was replaced in southern France by a new tool kit, the Gravettian, in which small, parallel-sided blades predominated and bone points were replaced by bone awls. Gravettian-era individuals buried in France, Germany, Belgium, Italy, and the Czech Republic are all very genetically similar despite their great geographic range. Most of the ancestry of the Gravettians descends from the same source as a 37-ka individual found in European Russia; this genetic source, in turn, is similar to that of the 35 ka Belgian associated with Aurignacian tools mentioned above. Over time, this Gravettian ancestry component spread from west to east, replacing the ancestry component found in the 35-ka individual associated with Aurignacian tools in Belgium. Thus, like the transition from IUP to Aurignacian culture, the transition from Aurignacian to Gravettian culture also appears to have come about via the spread of new people who largely replaced the previous residents, rather than via a long-standing resident population changing its culture over time.

Then about 21 ka, the Solutrean, with its beautiful leaf-shaped points, developed in Europe. And about 16 ka, the Solutrean gave way to the Magdalenian, a tool kit dominated by carved, decorated bone and antler points together with many microliths. The Magdalenian culture first appears in southwestern parts of Europe that were not covered in ice during the last glacial maximum, and then spreads across Europe as the ice sheets retreat. The oldest ancient DNA from an individual associated with Magdalenian tools is a 19,000-year-old individual from Spain. Over the next 5,000 years, the distinctive ancestry component associated with this 19-ka Spaniard spreads across Europe along with Magdalenian tools, replacing the ancestry component found in individuals associated with the preceding Gravettian tools. This Magdalenian-associated ancestry component is very similar to that of the 35-ka Belgian associated with Aurignacian tools. Thus, while people associated with Aurignacian tools were largely replaced by a population with distinctive ancestry and Gravettian tools, the descendants of these early Aurignacians persisted in southwest

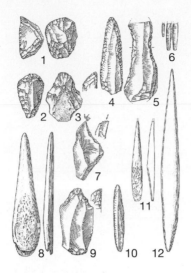

FIGURE 13.23

The Aurignacian tool kit, the earliest of several tool industries in the Upper Paleolithic of Europe, contains a variety of standardized tool types: 1, 2, 3, 9 = scrapers; 4, 5 = edged, retouched blades; 6, 10 = bladelets; 7 = a burin (a tool now used for engraving); 8, 11, 12 = bone points.

Europe, where they developed the Magdalenian tool kit and eventually replaced the people associated with Gravettian culture.

Finally, in a warming period following the end of the ice age around 14 ka, there was a major genetic turnover in Europe. Recall that although there are detectable genetic differences among hunter-gather populations living in Europe in the period 37 ka to 14 ka (these differences are what allow scientists to determine whether cultural changes in a given area occurred alongside population continuity or replacement), all populations living at this time were nevertheless descended from the same ancestral founder population and were genetically isolated from populations outside Europe. After 14 ka, European hunter-gatherers are more closely related to contemporary people living in the Near East, suggesting that some of the ancestors of Near Easterners began migrating into Europe around this time. This population movement was again associated with cultural change, with the Magdalenian giving way to different tool kits such as the Epigravettian and Azelian.

Upper Paleolithic peoples manufactured blade tools, which made efficient use of stone resources.

As we noted earlier, blades are stone flakes that look like modern knife blades: They are long, thin, and flat and have a sharp edge. Blades have a longer cutting edge than flakes do, so blade technology made more efficient use of raw materials than older tool technologies did. However, there was a cost: They also took more time to manufacture, requiring more preparation and more finishing strokes.

The Upper Paleolithic tool kit includes many distinctive, standardized tool types.

Upper Paleolithic peoples made many more kinds of tools than earlier hominins had made. Tools thought to be used as chisels, various types of scrapers, several kinds of points, knives, burins (pointed tools used for engraving), drills, borers, and throwing sticks are just some of the items from the Upper Paleolithic tool kit.

Even more striking, the different tool types have distinctive, stereotyped shapes. It is as if the Upper Paleolithic toolmakers had a sheaf of engineering drawings on which they recorded their plans for various tools. When toolmakers needed a new 10-cm (4-in.) burin, for example, they would consult the plan and produce one just like all the other 10-cm burins. Of course, these toolmakers didn't really use drawings as modern engineers do, but the fact that we find standardized tools suggests that they carried these plans in their minds. The final shape of Upper Paleolithic tools was not determined by the shape of the raw material; instead, the toolmakers seem to have had a mental model of what the tool was supposed to look like, and they imposed that form on the stone by careful flaking. The craft approach to toolmaking reached a peak in the Solutrean tool tradition, which predominated in southern France between 21 ka and 16.5 ka (**Figure 13.24**). Solutrean toolmakers crafted exquisite points shaped like laurel leaves that were sometimes 28 cm (about 1 ft.) long, 1 cm (about 0.5 in.) thick, and perfectly symmetric.

Upper Paleolithic people made tools from bones, antlers, and teeth. Earlier hominins had made limited use of bone, but Upper Paleolithic people transformed bone, ivory, and antler into barbed spear points, awls, sewing needles, and beads.

FIGURE 13.24

Long, thin, delicate points characterized the Solutrean tool tradition.

Stones and other raw materials for toolmaking were often transported hundreds of kilometers from their place of origin.

At Bacho Kiro, a 46,000-year-old Aurignacian site in Bulgaria, more than half of the flint used to make blades was brought from a source 120 km (about 75 mi.) away. Distinctive, high-quality flint quarried in Poland has been found in archaeological sites more than 400 km (about 250 mi.) away. Seashells, ivory, soapstone, and amber used in ornaments were especially likely to be transported long distances. By comparison, most of the stone used at a Mousterian site in France was transported less than 5 km (3 mi.).

Neanderthals and modern humans had similar subsistence economies.

The Upper Paleolithic spanned the depths of the last ice age. The area in Europe that contains the richest archaeological sites was a cold, dry grassland that supported large populations of a diverse assemblage of large herbivores, including reindeer (called caribou in North America), horse, mammoth, bison, woolly rhinoceros, and a variety of predators, such as cave bears and wolves.

Bones found at Upper Paleolithic sites indicate that large herbivores played an important role in the diets of Upper Paleolithic peoples. Everywhere, Upper Paleolithic peoples hunted herbivores living in large herds, fished, and hunted birds. In some places, they concentrated on a single species—reindeer in France, red deer in Spain (**Figure 13.25**), bison in southern Russia, and mammoths farther north and east. In some areas, such as the southern coast of France, rich salmon runs may have been an important source of food. There are also places where these peoples harvested several kinds of animals.

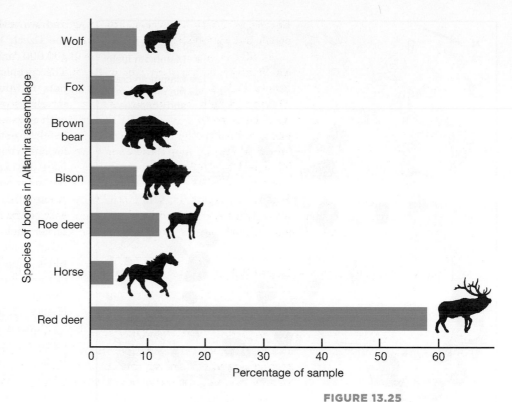

FIGURE 13.25

The bones of red deer dominate the assemblage at Altamira in northern Spain, as illustrated by these data, which come from a Magdalenian site dated to 15 ka to 13 ka.

Modern humans probably used plant foods extensively, but few sites preserve the remains from vegetation.

All modern foraging peoples rely on plant foods that they gather as well as animals that they hunt. Earlier hominins probably used plant foods as well, but the remains of these foods have not been preserved. Unusual preservation conditions at several sites provide glimpses into this part of the subsistence economy of Upper Paleolithic peoples. Dani Nadel of the University of Haifa and his colleagues have excavated a site on the shores of the Sea of Galilee that is dated to 19 ka. This site, which is called Ohalo II, contains more than 90,000 specimens of plants from 142 plant taxa, including wild varieties of barley, wheat, acorns, pistachios, olives, raspberries, figs, and grapes. Starch grains from barley have been found on a grinding stone, indicating that wild cereal grains were processed to make food.

The peoples of the Upper Paleolithic developed complex forms of shelter and clothing.

In what is now western and central Europe, Russia, and Ukraine, the remains of small villages have been found. Living on a frigid, treeless plain, Upper Paleolithic peoples hunted or scavenged mammoth and used the hairy beasts for food, shelter,

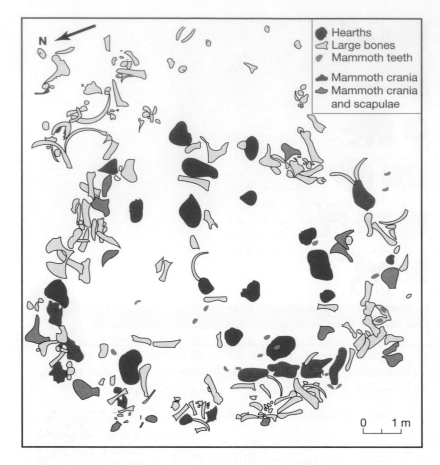

FIGURE 13.26

On the Russian plain, early humans had to cope with harsh climate conditions. They may have used mammoth bones to construct shelters. Mammoth bones and teeth litter this Upper Paleolithic site of Moldova. The presence of multiple hearths suggests that the site may have been occupied by several families.

Legend:
- Hearths
- Large bones
- Mammoth teeth
- Mammoth crania
- Mammoth crania and scapulae

0   1 m

and warmth. At the site of Předmostí in the Czech Republic, the remains of almost 100,000 mammoths have been discovered. The people who lived there constructed huts by arranging mammoth bones in an interlocking pattern and then draping them with skins. (Temperatures of about −45°C [−50°F] outside provided a strong incentive for chinking the cracks.) Huge quantities of bone ash found at these sites indicate that the mammoth bones were also used for fuel. A site about 470 km (about 300 mi.) southeast of Moscow contains the remains of even larger shelters. They were built around a pit about a meter deep and were covered with hides supported by mammoth bones. Some of these huts had many hearths, suggesting that several families may have lived together (**Figure 13.26**).

Modern foragers living in warm places often construct simple brush huts, using large branches to support the roof and smaller twigs and grasses to form the walls and cover the roof. Grasses may be gathered and spread on the floor and used as bedding. At the Ohalo II site, Nadel and his colleagues found kidney-shaped depressions that are about 5 m by 13 m (16 ft. by 43 ft.) in area. These depressions contain substantial pieces of tree branches that might have held up the roof of a hut as well as smaller twigs and stems that might have been part of a roof. In one of the depressions, archaeologists found grass stems that were loosely woven together around a mass of ash. It looks like the grasses may have formed a mat around a fire that people sat on as they cooked and on which they slept. If this interpretation is correct, then this site represents the oldest evidence for the construction of brush huts.

Several lines of evidence indicate that Upper Paleolithic peoples living in glacial Europe manufactured tailored fur clothing. When modern hunters skin a fur-bearing animal, they usually leave the feet attached to the pelt and discard the rest of the carcass. Many skeletons of foxes and wolves that are complete except for their feet have been found at several Upper Paleolithic sites in Russia and Ukraine, suggesting that early modern humans kept warm in sumptuous fur coats. Bone awls and bone needles are also common at Upper Paleolithic sites, so sewing may have been a common activity. Finally, three individuals at a burial site in Russia seem to have been buried in caps, shirts, pants, and shoes lavishly decorated with beads.

Upper Paleolithic peoples coped with their environment better than the Neanderthals did.

The richness of the fossil and archaeological record in Europe provides a detailed comparison of Neanderthal and Upper Paleolithic peoples. Three kinds of data suggest that Upper Paleolithic peoples were better adapted to their environment than Neanderthals:

1. Upper Paleolithic peoples lived at higher population densities in Europe than did Neanderthals. Archaeologists estimate the relative population size of vanished peoples by comparing the density of archaeological sites per unit of time.

Thus, if one group of people occupied a particular valley for 10,000 years and left 10 sites, and a second group occupied the same valley for 1,000 years and left 5 sites, then archaeologists would estimate that the second group had five times as many people. (In using this method, it is important to be sure that the sites were occupied for approximately the same length of time.) By these criteria, Upper Paleolithic peoples had far higher population densities in Europe than the Neanderthals.

2. Upper Paleolithic peoples lived longer than Neanderthals. Anthropologists Rachel Caspari of Central Michigan University and Sang-Hee Lee of the University of California, Riverside, have estimated that Upper Paleolithic peoples lived much longer than the Neanderthals. About a third of their sample of 113 Neanderthals reached age 30. In contrast, two-thirds of a sample of 74 Upper Paleolithic fossil individuals reached that age. The fact that Upper Paleolithic populations included a substantial fraction of older individuals, whereas Neanderthals were dominated by the young, may have allowed Upper Paleolithic peoples to retain and transmit more complex cultural knowledge than Neanderthals could.

3. Upper Paleolithic peoples were less likely to suffer serious injury or disease than were Neanderthals. In sharp contrast to the Neanderthals, the skeletons of Upper Paleolithic people rarely show evidence of injury or disease. Among the few injuries that do show up in the fossil record, there is a child buried with a stone projectile point embedded in its spine, and a young man with a projectile point in his abdomen and a healed bone fracture on his right forearm. Evidence of disease is slightly more prevalent than evidence of injury among the remains of Upper Paleolithic peoples; affected specimens include a young woman who probably died as the result of an abscessed tooth and a child whose skull seems to have been deformed by hydrocephaly (a condition in which fluid accumulates in the cranial cavity and the brain atrophies). Nonetheless, there is still less evidence for disease among these peoples than among Neanderthals.

There is good evidence for ritual burials during the Upper Paleolithic period.

Like the Neanderthals, Upper Paleolithic peoples buried their dead. Upper Paleolithic sites provide the first unambiguous evidence of both multiple burials and burials outside caves. Unlike the Neanderthals, Upper Paleolithic burials appear to have been accompanied by ritual. Upper Paleolithic burials are often associated with tools, ornaments, and other objects that suggest they had some concept of life after death. **Figure 13.27** shows the diagram of the grave of a child who died about 24 ka at the Siberian site of Mal'ta. The child was buried with several items, including a necklace, a crown (diadem), a figurine of a bird, a bone point, and stone tools.

Upper Paleolithic peoples were skilled artisans, sculpting statues of animals and humans and creating sophisticated cave paintings.

It is their art that distinguishes Upper Paleolithic peoples most dramatically from the hominins who

FIGURE 13.27

A child's grave from the Upper Paleolithic illustrating the rich collection of goods frequently included in Upper Paleolithic burials.

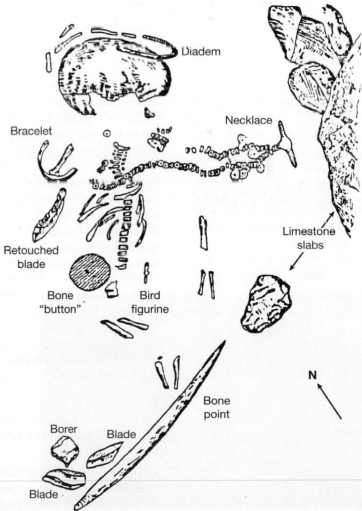

FIGURE 13.28

Small animal figures were carved from mammoth tusks during the Aurignacian period.

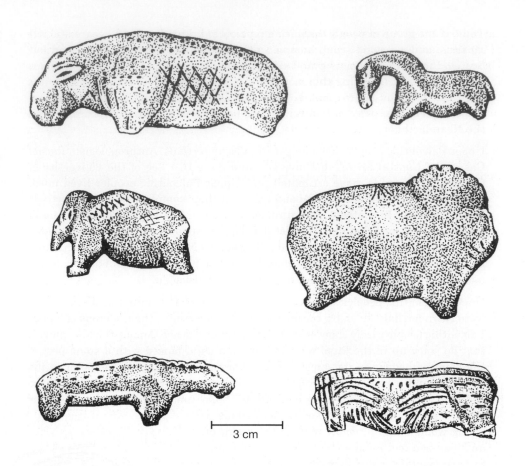

3 cm

preceded them. They engraved decorations on their bone and antler tools and weapons, and they sculpted statues of animals and female figures (**Figure 13.28**). The female statues are generally believed to be fertility figures because they usually emphasize female sexual characteristics. Upper Paleolithic peoples also adorned themselves with beads, necklaces, pendants, and bracelets and may have decorated their clothing with beads.

All of these artistic efforts are remarkable, but it is their cave art that seems most amazing now. Upper Paleolithic peoples painted, sculpted, and engraved the walls of caves with a variety of animal and human figures. They used natural substances that included iron oxides and manganese to create a variety of paint colors. They used their fingers, horsehair, and sticks to apply the paint. Their cave paintings frequently depict animals that they must have hunted: reindeer, mammoths, horses, and bison.

FIGURE 13.29

This image from Le Chauvet Cave in France was created about 30 ka. At the upper left, several lions are depicted. To the right are several rhinoceroses. At the top right, the multiple outlines suggest a rhinoceros in motion.

Some figures are half human and half animal. Sometimes the artists incorporated the natural contours of the cave walls in their work, and sometimes they drew one set of figures on top of others. Some of their work is spectacular with accurate perspective, lifelike characterization of behavior, and complex scenes. We don't know precisely why or under what circumstances these cave paintings were made, but they represent a remarkable cultural achievement that sets the people of this period apart from earlier hominins.

Techniques for dating pigments have allowed scientists to determine the age of the Upper Paleolithic cave paintings. Many of the famous caves, such as Lascaux, are dated to the end of the Upper Paleolithic, about 17 ka, but spectacular paintings in Le Chauvet Cave in France are about 36,000 years old (**Figure 13.29**). Most of the famous carved figurines are also less than 20,000 years old, but there are good examples of representational sculpture that date to the earliest Aurignacian. For example, **Figure 13.30** shows an ivory figure with a human body and a lion's head that was found at a site in Germany dated to 32 ka. There is also abundant evidence of the manufacture of beads, pendants, and other body ornaments at Aurignacian sites in France. The beads are standardized and often manufactured from materials that had been transported hundreds of kilometers. Archaeologists have found a well-preserved musical instrument that looks like a flute at an Aurignacian site in southwestern France. It is 10 cm (4 in.) long and has four holes on one side and two on the other. When this instrument was played by a professional flutist, it produced musical sounds.

The End of History?

Here we end our history of the human lineage. By about 100 ka, physically and behaviorally modern humans had evolved in Africa, and by the end of the Pleistocene about 10 ka, they had spread to every terrestrial habitat except Antarctica and a few remote islands. Of course, this was not the end of human history. There have been many changes in human life over the past 10,000 years. The warm and stable climates of this period allowed the development of agriculture, which led to large sedentary settlements, larger and larger polities, and immense technological progress. However, most of these changes resulted from cultural rather than genetic change and are not usually covered in biological anthropology courses like the ones this textbook is meant to serve. Of course, the new environments created by agriculture and human life have led to natural selection, which in turn has led to significant genetic change. But most of this change is in local adaptation, which creates human variation, and is the topic of next chapter.

FIGURE 13.30

This ivory figure depicts a human body with a lion's head. It comes from an Aurignacian site in southern Germany and is 32,000 years old.

CHAPTER REVIEW

Key Terms

Middle Stone Age (MSA) (p. 333)
blades (p. 334)
Mode 4 (p. 334)
pressure flaking (p. 334)

microlith (p. 334)
Mode 5 (p. 334)
most recent common ancestor (MRCA) (p. 339)
gene trees (p. 339)

haplogroup (p. 341)
microsatellite loci (p. 344)
haplotype (p. 346)
introgression (p. 346)
Upper Paleolithic (p. 352)

Initial Upper Paleolithic (p. 352)
Aurignacian (p. 353)
Châtelperronian (p. 353)
Uluzzian (p. 353)

Study Questions

1. What derived anatomic features distinguish modern humans from other hominins?

2. Where and when did modern humans originate? What is the evidence that supports your answer?

3. Describe the evidence that modern humans emerged from Africa more than 60 ka.

4. Explain why it is possible to make phylogenetic trees for mitochondrial DNA and the Y chromosome but not for other chromosomes. Why is it possible to make phylogenetic trees for individual genes on other chromosomes?

5. Explain why the mitochondrial DNA and Y chromosome trees are consistent with the hypothesis that modern humans evolved in Africa and then later spread across the rest of the globe. Explain why this hypothesis is consistent with the more recent data from whole genomes.

6. Suppose you were able to choose three fossils from which to extract DNA and that your goal was to

test the hypothesis that humans evolved in Africa. Which three would you pick? Explain why.

7. How do worldwide patterns of genetic variation support the idea that humans spread from Africa across Eurasia and Australia?

8. What is the evidence that there was interbreeding between modern humans and Neanderthals and Denisovans? What does this evidence tell us about where and when these matings took place?

9. Describe what the archaeological record tells us about the pattern of human behavior across the world between 100 ka and 30 ka. Which facts are widely accepted? Which are in dispute?

10. Describe the main differences between the tools of Upper Paleolithic peoples and those of their predecessors.

Further Reading

Groucutt, H. S., M. D. Petraglia, G. Bailey, et al. 2015. "Rethinking the Dispersal of *Homo sapiens* out of Africa." *Evolutionary Anthropology* 24: 149–164.

Klein, R. G. 2008. *The Human Career*. 3rd ed. Chicago: University of Chicago Press.

Malaspinas, A.-S., M. C. Westaway, C. Muller, et al. 2016. "A Genomic History of Aboriginal Australia." *Nature* 538: 207–214, doi:10.1038/nature18299.

Mallick, S., H. Li, M. Lipson, et al. 2016. "The Simons Genome Diversity Project: 300 Genomes from 142 Diverse Populations." *Nature* 538: 201–206, doi:10.1038/nature18964.

Marean, C. 2015. "An Evolutionary Anthropological Perspective on Modern Human Origins." *Annual Review of Anthropology* 44: 533–556, doi:10.1146/annurev-anthro-102313-025954.

McBrearty, S., and A. Brooks. 2000. "The Revolution That Wasn't: A New Interpretation of the Origin of Modern Human Behavior." *Journal of Human Evolution* 39: 453–563.

O'Bleness, M., V. B. Searles, A. Varki, P. Gagneux, and J. M. Sikelal. 2013. "Evolution of Genetic and Genomic

Features Unique to the Human Lineage." *Nature Reviews Genetics* 13: 853–866.

Pääbo, S. 2014. "The Human Condition—A Molecular Approach." *Cell* 157: 216–226, doi:10.1016/j.cell.2013.12.036.

Pagani, L., D. J. Lawson, E. Jagoda, et al. 2016. "Genomic Analyses Inform on Migration Events during the Peopling of Eurasia." *Nature* 538: 238–242, doi:10.1038/nature19792.

Reich, D. 2018. *Who We Are and How We Got Here*. New York: Random House.

Sankararaman, S., S. Mallick, M. Dannemann, et al. 2014. "The Genomic Landscape of Neanderthal Ancestry in Present-Day Humans." *Nature* 507: 354–357, doi:10.1038/nature12961.

Veeramah, K. R., and M. F. Hammer. 2014. "The Impact of Whole-Genome Sequencing on the Reconstruction of Human Population History." *Nature Reviews Genetics* 15: 149–162.

 Review this chapter with personalized, interactive questions through INQUIZITIVE.

 This chapter also features a Guided Learning Exploration on genetics and common ancestry.

PART
FOUR

EVOLUTION AND MODERN HUMANS

14

HUMAN GENETICS AND VARIATION

CHAPTER OBJECTIVES

By the end of this chapter you should be able to:

A Describe how humans differ genetically from other apes.

B Explain why variation in traits influenced by single genes is different from variation in traits affected by many genes.

C Describe how mutation and natural selection maintain differences at single genetic loci between people in the same population.

D Explain how genetic drift and natural selection create variation among populations in traits influenced by single genes.

E Describe how to measure genetic variation in complex traits.

F Assess why the existence of genetic variation in complex traits within populations does not imply that there is variation among populations.

G Evaluate the argument that common concepts of race do not correspond to any meaningful biological category.

363

Explaining Genetic Variation

Human beings vary in myriad ways. In any sizable group of people, there are differences in height, weight, hair color, eye color, food preferences, hobbies, musical tastes, skills, interests, and so on. Some of the people you know are tall enough to dunk a basketball, others have to roll up the hems of their pants; some have blue eyes and freckle in the sunshine, others have dark eyes and can get a terrific tan; some people have perfect pitch, others can't tell a flat from a sharp. Your friends may include heavy drinkers and teetotalers, great cooks and people who can't microwave popcorn, skilled gardeners and some who can't keep a geranium alive, some who play classical music and others who prefer heavy metal.

If we look around the world, we encounter an even wider range of variation. Some of the variation is conspicuous. Language, fashions, customs, religion, technology, architecture, and other aspects of behavior differ among societies. People in different parts of the world also look very different. For example, most of the people in northern Europe have blond hair and pale skin, and most of the people in southern Asia have dark hair and dark skin. Arctic peoples are generally stockier than people who live in the savannas of East Africa. Groups also differ in ways that cannot be detected so readily. For instance, the peoples of the world vary in blood type and the incidence of many genetically transmitted diseases. For example, Tay–Sachs disease, a recessive genetic disorder that usually kills children before the age of 4, is nearly 10 times more common among Ashkenazi Jews in New York than among other New Yorkers.

If you peer even farther out in the natural world, it is also easy to see that humans are different from other primates. We are comparatively hairless, walk bipedally, live in large cooperative societies, cook our food, depend on complex tools, and occupy every part of the globe. Our closest primate relatives do none of these things.

In this chapter, we consider how genetic differences lead to phenotypic differences among humans and between humans and other primates. We begin by discussing the genetic differences between humans and other primates. The sequencing of the great ape genomes has revealed much about this variation, and we are beginning to understand how apes and humans differ genetically. We then turn to the question of how people vary genetically within and among societies and the processes that create and sustain this variation. We begin by describing the nature of differences between people that are influenced by single genes that have large effects, and for which the differences are not influenced very strongly by the environment. We then consider variation in complex traits that are influenced by many genes and also substantially by the environment. As you will see, the methods used to assess variation in simple and complex traits are quite different. In both cases, we consider the processes that give rise to variation within and among populations. Finally, we will use our understanding of human genetic diversity to explore the significance and meaning of a concept that plays an important, albeit often negative, role in modern society: race. We argue that a clear understanding of the nature of human genetic variation indicates that there are real genetic differences between human populations, but that these do not map onto the common understanding of race.

How Humans Are Different from Other Apes

Sequencing the genomes of humans and other mammals provides new information about genetic differences between modern humans and other primates. In the past, humans were distinguished from other species mainly on the basis of morphology, and fossils were our only source of information about human evolutionary change. But today, molecular genetics provides us with another source of information about our evolutionary history. In this section, we summarize what is known about how modern humans differ genetically from our closest primate relatives.

For many years, our knowledge of the genome was indirect: We could study the protein products of genes, and we could use specialized molecular techniques to identify particular DNA segments called genetic markers. Lately, however, it has become possible to sequence large chunks of the genome. In 2002, a first cut at sequencing the entire human genome was announced with great fanfare. By 2013, the cost of sequencing had fallen by a factor of 100,000, and thousands of complete genomes had been sequenced, with various degrees of accuracy, including sequences from 2,504 people from 26 populations from around the world. More quietly, geneticists have worked on sequencing the genomes of dozens of other organisms, including yeast, fruit flies, mice, cats, dogs, and rhesus macaques. Of particular interest to us are the genomes of our closest primate relatives: chimpanzees, bonobos, and gorillas. These vast troves of genetic information, all published since 2005, give us new insights about the kinds of evolutionary changes that have occurred in the human lineage.

Human and chimpanzee genomes are very similar.

By aligning the human, chimpanzee, bonobo, and gorilla genomes and comparing the sequences nucleotide by nucleotide, geneticists have been able to measure the magnitude of the genetic differences between us and them. The Chimpanzee Sequencing and Analysis Consortium found that humans and chimpanzees differ by about 1.2%. This means that in 1.2% of the nucleotides, all chimpanzees have one nucleotide and all humans have a different nucleotide. A group of researchers led by Svante Pääbo of the Max Planck Institute for Evolutionary Anthropology in Leipzig, Germany, reported that the difference between humans and bonobos is about 1.3%, and a consortium led by Richard Durbin of the Wellcome Trust found that the difference between humans and gorillas is 1.75%. These percentages sound like minuscule differences, but remember that there are 3 billion bases in the human genome. A 1% difference represents differences in about 30 million nucleotides. There have also been approximately 5 million insertions and deletions of bits of DNA in or out of the human genome or the chimpanzee genome. Most of these insertions and deletions involve a few nucleotides, and most involve repetitive sequences or **transposable elements**, copies of DNA segments from one part of a genome that have been inserted somewhere else. Insertions and deletions contribute roughly another 3% to the overall difference between the genomes of humans and the great apes.

Most protein-coding genes differ between humans and chimpanzees.

Many people are puzzled by the small genetic difference between humans and chimpanzees. How could such a puny genetic difference produce the sizable phenotypic differences between humans and chimpanzees? The answer is that small differences in the sequence of nucleotides can lead to big differences in the phenotype because the percentage of DNA that differs is not the same as the percentage of genes that differ. To understand why this is true, consider two extreme possibilities. If DNA differences were distributed evenly across all the genes in the genome, then even a 1% difference between two species would cause every gene in the two species to differ. On the other hand, if the differences were clustered in certain parts of the genome, then a smaller proportion of the genes would differ. So clearly it is important to know something about the pattern of differences between chimpanzees and humans as well as the overall magnitude of differences.

This is where we benefit from having sequenced the complete genomes of humans and other ape species. Protein-coding genes (DNA sequences that code for proteins) can be identified in a DNA sequence by the "start" and "stop" codons that mark the beginning and end of each coding sequence. Using this approach, geneticists identified 13,454 homologous protein-coding genes in chimpanzees and in humans. Only 29% of these genes have the same amino acid sequences. Among those that differ, the median number of base substitutions is two. Thus, even though the DNA sequences of chimpanzees and humans differ by only a small percentage, 71% of the proteins produced by their genes differ.

Only a small fraction of protein-coding genes shows evidence of selection since the divergence of human and chimpanzee/bonobo lineages.

It seems plausible that many of the differences in morphology and behavior between humans and chimpanzees are the result of natural selection, which favored particular traits in each lineage. However, mutation and genetic drift could also create differences between the DNA sequences of humans and chimpanzees. Are the differences in the protein-coding genes of humans and chimpanzees due to natural selection or to nonadaptive processes such as mutation and genetic drift? To try to answer this question, geneticists make use of the fact that the DNA code is redundant (see Chapter 2). This redundancy means that some nucleotide substitutions (**synonymous substitutions**) do not produce any change in the amino acid sequence of the protein that results from the gene. By contrast, **nonsynonymous substitutions** alter the amino acid sequence of a protein. Directional selection typically favors a particular protein that produces a particular phenotype. Thus, structural genes that have been subjected to selection are expected to show fewer nonsynonymous substitutions than synonymous ones, whereas nonadaptive processes such as genetic drift are expected to affect synonymous and nonsynonymous substitutions to the same extent.

Once such **positively selected** genes have been identified, we want to determine when the change occurred. To do this, geneticists compare the sequences of humans and chimpanzees with that of a more distantly related species, called the out-group. Some studies use the mouse as an out-group, whereas others use the macaque. When the out-group and the chimpanzee are the same and humans differ, the out-group and chimpanzees probably share the ancestral DNA sequence and humans probably have the derived one. Similarly, when the out-group and humans are the same and chimpanzees differ, we assume that the chimpanzee sequence is derived.

Several research groups have used this technique to determine which of the protein-coding genes that differ between humans and chimpanzees have been subject to positive selection in one of the two lineages. Although results of these studies vary in detail, they all conclude that the percentage of positively selected genes is quite small. For example, the consortium of scientists responsible for sequencing the chimpanzee genome estimated that about 4.4% of the genes showed signs of positive selection. A more recent study by another research group places the estimate at 2.7%. These estimates represent upper bounds on the actual percentage of positively selected genes because a substantial number of cases would be expected to occur by chance alone. Thus, even though most coding sequences differ, only a small fraction of these differences seem to be functional. Again, this result is surprising given the magnitude of the phenotypic differences between humans and chimpanzees.

How can the apparent lack of genetic change be reconciled with the substantial amount of phenotypic change? One possibility is that measuring selection by comparing synonymous and nonsynonymous changes underestimates the amount of change due to selection. This method incorporates the assumption that the amount of evolution of a protein-coding gene is proportional to the number of nonsynonymous DNA bases that differ between two species. However, sometimes the change of one or two base pairs in a DNA sequence can strongly affect phenotype. For example, we will see later in this chapter that the FOXP2 gene in humans has a major impact on speech, even though the human and chimpanzee versions differ by only two substitutions. Such small sequence changes are not usually detected by counting synonymous and nonsynonymous substitutions (and the FOXP2 gene was not identified as a positively selected gene).

Second, many of the big differences in phenotype may involve traits that are affected by genes at many loci. We will see later in this chapter that human height is affected by genes at more than 3,000 loci, each with a small effect, and that differences in height between human populations are due to small changes in frequency at many loci. This means selection generating these changes at any particular locus is very weak and leaves little trace in the structure of the genome. Many of the important differences between humans and chimpanzees may be the result of selection at many loci and will be difficult to detect by these kinds of methods.

A final possibility is that most of the evolutionary changes are not the result of changes in protein-coding genes but changes in regulatory genes. Unfortunately, we cannot yet identify regulatory genes from DNA sequence data alone, so geneticists can't study the changes in regulatory genes in the same way that they study changes in structural genes. However, evidence from studies of gene expression suggests that regulatory changes may play a big role in shaping the differences between humans and other apes. Scientists at the Partner Institute for Computational Biology, established by the Chinese Academy of Sciences and the Max Planck Society, studied 184 genes that are expressed at the same time in the prefrontal cortex, the part of the brain involved in reasoning and decision making. As shown in **Figure 14.1**, these genes are expressed at earlier ages in chimpanzees and macaques than in humans.

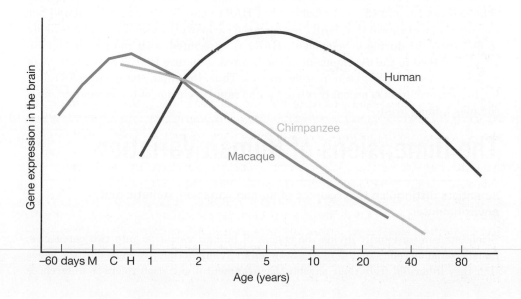

FIGURE 14.1

The level of expression of 184 genes in the prefrontal cortex at different ages for macaques, chimpanzees, and humans. Genes that are rarely expressed in adult macaques or chimpanzees are expressed in adult humans. M, C, and H = age of birth of individuals belonging to the three species; −60 days = time of conception.

This pattern of delayed maturation, called **neoteny**, is seen in many aspects of human development. It means that humans are, in a sense, apes who retain juvenile characteristics into adulthood. For brains, this means we retain neural plasticity longer, which is consistent with our greater behavioral plasticity. Such a coordinated shift in the expression of many genes is what would be expected to result from a shift in a regulatory cascade (see Chapter 2).

Noncoding sequences show evidence of positive selection since the human–chimpanzee split.

A team led by Katherine Pollard of the University of California, San Francisco, used a different method to identify regions that have experienced significant positive selection after human and chimpanzee lineages diverged. They searched the genomes of the mouse, rat, and chimpanzee to find DNA sequences that were at least 100 base pairs long and were at least 96% identical in all three taxa. Because the two rodent species are separated from chimpanzees by about 70 million years of independent evolution, Pollard and her colleagues reasoned that these sequences must be subject to strong **negative selection**—selection that favored the observed stable sequence over mutants that arose during these millions of years. They found about 35,000 negatively selected sequences. Then, for each of these regions, they compared the rate of change in the human lineage with the average rates of change in 12 other vertebrate species (not including the chimpanzee, rat, or mouse). The rates of change were significantly greater in the human lineage than in other lineages in 202 of these regions. The investigators ranked the regions by the rate of change and then assigned each region a label on the basis of the ranking. Thus, HAR1 was the fastest of the fast, and HAR202 was the slowest of the fast; *HAR* stands for "highly accelerated region." Almost all of these HAR segments are in noncoding regions. Even though these segments do not encode the structure of proteins, and for the most part, we don't know what they do, we do know that they have evolved very rapidly during human evolution. This suggests they have been shaped by natural selection, not genetic drift.

The fastest-changing region, HAR1, provides one example of how this might work. HAR1 is a 118-base-pair sequence on chromosome 20. In other vertebrates, HAR1 is extremely conservative, showing only two base pair changes between the chicken and the chimpanzee. If this rate of change had been continued in the human lineage, there would be only a 25% chance of having even one difference between the sequence in chimpanzees and that in modern humans. However, there have been 18 base pair changes, nearly an 80-fold increase in the rate of evolution. The HAR1 segment codes for a long noncoding RNA (lncRNA) molecule that folds itself into a stable structure. These kinds of RNA molecules often work with proteins to regulate gene expression, and this is probably what HAR1 does. Studies by Pollard and her co-workers have shown that HAR1 is expressed exclusively in the brain and is likely to be expressed during development. HAR1 is associated with the protein reelin, which is linked to the development of the layered structure that is characteristic of human brains but is not seen in other species. Thus, the rapid change in HAR1 during human evolution is probably related to the rapid evolution of the larger and more complex human brain.

The Dimensions of Human Variation

Scientists distinguish two sources of human variation: genetic and environmental.

Scientists conventionally divide the causes of human variation into two categories. **Genetic variation** refers to differences between individuals that are caused by the genes that they inherited from their parents. Environmental variation refers to differences between individuals caused by environmental factors affecting the organisms' phenotypes. Environmental factors include things like climate, habitat, and the presence of competing species, as well as things like the way that individuals were raised by their parents or birth order. For humans, culture is an important source of environmental variation.

A practical example—variation in body weight—will clarify this distinction. Many environmental factors affect body weight. Some factors, such as the availability of food, have an obvious and direct impact on body weight. Most people living under siege in Sarajevo in the mid-1990s were leaner than they were a decade earlier, when Sarajevo was a rich, cosmopolitan city. Other environmental effects are more subtle. For example, culture can affect body weight because it influences our ideas about what constitutes an appropriate diet and shapes our standards of physical beauty. In the United States, many young women adopt strict diets and rigorous exercise regimens to maintain a slim figure because thinness is considered desirable. But in several West African societies, young women are secluded and force-fed large meals several times a day for the express purpose of causing them to gain weight and become fat. In these societies, obesity is extremely desirable, and fat women are thought to be very beautiful. Body weight also appears to have an important genetic component. Recent research has shown that individuals with some genotypes are predisposed to be heavier than others, even when diet and levels of activity are controlled.

Genetic and environmental causes of variation may also interact in complicated ways. Consider two people who have inherited quite different genes affecting body weight. One is easily sated, whereas the other craves food constantly. Both individuals may be thin if they have to subsist on one cup of porridge a day, but only the one who craves food will gain weight when Big Macs and fries are readily available.

The relative importance of genetic and environmental influences for particular phenotypic traits is hard to determine.

It is often difficult to separate the genetic and environmental causes of human variation in real situations. The problem is that both genetic transmission and shared environments cause parents and offspring to be similar. Suppose we were to measure the weights of parents and offspring in a series of families living in a range of environments. The weight of parents and offspring (corrected for age) would probably be closely related. However, we would not know whether the association was due to the fact that they share genes or live in the same environment. Children might resemble their parents because they inherited genes that affect fat metabolism or because they learned eating habits and acquired food preferences from their parents.

Quite different processes create and maintain genetic and environmental variation among groups, and identifying the source of human differences will help us understand why people are the way they are. Genetic variation is governed by the processes of organic evolution: mutation, drift, recombination, and selection. Biologists and anthropologists know a great deal about how the various processes work to shape the living world and how evolutionary processes explain genetic differences among contemporary humans in particular cases.

Variation within human groups is distinct from variation among human groups.

Variation within groups refers to differences between individuals within a given group of people. In the Women's National Basketball Association, for example, Becky Hammon, 1.7 m (5 ft. 6 in.), competes alongside much taller players, such as Margo Dydek, 2.2 m (7 ft. 2 in.; **Figure 14.2**). **Variation among groups** refers to differences

FIGURE 14.2

Variation in stature within the Women's National Basketball Association is illustrated by the difference in height between Margo Dydek (2.2 m, or 7 ft. 2 in.) and Becky Hammon (1.7 m, or 5 ft. 6 in.).

FIGURE 14.3

Variation in stature between Olympic volleyball players and Olympic gymnasts is illustrated here by the difference in height between (*top*) Yuan Xinyue, 1.93 m (6 ft. 7 in.), and (*bottom*) Simone Biles, 1.4 m (4 ft. 8 in.).

between entire groups of people. For instance, as seen in **Figure 14.3**, the average height of Olympic volleyball players (exemplified by Yuan Xinyue, 1.93 m [6 ft. 7 in.]) is much greater than the average height of Olympic gymnasts (exemplified by Simone Biles, 1.4 m [4 ft. 8 in.]). Distinguishing these two levels of variation is important because, as we will see, the causes of the variation within groups can be very different from the causes of variation among groups.

Variation in Traits Influenced by Single Genes

By establishing the connection between particular DNA sequences and specific traits, scientists have shown that variation in some traits is genetic.

Although establishing the source of variation in human traits is often hard, sometimes we can be certain that variation arises from genetic differences between individuals. For example, recall from Chapter 2 that many people in West Africa suffer from sickle-cell anemia, a disease that causes their red blood cells to have a sickle shape instead of the more typical rounded shape. People with this debilitating disease are homozygous for a gene that codes for one variant of hemoglobin, the protein that transports oxygen molecules in red blood cells. Hemoglobin is made up of two protein subunits, labeled α (the Greek letter alpha) and β (the Greek letter beta). The DNA sequence of the most common hemoglobin allele, hemoglobin A, specifies the amino acid glutamic acid in the sixth position of the protein chain of the β subunit. But there is another hemoglobin allele, hemoglobin S, that specifies the amino acid valine at this position. People who suffer from sickle-cell anemia are homozygous for the hemoglobin S allele.

We can prove that traits are controlled by genes at a single genetic locus by showing that their patterns of inheritance conform to Mendel's principles.

We can sometimes distinguish between genetic and environmental sources of variation when traits are affected by genes at a single genetic locus. In such cases, Mendel's laws make detailed predictions about the patterns of inheritance (see Chapter 2). If scientists suspect that a trait is controlled by genes at a single genetic locus, they can test this idea by collecting data on the occurrence of the trait in families. If the pattern of inheritance shows a close fit to the pattern predicted by Mendel's laws then we can be confident that the trait is affected by a single genetic locus.

Research on the genetic basis of a language disorder called **specific language impairment (SLI)** illustrates the strengths and weaknesses of this approach. Children with SLI have difficulty learning to speak, and sometimes they have small vocabularies and make frequent grammatical errors as adults. SLI runs in families, but the genetic basis of the condition is usually unclear.

The pattern of inheritance of SLI in one family suggests that at least some cases of SLI are caused by a dominant allele at a single genetic locus. A group of researchers at the Wellcome Trust Centre for Human Genetics in Oxford, England, studied the expression of SLI in three generations of one family (known as the KE family; **Figure 14.4**). The members of this family who suffer from SLI have severe problems learning grammatical rules; they also have difficulty with fine-motor control of the tongue and jaws. The grandmother (shown as a blue circle at the top of the figure) had SLI, but her husband (shown as an orange triangle) did not. Four of her five children and 11 of her 24 grandchildren also had SLI. Suppose that SLI is caused by a dominant gene. Then because SLI is rare in the population as a whole, the Hardy–Weinberg equations tell us that almost all SLI sufferers will be heterozygotes. Of course, anyone

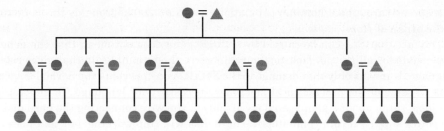

FIGURE 14.4

The pattern of specific language impairment (SLI) in the KE family tree suggests that some cases of SLI are caused by a single dominant gene. Circles represent women, triangles represent men, and blue represents people with SLI. If SLI is caused by a dominant gene, then, because SLI is rare in the population as a whole, we know from the Hardy–Weinberg equations that almost all people with SLI will be heterozygotes. Thus Mendel's laws tell us that, on average, half of the offspring of a mating between a person with SLI and a person without it will have SLI, and half of the offspring will have normal linguistic skills. Notice how well the family shown in this tree fits this prediction.

without SLI must be a homozygote for the normal allele at this locus. From Mendel's laws, on average half of the offspring of a mating between a person with SLI and one without the disorder will have SLI and half will have normal linguistic skills. The KE family fits this prediction very well. Both children of the son without SLI are normal, and the rest of the matings produced approximately equal numbers of normal and language-impaired children.

Although the pattern in the KE family is consistent with the idea that SLI is caused by a single dominant gene, it is possible that an environmental factor causes SLI to run in families and that the observed pattern arose by chance. Scientists search for two kinds of data to clinch the case. First, they collect data on more families. The more families that fit the pattern associated with the inheritance of a single-locus dominant gene, the more confident researchers can be that this pattern did not occur by chance. Second, researchers search for genetic markers (genes whose location in the genome is known) that show the same pattern of inheritance. Thus if every individual who has SLI also has a specific marker on a particular chromosome, we can be confident that the gene that causes SLI lies close to that genetic marker.

In 1998, the Wellcome Trust researchers demonstrated that SLI in the KE family is closely linked to a genetic marker on chromosome 7, and so SLI in this family is probably controlled by a gene closely linked to this marker. The later discovery of an unrelated person with the same symptoms allowed the researchers to identify the specific gene that causes the disorder in the KE family. The same allele of this gene, named FOXP2, was found in all affected members of the KE family and not in 364 unrelated people without SLI. It differs from the normal allele by a single nucleotide substitution.

Notably, molecular evidence suggests that the FOXP2 gene has undergone strong directional selection since the divergence of humans and chimpanzees. Different alleles of the FOXP2 gene are found in many animal species. The versions found in mice and humans differ by three amino acid substitutions, and the version found in chimpanzees differs from mice by only one of these substitutions. The last common ancestor of humans and mice lived more than 65 Ma, and the last common ancestor of humans and chimps lived only 6 Ma, which suggests that there has been rapid change in the FOXP2 gene during human evolution. Analyses of Neanderthal genomes show that the FOXP2 gene yields the same amino acid sequence as it does in modern humans. However, differences between Neanderthals and modern humans in one of the introns are related to the expression of FOXP2. Studies of variation in introns of the FOXP2 gene and surrounding noncoding sequences suggest that the normal version of the gene spread throughout hominin populations less than 200 ka, roughly the same time as the origin of *Homo sapiens*.

The fact that SLI can be caused by a single gene does not mean that the gene is responsible for all the psychological machinery in the human brain that gives rise to language. It means only that damage to the FOXP2 gene prevents the normal development of some of the psychological machinery necessary for language. To understand this idea, think about a simple analogy. If you cut the wire connecting the hard disk to the power supply in your computer, the hard disk will stop working, but that does not mean the wire contains all of the machinery necessary for operation of the hard disk. This argument is supported by the fact that the FOXP2 gene codes for a transcription factor belonging to a family of genes that play an important role in regulating gene expression during development. FOXP2 itself is strongly expressed in the brains of developing fetuses. By the same reasoning, SLI in other families may be caused by other genes whose expression is necessary for normal brain development, and geneticists later discovered three other genes that lead to SLI. Just as there are many ways to wreck your hard disk, there are probably many mutants at many loci that damage the parts of the brain necessary for language.

Causes of Genetic Variation within Groups

Mutation can maintain deleterious genes in populations, but only at a low frequency.

Many diseases are caused by recessive genes. For example, only people who are homozygous for hemoglobin S are afflicted with sickle-cell anemia. Other diseases caused by recessive alleles include phenylketonuria (PKU), Tay–Sachs disease, and cystic fibrosis. All these diseases are caused by mutant genes that code for proteins that do not serve their normal function, and all produce severe symptoms and sometimes death. Why haven't such deleterious genes been eliminated by natural selection?

One answer to this question is that natural selection steadily removes such genes, but mutation constantly reintroduces them. Very low rates of mutation can maintain recessive deleterious genes because most individuals who carry the gene are heterozygotes and do not suffer the consequences of having two copies of the deleterious gene. The observed frequency of many deleterious recessive genes is about 1 in 1,000. According to the Hardy–Weinberg equations, the frequency of newborns homozygous for the recessive allele will be $0.001 \times 0.001 = 0.000001$! Thus, only 1 in 1 million babies will carry the disease. This means that even if the disease is fatal, selection will remove only two copies of the deleterious gene for every 1 million people born. Because mutation rates for such deleterious genes are estimated to be a few mutations per million gametes produced, mutation will introduce enough new mutants to maintain a constant frequency of the gene. When this is true, we say that there is **selection–mutation balance**.

Selection can maintain variation within populations if heterozygotes have higher fitness than either of the two homozygotes.

Some lethal genes are too common to be the result of selection–mutation balance. In West African populations, for example, the frequency of the hemoglobin S allele is typically about 1 in 10. How can we account for this? The answer for hemoglobin S is that this allele increases the fitness of heterozygotes. It turns out that individuals who carry one copy of the sickling allele, *S*, and one copy of the normal allele, *A*, are partially protected against the most dangerous form of malaria, called **falciparum malaria** (**Figure 14.5**). Consequently, when falciparum malaria is prevalent, heterozygous *AS* newborns are about 15% more likely to reach adulthood than are *AA* infants.

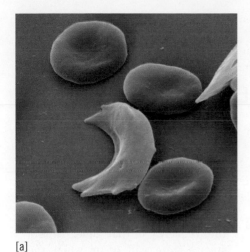

[a]

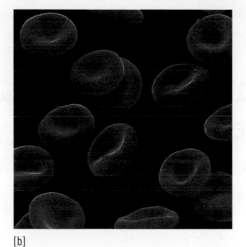

[b]

FIGURE 14.5

(a) Sufferers of sickle-cell anemia have abnormal red blood cells with a sickle shape. (b) Normal red blood cells are round. The sickling allele partially protects against falciparum malaria.

When heterozygotes have a higher fitness than either homozygote, natural selection maintains a **balanced polymorphism**, a steady state in which both alleles persist in the population. To see why balanced polymorphisms exist, consider what happens when the S allele is first introduced into a population and is very rare. Suppose that its initial frequency is 0.001. The frequency of SS individuals will be 0.001×0.001, or about 1 in 1 million, and the frequency of AS individuals will be $2 \times 0.001 \times 0.999$, or about 2 in 1,000. This means that for every individual who suffers the debilitating effects of sickle-cell anemia, there will be about 2,000 heterozygotes who are partially immune to malaria. Thus, when the S allele is rare, most S alleles will occur in heterozygotes, and the S allele will increase in frequency. However, this trend will not lead to the elimination of the A allele. To see why, let's consider what happens when the S allele is common and the A allele is rare. Now almost all the A alleles will occur in AS heterozygotes, making those individuals partially resistant to malaria. But almost all the S alleles will occur in SS homozygotes, and those individuals will suffer debilitating anemia. The A allele has higher fitness than the S allele when the S allele is common (**A Closer Look 14.1**). The balance between these two processes depends on the fitness advantage of the heterozygotes and the disadvantage of the homozygotes. Here, the equilibrium frequency for the hemoglobin S allele is about 0.1, approximately the frequency actually observed in West Africa.

Scientists suspect that the relatively high frequencies of genes that cause several other genetic diseases may also be the result of heterozygote advantage. The gene that causes Tay–Sachs disease has a frequency as high as 0.05 in some eastern European Jewish populations. Children who are homozygous for this gene seem normal for about the first 6 months of life. Over the next few years a gradual deterioration takes place, leading to blindness, convulsions, and finally death, usually by age 4. However, there is some evidence that individuals who are heterozygous for the Tay–Sachs allele are partially resistant to tuberculosis. Jared Diamond of the University of California, Los Angeles, points out that tuberculosis was much more prevalent in cities than in rural areas of Europe over the past 400 years. Confined to the crowded urban ghettos of eastern Europe, Jews may have benefited more from increased resistance to tuberculosis than did other Europeans, most of whom lived in rural settings.

Variation may exist because environments have recently changed and genes that were previously beneficial have not yet been eliminated.

Some genetic diseases may be common because the symptoms they create have not always been deleterious. One form of diabetes, **non-insulin-dependent diabetes (NIDD)**,

14.1 Calculating Gene Frequencies for a Balanced Polymorphism

It is easy to calculate the frequency of hemoglobin S when selection has reached a stable, balanced polymorphism. Suppose the fitness of AA homozygotes is 1.0, the fitness of AS heterozygotes is 1.15, and the fitness of SS homozygotes is 0, and let p be the equilibrium frequency of the allele S. If individuals mate at random, a fraction p of the S alleles will unite with another S allele to form an SS homozygote, and a fraction $1 - p$ will unite with an A allele to form an AS heterozygote.

Thus the average fitness of the S allele will be

$$0p + 1.15(1 - p)$$

By the same reasoning, the average fitness of the A allele will be

$$1.15p + 1(1 - p)$$

The relationship between the average fitness of each allele and the frequency of hemoglobin S, shown in **Figure 14.6**, confirms the reasoning given in the text. When the S allele is common so that p is close to 1, the average fitness of the S allele is close to 0; but when S is rare, its

average fitness is almost 1.15. If one gene has a higher fitness than the other, natural selection will increase the frequency of that gene. Thus a steady state will occur when the average fitnesses of the two alleles are equal; that is, when

$$1.15(1 - p) = 1.15p + (1 - p)$$

If you solve for p, you will find that

$$p = \frac{(1.15 - 1)}{(1.15 + 1.15 - 1)} = \frac{0.15}{1.3} \approx 0.1$$

which is about the observed frequency of the sickle-cell allele in West Africa.

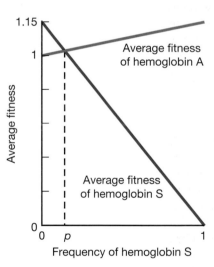

FIGURE 14.6

The average fitness of the S allele of hemoglobin S decreases as the frequency of S increases because more and more S alleles are found in SS homozygotes. Similarly, the average fitness of the A allele of hemoglobin A increases as the frequency of S increases because more and more A alleles are found in AS heterozygotes. A balanced polymorphism occurs when the average fitness of the two alleles is equal.

also called type 2 diabetes, may be an example of such a disease. **Insulin** is a protein that controls the uptake of blood sugar by cells. In NIDD sufferers, blood-sugar levels rise above normal levels because the cells of the body do not respond properly to insulin in the blood. High blood-sugar levels cause several problems, including heart disease, kidney damage, and impaired vision. NIDD also has a genetic basis. (The other form of diabetes, insulin-dependent diabetes, occurs because the insulin-producing cells in the pancreas have been destroyed by the body's own immune system. It is unlikely that insulin-dependent diabetes was ever adaptive.)

In some contemporary populations, the occurrence of NIDD is very high. For example, according to the International Diabetes Federation figures for 2015, on the Micronesian island of Nauru one in four adults between the ages of 20 and 79 has the disease. This is nearly three times higher than the world average. High rates of NIDD are a recent phenomenon, although the genes that cause the disease are not new. The late human geneticist James V. Neel suggested that the genes now leading to NIDD were beneficial in the past because they caused a rapid buildup of fat reserves during periods of plenty—fat reserves that would help people survive periods of famine in harsh environments. Traditionally, life on Nauru was very difficult. The inhabitants subsisted by fishing and farming. The islands of the Pacific are strongly affected by typhoons and volcanic activity; famine was common. NIDD was virtually unknown during this period. However, Nauru was colonized by Britain, Australia, and New Zealand in more recent times, and these influxes brought many changes in the residents' lives. They obtained access to Western food, and prosperity derived from the island's phosphate deposits allowed them to adopt a sedentary lifestyle. NIDD became very common in Nauru and other islands in the region. While there are other suggested explanations for the increase in NIDD, it is quite plausible that genes that formerly conferred an advantage for the residents of these small islands now lead to NIDD.

Causes of Genetic Variation among Groups

There are many genetic differences between groups of people living in different parts of the world. The existence of genetic variation among groups is intriguing because we know that all living people are members of a single species, and as we saw in Chapter 4, gene flow between populations within a single species tends to make them genetically uniform. In this section we consider several processes that oppose the homogenizing effects of such gene flow, thereby creating and maintaining genetic variation among human populations.

Selection that favors different genes in different environments creates and maintains variation among groups.

The human species inhabits a wider range of environments than any other mammal. We know that natural selection in different environments may favor different genes and that natural selection can maintain genetic differences in the face of the homogenizing influence of gene flow if selection is strong enough. Variation in the distribution of hemoglobin genes provides a good example of this process. Hemoglobin S is most common in tropical Africa, around the Mediterranean Sea, and in southern India (**Figure 14.7a**). Elsewhere it is almost unknown. Generally, hemoglobin S is prevalent where falciparum malaria is common, and hemoglobin A is prevalent where this form of malaria is absent (**Figure 14.7b**). Southeast Asia represents an exception to this pattern, and it is possible that hemoglobin E, a variant that is common in that region, also provides resistance to malaria.

The digestion of **lactose**, a sugar found in mammalian milk, provides another interesting example of genetic variation maintained by natural selection. Lactose is synthesized in the mammary glands and occurs in large amounts only in mammalian milk. Most mammals can digest lactose as infants but lose this ability after they are weaned. About two-thirds of humans follow the mammalian pattern, and synthesis of the necessary enzyme, **lactase-phlorizin hydrolase (LPH)**, ends after weaning. Such people are said to lack **lactase persistence**. When people who lack lactase persistence drink substantial amounts of fresh milk at one sitting, they suffer gastric distress that ranges from mild discomfort to severe pain. However, some people retain the ability

FIGURE 14.7

Hemoglobin S is common only in areas of the world where falciparum malaria is prevalent. (a) The colors show the frequency of hemoglobin S throughout the world. (b) The regions of Africa and Eurasia in which falciparum malaria is prevalent are in red.

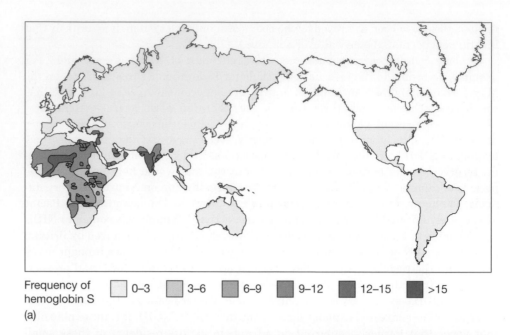

Frequency of hemoglobin S

| 0–3 | 3–6 | 6–9 | 9–12 | 12–15 | >15 |

(a)

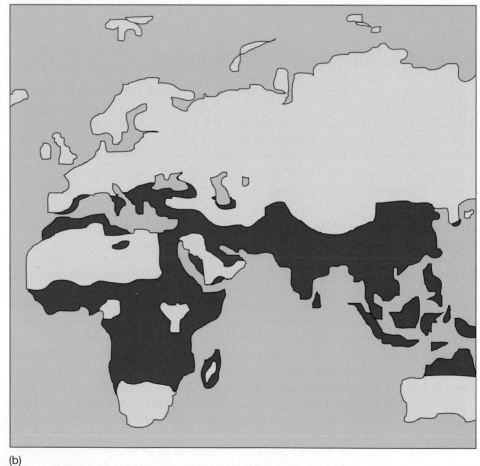

(b)

to digest lactose as adults and are said to have lactase persistence (**Figure 14.8**). The frequency of lactase persistence varies greatly among populations, ranging from 5% to almost 100%, with the highest frequencies found in people of northern European descent and in some populations from West Africa, East Africa, and the Middle East.

Molecular studies indicate that lactase persistence evolved independently in Africa and in Europe. In 2002, Nabil Enattah and a group of researchers from the

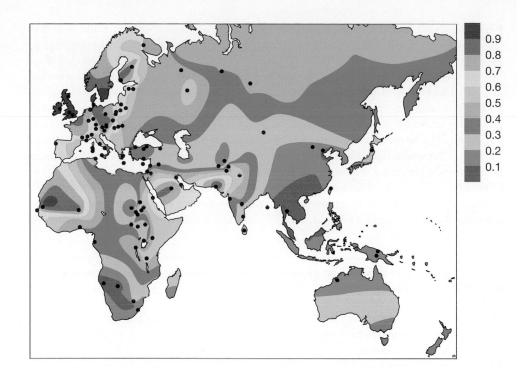

FIGURE 14.8

The distribution of lactase persistence in Africa and Eurasia interpolated from data given by the black dots. The numbers represent the fraction of the population who are lactase persistent.

University of Helsinki in Finland showed that the mutation of a single nucleotide from C to T in a noncoding region close to **LCT**, the structural gene that codes for LPH, is strongly associated with lactase persistence in Finnish populations. Later work showed that this mutant allele, labeled *T-13910*, is associated with lactase persistence elsewhere in northern Europe and southern Asia and that it regulates the expression of LCT. Thus, across all of Eurasia, lactase persistence is the product of a single gene change. The situation is different in East Africa, where four different mutations in areas of the genome close to *T-13910* have been found to be associated with lactase persistence. Lactase persistence is thus an example of convergent molecular evolution, in which different variants underlying the same phenotypic change arise in different populations. This parallel increase in the frequency of several different alleles in different populations suggests the same selective advantage (in this case the ability to absorb lactose) has led to these changes.

Additional research suggests that this common source of natural selection is the incorporation of milk in the diet. High frequencies of lactase persistence occur in populations with a long history of dairying, suggesting that the lactase-persistence gene evolved in response to this cultural practice. **Pastoralists** are people who keep livestock and do not farm. The ability to digest fresh milk is probably advantageous for pastoralists. This is clearly the case among desert pastoralists today. Gebhard Flatz of the Medizinische Hochschule Hannover in Germany studied the Beja, a people who wander with their herds of camels and goats in the desert lands between the Nile and the Red Sea (**Figure 14.9**). During the 9-month dry season, the Beja rely almost entirely on milk from their camels and goats. They drink about 3 liters of fresh milk a day, and they obtain virtually all their energy, protein, and water from milk.

The integration of archaeological and ancient DNA data provides further insight into the evolution of lactase persistence, particularly in Europe. Archaeological sites rich in the bones of domesticated cows, sheep, and goats indicate that pastoralism reached northern Europe about 6 ka. Pottery that dates to this

FIGURE 14.9

Pastoralists in northern Africa often herd camels. During part of the year, they obtain virtually all their nourishment from fresh milk, so many can digest lactose as adults.

Variation in Traits Influenced by Single Genes ■ **377**

period contains minute residues of chemical compounds found only in milk, suggesting that milk production played a significant role in these early pastoral economies. However, analyses of ancient DNA from people of this area at this time show that they did not carry the *T-13910* Eurasian lactase-persistence allele, although it is possible that it was present at very low frequencies and was not sampled. The *T-13910* allele was not present in appreciable frequency in Europe until 4,000 years ago and did not really become common until the Middle Ages. This delay suggests that milk consumption may have initially been restricted to children who could still digest lactose; limited to fermented products like cheese and yogurt that contain less lactose; or, if adults were consuming milk without having the *T-13910* allele, any side effects were simply tolerated. Indeed, experimental evidence indicates that many individuals who lack genetic adaptations to lactase persistence can avoid most of the negative consequences of drinking milk by drinking small amounts throughout the day and drinking milk along with food.

The sequencing of the human genome makes it possible to detect selection from DNA sequences.

Until recently, the only way to determine which human genes had been subject to selection was to guess which genetic loci might have been subject to natural selection and then to determine whether this was the case. For example, the relationship between the sickle-cell trait and malaria was initially detected by the correlation between the high frequency of the sickle-cell allele and the prevalence of falciparum malaria. As of 2006, about 90 genes have been shown to be subject to recent positive selection by using this "candidate gene" approach.

As we saw earlier, the availability of complete DNA sequences has allowed scientists to detect selection directly, without any information about the function of the gene or its prevalence in different populations. This is possible because positive selection leaves detectable patterns in the genome. Here we focus on one of these patterns, which is called a **selective sweep**. A selective sweep occurs when a beneficial mutation arises and then both the mutation and DNA linked to the mutation on the same chromosome spread through the population. This means selection leading to the spread of a favorable mutation can be detected by looking at regions of the genome in which identical long DNA sequences are common.

To understand why such sequences provide evidence for positive selection, assume for a moment that there is no crossing-over and, therefore, no recombination of genes carried on the same chromosome. Now suppose a favorable mutation arises in the population. Individuals carrying that mutation have higher fitness, and the mutation increases in frequency. If there were no recombination, all the DNA on the chromosome that contained the mutant would also spread, and this, in turn, would increase the frequency of all the alleles that have the good luck to be linked to the beneficial mutant. Eventually, the whole population would carry this particular sequence of genes. Of course, this does not really happen because crossing-over occurs and this shuffles the alleles at other loci. However, as we learned in Chapter 2, the rate at which this happens depends on how closely linked the loci are—the closer they are to the beneficial mutant, the less likely it is that recombination will separate the mutant from the allele that it was originally linked to. This means when a new favorable mutant initially spreads, it tends to be surrounded by a long chunk of DNA with the same sequence. As we saw in Chapter 13, sequences at adjacent loci on a chromosome that are inherited together are called haplotypes. Eventually, recombination breaks up the sequence, shortening the haplotype shared by the carriers of the beneficial allele, but this takes about 10,000 years. So genes that have been subject to recent selection are surrounded by a long haplotype, unlike alternative alleles that are not subject to selection. The Eurasian version of the lactase-persistence allele lies in the middle of a haplotype that is about 1 million base pairs long, suggesting that it has been subject to recent selection. Using this procedure,

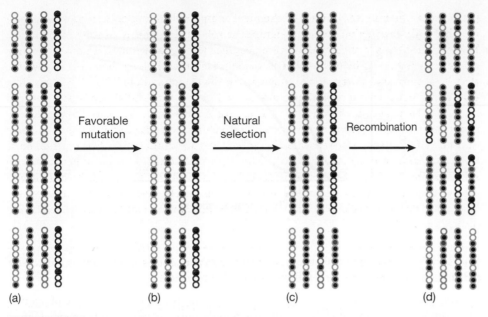

FIGURE 14.10

Why do selective sweeps lead to long haplotypes? (a) There are 16 chromosomes, each with 10 genetic loci. There are two alleles at each locus, black and white, and each has a frequency of 0.5. There are four kinds of chromosome, each with a different pattern of black and white alleles. These haplotypes are outlined in gray, blue, green, and purple. (b) A favorable mutation, colored red, arises at the third locus on one of the blue chromosomes. Notice that, by chance, this haplotype carries more black alleles than white. (c) Individuals carrying the mutation have higher fitness; thus chromosomes with the mutation increase in frequency. We assume that this happens so fast that there hasn't been any recombination. As a result, most of the chromosomes have the same sequence of alleles as the blue haplotype on which the mutation initially occurred. By searching DNA sequences for long common sequences of this type, geneticists can locate beneficial mutations that have recently spread. (d) Eventually, recombination shuffles genes between chromosomes, and, as a result, the blue haplotype is no longer common. However, notice that black alleles have become more common in the population because they have hitchhiked on the beneficial mutant. This means at all these loci there is one common allele, the black one, and a rare allele, the white one. We initially assumed that all the alleles were of equal frequency because they were not subject to selection. Such neutral alleles tend to occur with similar frequencies. Thus, by searching for chromosomal regions in which there is one very common allele at each locus, geneticists can find loci that have been subject to selection.

geneticists can identify sequences that have recently been subject to strong natural selection. **Figure 14.10** provides a simple example of this process.

Benjamin Voight and his colleagues at the University of Chicago used this technique to scan three human populations for signs of recent natural selection. They used DNA sequence data on 209 people: 89 from Tokyo and Beijing; 60 Yoruba-speaking individuals from Ibadan, Nigeria; and 60 people of northern and western European origin. These data, collected by a large consortium called the International HapMap Project, include about 800,000 polymorphic **single-nucleotide polymorphisms (SNPs)**. A SNP (pronounced "snip") is a location in the DNA sequence where individuals differ by a single base. In each population for each of these SNPs, researchers calculated the ratio of the length of the haplotype containing the more common allele and the length of the haplotype containing the less common allele. Large values of the index indicate a recent selective sweep. Using this approach, Voight and his colleagues identified 579 regions in which a sweep is likely to have occurred. Three-quarters of these sweeps occurred in only one of the three populations. This makes sense because most of the changes detected by this method occurred in the

FIGURE 14.11

If the ability to digest lactose as an adult leads to even as little as a 3% increase in fitness (*s*), then it is possible that the gene allowing the digestion of lactose has spread in the 7,000 years (300 to 350 generations) since the origin of dairying.

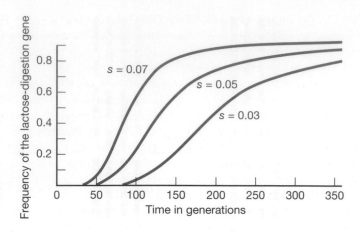

past 10,000 years, long after the populations of Asia, Africa, and Europe split. The coding genes in these regions fall into several categories:

- *Reproductive system.* This includes genes affecting the protein structure of sperm, sperm motility, gamete viability, and the female immune response to sperm. Genes in this category also show rapid evolution during the divergence of humans and chimpanzees and may reflect ongoing male–female conflict or selection for disease resistance.

- *Morphology.* Genes affecting skin color show evidence of strong selection among Europeans, and genes affecting bone development also show rapid evolution. This is consistent with the large amount of phenotypic variation seen in contemporary humans.

- *Digestion.* This includes genes affecting the metabolism of alcohol, carbohydrates, and fatty acids. These genetic changes may be due to changes in diet that followed the adoption of agriculture.

Many people think of evolution by natural selection as a glacially slow process that acts over millions of years. However, the length of the haplotypes surrounding these alleles suggests that they are only 6,000 to 9,000 years old. We don't need to imagine that selection has huge effects on fitness in order to account for such rapid evolution. **Figure 14.11** shows how fast the frequency of a new mutation would increase if it increased fitness by just 3%. As you can see, even this relatively small benefit could easily explain the spread of a new allele in less than 10,000 years.

Genetic drift creates variation among isolated populations.

In Chapter 3, we saw that genetic drift causes random changes in gene frequencies. This means if two populations become isolated, both will change randomly, and, over time, the two populations will become genetically distinct. Because drift occurs more rapidly in small populations than in large ones, small populations will diverge from one another faster than large ones will. Genetic drift caused by the expansion of a small founding population is sometimes called the **founder effect**.

Genetic differences among members of three religious communities in North America demonstrate how this process can create variation among human groups. Two of these groups are Anabaptist sects—the Old Order Amish (**Figure 14.12**) and Hutterites—and the third group is the Utah Mormons. Each of these groups forms a well-defined population. About 2,000 Mormons first arrived in the area of what is now Salt Lake City in 1847, and members of the church continued to arrive until 1890. Virtually all the immigrants were of northern European descent. At the turn of the twentieth century, there were about 250,000 people in this area, and about 70% of them belonged to the Mormon Church. In contrast, the two Anabaptist groups were

FIGURE 14.12

The Old Order Amish were founded by a group of 200 people. These Amish dress plainly and shun most forms of modern technology, including motor vehicles.

much smaller. The founding population of the Old Order Amish was only 200 people, and gene flow from outside the group has been very limited. Contemporary Hutterites are all descended from a population of only 443 people and, like the Amish, have been almost completely closed to immigration.

Researchers studying the genetic composition of each of these populations have found that Mormons are genetically similar to other European populations. Thus, even though Mormon populations have been partly isolated from other European populations for more than 150 years, genetic drift has led to very little change. This is just what we would expect, given the size of the Mormon population. In contrast, the two Anabaptist populations are quite distinct from other European populations. Because their founding populations were small and the communities were genetically isolated, drift has created substantial genetic changes in the same period.

Genetic drift can also explain why certain genetic diseases are common in some populations but not in others. For example, Afrikaners in what is now the Republic of South Africa are the descendants of Dutch immigrants who arrived in the seventeenth century. By chance, this small group of immigrants carried several rare genetic diseases, and these genes occurred at much higher frequency among members of the colonizing population than in the Dutch populations from which the immigrants were originally drawn. The Afrikaner population grew very rapidly and preserved these initially high frequencies, causing these genes to occur in higher frequencies among modern Afrikaners than in other populations. For example, sufferers of the genetic disease **porphyria variegata** develop a severe reaction to certain anesthetics. About 30,000 Afrikaners now carry the dominant gene that causes this disease, and every one of them is descended from a single couple who arrived from Holland in the 1680s.

Overall patterns of genetic variation mainly reflect the history of migration and population growth in the human species.

Much of the genetic variation among human groups reflects the history of the peoples of Earth. In Chapter 13 we explained that the geographic patterns of genetic variation indicate the human species underwent a population expansion in Africa about 130 ka and then throughout the world about 60 ka. There have been many population expansions since that time. The invention of agriculture led to expansions of farming peoples from the Middle East into Europe, from Southeast Asia into **Oceania** (the Pacific island groups of Polynesia, Melanesia, and Micronesia), and from west central Africa to most of the rest of the continent between 4,000 and 1,000 years ago. The domestication of the horse and associated military innovations led to several expansions of peoples living in the steppes of central Asia between 6,000 and 500 years ago, and improvements in ships, navigation, and military organization led to the expansion of European populations during the past 500 years.

As we saw in Chapter 13, worldwide patterns of genetic variation preserve a record of these expansions. As human populations expand, local populations become genetically isolated from one another and begin to accumulate genetic differences. Expanding local populations exchange genes with their neighbors and with other populations they encounter as they expand. This gene flow tends to blur the effects of the expansion. However, if there is not too much gene flow, the current patterns of genetic variation will reflect the pattern of past migrations.

This hypothesis is consistent with a variety of genetic data. For example, we have already seen how patterns of variation in Y chromosomes and mitochondrial DNA haplotypes can be used to infer the human expansions out of Africa. We have seen that genetic variation decreases along the likely path of human expansion from East Africa about 60 ka to the current location of contemporary populations (**Figure 14.13**). If contemporary patterns of human variation are due to the initial expansion of humans out of Africa, genetic distance should also be correlated with these distances, and that is exactly what the data show. Geographic distance is a very good predictor of genetic distance between human populations (**Figure 14.14**).

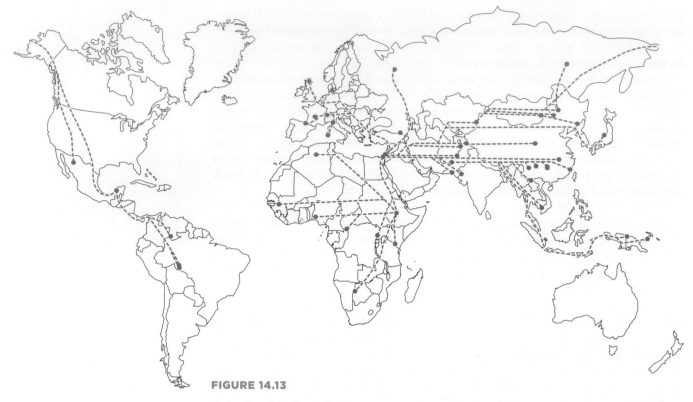

FIGURE 14.13

The location of the 51 populations in the Human Genome Diversity Project (*blue circles*) and the paths used to calculate distances in between these populations used on the x-axis in Figure 14.14 (and 13. 17).

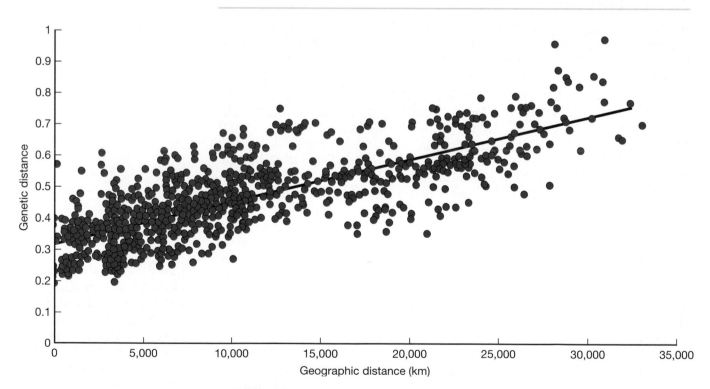

FIGURE 14.14

There are 51 populations in the Human Genome Diversity Project sample, so there are 1,275 pairs of populations. This graph shows the genetic distance between each pair plotted against the distance between the populations along the likely pathway of expansion of humans out of Africa. The two distances are strongly correlated, suggesting that the original expansion out of Africa accounts for much of the genetic variation among human populations.

Variation in Complex Phenotypic Traits

As we saw in Chapter 3, most human traits are influenced by many genes, each having a relatively small effect, and by environmental factors. Until recently it was impossible to detect the effects of single genes for such traits. Instead, geneticists developed statistical methods that enabled them to estimate the relative importance of genetic and environmental components of variation within groups. Because the relative importance of genetic variation and environmental variation will affect the resemblance between parents and offspring, the measure that computes the proportion of variation due to the effects of genes is referred to as the **heritability** of phenotypic traits. Technological advances are rapidly altering our understanding of the genetic basis of many traits.

This material will be easier to understand if we have a concrete example in mind. Height is an ideal trait: It is easy to measure, it varies within and between populations, it is relatively stable once individuals reach adulthood, and there is a wealth of data on height of individuals in different populations. In any moderately large sample of people, height will vary widely. For example, the data in **Figure 14.15** are taken from men who joined the British Army in 1939. Some of these young recruits were more than 2 m (around 7 ft.) tall; others were less than 1.5 m (5 ft.) tall.

Genetic Variation within Groups

Measuring the phenotypic similarities among relatives, particularly twins, can allow us to estimate the proportion of the variation within the population that is due to genes.

In Part One, we saw that the transmission of genes from parents to offspring causes children and parents to be phenotypically similar. If parents who are taller than average tend to have offspring who are taller than average, and parents who are shorter than average tend to have children who are shorter than average, you might think that variation in height is determined by variation in genes. The problem with this reasoning is that parents and offspring are not only more likely to share similar genes, they are also more likely to share similar environmental factors that could plausibly affect height, such as levels of nutrition and infectious diseases. Similarity between the environments of parents and their offspring is called **environmental covariation** and is a serious complication in computing heritability.

Data from studies of twins are particularly useful for separating the effects of genetic transmission from environmental covariation. The technique involves

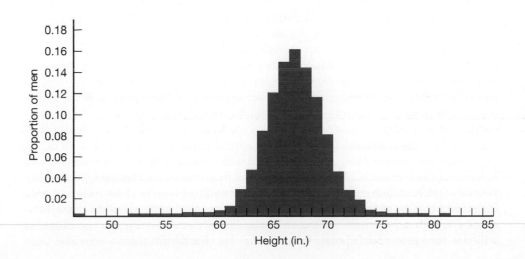

FIGURE 14.15

The heights of men joining the British Army in 1939 varied considerably, illustrating the range of variation in morphological characters within populations. The tallest men joining the army were more than 2 m (around 84 in., or 7 ft.) tall, and others were less than 1.5 m (60 in., or 5 ft.) tall.

comparing the similarity between monozygotic and dizygotic twins. Identical twins, or **monozygotic twins**, begin life when the union of a sperm and an egg produces a single zygote. Then, very early in development, this embryo divides to form two separate, genetically identical individuals. Fraternal twins, or **dizygotic twins**, begin life when two different eggs are fertilized by two different sperm to form two independent zygotes. Like other pairs of full siblings, they share approximately half of their genes, but unlike other full siblings, they are conceived at the same time and share the womb during gestation. To illustrate how twin studies work, imagine a large sample of twins, some identical and some fraternal, where each member of the twin pair grew up in the same family environment (that is, no cases where one twin was adopted and raised in a different family). Now for each twin pair we measure the height of twin 1 and twin 2. We then calculate the correlation of heights among twin pairs, essentially asking, "How well, on average, can we predict the height of this individual from the height of their twin?" Finally, we calculate the extent to which pairs of identical twins are more similar in height than are pairs of fraternal twins. Because each individual grows up in the same family as his or her twin, environmental covariance is less likely to affect the results. Differences in the level of similarity in pairs of identical twins versus pairs of fraternal twins is due to the greater genetic similarity of identical twins. These calculations performed on twins provide an estimate of the heritability of height in the population to which the twins belong.

Twin studies conducted in Europe and in America suggest that height is highly heritable, with about 80% of the variation due to genetic variation. Indeed, almost all human traits that have been studied in twins are at least partially heritable. A 2016 meta-analysis (a combination of data from multiple studies) involving more than 14 million twin pairs examined the heritability of more than 1,700 phenotypic traits, including morphological traits such as height, cognitive traits such as intelligence and personality, and social attitudes (for example, conservative vs. liberal). Across the full range of traits, the average value for heritability was 49%. Morphological traits tended to have higher heritability than traits such as social attitudes, but even for social attitudes about 30% of the variation was due to genetic variation.

Some critics are skeptical of these high heritability estimates, especially for cognitive and behavioral traits. Remember that twin studies rely on the "assumption of equal environments": that the extent to which monozygotic twins raised in the same family experience similar environments is no greater than the extent to which fraternal twins raised in the same family experience similar environments. But this assumption may be unfounded. For example, in the uterus, some monozygotic twins are more intimately associated than dizygotic twins are. And after birth, it is not uncommon to see monozygotic twins dressed in identical outfits or given rhyming names, and it is inevitable that their physical similarities to one another will be pointed out to them over and over (**Figure 14.16**).

Supporters of twin studies have responded to these criticisms by pointing to studies involving misclassified twins. A surprisingly large number of fraternal twins look so similar that they are mistaken for identical twins, and a surprisingly large number of identical twins look different enough that they are mistakenly considered to be fraternal twins. If the critics of twin studies are correct that the greater cognitive and behavioral similarity among pairs of identical twins than fraternal twins is due to identical twins being treated more alike by other people—rather than due to their having greater genetic similarity—then heritability should be lower when it is based on samples of misclassified identical twins than when based on samples of correctly classified twins. This is because misclassified identical twins will be treated more differently from one another than will correctly classified twins, and so their cognitive and behavioral similarity should be more like that of fraternal twins. However, heritability estimates derived from misclassified twins are actually as high as those from correctly classified twins. In addition, twins are not the only type of family members that can be studied to estimate heritability. For example, adoption studies examine the extent to which, among pairs of individuals raised in the same family, individuals resemble their

FIGURE 14.16

Identical, or monozygotic, twins are produced when an embryo splits at an early stage and produces two genetically identical individuals.

biological siblings more than their adopted siblings. Adoption studies are less common than twin studies but generally give similar heritability estimates as twin studies.

Heritability is a concept that is often misunderstood. There are two key points to remember about heritability.

First, low heritability does not mean that genes have little to do with the development or expression of a trait in an individual. For example, in all human populations, the vast majority of people have two eyes, and environmental rather than genetic causes are responsible for the few that don't (that is, some people lose an eye in an accident). Because genetic differences explain so little of the variation in eye number, heritability of eye number is close to zero. But it would be incorrect to conclude that genes have nothing to do with why most people have two eyes.

Second, high heritability does not mean that the phenotypic trait is determined by genes and can't be changed by the environment. Rather, high heritability only means that in this particular population, variation in environmental conditions doesn't explain much of the phenotypic differences among individuals; as a consequence, most of the phenotypic differences among individuals are explained by their genetic differences. In contemporary Western societies, the vast majority of people receive adequate nutrition, so most differences in height are explained by genetic differences, and heritability is high (about 80%). The heritability of height in Western societies would have been lower 200 years ago, because more of the variation in height would have been explained by differences in nutrition among individuals. This logic also applies to comparisons between contemporary societies: In areas of the world today where individuals vary more dramatically in nutrition, such as Africa and India, heritability of height is lower, at around 60%.

Genomewide association studies confirm that many genes, each with a small effect, affect height.

Drops in the cost of assessing an individual's genotype at many loci have allowed geneticists to conduct **genomewide association studies (GWAS)**. These studies make use of what are called SNP chips to examine the genetic basis of complex traits. Hundreds of thousands of short DNA segments are bound to each silicon SNP chip. Each of these segments matches the sequence of bases surrounding one allele of a known SNP. Then a sample of an individual's DNA is chemically chopped into small pieces and applied to the chip; the DNA bits bind to the matching segments on the chip. In this way, molecular biologists can assess an individual's genotype at half a million SNPs at the same time. Because the cost is much lower than whole-genome sequencing, researchers can afford to evaluate the genotypes of many individuals.

The basic idea behind GWAS is very simple. To determine which SNPs are associated with variation in height, researchers sample many individuals. They measure the height of each individual and assess his or her genotype for many SNPs. Then they determine which SNPs are more often found in tall people and which are more often found in short people. An important point to remember about GWAS is that the SNPs they assess do not actually directly affect height themselves. Instead, SNPs are associated with variation in height because they are on the same chromosome and in close proximity to genes that *do* play a causal role. This phenomenon of **linkage disequilibrium** means that, for example, if you have an "A" nucleotide at a particular SNP, you are also likely to have an allele variant at some nearby gene that, say, increases bone growth. However, the relationship between the SNP and the phenotype may also be more indirect than this. For example, another SNP may be located near a gene that affects appetite, which leads to increased height through better nutrition.

In the largest GWAS of height yet conducted, a 2018 study identified 3,290 SNPs associated with variation in height among ~700,000 Europeans. To illustrate how complex the genetic causes of height variation are, no single SNP itself was strongly associated with differences in height—individuals who had the particular SNP that most strongly predicted height were on average only 4 mm (0.16 in.) taller than individuals who lacked this SNP. Furthermore, if you add up the effects of all 3,290 of these SNPs, together they still only account for about 25% of the variation in height. Recall that estimates of heritability derived from studying twins in European populations are much higher, at around 80%. The lower heritability estimates from GWAS compared to twin studies has been found not only for height but also for many other phenotypic traits and has been termed "the missing heritability problem." Does the missing heritability problem indicate that critics of twin studies were correct, and that twin studies do indeed overestimate heritability in height and other traits?

While this is a possibility, there are also several reasons to draw the opposite conclusion—that GWAS *underestimate* the true heritability. First, many of the SNPs associated with phenotypic variation have not yet been found, as their association with height is so small that even larger sample sizes will be required to statistically detect them. Indeed, since GWAS studies began in the mid-2000s, they have included larger and larger sample sizes over time, which has led to the identification of increasing numbers of SNPs that explain more and more phenotypic variation. Second, much of the genetic variation associated with phenotypic variation might not actually reside in SNPs. SNP chips only evaluate **common variation**, positions in the genome where the rarest allele in a population has a frequency of at least 1%. Height-associated SNPs that are more rare than this will not be detected in SNP-based GWAS, as SNP chips do not even attempt to detect very rare variants. The first large-scale GWAS study based on whole-genome sequencing, which assesses every position in the genome rather than just common variants, was conducted in 2019. It showed that when particularly rare variants are accounted for, genetic variation accounts for 79% of the variation in human height in Europeans, in line with the heritability estimates from twin studies.

Genetic Variation among Groups

Stature varies among human populations.

Just as people within groups vary, groups of people collectively vary in certain characteristics. For example, there is a considerable amount of variation in average height among populations. People from northwestern Europe are tall, averaging about 1.75 m (5 ft. 9 in.). People in Italy and other parts of southern Europe are about 12 cm (5 in.) shorter on average. African populations include very tall peoples such as the Nuer and the Maasai and very short peoples such as the !Kung. Within the Western Hemisphere, Native Americans living on the Great Plains of North America and in Patagonia are relatively tall, and peoples in tropical regions of both continents are relatively short.

Some of the variation in overall body size among human groups appears to be adaptive.

In Chapter 12 we learned that larger body size is favored by natural selection in colder climates. Large-bodied peoples, such as the indigenous peoples of Patagonia in South America and the Great Plains in North America, usually live in relatively cold parts of the world, but small-bodied peoples typically live in warmer areas, such as southern Europe or the neotropics. This pattern suggests that variation in body size among groups may be adaptive, a conjecture that is borne out by data on the relationship between body size and climate from many human groups (**Figure 14.17**). Thus, at least some fraction of the variation in body size among groups is adaptive.

FIGURE 14.17

People living in cold climates have larger bodies than those living in warm climates. The vertical axis plots mean chest girth for many human groups, and the horizontal axis plots the mean yearly temperature in the regions in which each group lives. Because chest girth is a measure of overall body size (meaning body volume), these data show that people living in colder climates have larger bodies.

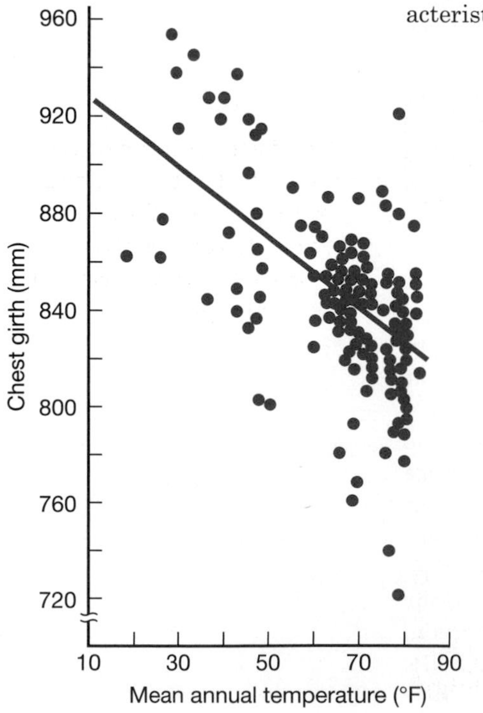

The fact that variation within populations has a genetic component does not mean that differences between groups are caused solely by genetic differences.

We have seen that a significant proportion of the variation in height and body size within some populations reflects genetic variation. We also know that the variation in height and body shape among populations seems to be adaptive. From these two facts, it might seem logical that the variation in body size among populations is genetic and that this variation represents a response to natural selection. However, this logic is wrong. All the variation in height within populations could be genetic *and* all the variation in height among populations could be adaptive, but this would not mean that there is variation among populations in the distribution of genes that influence height. In fact, all the observed variation in height among groups could be due solely to differences in environmental conditions.

This point is often misunderstood, and this misunderstanding leads to serious misconceptions about the nature of genetic variation among human groups. A simple example may help you see why the existence of genetic variation within groups does not imply the existence of genetic variation among groups. Suppose Rob and his next-door neighbor Pete both set out to plant a new lawn. They go to the garden store together, buy a big bag of seed, and divide the contents evenly. Pete, an avid gardener, goes home and plants the seed with great care. He fertilizes, balances the soil acidity, and provides just the right amount of water at just the right times. Rob scatters the seed in his backyard, waters it infrequently and inadequately, and never even considers fertilizing it. After a few months, the two yards are very different. Pete's lawn is thick, green, and vigorous, but Rob's lawn hardly justifies the name (**Figure 14.18**). We know that the difference between the two lawns cannot be due to the genetic characteristics of the grass seed because Rob and Pete used seed from the same bag. At the same time, variation in the height and greenness of the grass within each lawn might be largely genetic because the seeds within each lawn experience very similar environmental conditions. All the seeds in Pete's lawn get regular water and fertilizer, whereas all the seeds in Rob's lawn are neglected to an equal extent. Thus, if the seed company has sold genetically variable seed, differences between individual plants *within* each lawn could be due mainly to genetic differences between the seeds themselves.

The same argument applies to variation in any variable human trait, including height. Modernization has had a striking effect on the average heights of many peoples. For example, **Figure 14.19** plots the heights of several groups of English boys between

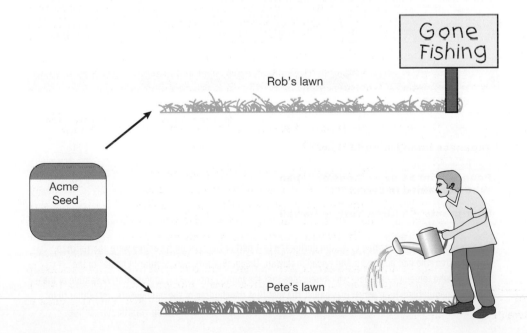

FIGURE 14.18

The differences between two lawns planted from the same bag of seed must be environmental. However, if the seed used was genetically variable, the differences within each lawn could be genetic.

FIGURE 14.19

Height increases with time in English populations, but at any given time richer people are taller. Note that English *public* schools are the equivalent of American *private* schools.

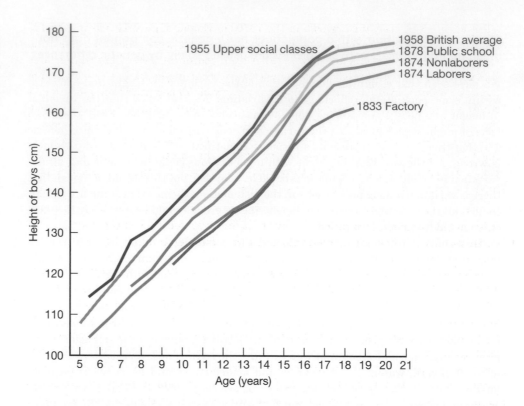

the ages of 5 and 21 in the nineteenth and twentieth centuries. In 1833, 19-year-old factory workers averaged about 160 cm (about 5 ft. 3 in.). In 1874, laborers of the same age averaged about 167 cm (5 ft. 6 in.). In 1958, the average British 19-year-old stood about 177 cm (5 ft. 9 in.). Similar increases in height over the past hundred years can be seen among Swedish, German, Polish, and North American children. These changes have occurred very rapidly, probably too fast to be due to natural selection.

There also have been substantial changes in height among immigrants to the United States over a few generations. During the first part of the twentieth century, when Japan had only begun to modernize, many Japanese came to Hawaii to work as laborers on sugar plantations. The immigrants were considerably shorter than their descendants who were born in Hawaii (**Table 14.1**). This change in height among immigrants and their children was so rapid that it cannot be the result of genetic change. Instead, it must be due to some environmental difference between Japan and Hawaii in the early twentieth century. The underlying cause of this kind of environmental

TABLE 14.1

	Average Height (cm)	Sample Size
Japanese Immigrants to Hawaii	158.7	171
People from Same Regions of Japan Who Remained in Japan	158.4	178
Immigrants' Children Born in Hawaii	162.8	188

Japanese men who immigrated to Hawaii during the first part of the twentieth century were shorter than their children who had been born and raised in Hawaii. The immigrants were similar in height to the Japanese who remained in Japan, which indicates that the immigrants were a representative sample of the Japanese populations from which they came. The fact that the children of the immigrants were taller than their parents shows that environmental factors play an important role in creating variation in stature.

effect is not completely understood. In the 1870s in England, poverty was involved to some extent because relatively wealthy public-school boys were taller than less affluent nonlaborers, and nonlaborers were taller than poorer laborers. Observations like this have led some anthropologists to hypothesize that increases in the standard of living associated with modernization improve early childhood nutrition and increase children's growth rates. However, this cannot be the full explanation because even the richest people in England 150 years ago were shorter than the average person in England today. Because it seems unlikely that wealthy Britons were malnourished in the 1870s, other factors must have contributed to the increase in height during the past 150 years. Some authorities think that the control of childhood diseases may have played an important role in these changes.

It is very difficult to determine if genetic differences and natural selection contribute to differences between groups for complex phenotypic traits.

Recall that it is fairly straightforward to determine if genetic differences contribute to population differences in a phenotypic trait that is controlled by variation at just one (or a few) loci, and even to determine that natural selection, rather than just random genetic drift, is responsible for the population difference. For example, we know that individual differences in lactase persistence have a genetic cause because there is a near one-to-one correspondence between genotype and the ability to drink milk: For the most part, if you have a certain allele at a particular locus, you will be able to comfortably drink milk as an adult no matter what environment you experienced growing up. And we know that populations differ dramatically in the frequency of alleles conferring lactase persistence. Finally, we know these population differences in allele frequencies are due at least in part to differences in natural selection: The functional alleles are much more common in populations with a long evolutionary history of milk-drinking, and they are surrounded by long-shared haplotypes, which reveal that the allele has recently quickly risen to high frequency in these populations.

But how can we determine the role of genetics and natural selection in generating differences between populations for complex phenotypic traits that are highly polygenic, affected by variation at thousands of loci, and also more substantially affected by environmental variation?

Earlier we saw that northern Europeans are taller on average than are southern Europeans. And remember that genomewide association studies have identified many loci that affect stature. At each of these loci there are alleles that are more common in taller people and others that are more common in shorter people. Neil Hirschhorn and his colleagues at Harvard University recently conducted a study that showed, on average, "tall" alleles are a little more common in northern European populations and the "short" alleles are a little more common in southern European populations, suggesting that the difference in stature between these two populations is in part due to small genetic differences at many loci. Additional analyses suggested that these genetic differences in height were driven in part by natural selection. Because each of the thousands of genes that affect height have only a tiny effect, selection on any individual gene will be very weak; therefore, we should not expect to observe the signs of a selective sweep at any individual SNP associated with height. However, the allele frequency difference between southern and northern European populations was significantly higher for height-associated SNPs than for SNPs across the genome, and this suggests natural selection played a role in generating these genetic differences in height between populations.

Genomewide association studies are less useful for comparisons of distantly related populations.

To understand why GWAS are less useful in comparing distantly related populations, remember that these studies use SNP chips that are developed in particular populations. The SNPs that they identify do not necessarily affect the trait themselves, but are in linkage disequilibrium with nearby genes that do directly affect the trait. The more distantly related two populations are, the more different their patterns of linkage disequilibrium will be, and the worse a SNP chip developed in one population will be at predicting height in the other population. For example, SNP chips developed and used in GWAS of height in Europeans have been used to genotype West African populations. Per this method, alleles associated with tallness in Europeans are less common in West Africans than in Europeans, which might lead one to predict that West Africans, on average, would be shorter than Europeans. But this prediction is not correct: Robust anthropological evidence shows that West Africans are as tall as Europeans on average. GWAS of many other phenotypic traits have also been shown to suffer from what geneticists call **poor portability**. Unfortunately, about 80% of the individuals who have participated in GWAS studies are of European descent. This bias limits our ability to disentangle the role of genetic versus environmental causes of phenotypic differences between populations for traits that are affected by many genes.

The Race Concept

Widely held views of race are bad biology.

Race is an important part of everyday life. Today, race affects how we see the world and how the world sees us; it affects our friendships, our choice of marriage partners, our employment prospects, and what happens to us at a traffic stop. In other times and places, racial differences have been used to justify slavery and genocide.

But what *is* race? The phenomenon of people identifying with large groups is found in almost every culture and is often associated with the idea that other groups are dirty or inferior. In small-scale societies, group members often share a language and cultural norms and live in a contiguous territory. In complex, urban societies, groups can be based on cultural traits such as language or religion or physical traits such as skin color or facial morphology (**Figure 14.20**). In many cultures, people acquire the identity of their parents. Often this is associated with the belief that children acquire an "essence" from their parents that makes them who they are and determines what they are like. Cross-cultural experiments indicate that many people think children acquire their parents' essence, even when the child is raised by members of a different group, suggesting that these beliefs are rooted in pre-scientific folk biology. In actual fact, people often manage to change their group identity for a variety of reasons.

FIGURE 14.20

This map shows contours in skin color. Notice that there are smooth gradients away from the equator.

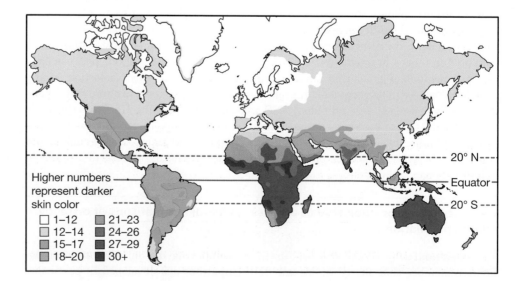

Higher numbers represent darker skin color

- 1–12
- 12–14
- 15–17
- 18–20
- 21–23
- 24–26
- 27–29
- 30+

20° N
Equator
20° S

The word *race* entered English in the late Renaissance and derives from the Italian word *razza*, which in Renaissance Italy was used to refer only to a breed or type of animal. Colonial expansion brought Europeans into contact with people in other parts of the world, and they gradually applied the word to describe the different groups that they encountered. In the eighteenth and nineteenth centuries, the development of population biology led many biologists and anthropologists to argue that races—generally referring to a small number of continent-scale groups such as sub-Saharan Africans, Western Eurasians, and Eastern Eurasians—were long-isolated biological populations that had evolved different attributes over tens or hundreds of thousands of years. In the early twentieth century this view of race was taught by most physical anthropologists as scientific fact. Developments in genetics, coupled with events such as the Nazi Holocaust and the American struggles over civil rights, led many to reject this conception of race in the latter half of the twentieth century. Today, some have extended the rejection of the race concept to a rejection of the possibility that there are any significant genetic differences at all between human groups.

In this section, we will present evidence that neither of these views is correct. The folk view that racial groups represent sharply bounded, genetically distinct, long-isolated populations is simply wrong. Race in this sense does not exist. However, it is also inaccurate to maintain that there are no genetic differences between human populations. Rather, contemporary data support the following propositions:

1. There are genetic differences between groups of people living in different parts of the world, and people from nearby populations are more similar genetically than they are to people from distant populations.

2. Most of the genetic variation is within populations. This means two people drawn from two different populations are often more similar at a single genetic locus than are two people drawn from the same population.

3. Nonetheless, genetic variation can be used to identify a person's geographic origin with some precision, and statistical techniques can be used to classify people into geographically based groups on the basis of these genetic data.

4. However, there are no sharp boundaries separating groups. Instead, most variation between groups is gradual.

5. Most groups have not been isolated for long periods of time. People have been on the move for as far back as we can see, and as a consequence, contemporary groupings are the result of admixture in the relatively recent past.

Measuring the amount of variation within and between populations.

Geneticists use a statistic labeled F_{ST} to measure the amount of variation between populations. F_{ST} is the proportion of the total genetic variance contained in a particular population relative to the total genetic variance in the global population. To understand what this means, let's use a simple example. Imagine we genotype a large number of people from across the globe at a single genetic locus and calculate the probability that any two individuals that are randomly drawn from this sample have different alleles at this locus. This is the total genetic diversity. Then, we separate our worldwide sample of people into some number of groups; for example, each group might consist of people living on the same continent. Then, we calculate genetic diversity again for each of these groups and compute the average of these population values; we call this the average within-population genetic diversity. To calculate the F_{ST} value we use the following formula:

$$\frac{\text{(total genetic diversity} - \text{average within-population genetic diversity)}}{\text{total genetic diversity}}$$

F_{ST} values can range from 0 to 1. If most of the genetic variation is within groups, then the F_{ST} value will be low (closer to 0). In contrast, if most of the variation is between groups, the F_{ST} value will be high (closer to 1). F_{ST} values are calculated for multiple loci across the genome, and the average F_{ST} value provides us an overall estimate of how much of the total variation in genetic diversity is due to variation within groups. Put simply: A low average F_{ST} value would suggest that the way people were grouped—such as by their perceived race—doesn't strongly correlate with how they differ genetically.

F_{ST} values for continental groups are low.

Multiple studies have examined F_{ST} values for continental populations that correspond to what most people think of as human races (that is, Europeans, Africans, East Asians, Native Americans). These studies vary in the type of genetic markers they examine and in the precise ways in which they divide humans up into a few big populations, but all find that F_{ST} is about 0.11 to 0.18. These values indicate that if you examine any individual locus, the average probability of drawing two identical alleles is increased by only 11% to 18% if you sample two individuals from the same continental population versus a sample of individuals from the entire species. To put it another way, suppose that a malevolent extraterrestrial wiped out the entire human species except for one of the continental-level populations, which it preserved in an extraterrestrial zoo. The members of the surviving continental population would still contain, on average, 82% to 89% of the total genetic variation that is present in the entire human species.

Although F_{ST} values for continental groups are low, this variation is not patterned randomly.

F_{ST} values are based on averages of single loci. Loci are combined in genomes, so we can also examine how combinations of alleles across loci are patterned. To illustrate how this works, imagine that we genotype 100 loci in three populations. Each locus has only two alleles, *A* and *B*. Allele frequencies only vary slightly between populations: At each of the 100 loci, the frequency of the *A* allele is 55% in population 1, 50% in population 2, and 45% in population 3. If you try to predict what population an individual comes from on the basis of that individual having the *A* allele at any given locus, you will indeed do only slightly better than chance by picking population 1. However, if you investigate all 100 loci simultaneously, you will do much better, as an individual that has an *A* allele at all 100 loci is much more likely to come from population 1 than from population 2 or especially population 3. This phenomenon, termed **correlation structure**, can also be illustrated with an example that uses phenotypic instead of genetic variation: the facial characteristics of people of different sexes. While males tend to have squarer jaws, smaller eyes, and bigger browridges than females, for each of the traits the variation within the sexes is larger than the variation between the sexes. If you use any one of these traits in isolation to decide if an individual is male or female, you won't do much better than chance. However, if you consider all three of these traits simultaneously (that is, this individual has a rounded jaw *and* large eyes *and* a small browridge; therefore, I think she is female), your odds of correct classification will improve. And the more traits you simultaneously examine, the better your odds will be. With genetic data, we can examine combinations of hundreds or even millions of variable loci whose allele frequencies only vary slightly among populations from different continents. A number of studies have shown that using 1,000 SNPs, individuals can be correctly assigned to their continental population with a high degree of accuracy.

Patterns of genetic variation can be used to classify people into groups, but such groupings don't generally align with conventional notions of race.

The calculations of F_{ST} values that we have described are based on a predetermined category: the continent that people live on. But that is not the only way we could divide people up. We could divide them by the country they live in, the language they speak, or their distance from Africa. Geneticists have developed a set of methods that allow them to generate groupings that emerge from the patterns in the data, rather than from the predetermined decisions of the researcher. That is, in conducting these analyses we don't start with a set number of populations in our mind, but instead assume that individuals come from some fixed but unknown number of populations, and then test which number of clusters best fit the pattern of variation in the data. These methods, called unsupervised clustering algorithm studies, are designed to maximize genetic similarities among individuals that are assigned to the same cluster and to minimize genetic similarities among individuals assigned to different clusters. In one study using these methods, geneticist Sarah Tishkoff and colleagues found that the best estimate of the number of clusters was actually 14, six of which are in Africa. The other clusters mainly corresponded to continental populations. However, there were other clusterings that were nearly as effective, which reflects the facts that the boundaries between clusters are not very sharp and that there is considerable overlap in the distribution of genes across clusters.

While clusters are statistically detectable, most genetic variation is gradual.

Recall from Figure 14.14 that the geographic distance between populations is a good predictor of their genetic distance. What this means is that in general, allele frequencies change very gradually with geography, and that there are few areas of the world where you see large differences in the allele frequencies between neighboring populations. Genetic variation changes gradually with geographic distance even among populations that are assigned to the same cluster in unsupervised clustering algorithm studies. For example, among West Eurasians (the label frequently given to populations from Europe, the Near East, and central Asia that often group together as a cluster in unsupervised clustering algorithm studies), allele frequencies change gradually from northern Europe to central Asia. But if most genetic variation changes smoothly with geographic distance, how is it that unsupervised clustering algorithms can reliably assign populations to clusters? Wouldn't such a task be like looking at a color spectrum, and trying to determine where blue ends and green begins (Figure 14.21a)?

[a]

[b]

[c]

FIGURE 14.21

Unlike the colors of the spectrum (a), where variation is so perfectly smooth and gradual that it is impossible to determine where one color ends and another begins, there are some discontinuities in human genetic variation. But you shouldn't think of these discontinuities as leading to the existence of distinct, clearly demarcated populations, as shown in (b) with three homogenous, clearly demarcated colors. Rather, the true picture of human genetic variation is one that is mostly clinal (c), but with a very small amount of discontinuity that allows one to see fuzzy, indistinct clusters. The degree of discontinuity shown in (c) is much greater than the discontinuities in human variation.

The answer to this question is that clusters are statistically detectable only because there are some areas of the world where allele frequencies change just a bit more sharply than one would expect based on geographic distance alone. For example, a central Asian population will have allele frequencies that are just a bit more similar to those of another West Eurasian cluster population than they will be to an equally distant population assigned to a different cluster, such as East Asian. To put into perspective just how much of human worldwide genetic variation is explained by gradual change versus clustered patterns, consider a 2005 study by Stanford geneticist Noah Rosenberg and colleagues. Using 993 genetic loci typed in 1,048 individuals from different populations across the world, this study used unsupervised clustering algorithms to assign populations to clusters. They then measured genetic distance between each population pair and correlated these values with the corresponding geographic distances between population pairs. A correlation value of 1 would indicate that you could predict with certainty the genetic distance between a pair of populations by knowing the geographic distance between them. The researchers found a fairly high but not perfect correlation value, 0.69. Next they added to their statistical model information about whether a population pair was assigned to the same cluster or to different clusters by the unsupervised clustering algorithm. Finally, they determined how this additional information of cluster membership improved the prediction of genetic distance between population pairs, beyond what was already predicted by their geographic distance. The answer was basically "not very much," as the correlation only increased to 0.729. Thus, while genetic variation is not completely gradual, it is certainly *mostly* gradual.

So genetic variation is not exactly like a color spectrum, where completely smooth and gradual variation makes any objective grouping totally impossible (see Figure 14.21a). But neither should genetic clusters be thought of as analogous to clearly discrete colors such as red and green (**Figure 14.21b**). Instead, perhaps the best color-related analogy for human genetic variation is a very slightly lumpy spectrum, where most change is gradual, but with some minor discontinuities as well (**Figure 14.21c**).

The clusters identified through modern methodologies are the result of a history of genetic admixture, not long isolation.

The kinds of methods discussed above provide a description of patterns of genetic variation among contemporary populations; they don't provide much insight into the processes that led to those patterns of variation. For example, some early anthropologists speculated that the big continental "races" were descended from the original inhabitants of the continents, largely genetically isolated and independently evolving since the major out-of-Africa dispersal. Recent research combining ancient DNA and archaeology indicates that this speculation was completely wrong. For almost every region of the world where this has been examined, the people living there now are a poor guide to who was living there in the past. Moreover, current populations are the end product of multiple **admixture** (interbreeding) events between different populations that themselves do not exist in unadmixed form today.

Three major events led to the genetic and phenotypic variation characteristics of Europeans today. (As elsewhere in this text, when we refer to "Europeans" or any other contemporary population, we are referring to individuals for whom the vast majority of their ancestors lived in that area within the past 500 years. Of course, this excludes many individuals who migrated into the area more recently but who would nevertheless justifiably consider themselves as "European.") As discussed in the previous chapter, the first inhabitants of Europe were hunter-gatherers, who experienced a series of population transformations involving various degrees of admixture and replacement throughout the Upper Paleolithic. The most recent of these events was a massive expansion of hunter-gatherers out of the Near East around 14 ka that displaced most of the ancestry of the hunter-gatherers resident in Europe prior to this time. Alleles that are associated with blue eyes, dark hair, and dark skin in contemporary humans were present among these European hunter-gatherers, suggesting an unusual combination of phenotypic features that is rare in any extant human population.

Next, farmers from Anatolia (the western part of modern-day Turkey) entered and rapidly spread across Europe about 8,500 years ago. These early European farmers retained about 90% of their distinctive Anatolian ancestry, which indicates a fairly low amount of admixture with the hunter-gatherers they encountered. Indeed, isolated populations of hunter-gathers, especially in northern Europe where farming was more difficult, persisted for several thousand years after the arrival of the earliest European farmers. Beginning about 6,500 years ago, there was an increase in admixture between hunter-gathers and farmers. Compared to the earliest farmers, farmers after this time have about 20% more hunter-gatherer ancestry, and hunter-gatherer populations with no trace of Anatolian farmer ancestry largely disappear. Allelic variation suggests that European farmers had dark hair like the European hunter-gatherers, but they had light skin and brown eyes.

The third and final major event in the peopling of Europe was the large-scale immigration of sheep and goat herders associated with the Yamnaya archaeological culture from the grassland steppes of present-day southern Russia and Georgia. The expansion of the Yamnaya into Europe (as well as eastward) was fueled by the domestication of the horse and acquisition of the wheel. Yamnaya ancestry appeared in southeastern Europe by 6 ka, in northeastern Europe around 5 ka, and in central Europe at the time of the Corded Ware complex around 4.6 ka. In each of these areas, there is no evidence of Yamnaya-like ancestry just before their culture appears in the archeological record but high levels of Yamnaya ancestry thereafter. For example, individuals that are found associated with the Corded Ware cultural complex (which combines cultural elements from the Yamnaya and earlier European farmers) have on average about three-quarters Yamnaya and one-quarter Anatolian farmer ancestry. Allelic variation at functional loci suggests that like the farmer migrants that preceded them, Yamnaya pastoralists brought alleles for light skin and brown eyes to Europe. However, they also introduced allelic variation for blond hair to Europe for the first time.

All Europeans today are a mix of these three major sources of ancestry: On average, about half of the ancestry of contemporary Europeans can be traced to the Yamnaya-like component, half to the Anatolian farmers, and just 1% to 2% to hunter-gatherers. This average, however, conceals large regional variation: There is more Yamnaya ancestry in northern Europe, more Anatolian farmer ancestry in southern Europe (especially Sardinia and Sicily), and more hunter-gatherer ancestry in the Baltics and eastern Europe. The stereotypically northern European phenotype of blue eyes, light skin, and blond hair, which featured prominently in Nazi propaganda of a "pure" Aryan race, was a recent product of three major waves of migration and admixture, each of which contributed a unique component. And each of the three major sources of ancestry in modern Europeans were themselves the product of previous admixture events between divergent populations; the Yamnaya population, for example, appears to have arisen around 7 ka to 5 ka as populations related to ancient and present-day Armenians and Iranians mixed when they migrated from the south, passing between the Black Sea and the Caspian Sea and out onto the steppe.

Recent ancient DNA and archaeological research has also shown that other contemporary populations have formed as a result of recent admixtures between formerly isolated populations. For example, the present-day native people of North America descend from at least four distinct streams of ancestry. First, people bearing a distinctive ancestry component related to present-day East Asians moved into North and South America across the Bering land bridge approximately 15,000 years ago. Labeled *First Peoples*, this ancestry component is the major source of ancestry in all contemporary Native Americans. Second, people with a higher degree of genetic relatedness to contemporary Australasians, termed *Population Y*, contributed a small amount of additional distinct ancestry to indigenous groups from Amazonia. Third, a stream of ancestry that relates to Paleo-Eskimos spread throughout the Arctic after about 5,000 years ago. Paleo-Eskimo ancestry is particularly widespread today in Na-Dene language speakers, which includes Athabaskan and Tlingit communities from Alaska and northern Canada, the West Coast of the United States, and the Southwest of the United States. A fourth ancestry component, termed *Neo-Eskimo*, spread with the Thule and related archaeological cultures throughout the Arctic only about 800 years ago and is today present in Yup'ik and Inuit groups of the Arctic, who also have Paleo-Eskimo ancestry in addition to their First Peoples' ancestry.

And as in Europe, the distinctive ancestry components detectable in today's Native Americans were themselves the products of admixture. One pattern that puzzled geneticists when they only had access to DNA from living populations was the fact that Europeans, in particular northern Europeans, are more closely related to Native Americans than to East Asians, a pattern that would not be expected based on geographic proximity. When ancient DNA became available from European hunter-gatherers, the puzzle deepened, as these people lacked the signal showing closer affinity to Native Americans than to East Asians. The puzzle was solved with the publication of more ancient DNA from other time periods. In 2013, Eske Willerslev and his colleagues at the University of Copenhagen published genomewide data from the bones of a boy who had lived at the Mal'ta site in Siberia around 24 ka. The Mal'ta boy belonged to a population that geneticists now label the *Ancient North Eurasians*. Some people from this population migrated east across Siberia and contributed some of the ancestry to the population that crossed the Bering land bridge and gave rise to the First Peoples in the Americas. Others migrated west and contributed to some of the ancestry of living Europeans. But this westward movement did not occur until the big expansion of Yamnaya populations beginning some 6,000 years ago, which explains why European hunter-gatherers living in Europe before this time, in contrast to contemporary Europeans, were not more closely related to Native Americans than to East Asians. Populations bearing Yamnaya ancestry (and thus Ancient North Eurasian ancestry) also expanded to East Asia at this time but were later largely replaced by the ancestors of contemporary East Asians.

Compared to Europe and the Americas, the stories of the peopling of other parts of the world are less well-known because hot and moist climates are not good for the preservation of ancient DNA. However, the data that are available suggest that most of today's populations are not the descendants of the population that lived in that location 10,000 or more years ago. Just as ancient DNA showed us that a tree is not a good metaphor for the relationships among distantly related populations that some might designate as separate species (that is, Neanderthals, Denisovans, and modern humans), it has also demonstrated that a tree is not a good metaphor for the relationships among the more closely related populations that we observe in humans today. Unlike real tree limbs, which do not grow back together after they branch, mixtures of divergent human populations have occurred repeatedly in hominin and human evolution. Rather than a tree, a better metaphor may be a braided stream. Just like human populations have a recurring pattern of divergence and remixing extending back from the present far into the past, so too do the branches of a stream that pass over a vast delta separate only to be reunited (**Figure 14.22**).

So, what do these results about worldwide patterns of genetic variation tell us about the scientific validity of race in humans?

The data support the propositions that we outlined at the beginning of this section. The common notion that races represent long-isolated biological populations that have evolved different attributes over tens or hundreds of thousands of years is not supported. The data tell us that some genetic differences exist among groups of people living in different parts of the world today. However, at most genetic loci there is considerably more genetic variation within populations than between them.

FIGURE 14.22

The history of a group of related species is like a branching tree. Once two species split there is little genetic interchange between them. The history of populations within a single species is more like a braided river. Streams divide and rejoin in a complex web as they flow to the sea.

If we know enough about a person's genome, we'll be able to identify that person's geographic origin with some precision, and this makes it possible to use statistical techniques to classify people into geographically based groups. But these groups do not have sharp boundaries, and there is considerable overlap in the genetic composition of different groups. This variation is largely gradual, so populations that are closer together are more similar to one another than are populations that are farther apart. Human variation largely reflects the movements and mating history of people across the globe, with a repeated divergence and remixing extending back from the present far into the past.

The question of the scientific validity of races in humans is not just an academic one. Health-care providers have long struggled with the utility of race/geographic ancestry in medicine. The ultimate goal of **personalized** (or **precision**) **genetic medicine** is to identify specific genes and their allelic variants that influence either the risk of disease, a quantitative outcome of medical relevance (for example, blood pressure or lung capacity), or response to a particular treatment or drug. Calculating genetic risk for monogenic disorders such as sickle-cell anemia, cystic fibrosis, Huntington's disease, and Duchenne muscular dystrophy is relatively straightforward and a routine part of contemporary medical care. Many other medical conditions, however, such as coronary heart disease, hypertension, and diabetes, are caused by variation of genes at many loci. **Polygenic risk scores** use the statistical information gleaned from GWAS and an individual's own genotype to predict how likely a person is to have some phenotypic outcome, such as the onset of a particular disease. For example, it is now possible to use polygenic risk scores to help identify individuals who are at risk of developing coronary heart disease. Such information might be useful to doctors in deciding which individuals should be screened and to patients in deciding which lifestyle choices they should make in order to reduce their chances of actually developing the disease. While polygenic risk scores currently don't explain much of the variation in any given phenotypic trait, they will likely become more accurate in the future as GWAS studies include more people and use whole-genome sequencing data. As we discussed earlier, the poor portability of GWAS results means that personalized genetic medicine is currently mainly useful to people of European ancestry. Many researchers and clinicians are optimistic that in the not-too-distant future, better-powered GWAS studies from more non-European populations will eventually result in personalized genetic medicine that will help people regardless of their genetic ancestry (**A Closer Look 14.2**).

14.2 Direct-to-Consumer Genetic Testing

While genetic testing is often done in the health-care system, in the past decade dozens of companies have begun to offer direct-to-consumer genetic testing. Most direct-to-consumer genetic tests are based on SNP chips, although whole-genome sequencing services have recently become available and will be more common in the future. Direct-to-consumer genetic testing provides information in three major areas, which we discuss in the following.

Disease Risk and Health

Direct-to-consumer genetic testing can provide estimates of an individual's genetic risk of developing several common diseases, such as celiac disease, Parkinson's disease, and Alzheimer's disease, or other health-related phenotypes, such as weight. Some of these risk estimates are derived from polygenic risk scores, while others are based on individual loci. The models that the companies use to calculate a customer's risk scores are based on a combination of prior scientific research as well as the phenotypic data that the customer chooses to share with the company. These companies do not test all diseases that have a known genetic basis, and the tests that they do provide are not meant to be diagnostic, as they do not assay all of the genetic variation that is known to be associated with the disease from prior genetics research. Direct-to-consumer genetic status can also inform individuals who are carriers of specific recessive genetic conditions, which may be useful to people contemplating having children.

Genealogy

In common parlance, the word *relative* refers to individuals who share a recent common ancestor. But not all relatives in this genealogical sense are also relatives in a more genetic sense, in that two individuals actually share DNA that they both inherited from this recent common ancestor. Each generation you go back, your number of ancestors doubles. And although you inherit half of your DNA from your mother and half of your DNA from your father, chance events of inheritance mean that your four grandparents will vary in how much genetic material they will contribute to you. Although the odds are nearly certain that you inherited at least some genetic material from all four of your grandparents, the more generations you go back in time, the less likely this is to be true. University of California geneticist Graham Coop used simulations of known recombination rates to show that the odds of you inheriting any genetic material from a randomly chosen ancestor that existed only 14 generations back is almost zero.

Direct-to-consumer genetic tests search their databases to find individuals who share a recent common ancestor on the basis of the number and length of shared DNA segments—the higher both of these values are, the more recent the shared common ancestor and the more confidence there is in the assignment. For example, research using the 23andMe SNP chip platform shows that first cousins (sharing a grandparent, two generations in the past) and closer relatives can be detected with near 100% accuracy, while fifth cousins will be detected only about 15% of the time. However, despite the difficulty of identifying distant cousins, participants in direct-to-consumer genetic testing typically will have many identified,

But until such an idealized future of personalized genetic medicine arises, what role should race/geographic ancestry play in medicine? Everyone agrees that self-identified race correlates with geographic ancestry, that geographic ancestry correlates with genetic variation, and that genetic variation correlates with many medically relevant phenotypes (for example, probability of having a particular disease or response to a particular treatment). Proponents of racially based genetic medicine argue that although self-identified race is far from a perfect indicator of an individual's medically relevant genetic variation, it does have some utility. Many different as everyone has an extremely large number of distant cousins.

The growth of direct-to-consumer genetic testing has also fueled a revolution in forensic genetics. Traditionally, forensic genetics has been limited to determining if a sample collected at the scene of a crime came from the same individual as a particular suspect or one of the 16.5 million individuals included in governmental forensic genetic databases. However, forensic genetics was dramatically changed in 2018 with the case of the Golden State Killer, the name given to a single man suspected to have killed 12 people and raped 45 women across California between 1976 and 1986. Police compared the DNA collected from crime scenes linked to the Golden State Killer to millions of genotypes at GEDmatch, a free service that allows customers from multiple direct-to-consumer genetic services to upload their genetic data and search for relatives. As with the governmental forensic genetics database, investigators found no direct match to the Golden State Killer's genotype in this public database. However, they did find a distant relative, which narrowed down the list of potential suspects from possibly millions to a single family. The pool of suspects within this family was then narrowed down further using traditional, nongenetic investigative techniques and eventually confirmed when the DNA of one of the suspects that was obtained from a coffee cup revealed a perfect match to a sample obtained from the original crime scene. Since this case, similar relative searches of publicly accessible databases have been used to solve dozens of cold cases.

Geographic Ancestry

Direct-to-consumer genetic tests also provide customers with estimates of the percentage of their genome that derives from particular ancestral populations. Using various techniques related to those discussed in the section on clustering algorithms, the companies begin with a reference set of genotypes from customers of known ancestry (that is, those having all four grandparents in a particular geographic area). When a segment of customer DNA closely matches the DNA from one of the reference populations, that segment is assigned to that reference population. After assigning each individual segment, these values are summed up to derive the composite picture of percentage ancestry in each reference population. Some segments of DNA are shared across many reference populations and can be assigned only very broadly (for example, "Northwest European") rather than more specifically (for example, "German") or not even assigned to any area at all. In general, the algorithms are designed to be conservative, in the sense that they are biased toward false negative errors rather than false positive errors. An example of a false negative error would be leaving a truly German segment as unassigned, while a false positive error would be assigning a truly German segment as Italian.

Because each company has its own unique set of reference populations and algorithms, the same customer may get different results from different companies. The same individual may even get different results when tested twice by the same company, as there is some degree of error in the genotyping process. There is not much research on the accuracy of direct-to-consumer genetic testing. However, a 2019 study compared the estimates of ancestry for 21 pairs of identical twins. Identical twins have nearly identical DNA, so estimates of their ancestry should be highly concordant. This was largely the case for estimates produced by the same company, where the mean percentage agreement ranged from 95% to 99%. However, when the estimates of genetic ancestry were compared between different companies, the mean percentage agreement dropped to between 53% and 84%. This highlights the fact that genetic ancestry estimates depend, in part, on the reference populations and algorithms companies use.

disorders have similar symptoms, and a patient's self-identified race can be used to prioritize diagnostic tests. Critics of racially based medicine, however, argue that in practice racially based medicine often leads to underdiagnosis of disorders in individuals who come from ancestries where the genetic basis for the disorder is present but less common. Critics also point out that many health disparities between races or ethnicities have no known genetic basis and fear that a focus on the genetic causes for discrepancies can distract from a focus on socioeconomic and other environmental causes.

CHAPTER REVIEW

Key Terms

transposable elements (p. 365)

synonymous substitutions (p. 366)

nonsynonymous substitutions (p. 366)

positively selected (p. 366)

neoteny (p. 368)

negative selection (p. 368)

genetic variation (p. 368)

variation within groups (p. 369)

variation among groups (p. 369)

specific language impairment (SLI) (p. 370)

selection–mutation balance (p. 372)

falciparum malaria (p. 372)

balanced polymorphism (p. 373)

non-insulin-dependent diabetes (NIDD) (p. 373)

insulin (p. 374)

lactose (p. 375)

lactase-phlorizin hydrolase (LPH) (p. 375)

lactase persistence (p. 375)

LCT (p. 377)

pastoralists (p. 377)

selective sweep (p. 378)

single-nucleotide polymorphisms (SNPs) (p. 379)

founder effect (p. 380)

porphyria variegata (p. 381)

Oceania (p. 381)

heritability (p. 383)

environmental covariation (p. 383)

monozygotic twins (p. 384)

dizygotic twins (p. 384)

genomewide association studies (p. 385)

linkage disequilibrium (p. 385)

common variation (p. 386)

poor portability (p. 390)

correlation structure (p. 392)

admixture (p. 394)

personalized (or precision) genetic medicine (p. 397)

polygenic risk scores (p. 397)

Study Questions

1. How can we reconcile the fact that humans and other apes differ only by 1% or 2% of their genomes and at the same time there are huge phenotypic differences between humans and apes?

2. What sources of human variation are described in this chapter? Why is it important to distinguish among them?

3. Consider the phenotype of human finger number. Most people have exactly five fingers on each hand, but some people have fewer. What is the source of variation in finger number?

4. How can natural selection maintain genetic variation within human populations?

5. What is the evidence that selection has generated genetic variation between human groups?

6. Why is it hard to determine the source of variation in human phenotypes? Why might it be easier to determine the source of variation for other animals?

7. Explain how studies of human twins allow researchers to estimate how genetic and environmental differences affect phenotypic traits.

8. Suppose you are told that differences in IQ scores among white Americans have a genetic basis. What would that tell you about whether differences in average scores between white Americans and Americans belonging to other ethnic groups had an environmental or genetic basis? Why?

9. Explain why genowide association studies show race is not a biologically meaningful category of classification.

10. Imagine that in the archaeological record of a given area you observed a change in pottery styles over time. How could you use ancient DNA to determine whether this change occurred in the context of (a) population continuity, with the local population changing their own culture, perhaps by borrowing ideas that they observed in their neighbors, or (b) population replacement, with the local population being largely displaced by people who brought a new culture into the area with them?

Further Reading

Falconer, D. S., and T. F. C. Mackay. 1996. *Introduction to Quantitative Genetics*. 4th ed. Essex, UK: Longman.

Gerbault, P., A. Liebert, Y. Itan, A. Powell, M. Currat, J. Burger, D. M. Swallow, and M. G. Thomas. 2013. "Evolution of Lactase Persistence: An Example of Human Niche Construction." *Philosophical Transactions of the Royal Society* 366: 863–877.

Jobling, M. A., E. Hollox, M. Hurles, T. Kivisild, and C. Tyler-Smith. 2014. *Human Evolutionary Genetics*. 2nd ed. New York: Garland Science.

Mielke, J. H., L. W. Konigsberg, and J. H. Relethford. 2010. *Human Biological Variation*. New York: Oxford University Press.

O'Bleness, M., V. B. Searles, A. Varki, P. Gagneux, and J. M. Sikela. 2012. "Evolution of Genetic and Genomic Features Unique to the Human Lineage." *Nature Reviews Genetics* 13: 853–866.

Reich. D. 2018. *Who We Are and How We Got Here: Ancient DNA and the New Science of the Human Past*. Pantheon, NY.

Rosenberg, N. A., J. K Pritchard, J. L. Weber, H. M. Cann, K. K. Kidd, L. A. Zhivotovsky, and M. W. Feldman. 2002. Genetic Structure of Human Populations, *Science* 298, 2381–2385.

 Review this chapter with personalized, interactive questions through INQUIZITIVE.

 This chapter also features a Guided Learning Exploration on modern human variation and "race."

15

EVOLUTION AND HUMAN BEHAVIOR

CHAPTER OBJECTIVES

By the end of this chapter you should be able to:

A Evaluate the argument that the application of evolutionary reasoning to understanding contemporary human behavior does not entail genetic determinism.

B Explain why evolutionary thinking helps us understand how people learn.

C Discuss reasons why people usually do not mate with close relatives.

D Explain how natural selection helps us understand why men and women both value good character in a marriage partner, and why men usually care more about youth and women more about control of resources.

403

Why Evolution Is Relevant to Human Behavior

The great geneticist Theodosius Dobzhansky once said, "Nothing in biology makes sense except in the light of evolution." The theory of evolution is at the core of our understanding of the natural world. By studying how natural selection, recombination, mutation, genetic drift, and other evolutionary processes interact to produce evolutionary change, we come to understand why organisms are the way they are. And since we are biological creatures, it stands to reason that evolutionary principles can be extended to understand ourselves.

However, there has been considerable resistance to this idea, even among those that are perfectly comfortable with everything that we have discussed in the previous chapters of this book. This is because many social scientists think evolutionary analyses imply that behavior is genetically determined and that differences in behavior must be the product of genetic differences between individuals. Both of these objections are based on fundamental misunderstandings about how evolution works.

All phenotypic traits, including behavioral traits, reflect the interaction between genes and the environment.

Many people have the mistaken view that behaviors are either genetic or learned. They believe that behaviors that are genetic must be fixed and unchangeable (like reflexes), while behaviors that are learned (like language) are controlled entirely by environmental contingencies. This assumption about the relationship between genes and behavior lies at the heart of the *nature–nurture debate* that has plagued the social sciences for many decades, and the assumption is fundamentally incorrect.

Many people think that genes are like engineering drawings for a finished machine, and that individuals vary simply because their genes carry different specifications. For example, they imagine that Kevin Durant is tall purely because he carries genes that specify an adult height of 2 m (6 ft. 10 in.). As we explained in Chapter 14, height is partly heritable, and this means the genes that you inherited from your parents have some effect on your height. But genes are not like blueprints that specify phenotype. Every trait results from the *interaction* of a genetic program with the environment. Thus, genes are more like recipes in the hands of a creative cook—sets of instructions about how to build an organism using materials available in the environment. At each step, this very complex process depends on the nature of local conditions. The expression of any genotype always depends on the environment. Your height is shaped by the genes you inherited from your parents, how well-nourished you were when you were growing up, and the nature of the diseases you were exposed to during childhood.

Natural selection can shape developmental processes so that organisms develop different adaptive behaviors in different environments.

Some people believe that natural selection cannot create adaptations unless behavioral differences between individuals are caused by genetic differences. If this were true, then evolutionary explanations of human behavior would be invalid because there is no doubt that most of the variation in behavioral traits, such as the languages people speak, the ways they make a living, and their religious beliefs, are the product of learning and culture, not genetic differences between individuals.

However, differences in behavior are not necessarily based on genetic differences between individuals. Natural selection shapes learning mechanisms so that organisms adjust their behavior to local conditions in an adaptive way. Recall from Chapter 3 that this is exactly what happens with soapberry bugs. Male soapberry bugs in Oklahoma guard their mates when females are scarce, but not when females are abundant. Individual males vary their behavior adaptively in response to the local sex ratio. For this kind of flexibility in male behavior to evolve, there had to be small genetic differences in the male propensity to guard a mated female and small genetic differences in how mate guarding is influenced by the local sex ratio. If such variation exists, then natural selection can mold the responses of males so that they are locally adaptive. In any given population, however, most of the observed behavioral variation occurs because individual males respond adaptively to environmental cues.

Behavior in the soapberry bug is relatively simple. Human learning and decision making are immensely more complex and flexible. We know much less about the mechanisms that produce behavioral flexibility in humans than we do about the mechanisms that produce flexibility in mate guarding among soapberry bugs. Nonetheless, such mechanisms must exist, and it is reasonable to assume that they have been shaped by natural selection. (Note that we are not arguing that all behavioral variation in human societies is adaptive. We know that evolution does not produce adaptation in every case, as we discussed in Chapter 3.) The crucial point here is that evolutionary approaches do not imply that differences in behavior among humans are the product of genetic differences between individuals.

In this chapter and the next, we consider how evolutionary theory can be used to understand how we think and how we behave. Our goal is to highlight some of the insights that researchers from various disciplines (especially psychology, anthropology, and economics) have gained from investigating human minds and behavior from an evolutionary perspective. As you will see, there are some forms of behavior that we share with other primates (like our avoidance of inbreeding) and others that are derived (like our capacity for large-scale cooperation).

Understanding How We Think

Evolutionary analyses provide important insights about how our brains are designed.

The adaptation that most clearly distinguishes humans from other primates is our large and very complex brain. Natural selection hasn't just made our brains bigger; it has shaped our cognitive abilities to solve particular kinds of problems. As we explained in Chapter 3, these are not necessarily the kinds of problems we face today. That is because for most of the history of our species, humans lived in small-scale foraging societies in which people relied on food that they hunted or gathered (**Figure 15.1**). Stratified societies with agriculture and high population density have existed for only a few thousand years (**Figure 15.2**). Although this

FIGURE 15.1

Evolutionary psychologists believe that the human mind has evolved to solve the adaptive challenges that confront food foragers because this is the subsistence strategy that humans have practiced for most of our evolutionary history.

FIGURE 15.2

Indigenous peoples of the Mississippi delta constructed these mound structures 2,500 to 1,300 years ago. Monumental architecture like this is based on the ability of one group of people to control the labor of others, a signal of social stratification.

FIGURE 15.3

Rats initially sample small amounts of unfamiliar foods, and if they become ill soon after eating something, they will not eat it again.

may have been enough time to evolve some fairly simple adaptations, such as the ability to digest lactose or resist the effects of malaria, it has not been enough time to have had much impact on the evolution of more complex traits that rely on thousands of genes, such as our brains. This has led John Tooby and Leda Cosmides of the University of California, Santa Barbara, to argue that our brains are designed for life in foraging societies and to solve a very specific set of adaptive problems that our ancestors faced repeatedly.

Natural selection has favored a suite of special-purpose learning mechanisms.

Psychologists once thought that animals had a few general-purpose learning mechanisms that allowed them to modify any aspect of their phenotypes adaptively. However, we now know that animals are predisposed to learn some things and not others. Rats live in a very wide range of environments where they frequently encounter new things. How do rats know whether something new is safe for them to eat? They conduct an experiment: When they encounter a new item, they nibble on it, and then wait for several hours. If they become ill in the interval, they do not eat the new item again. Rats pay attention to the taste and smell of foods instead of color or other features, perhaps because when they are foraging it is often too dark to see what they are eating (**Figure 15.3**). This taste-and-wait strategy makes sense for rats and has enabled them to become an extremely successful species. However, there are limits to the flexibility of their food-learning mechanisms. There are certain items that rats will never sample, and in this way their diet is rigidly controlled by genes. Moreover, the learning process is not affected equally by all environmental contingencies. For example, rats are affected more by the association of new tastes with gastric distress than they are by other possible associations.

Learning What to Eat and How to Avoid Danger

Human children grow up slowly and acquire much of the knowledge that they need to survive through social learning.

Just as rats need to learn about unfamiliar foods as they venture into new habitats, children have to learn about the world as they grow up. It is important for them to avoid mistakes that will have costly consequences, and so we might expect selection to favor learning mechanisms that help prevent them from making costly mistakes. Researchers have explored how children learn about some of the things that can be dangerous to them: plants and predators.

Plants are important to human subsistence, but not all plants are safe to eat.

As we saw in Chapter 11, plants provide a substantial fraction of the calories that foragers consume. Foragers have highly sophisticated knowledge of what foods are edible, how foods need to be processed, and so on. They need this knowledge because plants can be very dangerous. This may surprise you. It's easy to imagine dangerous animals—think of the endless list of monsters that populate legend and cinema. Plants, not so much. But they can be just as deadly. As we explained in Chapter 5, many plants are loaded with poison meant to deter insect predators, and many of these poisons are toxic to humans as well. To get a feeling for how many plants are toxic, look up online a "list of poisonous plants." Such a list will include 13 domesticated plants that we eat, such as potatoes and cassava (also called manioc or tapioca), and others that we plant in our gardens, such as oleander, lupine, holly, and ivy. Poisonous plants come in all shapes and sizes, and there are no helpful cues to tell us which plants are safe to eat. Trial-and-error learning can be very dangerous. For example, a single seed from the "suicide tree," common in the southern Indian state of Kerala, contains enough toxin to stop the heart.

Children have psychological learning mechanisms that help them avoid dangers from unfamiliar plants and learn which plants are edible.

Annie Wertz, a developmental psychologist now at the Max Planck Institute for Human Development in Berlin, realized that learning which plants are safe to eat and which ought to be avoided is an important adaptive problem for children. If they accidentally eat something that is toxic, they could become very sick or die. So she predicted that young children would be cautious about unfamiliar plants. She designed a series of experiments in which she presented 6- to 18-month-old children with a real plant, an artificial plant, novel artifacts that look or feel like plants, familiar artifacts, and natural objects. Then the experimenters measured how long it took for the children to reach out and touch each of the objects. (Parents were instructed to close their eyes, so they couldn't see what items were being presented and could not provide cues to the children.) It turns out that children are very suspicious of plants (**Figure 15.4**). They wait much longer before they touch real plants or artificial plants than before they touch other things, even things that they have never seen before.

The reluctance to touch plants may protect children from coming into contact with things that are dangerous to them, but they need to learn that they don't have to avoid all plants. This is especially important for forager children who often begin to gather some of their own food at a very young age. Wertz and her colleagues have shown that children have a number of mechanisms (or learning rules) that help them figure out which plants are edible. For example, children are more likely to look at their parent when they are presented with a plant than when they are presented with other objects, suggesting that they are looking for cues about whether the plant is dangerous. In another experiment, Wertz and her colleagues showed that 18-month-old children are particularly attuned to where foods come from and whether adults eat them

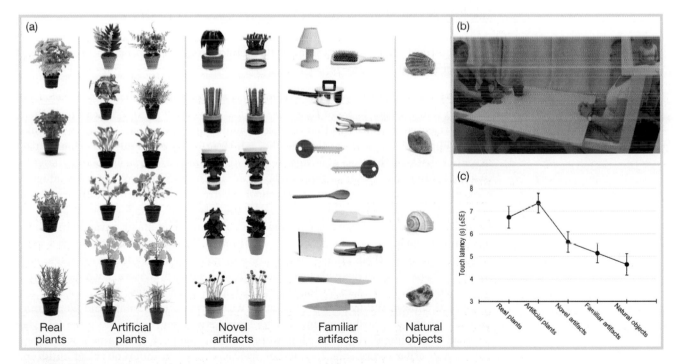

FIGURE 15.4

(a) Five sets of objects presented to the children: real plants, artificial plants, novel artifacts, familiar artifacts, and natural objects. The novel artifacts are matched to the color, shape, or texture of the plants, and the familiar artifacts include things children are typically allowed to touch and things that they are not. (b) The experimental setup: the parent, child, and stimulus object on the left and the experimenter on the right. The parent's eyes are closed to prevent cueing the child. (c) The latency to touch real and artificial plants (the interval between presentation and first contact) is much larger than the latency to touch other objects.

For this experiment, children were presented with two videos. In both videos, an adult sits between a plant with a fruit affixed to it and an unfamiliar artifact with a different fruit affixed to it. In one video, the adult puts the fruit from the plant in his mouth and says "hmm" in a neutral tone; he then does the same with the fruit from the unfamiliar artifact. In the other video, everything is exactly the same, but the adult puts the fruits behind his ear, instead of in his mouth. Children are shown one of these videos, and are then presented with the two fruits and asked which of the fruits they can eat. Children that see the video in which the adult puts the fruits in his mouth are much more likely to say that the fruit from the plant is edible than to say that the fruit from the unfamiliar artifact is edible. But children who see the video in which the adult puts the fruits behind his ear make no distinction between the fruit that comes from the plant versus that from the unfamiliar artifact. Experiments based on looking times suggest that even 6-month-old infants are surprised when they see adults put the fruit from the unfamiliar artifact into their mouths.

These experiments suggest that selection has prepared children to learn about plants in an adaptive way. They have a set of learning rules, or guidelines. First, be particularly careful about plants, as they can be more dangerous than other kinds of objects. Second, food is more likely to come from plants than from other kinds of things. Third, pay attention to what adults are doing; things that adults eat are likely to be safe.

The inference that these plant learning guidelines are part of our innate psychological machinery relies on the fact that the children who served as subjects of Wertz's experiments were very young and had little direct experience with plants or choosing their own foods. While this is plausible, it is also possible that the children's responses reflect specific elements of the culture in which they were raised. Evidence that children raised in very different environments show similar kinds of responses are needed to rule out this possibility. In the next section, we discuss a body of work that provides this kind of evidence.

Cross-cultural experiments indicate that children are also prepared to learn whether animals are dangerous.

Clark Barrett of the University of California, Los Angeles, has investigated whether children find it easier to learn that animals are dangerous than to learn about their other attributes. He reasoned that selection should have shaped our psychology so children find it easier to learn and remember which animals are dangerous than which are edible because the costs of failing to learn about danger are much greater than the costs of failing to learn something can be eaten. Barrett and his colleagues conducted a parallel set of experiments with preschool children living in urban California and children living in remote Shuar villages in the forests of Ecuador. The children in California interact with pets, see animals in zoos, and read about animals in books, but rarely come into contact with dangerous animals. The Shuar children live in a very different world. In their villages there are scorpions, highly venomous snakes, and they are more likely to hear about encounters with large predators, such as jaguars. Shuar men hunt for food and bring their kills back to their homes.

Barrett showed children a series of cards with pictures of animals that were unfamiliar to them, one card at a time. In one treatment, a set of children were told the name of each animal, and then they were told that some of the animals were dangerous and others were safe. In another treatment with a different set of children, children were told the name of each animal, and the animals were identified as edible or not edible, and in a control treatment, a third set of children were just told the name of each animal. Once a child had been shown all of the cards of a treatment, she was shown them again and asked the animal's name and whether it was dangerous or safe, edible or inedible, depending on the treatment. A week later, the children were asked again whether or not the animals were dangerous or edible. As shown in **Figure 15.5**, children in both cultures were more likely to remember which animals

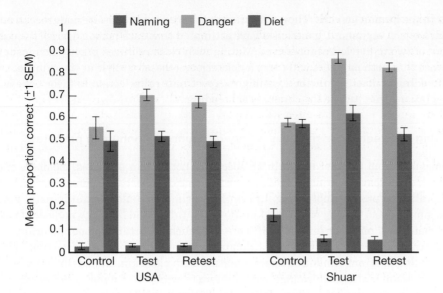

FIGURE 15.5

Both American and Shuar children are more likely to remember unfamiliar animals that are dangerous than unfamiliar animals are that are edible.

were dangerous than which animals were edible. The Shuar children did somewhat better on the recall tests than the American children did, perhaps because the Shuar sample included more older children than the American sample did.

The Psychology of Reproductive Behavior

As we explained in Chapter 6, reproduction is the central act in the life of every living thing, and so we might expect natural selection to have shaped our mating psychology in adaptive ways.

Choosing the right mate can have important effects on the fitness of individuals, and natural selection has shaped mate preferences in other animals in adaptive ways. For example, in some bird species, females prefer males with high-quality territories over males with poor territories. Female pea hens prefer male peacocks with more eyes in their trains, because this is a signal of male genetic quality. Martin Muller of the University of New Mexico and his colleagues have found that male chimpanzees prefer older females over younger females, probably because older mothers are more successful in rearing offspring than are young mothers. Evolutionary theory provides some insight about human mate preferences in terms of both who we should avoid mating with and who we should be attracted to.

Inbreeding Avoidance

The offspring of genetically related parents have lower fitness than that of the offspring of unrelated parents.

Matings between relatives are referred to as **inbred matings**, or inbreeding, and matings between unrelated individuals are referred to as **outbred matings**, or outbreeding. The offspring of inbred matings are much more likely to be homozygous for deleterious recessive alleles than are the offspring of outbred matings. As a consequence, inbred offspring are less robust and have higher mortality than the offspring of outbred matings. In Chapter 14, we discussed several genetic diseases, such as phenylketonuria, Tay–Sachs disease, and cystic fibrosis, that are caused by a recessive gene. People who are heterozygous for such deleterious recessive alleles are unaffected, but people who are homozygous suffer severe, often fatal consequences. Recall that such alleles occur at low frequencies in most human populations. However, there are many loci in the human genome. Thus, even though the frequency of deleterious recessives at each locus is very small, geneticists have estimated that each person carries the equivalent of two to five lethal recessives. Mating with close relatives greatly increases the chance that both partners will carry a deleterious recessive allele at the same locus. If inbreeding is deleterious, then we might expect natural selection to favor behavioral adaptations that reduce the chance of inbreeding.

Mating between close relatives is very rare among nonhuman primates.

Remember from Chapter 6 that in all species of nonhuman primates, members of one or both sexes leave their natal groups near the time of puberty. Dispersal is probably an adaptation to prevent inbreeding. In principle, primates could remain in natal groups and simply avoid mating with close kin. However, this would limit the number of potential mates and might be unreliable if there was much uncertainty about paternity.

Natural selection has provided primates with another form of protection against inbreeding: a strong inhibition against mating with close kin. In matrilineal macaque groups, some males acquire high rank and mate with adult females before they disperse from their natal groups. However, matings among close kin are extremely uncommon. Although most female chimpanzees disperse from their natal groups, some remain, and this means that some adult females live in the same groups as their fathers and have opportunities to mate with them (**Figure 15.6**). However, they rarely do. Female chimpanzees seem to have a general aversion to mating with males much older than themselves, and males seem to be generally uninterested in females much younger than themselves. These mechanisms may protect females from mating with their fathers, and vice versa.

Humans rarely mate with or marry close relatives.

The genetic disadvantages of inbreeding apply to humans as well as other animals, and inbreeding avoidance is a universal feature of modern human societies. There is not a single ethnographically documented case of a society in which brothers and sisters regularly marry or one in which parents regularly mate with their own children. The pattern for more distant kin is much more variable. Some societies permit both sex and marriage with nieces and nephews or between first cousins; other societies prohibit sex and marriage among even distant relatives.

Adults are not sexually attracted to the people with whom they grew up.

The fact that inbreeding avoidance is very common among primates suggests that our human ancestors probably also had psychological mechanisms preventing them from mating with close kin. These psychological mechanisms would disappear during human evolution only if they were selected against. However, mating with close relatives is deleterious in humans, as it is in other primates. Thus, both theory and data predict that modern humans will have psychological mechanisms that reduce the chance of close inbreeding, at least in the small-scale societies in which human psychology was shaped.

There is evidence that such psychological mechanisms exist. In the late nineteenth century, the Finnish sociologist Edward Westermarck speculated that childhood propinquity stifles desire. By this he meant that people who live in intimate association as small children do not find each other sexually attractive as adults. Several case studies provide support for Westermarck's hypothesis:

- *Minor marriage.* Until recently, an unusual form of marriage was widespread in China. In **minor marriages**, children were betrothed and the prospective bride was adopted into the family of her future husband often at a young age. There, the betrothed couple grew up together like brother and sister. Arthur Wolf, an

FIGURE 15.6

Female chimpanzees avoid mating with closely related males. Although mothers have close and affectionate relationships with their adult sons, matings between mothers and sons are quite uncommon.

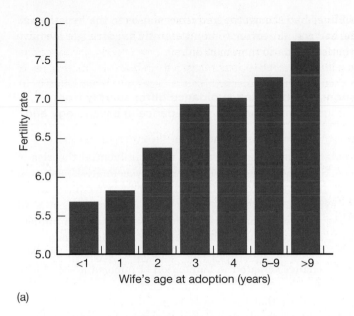

(a)

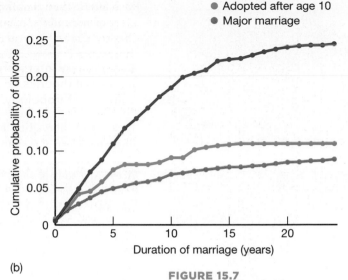

(b)

FIGURE 15.7

In minor marriages, the age of the wife when she arrives in her future husband's household (age at adoption) affects both fertility and the likelihood of divorce. (a) The fertility of women adopted at young ages is depressed. (b) The younger a woman is when she arrives in her husband's household, the less likely it is that the marriage will survive

anthropologist at Stanford University, conducted research on minor marriage in Taiwan. His informants told him that partners in minor marriages found each other sexually unexciting. Sexual uninterest was so great that fathers-in-law sometimes had to beat the newlyweds to persuade them to consummate their marriage. Wolf's data indicate that minor marriages produced about 30% fewer children than other arranged marriages did (**Figure 15.7a**) and were much more likely to end in separation or divorce (**Figure 15.7b**). Infidelity was also more common in minor marriages. When modernization reduced parental authority, many young men and women who were betrothed in minor marriages broke their engagements and married others.

- *Kibbutz age-mates.* Before World War II, many Jewish immigrants to Israel organized themselves into utopian communities called *kibbutzim* (plural of **kibbutz**). In these communities, children were raised in communal nurseries, and they lived intimately with a small group of unrelated age-mates from infancy to adulthood. The ideology of the kibbutzim did not discourage sexual experimentation or marriage by children in such peer groups, but neither occurred. The Israeli sociologist Joseph Sepher, himself a kibbutznik, collected data on 2,769 marriages in 211 kibbutzim. Only 14 of them were between members of the same peer group, and in all these cases one partner joined the peer group after the age of 6. From data collected in his own kibbutz, Sepher found no instances of premarital sex among members of the same peer group.

- *Third-party attitudes toward incest.* As you may have realized already, aversions to inbreeding extend beyond our attitudes toward our own mating behavior to include strong beliefs about appropriate mating behavior by other individuals. We are disgusted not only by the idea of having sex with our parents or our own children but also by the idea of other people having sex with their children. Westermarck hypothesized that co-residence during childhood generates sexual aversions to particular partners. If that is the case, then the extent of exposure to siblings of the opposite sex during childhood might also be linked to the strength of feelings about one's own behavior and the strength of feelings about the behavior of others. These predictions have been tested in two experimental studies. One was conducted by Daniel Fessler of the University of California, Los Angeles, and Carolos Navarette of Michigan State University. The other was conducted by Debra Lieberman at the University of Miami, John Tooby, and Leda Cosmides. In both studies, subjects (university undergraduates) were asked to contemplate hypothetical cases of consensual sexual relationships involving adult siblings. Those who had grown up with opposite-sex siblings had stronger negative responses to the hypothetical scenario than those who had not. Moreover, women generally had stronger aversive responses to the hypothetical scenario than men did.

Evolutionary interpretations of inbreeding avoidance differ sharply from influential theories about incest and inbreeding avoidance in psychology and cultural anthropology.

Incest and inbreeding avoidance play a central role in many influential theories of human society. For example, Sigmund Freud (the founder of psychoanalysis) asserted that people harbor a deep desire to have sex with members of their immediate family. According to this view, the existence of culturally imposed rules against incest is all that saves society from these destructive passions. Claude Levi-Strauss, an influential anthropological theorist, argued that prohibitions against matings among close kin are distinctly cultural phenomena that encourage exchanges of women between groups, and that reciprocal exchanges create the foundation of important social alliances. Neither of these arguments is very plausible from an evolutionary perspective. Both theory and observation suggest that the family is not the focus of desire; it's a tiny island of sexual indifference. There are compelling theoretical reasons to expect that natural selection will erect psychological barriers to incest and good evidence that inbreeding avoidance evolved well before modern humans appeared.

However, the evolutionary analysis we have outlined here is not quite complete. For example, we might expect the Westermarck effect and egocentric empathy to produce an aversion to minor marriage in China. Yet this practice has persisted for a long time. It is possible that psychological mechanisms are supplemented or perhaps superseded by conscious reasoning. People in many societies believe that incest leads to sickness and deformity, and their beliefs may guide their behavior and shape their cultural practices. Finally, it seems clear that attitudes about incest are not based solely on the deleterious effects of inbreeding. If they were, then all societies would have the same kinds of rules about who can have sexual relationships. Instead, we find considerable variation. For example, some societies encourage first cousins to marry, whereas others prohibit them from doing so.

Human Mate Preferences

Marry

Children—(if it Please God)—Constant companion, (& friend in old age) who will feel interested in one,—object *to be* beloved and played with. better than a dog anyhow.—Home, & someone to take care of house—Charms of music & female chit-chat.—These things good for one's health.—*but terrible loss of time.*—

My God, it is intolerable to think of spending one's whole life, like a neuter bee, working, working, & nothing after all.—No, no won't do.—Imagine living all one's day solitary in smoky dirty London house.—Only picture to yourself nice soft wife on a sofa with good fire, & books, & music perhaps—Compare this vision with the dingy reality of Grt. Marlbro St.

Marry—Mary—Marry Q.E.D.

Not Marry

Freedom to go where one liked—choice of Society & *little of it.*—Conversation of clever men at clubs—Not forced to visit relatives, & to bend in every trifle.—to have the expense & anxiety of children—perhaps quarrelling—**Loss of time.**—cannot read in the Evenings—fatness & idleness—Anxiety & responsibility—less money for books & c—if many children forced to gain one's bread.—(But then it is very bad for one's health to work too much)

Perhaps my wife won't like London; then the sentence is banishment & degradation into indolent, idle fool (Burkhardt and Smith, 1986, p. 444).

These are the thoughts of 29-year-old Charles Darwin, recently returned from his 5-year voyage on HMS *Beagle*. Soon after writing these words, Darwin married his cousin Emma, the daughter of Josiah Wedgwood, the progressive and immensely wealthy manufacturer of Wedgwood china (**Figure 15.8**). By all accounts, Charles and Emma were a devoted couple. Emma bore 10 children and nursed Charles through countless bouts of illness. Charles toiled over his work and astutely managed his investments, parlaying his modest inheritance and his wife's more substantial one into a considerable fortune.

Darwin's frank reflections on advantages and disadvantages of marriage were very much those of a conventional, upper-class Victorian gentleman. But people of every culture, class, and sex have faced the problem of choosing mates. Sometimes people choose their own mates, and other times parents arrange their children's marriages. But everywhere, people care about the kind of person they will marry and mate with.

FIGURE 15.8

Charles Darwin courted and married his cousin, Emma Wedgwood. This portrait was painted when Emma was 32, just after the birth of her first child.

Evolutionary theory generates some testable predictions about the psychology of human mate preferences.

David Buss, a psychologist at the University of Texas at Austin, was among the first to examine the evolutionary logic underlying human mating preferences and tactics. Buss reasoned that for much of their evolutionary history, humans have lived in foraging societies. The adaptive challenges that men and women face in these kinds of societies are likely to have shaped their mating preferences. For women, this might have meant choosing men who would provide them with access to resources. Recall from Chapter 11 that there is considerable interdependence between men and women in foraging societies. Women mainly gather plant foods, and men are mainly responsible for hunting. Women's consumption exceeds their production for much of their reproductive life. Children do not begin to provide substantial amounts of their own food until they reach adolescence. Thus, for women, it might be important to choose a mate who will be a good provider. Humans tend to form lasting reproductive bonds, so men's reproductive success depends largely on the fertility of their mating partners. Women's fertility is highest when they are in their twenties and declines to zero when women reach menopause, at about 50 years of age (**Figure 15.9**). Thus, selection should

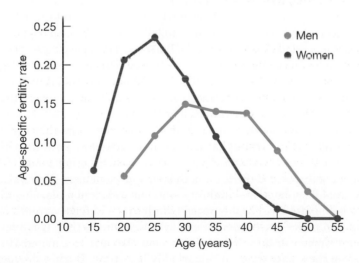

FIGURE 15.9

Age-specific fertility rates give the probability of producing a child at particular ages. !Kung women have their first child between the ages of 15 and 19 and have the highest fertility rate in their twenties. Women's fertility falls to zero by age 50. !Kung men do not begin to reproduce until their early twenties, and their fertility rates are fairly stable in their thirties and forties, dropping to low levels in their fifties.

FIGURE 15.10

People in 33 countries (red) were
surveyed about the qualities
of an ideal mate.

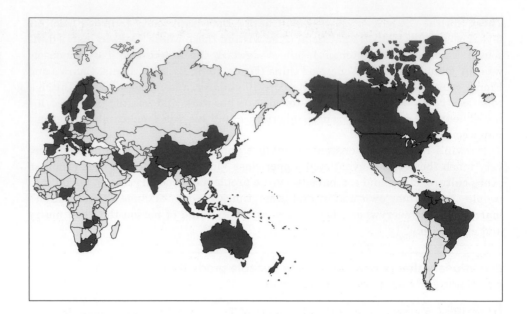

have favored men who chose young and healthy mates. Picture IDs were scarce in the Pleistocene, so selection may have shaped men's psychology so that they are attracted to cues that reliably predict youth and health, such as smooth skin, good muscle tone, shiny hair, and symmetric features. Thus, Buss predicted that there would be substantial sex differences in mate preferences. Women should be more concerned about their mate's ability to provide resources than men are, and men should be more concerned about their mate's looks than women are.

If evolution has shaped the psychology of human mating strategies, then we would expect to find common patterns across societies.

To test his hypotheses about sex differences in mate preferences, Buss used a standardized questionnaire that had been developed by Reuben Hill, a sociologist, in the 1930s. In the questionnaire, subjects are asked to rate 18 traits of potential mates, including attractiveness, financial prospects, compatibility, and chastity, according to their desirability. Buss enlisted colleagues in 33 countries around the world to administer the survey and generated a sample of more than 10,000 men and women (**Figure 15.10**). The results strongly supported Buss's hypothesis. Buss found that people's sex had the greatest effect on their ratings of the following traits: "good financial prospect," "good looks," "good cook and housekeeper," and "ambition and industriousness." As the evolutionary model predicts, women value good financial prospects and ambition more than men do, and men value good looks more than women do.

Although Buss found the expected sex differences, the traits that generate these sex differences are not the traits that matter most to people. Both men and women around the world rate mutual attraction or love above all other traits (**Table 15.1**). The next most highly desired traits for both men and women are personal attributes, such as dependability, emotional stability and maturity, and a pleasing disposition. Good health is the fifth most highly rated trait for men and the seventh for women. These traits also make sense from an evolutionary perspective. It is important for both men and women to find mates with whom they can get along. Human children depend on their parents for a remarkably long time. During this period, both parents provide food and shelter for their children. Because parental investment lasts for many years, it makes sense that both men and women would value traits in their partners that help them sustain their relationships. Both partners are likely to value personal qualities such as compatibility, agreeableness, reliability, and tolerance.

TABLE 15.1

Trait	Ranking of Trait by:	
	Males	**Females**
Mutual attraction/love	1	1
Dependable character	2	2
Emotional stability and maturity	3	3
Pleasing disposition	4	4
Good health	5	7
Education and intelligence	6	5
Sociability	7	6
Desire for home and children	8	8
Refinement, neatness	9	10
Good looks	10	13
Ambition and industriousness	11	9
Good cook and housekeeper	12	15
Good financial prospect	13	12
Similar education	14	11
Favorable social status or rating	15	14
Chastity*	16	18
Similar religious background	17	16
Similar political background	18	17

Men and women from 33 countries around the world (shown in Figure 15.10) were asked to rate the desirability of a variety of traits in prospective mates. The rankings of the values assigned to each trait, on average, are given here. Subjects were asked to rate each trait from 0 (irrelevant or unimportant) to 3 (indispensable). High ranks (low numbers) represent traits that were generally thought to be important.

*Chastity was defined in this study as having no sexual experience before marriage.

Culture predicts people's mate preferences better than their gender does.

Buss's data set reveals considerable variation from country to country. Buss and his colleagues found that the country of residence has a greater effect than a person's gender on variation in all of the 18 traits in Table 15.1, except for "good financial prospect." This means that knowing where a person lives tells you more about what he or she values in a mate than knowing the person's gender. Of the 18 traits, chastity (defined in Buss's study as no sexual experience before marriage) shows the greatest variability among populations. In Sweden, men rate chastity at 0.25, and women rate it at 0.28 on a scale of 0 (irrelevant) to 3 (indispensable). In contrast, Chinese men rate chastity 2.54, and Chinese women rate it 2.61 (**Figure 15.11**), indicating that there is more similarity between men and women from the same population than there is among members of each gender from different populations.

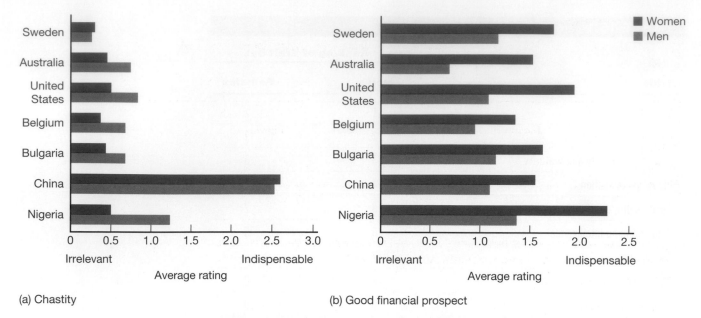

(a) Chastity

(b) Good financial prospect

FIGURE 15.11

Culture accounts for substantial variation in mate preferences. The average ratings given by men and women in several countries surveyed are shown for the trait with (a) the highest interpopulation variability ("chastity") and (b) the lowest interpopulation variability ("good financial prospect").

Patterns of jealousy reflect the complex interaction between our evolved psychology, culture, and environment.

Although we usually think that jealousy is a negative emotion, David Buss and his colleagues have suggested that jealousy may serve a useful evolutionary function because it reduces the risk of infidelity. Infidelity can reduce the fitness of both men and women. Women may suffer fitness costs if their partners form relationships with other women and divert resources to them. Men may suffer fitness costs if their partners become pregnant by other men, and they inadvertently invest in children they have not sired. Although both men and women may be affected by their partners' infidelity, Buss and his colleagues thought that there might be gender differences in the focus of their concern. So they designed a study in which they asked whether subjects would be more concerned about their partner having sex with someone else (sexual infidelity) or their partner falling in love with someone else (emotional infidelity). They found that men were more concerned about sexual infidelity than emotional infidelity, while women were more concerned about emotional infidelity than sexual infidelity. This basic pattern of results has been replicated in a number of studies conducted in industrialized countries, mainly with student populations.

Do the same patterns characterize a wider range of human societies? To answer this question, Brooke Scelza of the University of California, Los Angeles, assembled a team of researchers to conduct parallel studies of partner jealousy in eight small-scale societies around the world and in three urban areas. In each society, they asked people to rate infidelity by men and women on a 5-point scale from very good to very bad. They also asked whether sexual infidelity or emotional infidelity would be more distressing. Overall, men rated sexual infidelity more negatively than women did. And both men and women agreed that sexual infidelity by women was worse than sexual infidelity by men. However, there was considerable variation in how bad people thought that infidelity was. There was also considerable variation across societies about whether sexual infidelity or emotional infidelity would be more upsetting (**Figure 5.12**). In six of the eight small-scale societies and one of the

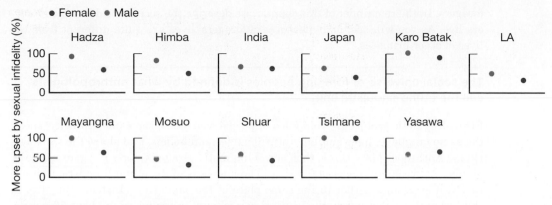

FIGURE 15.12

Men and women were asked whether they thought sexual or emotional infidelity would be more upsetting. There was considerable variation among societies about whether sexual infidelity would be more upsetting than emotional infidelity.

three urban societies, a majority of women thought that sexual infidelity was more distressing than emotional infidelity. As we saw in Buss's study of mate preferences, knowing where someone comes from tells you more about their feelings about infidelity than their gender does.

Scelza and her colleagues also wanted to understand the factors that might contribute to the variation that they observed across societies. They reasoned that if concerns about partner fidelity are based on paternity uncertainty and the risk of misallocating investment, then there ought to be a link between the extent of paternal investment and the extent of concern about infidelity. After all, if males invest very little in their offspring, then the costs of misallocating investment will be low, and vice versa. So members of Scelza's team put together information for each society about the amount of direct care that men provided for children (holding, grooming, feeding, play) and how much men contributed to their own children's subsistence through food that they provided, as well as whether they provided resources (money, land, livestock) and paid costs associated with their children's marriage. As they predicted, the extent of concern about infidelity was substantially higher in societies in which men invested more heavily in their children than in societies in which men invested less.

These studies of inbreeding aversion, mate preferences, and partner jealousy illustrate an important point: Evolutionary explanations that invoke an evolved psychology and cultural explanations that are based on the social and cultural milieu are not mutually exclusive. The cross-cultural data suggest that some uniformities among people are the result of evolved psychological mechanisms. But this is not the whole story. The cross-cultural data suggest that people are also strongly influenced by the cultural environment in which they live.

Human Social Organization, Mating Systems, and Family Structure

For much of human evolutionary history, people made a living by gathering wild plants, collecting honey, and hunting wild game.

Research on how children learn about the world and studies of human mating psychology share the assumption that our minds have been shaped to solve the kinds of problems that our ancestors faced during the thousands of years that humans lived as foragers. In the remainder of this chapter, we describe the social world in which foragers live. As you will see, many features of foragers' lives are quite different from the lives of other primates.

The social universe of foraging peoples is defined by what anthropologists call the ethno-linguistic unit.

Ethno-linguistic groups are composed of people who share the same ethnicity, speak the same language, have similar cultural values and beliefs, and share the same territory. Anthropologists think that ethno-linguistic groups among foragers typically number between 500 and 1,000 people. The members of the ethno-linguistic unit are rarely, if ever, all together in the same place at the same time. Instead, they live in smaller units, called bands or camps, that are usually composed of about 30 people. Bands typically include several nuclear families, as well as some unmarried adults. Within bands, people form households, and sometimes a few households are clustered together in space. Recall from Chapter 5 that no other primates form groups composed of multiple bonded pairs, although a number of primates, including geladas and hamadryas baboons, live in multilevel societies.

Membership in forager bands is quite fluid, as bands may split apart or two bands may merge together. Individuals also move regularly from one band to another. For example, studies of Hadza band membership led by Coren Apicella of the University of Pennsylvania show that people continue to live with only 20% of their bandmates from one year to the next. In most forager societies, camps shift from one location to another several times a year, as resource availability changes. High levels of mobility are possible because foragers in these societies do not build elaborate structures or accumulate many material possessions. However, in some places of the world, such as the northwest coast of North America, past foraging peoples had access to high-density marine resources and established permanent settlements, lived in larger groups, and built more elaborate structures.

Foragers' mating arrangements also differ from the mating arrangements of other primates.

People don't just mate, they marry. Marriages are socially recognized unions, which serve romantic, reproductive, economic, and social functions. A review of marriage practices by Menelaos Apostolou of the University of Nicosia in Cyprus indicates that marriages are arranged by parents or other close relatives in 85% of forager societies. This does not mean parents are entirely indifferent to their children's wishes, but they may take advantage of opportunities to establish connections to other families that create social or economic benefits. Marriage creates ongoing connections between families because people maintain ties to their natal families after they marry and also recognize kin connections through marriage (**affinal kinship**). Affinal kin are what we call our in-laws. Polygyny is permitted in most forager societies, but it is not common in practice. The late Frank Marlowe estimated that about 10% to 20% of men in forager groups have more than one wife.

There is considerable variation (and flexibility) in marital residence patterns in forager societies, as couples in some societies live with the wife's kin after marriage, couples in other societies live with the husband's kin, and couples in still other societies are equally likely to live with the husband's kin or the wife's kin. In addition, couples may move back and forth between their families for the duration of their marriage. An analysis of band composition in 32 forager groups led by Kim Hill of Arizona State University shows that one of the consequences of this variation in marital residence patterns is that foragers are equally likely to live with the husband's parents as they are to live with the wife's parents. And because men and women are equally likely to remain in their parents' bands in many forager societies, it is fairly common for men to live in the same group as their adult sisters and for women to live in the same group as their adult brothers. Co-residence of adult opposite-sex siblings is very rare in other primates.

The composition of forager bands reflects the overlapping connections between individuals, their own kin, and their spouses' kin.

Hill and his colleagues were able to reconstruct the complete composition of 58 Aché bands in the Amazon basin (**Figure 15.13**) and six Ju/'hoansi bands in southern Africa. As Figure 15.13 shows, people in these societies live with relatively few of their own primary kin (parents, siblings, offspring). However, they are connected to about 75% of the other adults in their groups through blood or marriage.

High levels of residential mobility and mechanisms for organizing a complex web of connections enable foragers to build extensive social networks. Kim Hill and Brian Wood, the latter now at the University of California, Los Angeles, conducted parallel sets of interviews among the Aché and the Hadza in which they asked people whether they had interacted with other individuals in particular ways. They were able to extrapolate from these data how many different people men hunted with, shared food with, observed making tools, joked with, and so on each year. According to their calculations, each Aché and Hadza man would hunt with about 10% of the men in his ethnic group each year, and over the course of his lifetime each man would have hunted with 200 to 300 different men. They estimated that each adult is connected through these kinds of interactions to about 1,000 people, a number that greatly exceeds the expected number for any other primate.

People rely on kinship to organize their social world.

The combination of reasonably high paternity certainty provided by pair bonding and the maintenance of natal kin connections enables people to track kinship through both maternal and paternal lines. All modern societies, including the societies of foraging peoples, have kinship systems that provide linguistic terms that distinguish different categories of kin and specify social norms that regulate the nature of rights and obligations among them. These kinship systems help people map their social universe.

There is considerable variation across cultures in how particular types of relatives are distributed into kin categories. For example, we use the term *sister* to refer to our female siblings who share the same mother and father, and we use the term *first cousin* to refer to our parents' siblings' male and female offspring. However, there are some kinship systems in which sisters and female first cousins would be lumped together in a single category. People everywhere understand the difference between what we call sisters and cousins, but these are not socially relevant distinctions in some societies. In some kinship systems, distinctions are made between maternal and paternal kin (**Figure 15.14**), although our kinship system does not highlight that distinction.

Kin terms help people keep track of their connections to others. Other primates rely on their own direct experience or observations of others to identify kin, but kin terms enable us to learn about a much wider range of relatives—even those that we have never met. This enables people to construct extensive family trees that extend across generations. Kin terms also make it much easier to talk about kin connections. Referring to your brother-in-law's cousin is much simpler than saying "my sister's husband's parent's sibling's child," and your listener doesn't need to do the mental gymnastics to trace this kin connection through its many steps. Moreover,

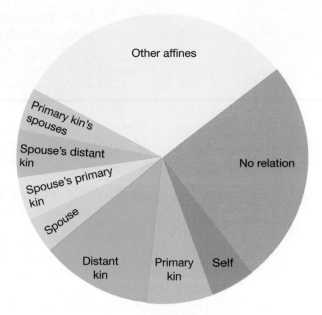

FIGURE 15.13

This figure shows the distribution of various types of kin living in Aché bands. Primary kin include parents, siblings, and offspring. Affines are relatives through marriage. The category "other affines" includes spouses of distant kin, distant kin of spouse, and affines of affines.

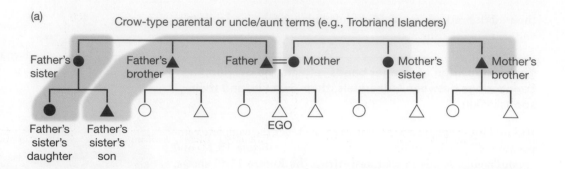

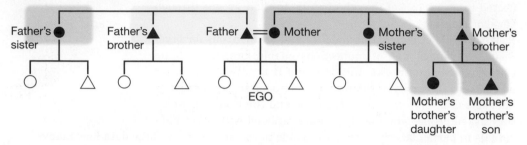

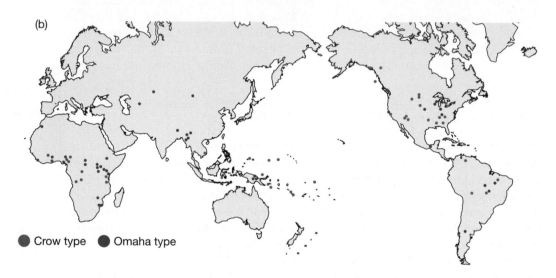

FIGURE 15.14

There are about six major types of kinship systems across the world. (a) The Crow-type system, named after the Crow people of North America, maps kinship along matrilineal lines, and the Omaha-type system, named after another group of North American people, maps kinship along patrilineal ones. Note that these systems are mirror images of each other. (b) Both systems are found across the world.

your listener doesn't need to know that your sister's name is Meghan, and her husband's name is Harry, and so on.

Cooperative networks extend beyond networks of kin and reciprocating partners.

As we explained earlier, cooperation plays an important role in the lives of foragers. Sexual division of labor and food sharing are universal features of foraging societies.

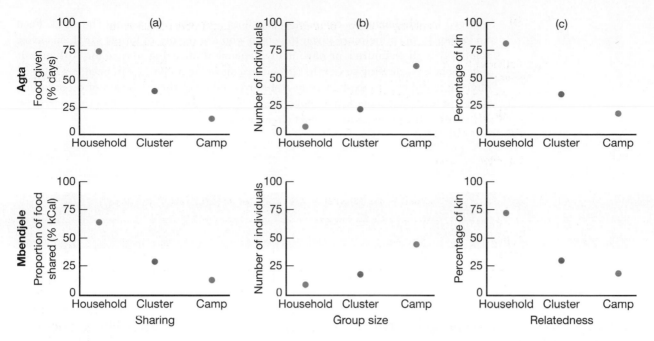

FIGURE 15.15

In (a), the distribution of food within households, between households within the same household cluster, and to other members of the camp is shown for the Agta (top) and Mbendjele (bottom). In (b), the group size for each of these units is shown, and in (c) the degree of relatedness within each of these units is shown.

Much of the food that foragers bring back to camp, particularly meat, is consumed by members of other households. A collaborative study of food sharing in two forager societies, the Agta of the Philippines and the Mbendjele of central Africa, conducted by researchers from University College London, documented how food was distributed within camps. In both cases, the most food is reserved for the forager's own household, but substantial amounts are given to other households within the same neighborhood cluster, and some is distributed to camp members outside the neighborhood cluster (**Figure 15.15**). Data on the degree of relatedness within households, clusters, and camps suggest that kinship and reciprocity have an important impact on food distribution among the Agta and Mbendjele, but sharing also extends to people that are not related and are not very likely to reciprocate.

Need also plays an important role in food sharing in forager societies. For example, the researchers who were studying food distribution among the Agta conducted an experiment in which people were presented with pictures of themselves and 10 randomly selected residents of their camps. They were given a number of tokens, each worth about 125 g (4.4 oz.) of rice, and asked to place the tokens on the photos of the people they wanted to give rice to (including themselves). People gave more to close kin than to distant kin, more to people who also allocated resources to them, and more to people who were needy. And when people were asked to explain the reasons for their allocations, recipient need was a major factor (**Figure 15.16**).

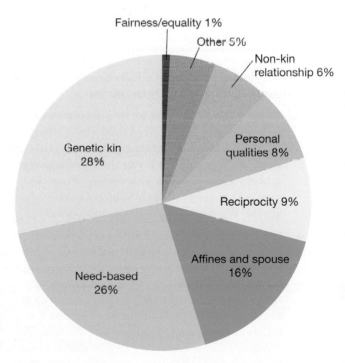

FIGURE 15.16

This pie chart shows responses of the Agta when they were asked about the reasons for their allocations of rice. Genetic or affinal kinship was mentioned about half of the time, and recipient need was mentioned about one-quarter of the time.

Need-based sharing is common among foragers, not just among the Agta. Food and other forms of help are given to people who are too old to forage for themselves, temporarily ill or injured, or have many dependent offspring to care for. Although it may seem unsurprising to you that people are moved to help those in need, need-based sharing is difficult to explain from standard models of the evolution of cooperation, which predict that cooperation will be limited to close kin and reciprocating partners. We turn to this problem in the next chapter.

CHAPTER REVIEW

Key Terms

inbred matings (p. 409)   minor marriages (p. 410)   kibbutz (p. 411)   affinal kinship (p. 418)
outbred matings (p. 409)

Study Questions

1. Much of the behavior of all primates is learned. Nonetheless, we have suggested many times that primate behavior has been shaped by natural selection. How can natural selection shape behaviors that are learned?

2. What is the nature–nurture debate? Explain why this debate is based on a flawed understanding of how evolution shapes behavior.

3. Many of the things that we do are consistent with general predictions derived from evolutionary theory. We love our children, help our relatives, and avoid sex with close kin. But there are also many aspects of the behavior of members of our own society that seem unlikely to increase individual fitness. What are some of these behaviors?

4. In Chapter 6, we said that the reproductive success of most male primates depends on the number of females with which they mate. Here we discussed Buss's argument that a man's reproductive success will depend mainly on the health and fertility of his mate. Why are humans different from most primates? Among what other primate species should we expect males to attend to the physical characteristics of females when choosing mates?

5. How do data from cross-cultural studies of sexual jealousy illustrate the complex interactions of biology, environment, and culture?

6. Some people who read about evolutionary analyses of human mating patterns might come to the conclusion that it is justifiable for men to prefer younger partners over older ones or for women to prefer wealthier partners over poorer ones. Assess this conclusion in light of the naturalistic fallacy we discussed in Chapter 6.

7. Affines (or in-laws) play an important role in human societies, which seems to fit with predictions derived from kin selection. In fact, it seems surprising that affinal connections do not play a bigger role in other primate species. Why do you think that these kinds of kin ties have come to play a larger role in human societies than in the societies of other primates?

8. In the text, we make a distinction between mating and marriage. What is the basis of this distinction? How does marriage function in human societies?

9. Some researchers believe that language has played an important role in the development of human social organization and the development of extensive kin networks. This suggests that language shapes the way we think about the world. Can you think of other ways in which the way we think is shaped by language?

10. In some societies, help is allocated on the basis of the need of recipients. Why is need-based sharing a puzzle from an evolutionary perspective?

Further Reading

Barkow, J. H., L. Cosmides, and J. Tooby, eds. 1995. *The Adapted Mind: Evolutionary Psychology and the Generation of Culture*. New York: Oxford University Press.

Barrett, L., R. Dunbar, and J. Lycett. 2002. *Human Evolutionary Psychology*. Princeton, N.J.: Princeton University Press.

Buss, DA. 2019. *Evolutionary Psychology: The New Sciences of the Mind*. New York: Routledge.

Hill, K. R., R. S. Walker, M. Božičević, J. Eder, T. Headland, B. Hewlett, A. M. Hurtado, F. Marlowe, P. Wiessner, and B. Wood. 2011. "Co-residence Patterns in Hunter-Gatherer Societies Show Unique Human Social Structure." *Science*, 331(6022), pp. 1286–1289.

Smith, EA and B. Winterhalder. 2017. *Evolutionary Ecology and Human Behavior*. New York: Routledge.

Wertz, A. E. 2019. "How Plants Shape the Mind." *Trends in Cognitive Sciences*, 23(7), 528–531.

Review this chapter with personalized, interactive questions through INQUIZITIVE.

16

CULTURE, COOPERATION, AND HUMAN UNIQUENESS

CHAPTER OBJECTIVES

By the end of this chapter you should be able to:

A Describe how cumulative cultural adaptation allows humans to evolve more rapidly to a wider range of habitats than other mammal species can.

B Assess how different learning mechanisms can sustain cultural traditions.

C Assess possible reasons why, despite cultural traditions being common in other species, cumulative cultural adaptation is very rare.

D Discuss why adaptive modes of cultural learning can lead to maladaptive behavior.

E Compare the pattern and scope of cooperation in humans to that of other mammal species.

F Explain why the pattern and scale of human cooperation are puzzling from an evolutionary perspective.

As we conclude this book, we turn to a final question: What has made humans a unique species? Some readers will think that this is a trivial question; others, a controversial one. It may seem trivial because every species is unique, just as every snowflake is different from every other snowflake. But others may think the question is controversial because humans are products of the same evolutionary processes that have shaped all other forms of life on the planet. Each one of us is descended from a tiny shrewlike insectivore that lived among the dinosaurs more than a hundred million years ago. That small creature was gradually transformed by natural selection into a monkey-like animal clambering through the Oligocene forests of Africa; then to one of the many Miocene apes; then to a bipedal australopith in the woodlands of East Africa 3 Ma; and then to the genus *Homo*, which was the first of our ancestors to venture out of Africa; and finally to *Homo sapiens*, the brainy tool-addicted creature that now lives in practically every part of the world. We share approximately 96% of our genome with chimpanzees and bonobos, our physiology and morphology are only slightly modified versions of the standard primate model, and much of our behavior and psychology can be understood in the same terms as the behavior and psychology of other animals. Many would argue that the claim that humans are unique denies our knowledge of our evolutionary origins and obscures our place in nature.

But humans are an outlier in the natural world. Contemporary human biomass (the sum of all our weights) is eight times the biomass of all other wild terrestrial vertebrates combined and equals the biomass of all of the more than 14,000 species of ants. This is not just a consequence of agricultural and modern industrial technology. Human hunter-gatherers were outliers in the natural world even before the origin of agriculture. As we learned in Chapter 13, modern humans left Africa about 60,000 years ago and by 12,000 years ago they occupied every terrestrial habitat on Earth except Antarctica and a few remote islands. Their geographic and ecological range was larger than that of any other creature. As we have seen, most primates are limited to a narrow range of habitats on a single continent. We find chimpanzees in central African forests, baboons in African woodlands and savannas, and capuchins in the forests of Central and South America. The animals with the largest ranges are big predators such as wolves and lions, but their ranges are still much smaller than the range of human foragers 12,000 years ago. Foragers were able to accomplish this because they were better at rapidly adapting to a wide range of environments than any other creature.

Our goal in this chapter is to explain how and why this happened.

One reason humans have become so successful is that we are smarter than other animals. Over the past 2 million years human brains have become about three times the size of chimpanzee brains, and as we saw in Chapter 8, larger brains seem to lead to more complex cognition. We are better at causal reasoning, theory of mind, and other reasoning tasks that help us learn how to solve new problems. And this would have helped humans make a living in new environments. However, we want to convince you that although we are smart, we are not nearly smart enough to solve the problems humans need to solve to survive and thrive in such a wide range of habitats. Two more ingredients are essential parts of the human recipe. The first is culture. Unlike other creatures, people can learn from one another in a way that leads to the accumulation of locally adaptive knowledge, tools, and social institutions, and this allows humans to solve adaptive problems collectively that are too hard for individuals to solve on their own. The second ingredient is cooperation. People cooperate far more than any other mammal. This allows for specialization, exchange, and division of labor, which in turn vastly amplify the ability of people to extract resources from their environments. The three Cs—cognition, culture, and cooperation—have made humans a runaway ecological success.

Evolution and Human Culture

Foraging populations solve problems that are beyond the inventive capacity of individuals.

You learned in Chapter 13 that by 45 ka, modern people were living above the Arctic Circle near the Yenisei River. The people who left Africa 60 ka were tropical foragers living in a hot, dry coastal environment. To adapt to the high Arctic they had to create an entirely new way of life. We don't know very much about the Yana River people (see Chapter 13, p. 350), but we do know a lot about the Central Inuit, foragers who lived at about the same latitude in the Canadian Arctic.

The Central Inuit lived in small groups and made a living mainly by hunting and fishing. They depended on a tool kit crammed with complex, highly refined, and well-designed implements. Winter temperatures average about −25°C (−13°F), so survival required warm clothes. The Central Inuit made cleverly designed clothes, mainly from caribou skins that were both light and warm. To make these kinds of clothes, you need a host of complex skills: You must know how to cure and soften hides, spin thread, carve needles from bone, and cut and stitch well-fitting garments. But even the best clothing is not enough during winter storms; shelter is mandatory. The Central Inuit made snow houses that were so well designed that interior temperatures were about 10°C (50°F).

There is no wood in these environments, so people carved soapstone lamps and filled them with rendered seal fat to light their homes, cook their food, and melt ice for drinking water. During the winter, the Central Inuit hunted seals with multipiece toggle harpoons, mainly by ambushing them at their breathing holes, and moved their camps by using dogsleds. During the summer, they used the leister—a three-pronged spear with a sharp central spike and two hinged, backward-facing points—to harvest Arctic char caught in stone weirs (**Figure 16.1**). They also hunted seals and walrus in open water from kayaks. Later in summer and into the fall, the Central Inuit shifted to caribou, which they hunted with sophisticated composite bows made from driftwood and sinew. And these items and technologies are only part of the Central Inuit tool kits.

Having an extensive tool kit, however, is still not enough to survive in the Arctic environment; you also need a vast amount of knowledge. You need to know the habits of the animals that you hunt, how to move on ice, how to judge the weather, and where food can be found as the seasons change. You also need social rules and customs that allow groups of people to work together in such difficult conditions.

Do you think that you could acquire all the local knowledge necessary to live in the Arctic on your own? You are smart and probably did well on the SAT. You can probably drive a car, operate a computer, and understand something about physics. So if individual cognition *alone* is the key to the human ability to adapt to a wide range of environments, you should be able to figure out how to survive in the Arctic. This is exactly the way that other animals learn about their environments—they rely mainly on information encoded in their genes and personal experience to figure out how to find food, make shelter, and in some cases make tools.

We're pretty sure you'd fail because this experiment has been repeated many times, and the outcome is almost always the same. We think of it as "the lost European explorer experiment." Over the past couple of centuries, various European explorers have become stranded in unfamiliar habitats. Despite desperate efforts and ample learning time, these hardy men and women suffered or died because they could

FIGURE 16.1

The Inuit use many specialized tools to make a living in the Arctic. Here a man holds a leister, a specialized fishing spear.

FIGURE 16.2

Sir John Franklin, member of the Royal Society and leader of an expedition lost in the northern reaches of North America.

not figure out how to adapt to the habitat they found themselves in. The Franklin Expedition of 1846 illustrates this point. Sir John Franklin, a Fellow of the Royal Society and an experienced Arctic traveler, set out to find the Northwest Passage and spent two ice-bound winters in the Arctic. Everyone on his team eventually perished from starvation and scurvy (**Figure 16.2**). Their fate was tragic but also instructive. Members of the expedition spent their second winter on King William Island. The Central Inuit have lived around King William Island for at least 700 years, and this area is rich in animal resources. But the British explorers starved because they did not have the necessary local knowledge to make a living. Even though they had the same basic cognitive abilities as the Inuit and they had 2 years to use those abilities to figure out how to survive, they failed to acquire the skills necessary to subsist in the northern habitat.

Results from this version of the lost European explorer experiment and many others suggest that the technologies of foragers and other relatively "simple" societies are way beyond the inventive capacity of individuals. It's not hard to see why. Kayaks, bows, and dogsleds are complicated artifacts with multiple interacting parts made of many materials. Working out the best design or even a workable design for something like this from scratch is very hard to do. The Inuit could make the tools that they needed and master all the tasks that they needed to stay alive in the Arctic because they could draw on a vast pool of information that was known by other people in their population. They could gain access to this information by watching them, asking questions, or being taught. That is, unlike other organisms, humans rely on culturally acquired information, and it is the ability to make use of culturally acquired information that has made our species such a spectacular evolutionary success.

Humans rely on the accumulation of culturally acquired information to survive.

For many anthropologists, culture is what makes us human. Each of us is immersed in a cultural milieu that influences the way we see the world, shapes our beliefs about right and wrong, and endows us with the knowledge and technical skills to get along in our environment. Despite the central importance of culture in anthropology, there is little consensus about how or why culture arose in the evolution of the human lineage. In the discussion that follows, we present a view of the evolution of human culture that one of us (R.B.) developed with Peter Richerson at the University of California, Davis. Although we believe strongly in this approach to understanding the evolution of culture, there is not a broad consensus among anthropologists that this or any other particular view of the origins of culture is correct.

There are many definitions of **culture**. When thinking about the role of culture in human evolution, we think it is useful to define culture as *information acquired by individuals through some form of social learning*. For example, a child may learn that it is important to defer to her elders from watching her parents interact with her grandparents. She may also be corrected if she fails to behave appropriately. When individuals acquire different behaviors as a result of some form of social learning, then we observe cultural variation. The properties of culture are sometimes quite different from the properties of other forms of environmental variation. If people acquire behavior from others through teaching or imitation, then different populations living in similar ecological environments may behave very differently because they acquire different behaviors from members of the previous generation.

Culture is common among other animals, but cumulative cultural evolution is rare.

Over the past few decades, primatologists have documented a great deal of behavioral variation across groups in various species, most notably chimpanzees, orangutans, and capuchins. For example, chimpanzees living on the western shores of Lake Tanganyika raise their arms and clasp hands while they groom (**Figure 16.3**), but chimpanzees living on the eastern shore of the lake don't do this. Orangutans in some areas use sticks to pry seeds out of fruits, but orangutans at other sites have not mastered this technique and cannot extract the seeds. Capuchins show considerable variation in foraging techniques and social conventions. For example, capuchins at some sites participate in long bouts of mutual hand sniffing, but capuchins at other sites never display this behavior. In some cases, scientists have documented the appearance, diffusion, and eventual extinction of behavioral variants. There are also examples of cultural traditions in a variety of species outside the primates, such as fish, birds, cetaceans, meerkats, and rodents. These traditions encompass a diverse array of ecologically significant behaviors, including food preferences, foraging techniques, and alarm calls.

In human populations, culturally transmitted adaptations can gradually accumulate over many generations (**Figures 16.4** and **16.5**), resulting in complex behaviors that no individual could invent on his or her own. In other animals, there are very few examples of this kind of cumulative cultural evolution. The best-documented case of cumulative cultural evolution comes from studies of song dialects in songbirds, such as cowbirds. Cowbirds lay their eggs in the nests of other bird species (a good trick in itself), so chicks don't hear the songs of their own parents when they are growing up. Once they leave the nests of their foster parents, young birds begin to imitate the songs of cowbirds in the neighboring area. The form of the song in each local population gradually changes, and dialect variation among populations can be used to create song trees just as genetic variation can be used to construct gene trees. The songs of whales and other cetaceans seem to evolve in the same way as birdsong. This is impressive, but in these cases, cumulative cultural evolution is limited to a single domain, song dialect, and other behaviors are not culturally transmitted. Humans are an outlier in the extent of cumulative cultural evolution.

FIGURE 16.3

Chimpanzees display a variety of behaviors that seem to vary from group to group. Here, chimpanzees raise their arms while grooming.

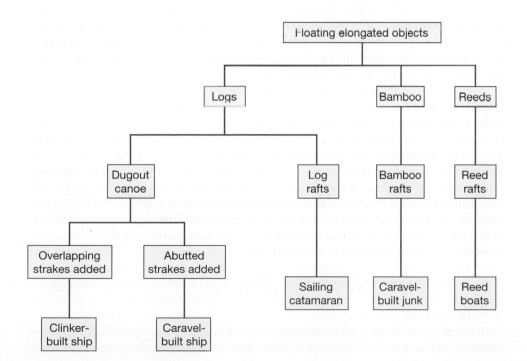

FIGURE 16.4

We can trace how certain technological innovations have developed. In China, the first boats were elongated structures that floated on the water. Some were made of logs, others of bamboo or reeds. Floating logs were transformed into canoes with the addition of a keel. Other types of ships, including Chinese junks, have a square hull and no keel.

Evolution and Human Culture ■ **429**

FIGURE 16.5

(a) Bamboo rafts may have been the precursors of (b) the great Chinese junks.

[a]

[b]

Why culture in other species does not accumulate is not clear.

Social learning creates traditions because experienced individuals do something (perform a behavior, make a tool, vocalize), and this makes it more likely that naive individuals will do something similar. A variety of social learning mechanisms lead to traditions, and it is useful to think of these mechanisms as ranging along a continuum. At one end of the continuum are mechanisms that do not preserve innovations and, therefore, cannot lead to cumulative cultural evolution. For example, **social facilitation** occurs when the activity of one animal increases the chance that other animals will learn the behavior on their own. Social facilitation could account for the persistence of tool use in the following scenario: Young chimpanzees accompany their mothers while the mothers are foraging. In populations in which females use stones to break open nuts, infants and juveniles spend a lot of time around nuts and hammer stones. Young chimpanzees fool around with stone hammers and anvils until they master the skill of opening the nuts. They do not learn the skill by watching their mothers. This means if a talented (or lucky) individual finds a way to improve nut cracking, the innovation will not spread to other members of the group.

At the other end of the continuum are mechanisms that preserve innovations. For example, **observational learning** (sometimes called imitation) occurs when naive animals learn how to perform an action by watching the behavior of experienced, skilled animals. If a chimpanzee female invents a new nut-cracking technique, then the innovation will be preserved if her offspring imitate it. Imitation allows innovations to persist because unskilled individuals can acquire new or improved techniques by observing the actions of others. Cumulative cultural evolution may occur if a series of innovations arises, gets copied, and spreads throughout the group.

In between social facilitation and observational learning are other mechanisms, such as **emulation**. Emulation occurs when naive individuals learn the end state of the behavior (a cracked nut) but not the behavior that generated that end state (pounding with a hammer stone). This can lead to the spread of innovations when individuals can learn on their own how to produce the end state, but not if the end state is difficult to achieve. For example, suppose one chimpanzee absconds with a can of tuna from the storage tent at the research camp and manages to open the can by using the same hammer-and-anvil technique that she uses to crack open nuts. The innovation (opening tuna cans) could spread once chimpanzees learn the goal (tuna) because the chimpanzees already know how to pound open nuts. However, if a new technique is needed to open the can, say, prying open the lid with a sharp stone, then emulation would not preserve the innovation.

Several recent studies suggest that monkeys and apes are capable of a form of observational learning. For example, chimpanzees at Gombe strip the leaves from slender twigs and use the twigs as probes to fish for termites. Elizabeth Lonsdorf of Franklin and Marshall College and her colleagues videotaped young chimpanzees while their mothers were fishing for termites. She found that young females watched their mothers carefully as they fished for termites, but young males were considerably less attentive (**Figure 16.6**). Lonsdorf also discovered that not all females used the same fishing techniques; some females consistently used longer twigs than others did. Young females tended to use the same kinds of tools that their mothers used, but young males did not adopt the techniques that their mothers used.

White-faced capuchins also seem to learn some foraging techniques by observation. The monkeys feed on seeds of *Luhea* fruits, which they obtain by pounding the fruits against a hard surface or scrubbing the fruit along a rough surface. The two techniques seem to be equally effective, but adults tend to use one technique or the other, not both. Susan Perry of the University of California, Los Angeles, monitored the development of *Luhea* processing techniques among immature monkeys. She found that the juveniles try out both techniques when they are young but eventually settle on only one. Those that associate most often with pounders tend to adopt the pounding technique, and those that associate most with scrubbers tend to adopt the scrubbing technique. Thus capuchins seem to learn *Luhea* foraging techniques through observation.

More evidence of observational learning comes from a set of experiments on captive chimpanzees conducted by Andrew Whiten of the University of St. Andrews and his colleagues. In these experiments, the animals are presented with a task, such as extracting a reward from a box that can be opened in two ways. Naive chimpanzees in one group observe the behavior of a group member that has been trained to open the box one way, naive chimpanzees in a second group learn a different technique (**Figure 16.7**). Chimpanzees tend to use the technique that they have seen demonstrated, suggesting that they must have learned the technique through imitation. Similar types of experiments provide evidence of social learning in vervet monkeys, capuchin monkeys, and lemurs.

So why don't chimpanzees make stone tools, bows, and arrows or build canopies over their nests to shelter them from the rain? We are not sure, but there are several possibilities. First, although the naturalistic data and experiments show that chimpanzees and capuchins can learn by observing others, the process is not very accurate. Most copy the behavior of demonstrators, but some don't. Repeated over generations, inaccurate social learning would rapidly degrade the innovations. Second, once chimpanzees have learned one way of getting inside the box, they are not inclined to learn another way, even if it is more efficient. Similarly, capuchins seem to settle on one technique for extracting seeds from *Luhea* fruits. This limits their ability to acquire progressively better skills and technology.

The third factor that may limit the development of complex culture repertoires in chimpanzees (and by extension other primates) is that chimpanzees do not blindly copy all the details of behaviors that they observe. Oddly enough, blind copying may be an important requirement for cumulative cultural evolution. In one set of experiments, chimpanzees observed human experimenters open the boxes in a way that included irrelevant, nonfunctional behaviors along with those required to get inside the box. Chimpanzees tend to acquire only behaviors necessary to actually open the box. In a parallel experiment conducted with people, the subjects faithfully copied all the irrelevant behaviors as well as the relevant ones. Faithful copying may be

FIGURE 16.6

Young female chimpanzees carefully watch their mothers fish for termites, and they tend to acquire the same kinds of techniques that their mothers use. Males are much less attentive to their mothers and do not match their mothers' techniques.

FIGURE 16.7

A young chimpanzee is trying to open a box in one of Whiten's experiments.

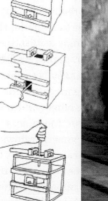

important for cumulative cultural evolution because many of the things that we learn are complicated and difficult to understand. (For example, why do you need to beat the eggs before adding them to the flour when you make a cake?) If people, like chimpanzees, copied only what they could understand, complicated tools and behavioral routines could not evolve. Doing something because you have seen others do it, even if you don't understand why they did it, may be important. But clearly, this could lead to unfortunate consequences.

Suppose that you are trying to learn how to fletch an arrow. Your mentor stops to scratch an itch or swat away a fly. The learner has to separate these irrelevant actions from the relevant ones. Psychologists George Gergely and Gergely Csibra of the Central European University think that learners can solve this problem only if demonstrators provide cues about which components of the behavior are important and which are not. Doing so need not involve overt verbal instruction; instead, demonstrators can use subtle cues such as the direction of their gaze or pointing.

Culture Is an Adaptation

Cumulative cultural adaptation is not a by-product of intelligence and social life.

Chimpanzees and capuchins are among the world's cleverest creatures. In nature, they use tools and perform many complex behaviors; in captivity, they can be taught very complicated tasks. Chimpanzees and capuchins live in social groups and have ample opportunity to observe the behavior of other individuals, and yet the best evidence suggests that neither chimpanzees nor capuchins make much use of observational learning in their daily lives. Thus the learning mechanisms that allow cumulative cultural adaptation, whatever they are, may not simply be a by-product of intelligence and opportunities for observing others.

This conclusion suggests, in turn, that the psychological mechanisms that enable humans to learn in a way that gives rise to cumulative cultural evolution are adaptations that have been shaped by natural selection because culture is beneficial (**Figure 16.8**). Of course, this need not be the case. These mechanisms may be by-products of some other adaptation that is unique to humans, such as language. But given the great importance of culture in human societies, it is important to think about the possible adaptive advantages of culture.

Culture allows humans to exploit a variety of environments by using a universal set of mental mechanisms.

Humans can live in a wider variety of environments than other primates because culture allows us to accumulate better strategies for exploiting local environments much more rapidly than genetic inheritance can produce adaptive modifications. Animals such as baboons adapt to different environments by using various learning mechanisms. For example, they learn how to acquire and process the food they eat. Baboons in the lush wetlands of the Okavango delta of Botswana learn how to harvest roots of water plants and how to hunt young antelope. Baboons living in the harsh desert of nearby Namibia must learn how to find water and process desert foods. All such learning mechanisms require prior knowledge about the environment: where to search for food, what strategies can be used to process the food, which flavors are reinforcing, and so on. More detailed and more accurate knowledge allows more accurate adaptation because it allows animals to avoid errors and acquire a more specialized set of behaviors.

In most animals, this knowledge is stored in the genes. Imagine that you captured a group of baboons from the Okavango Delta and moved them to the Namibian desert. It's a very good bet that the first few months would be tough for the baboons, but after a

FIGURE 16.8

Infants are prone to spontaneous imitation of the behaviors they observe. Here, a 13-month-old infant flosses her two teeth.

relatively short time, the transplanted group of baboons would probably be hard to tell from their neighbors. They would eat the same foods, have the same activity patterns, and have the same kinds of grooming relationships. The transplanted baboons would become similar to the local baboons because they acquire a great deal of information about how to be a baboon genetically; it is hardwired. Of course, the transplanted baboons would have to learn where to find water, where to sleep, which foods are edible, and which foods are toxic, but they would be able to do this without contact with local baboons because they have a built-in ability to learn such things on their own.

Human culture allows accurate adaptation to a wider variety of environments because *cumulative* cultural adaptation provides more accurate and more detailed information about the local environment than genetic inheritance systems can provide. The Inuit could make kayaks and do all the other things they needed to do to stay alive in the harsh environment of the Arctic because they could make use of a vast pool of useful information stored in the minds of other people in their population. The information contained in this pool is accurate and adaptive because the combination of individual learning and human social learning leads to rapid, cumulative adaptation. Even if most individuals blindly imitate the behavior of others, some individuals may occasionally come up with a better idea, and this will nudge traditions in an adaptive direction. Observational learning preserves the many small nudges and exposes the modified traditions to another round of nudging. This process generates adaptation more quickly than genetic inheritance does. The complexity of cultural traditions can explode to the limits of our capacity to learn them.

Culture can lead to evolutionary outcomes not predicted by ordinary evolutionary theory.

The importance of culture in human affairs has led many anthropologists to conclude that evolutionary thinking has little to contribute to understanding human behavior. They argue that evolution shapes genetically determined behaviors but not behaviors that are learned, and so culture is independent of biology. This argument is a manifestation of the nature–nurture controversy, and we explained at the beginning of the past chapter why this reasoning is flawed. Although many anthropologists have rejected evolutionary thinking about culture, many evolutionists have made the opposite mistake. They reject the idea that culture makes any *fundamental* difference in the way that evolution has shaped human behavior and psychology. If natural selection shaped the genes underlying the psychological machinery that gives rise to human behavior, the machinery must have led to fitness-enhancing behavior, at least in ancestral environments. If the adaptation doesn't enhance fitness in modern environments, that's because our evolved psychology is designed for life in a different kind of world.

We think both sides in this argument are wrong. Humans cannot be understood without the complex interplay between biology and culture. This is because cumulative cultural evolution is rooted in a new evolutionary trade-off between benefits and costs. Human social learning mechanisms are beneficial because they allow humans to accumulate vast reservoirs of adaptive information over many generations, leading to the cumulative cultural evolution of highly adaptive behaviors and technology. Because this process is much faster than genetic evolution, it allows human populations to develop cultural adaptations to local environments: kayaks in the Arctic and blowguns in the Amazon. The ability to adjust rapidly to local conditions was highly adaptive for early humans because the Pleistocene was a time of extremely rapid fluctuations in world climates. However, the psychological mechanisms that create this benefit come with a built-in cost. Remember that the advantage of learning from others is that it avoids the need for everyone to figure out everything for themselves. We can simply do what others do. But to get the benefits of social learning, people have to be credulous, generally accepting that other people are doing things in a sensible and proper way.

This credulity helps us learn complicated things, but it also makes us vulnerable to the spread of maladaptive beliefs and behaviors. If everyone in our community believes that it's beneficial to bleed sick people or that it's a good idea to boil polluted water before drinking, we believe that, too. This is how we get wondrous adaptations such as kayaks and blowguns. But we have little protection against the perpetuation of maladaptations that somehow arise. Even though the capacities that give rise to culture and shape its content must be (or at least must have been) adaptive on average, the behavior observed in any particular society at any particular time may reflect evolved maladaptations. Examples of these sorts of maladaptations are not hard to find.

Maladaptive beliefs can spread because culture is not acquired just from parents.

The logic of natural selection applies to culturally transmitted information in much the same way that it does to genes. Beliefs compete for our memory and our attention, and not all beliefs are equally likely to be learned or remembered. Beliefs are heritable, often passing from one individual to another without major change. As a result, some beliefs spread, and others are lost. However, the rules of cultural transmission are different from the rules of genetic transmission, so the outcome of selection among beliefs can be different from the outcome of selection on genes. The basic rules of genetic transmission are simple. With some exceptions, every gene that an individual carries in his body is equally likely to be incorporated into his gametes, and the only way that those genes can be transmitted is through his offspring. Thus only genes that increase reproductive success will spread. Cultural transmission is much more complicated. Beliefs are acquired and transmitted throughout an individual's life, and they can be acquired from grandparents, siblings, friends, co-workers, teachers, and even completely impersonal sources such as books, television, and now the Internet.

Most important, ideas and beliefs can spread even if they do not enhance reproductive fitness. If ideas about dangerous activities such as rock climbing and heroin use spread from friend to friend, these ideas can persist even though they reduce survival and individual reproductive success (**Figure 16.9**). Beliefs about heaven and hell can spread from priest to parishioner, even if the priest is celibate and has no offspring of his own to influence. Moreover, cultural variants may accumulate and be transmitted within groups of people who form clans, fraternities, business firms, religious sects, or political parties. This process can generate groups that are defined by cultural values and traditions, not by genetic relatedness.

FIGURE 16.9

Cultural evolution may permit the spread of ideas and behaviors that do not contribute to reproductive success. Dangerous sports, such as rock climbing, may be examples of such behaviors.

Culture is part of human biology, but culture makes human evolution qualitatively different from that of other organisms.

The fact that culture can lead to outcomes not predicted by conventional evolutionary theory does not mean that human behavior has somehow transcended biology. The idea that culture is separate from biology is a popular misconception that cannot withstand scrutiny. Culture is generated from organic structures in the brain that were produced by the processes of evolution. However, cultural transmission leads to new evolutionary processes. Thus to understand the whole of human behavior, evolutionary theory must be modified to account for the complexities introduced by these poorly understood processes.

The fact that culture can lead to outcomes that would not be predicted by conventional evolutionary theory does not mean that ordinary evolutionary reasoning is useless. The fact that there are processes that lead to the spread of risky behaviors such as rock climbing does not mean that these are the only processes that influence cultural behavior. In the last chapter we saw that many aspects of human psychology probably have been shaped by natural selection so that people learn to behave adaptively. We love our children and feel strong aversions to mating with close relatives. There is every reason to suspect that these predispositions play an important role in shaping human cultures. As long as this is the case, ordinary evolutionary reasoning will be useful for understanding human behavior.

Cooperation

Humans are more cooperative than other mammals.

Most mammals live solitary lives, meeting only to mate and raise their young. Among social species, cooperation is limited to relatives and, perhaps, small groups of reciprocators. After weaning, individuals acquire virtually all the food that they eat themselves. There is little division of labor, no trade, and no large-scale conflict. The sick, hungry, and disabled must fend for themselves. The strong take from the weak without fear of sanctions by third parties. Seventeenth-century philosopher Thomas Hobbes (**Figure 16.10**) famously described life in the state of nature as "the warre of all against all." Amend Hobbes to account for nepotism, and his picture of the state of nature is not so far off for most mammals.

In stark contrast, cooperation is an essential component of the economies of all foraging societies. Arizona State University anthropologist Kim Hill illustrates the difference with the following anecdote. Human hunter-gatherers and nonhuman primates both forage for fruit in trees. When a party of chimpanzees comes upon a fruiting tree, they all climb the tree and gather as much fruit as they can. Human hunter-gatherers send a couple of young guys up into the tree to shake the branches so that the fruit falls to the ground where everybody can easily harvest it. The young men are willing to go up into the tree because they know there will be fruit waiting for them when they get down, and this cooperative arrangement allows people to harvest the fruit more efficiently.

This kind of cooperative activity pervades human hunter-gatherer societies. Hill has studied the Aché, a hunter-gatherer group living in the forests of Paraguay, for more than 20 years (**Figure 16.11**). Here is a list of the cooperative foraging behaviors he recorded:

> Cuts a trail for others to follow; makes a bridge for others to cross a river; carries another's child; climbs a tree to flush a monkey for another hunter; allows another to shoot at prey when ego has first (best) shot; allows another to dig out an armadillo or extract honey or larva when ego encountered it; yells whereabouts of escaping prey; calls the location of a resource for another individual to exploit while ego continues searching; calls another to come to a pursuit of a peccary, paca, monkey, or coati; waits for others to join a pursuit, thus lowering own return rate; tracks peccaries when ego has no arrows (for other men to kill); carries game shot by another hunter; climbs fruit trees to knock down fruit for others to collect; cuts down palms (for others to take heart or fiber); opens a "window" in a tree to test for palm starch (for others to come take); carries the palm fiber others have collected; cuts down fruit trees for others to collect the fruit; brings a bow, arrow, ax, or other tool to another in a pursuit; spends time instructing another on how to acquire a resource; lends bow or ax to another when it could be used by ego; helps to look for another's arrows; prepares or repairs another man's bow and arrows in the middle of a pursuit; goes back on the trail to warn others of wasp nest; walks toward other hunters to warn of fresh jaguar tracks or poisonous snakes; removes dangerous obstacles from the trail before others arrive (Hill, 2002, pp. 113–114).

In each case, one individual helps another individual and incurs some cost in doing so. Such helping behavior is not structured by kinship among the Aché. Only a very small percentage of helping behavior is directed toward kin. Men help unrelated men, and women tend to help their husbands. Food produced by cooperative foraging is shared throughout the band.

FIGURE 16.10

The English political philosopher Thomas Hobbes.

FIGURE 16.11

The Aché are a group in Paraguay who lived solely by hunting and gathering until the 1970s and still acquire much of their food by foraging. Here, two Aché women cooperate to extract starch from a palm tree.

Human cooperation increases our ability to adapt.

Although the members of other mammal species don't engage in the division of labor, trade, mutual aid, and the construction of large-scale capital facilities, there are animals that do all these things, and they have been spectacular ecological successes and have radiated into a vast range of habitats. Multicellular organisms arose when groups of single-celled creatures evolved specialization and exchange, and multicellular organisms have been able to occupy a dramatically large number of niches. Their success indicates that the benefits of cooperation among cells were present in niches as different as those occupied by plants and animals; ecologies as different as aquatic, terrestrial, and subterranean habitats; and climates as varied as tropical and tundra. Similarly, eusocial insects have a very wide range of lifeways—some ant species herd aphid "cows," protect their herds from predators, and subsist on sweet "honeydew" produced by their carefully tended domesticates. Others are like farmers, carefully tending and fertilizing fungus gardens. Army ants, which have several castes of workers specialized for different tasks, can work together to build bridges, defend the colony, and manage traffic. Like humans, the eusocial insects have been a spectacular ecological success. Ants, for example, make up 2% of insect species but more than a third of insect biomass; in tropical forests, ants outweigh all vertebrates combined.

We believe that cooperation has played a similar role in the human expansion across the globe. Specialization is beneficial because it is efficient to subdivide labor among individuals who specialize in one or a few specific tasks. Exchange allows the output of efficient production to be shared. If one individual specializes in building houses, a second in farming, and a third in making music, and they trade their products, all three will typically enjoy better housing, food, and music than if they tried to produce everything themselves. The same goes for mutual aid. When an individual is sick and cannot forage, others can greatly improve her fitness by providing food at relatively small cost to themselves. Cooperative child care can greatly increase the ability of parents to produce food and other resources.

Humans cooperate in large groups of unrelated individuals.

One of the most striking differences between people and other social mammals is the scale on which humans cooperate. In most other mammals, cooperation is limited to small groups of kin and reciprocators. The most striking exception is a spectacularly homely subterranean African rodent called the naked mole rat, which lives in underground colonies numbering about 80 individuals (**Figure 16.12**). Colonies of naked mole rats work much like the colonies of ants or termites. There is a single reproducing female, and colony members forage cooperatively, maintain the burrow, and defend the colony. The members of a colony are closely related to one another. Other mammals that cooperate in sizable groups, such as African wild dogs, also are closely related. Human societies differ from other cooperative species because they can mobilize many *unrelated* individuals for collective enterprises. This is obviously true of modern societies in which government institutions such as courts and police regulate behavior. But, as it turns out, societies without such institutions also can mobilize many cooperators.

The joint production of capital facilities, such as roads and bridges, relies on large-scale cooperative behavior. Each worker invests time and labor, but all members of the community will travel on the road or use the bridge to cross the river. In modern societies, contracts enforced by governments mean that workers are guaranteed to be compensated directly. But in small-scale societies, in which workers don't receive wages for their labor or sign legally binding contracts,

FIGURE 16.12

The naked mole rat is a subterranean rodent that lives in large cooperative colonies with a single reproducing female. Individuals within a colony are closely related.

communities can also organize large-scale construction projects. For example, before the twentieth century, there were massive runs of salmon up the Trinity, Klamath, and other western coastal rivers. The Native American groups living along these rivers constructed large weirs, which act like fences, across the rivers to harvest the salmon as they swam upstream to spawn (**Figure 16.13**). The Yurok constructed a weir across the Klamath River at Kepel. Cutting the wood for this weir required the labor of hundreds of men from several villages, and the construction involved 70 workers over an extended period. During the 10 days of the salmon run, huge numbers of salmon were collected, dried, and shared among members of the tribe.

FIGURE 16.13

A salmon weir built across the Trinity River in northern California by members of the Hupa tribe in the early twentieth century. The weir remained in place for a short period during the early summer run of king salmon. By blocking the path of the migrating salmon, it allowed the Native Americans to harvest many salmon by using nets. Many people cooperated in constructing the weir and harvesting the salmon.

Warfare is a particularly interesting case of large-scale cooperation for two reasons. First, warfare has played an undeniably important role in human history. Second, our ability to wage war is remarkable because war creates an especially high-stakes collective action problem. Individual warriors risk injury or death, whereas victorious military actions benefit all group members. People engage in armed conflict with neighboring groups in almost all human societies. In foraging societies, the size of warring groups is typically small, but even societies that have no formal institutions can mobilize sizable war parties under the right circumstances. A recent study of warfare among the Turkana, an African pastoralist society, conducted by Sarah Mathew of Arizona State University provides a good example.

The Turkana herd cows and sheep in the arid savanna of northwestern Kenya (**Figure 16.14**). They live in mobile settlements numbering a few hundred people. The Turkana are divided into approximately 20 territorial sections—geographic regions within which herdsmen from each territory are free to graze. Men also belong to age groups, which are composed of similar-aged men who tend to herd and fight together. There is no recognized political or military authority; no elected officials, official police, or fighting forces. The Turkana often engage in armed combat with members of other ethnic groups that live just outside the border of Turkana territory. Victors may acquire livestock to supplement their herds and new grazing land, and they may also deter attacks by other groups. However, going to war has sizable costs for individuals: Mathew's data indicate that warriors have a 1% chance of dying each time they go on a raid. The Turkana cooperate in large numbers; on average, 300 warriors are mobilized for each raid (**Figure 16.15**), and the warriors come from several settlements, territorial sections, and age groups (**Figure 16.16**). This means most of the men in these large raiding parties are unrelated to one another, and many members of the war party barely know one another.

FIGURE 16.14

The Turkana are a group of nomadic herders who live in northwestern Kenya.

Notice that cooperation does not always produce nice or socially desirable outcomes. When the Yurok construct a weir, they catch more salmon, and everyone in the group gets more food. When other group members give food to a woman too sick to forage, she and her children are better off. When the Turkana go off on a raid, some are wounded, suffer, and may die; the threat of attacks and counterattacks forces everyone to put effort into guarding herds, wasting resources that could be used for more productive purposes.

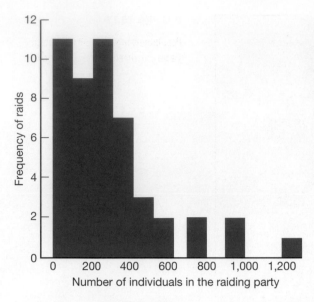

FIGURE 16.15

The distribution of the size of raiding parties among the Turkana. The horizontal axis gives the number of warriors participating in a raid, and the height of the vertical bars gives the fraction of the sample of raids that had that number of warriors. The average size was about 300 fighters, and the largest parties consisted of more than 1,000 individuals. This distribution means that a single individual has only a small effect on the success of the raid.

Human cooperation is regulated through prosocial sentiments and culturally transmitted norms enforced by rewards and punishments.

People are much more cooperative than other mammals, but they are not angels. Like other organisms, people are motivated by their own well-being and the welfare of their kin. The effect of each individual's contribution to large-scale collective enterprises is small. Selfish motives will tempt people to free ride; that is, to hide behind a tree when the shooting starts or feign illness when it is time to cut timber for the weir. If people succumb to these motives, there will be no cooperation. So what prevents free riding and sustains cooperation?

For most animals the answer is some combination of kinship and reciprocity. When individuals cooperate with kin, free riding can reduce their inclusive fitness. A naked mole rat who shirks its duties reduces the fitness of kin, and because relatedness is high, this effect can be enough to prevent free riding. In some primates, reciprocity plays an important role. Here, free riders are punished by retaliation by injured parties. A baboon who fails to reciprocate grooming may not get groomed by its partner next time.

While similar motives undoubtedly play a role in human cooperation, especially on smaller scales, they are not the whole story. Prosocial sentiments and the enforcement of culturally evolved moral norms by third parties play a crucial role in sustaining human cooperation.

Humans are not just exceptionally clever and cooperative creatures; we are also unusually nice ones. We donate to charity, give blood, return lost wallets, and give directions to bewildered tourists. As we noted earlier, individuals go to war, risking their lives to gain rewards that will mainly benefit others. Empathy motivates us to feel compassion for others, even people we don't know and will never meet. We have prosocial sentiments, such as generosity and a sense of fairness, and feel concern for the welfare of others. Such sentiments may motivate us to perform altruistic acts. As Abraham Lincoln once said, "When I do good, I feel good. When I do bad, I feel bad. That's my religion."

However, some researchers believe that people perform these kinds of acts for largely selfish reasons. They point out that heroes get to ride in parades, crusaders for justice become famous, and generous donors get their names on brass plaques. And, in some cases, we may expect recipients to reciprocate in the future. In an effort to get at the nature of people's social preferences, behavioral economists have designed a set of simple games in which individuals are faced with decisions that will affect their own welfare and the welfare of others. For example, in the dictator game, one player (the proposer) is given a sum of money. The proposer can keep all the money or can allocate some amount to another player. In the standard form of the game, the offer is

FIGURE 16.16

Membership in Turkana raiding parties is drawn from several (a) settlements, (b) age groups, and (c) territorial sections. This distribution means that warriors fight with many individuals they do not know well, suggesting that cooperation is not maintained by reciprocity.

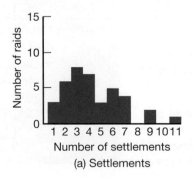

(a) Settlements

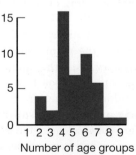

(b) Age groups

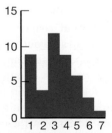

(c) Territorial sections

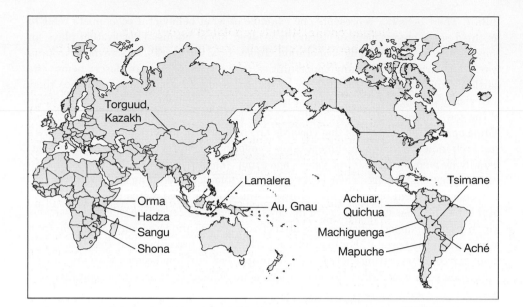

relayed anonymously; the two players never meet and never interact again. This is meant to eliminate the possibility that proposers will take advantage of opportunities to gain reputational benefits or expectations on the basis of reciprocity. Although proposers are free to keep all the money for themselves, not everyone does this. In fact, proposers typically allocate 20% to 30% of their endowments to the other player.

The ultimatum game adds a second step to the dictator game. As before, there are two players, a proposer and a recipient. The proposer is given a monetary endowment and proposes an anonymous allocation. But now the recipient decides whether to accept or reject the proposer's offer. If the recipient accepts the offer, each player gets the proposed amount; if the recipient rejects the offer, neither one gets any money. A recipient who rejects an offer above zero will actually lose money. Nonetheless, recipients typically reject offers of less than 20%. This is striking because recipients incur a cost when they reject a low offer, and they seem willing to punish proposers who make low offers, even though they are strangers and will not interact again.

You might think that these results reflect some peculiarity of industrialized societies or of college students, who are usually the subjects of such experiments. But the ultimatum game has now been played by thousands of people in dozens of countries all over the world. Joseph Henrich, now at Harvard University, coordinated a project in which the ultimatum game was played in 16 societies (**Figure 16.17**), including hunter-gatherers, nomadic herders, forest farmers, and urban people in less developed economies all outside of the United States. Henrich and his colleagues found considerable variation across societies in the size of offers and the likelihood that low offers would be rejected (**Figure 16.18**). However, in all societies, the more unequal an offer was, the more likely it was that it would be rejected.

Third-party enforcement of culturally evolved moral norms sustains human cooperation.

In the ultimatum game, recipients are often willing to punish proposers who make low offers even when they don't know their identity and won't interact again. They seem to be motivated by a sense of what constitutes an acceptable offer and what is simply

FIGURE 16.18

Results from cross-cultural ultimatum game experiments. The horizontal axis gives the fraction of the total monetary endowment offered by the proposer. The yellow bar gives the mean offer in each society, and the diameter of the blue circles gives the fraction of proposers who offered that fraction of the endowment. So, for example, most Lamalera participants offered half of the endowment, and the mean offer was slightly greater than a half. Among the Quichua, the most common offer was about 25% of the endowment, and the mean offer was about the same.

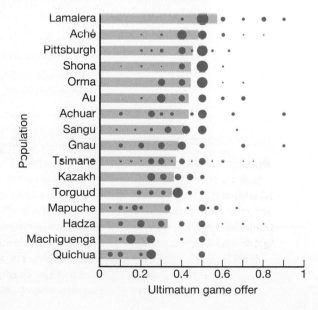

unfair. Their behavior represents one example of what behavioral economists refer to as third-party enforcement of culturally evolved moral norms. That system plays an important role in sustaining cooperation. To see how this works, let's revisit Turkana warfare. Warriors have many opportunities to free ride. They can desert before the battle begins. During the battle they can lag behind, hide, or otherwise reduce their own risk of getting killed. After the battle they can make off with more than their share of the cows. Moreover, some free riding occurs in about 50% of raids. Why doesn't free riding spread? The answer is that cowards and other free riders are punished. Mathew's research indicates that punishment takes two forms: First, there is direct punishment, usually by members of the free rider's age group. The first violation usually results in verbal sanctions; the free rider is ridiculed and told not to do it again. Further violations elicit corporal punishment and fines—the violator's age-mates tie him to a tree, beat him, and then slaughter one of his cows for a feast. Second, free riders lose various kinds of social support. They are less attractive as mates and less likely to get help from others when they need it. For example, a Turkana man traveling away from his settlement can count on getting shelter and food from other Turkana because they are obliged to provide such hospitality. However, there is no obligation to provide hospitality to a man who has a reputation for cowardice.

Norms enforced by third-party sanctions regulate a variety of behaviors in even the simplest foraging societies. Anthropologist Kim Hill surveyed the ethnographic literature and compiled a long list of norms that regulate behavior:

- *Marriage.* Whom you can marry based on age, kin relationship, or ritual group membership and whether it is permissible to have more than one wife or husband.

- *Food production.* What land is yours to exploit, what kinds of plants and animals you may harvest, and what economic activities are permissible.

- *Food sharing.* Whom you must share with, how much they receive, and who receives which cuts of meat.

- *Food consumption.* What kinds of food you may eat based on your age, sex, reproductive status, and ritual group membership.

- *Display rights.* What kinds of rituals you may participate in.

- *Residence.* Where you may live and with which people, again based on your sex, age, reproductive status, and ritual group membership.

- *Politics.* Who has political power and who can be a leader based on kinship, ritual membership, sex, age, and other factors.

- *Conflict.* Who is a legitimate opponent in ritual dueling and divining, what kinds of conflict are just and what kinds are not, and whether you are obligated to participate in conflicts with other groups.

- *Life history.* When you can have sex and who must invest in children.

- *Pollution.* Where and when you can relieve yourself, waste disposal, and other potentially polluting activities.

Notice that some norms regulate victimless crimes. For example, it is very common for norms to prohibit sex among siblings or parents and offspring. These behaviors don't injure third parties, yet third parties sometimes go to great efforts to suppress such behaviors.

It seems likely that third-party enforcement of norms makes it easier to maintain cooperation than does simple reciprocity for several reasons. First, it can increase the magnitude of penalties imposed on free riders. With reciprocation, a man who cheats his partner will lose the benefits of that relationship. With community-enforced norms, the violator may face more severe punishment and the loss of social support from nearly everyone in the community. Second, third-party norm enforcement can increase the chance that violators are detected. Without community monitoring and enforcement, a child who tells her mother that she is too sick to gather firewood or mind the goats can go out and play with her friends when her mother is out of camp. With community monitoring and enforcement, the cost of malingering will be much greater—she can't play when anyone is around the camp.

The extent of human cooperation is an evolutionary puzzle.

All the available evidence suggests that the societies of our Pliocene ancestors were like those of other social primates. Sometime during the past several million years, important changes occurred in human psychology that supported larger, more cooperative societies. Given the magnitude and complexity of the changes in human societies, the most plausible hypothesis is that they were the product of natural selection. However, the standard theory of the evolution of social behavior is consistent with Hobbes's vision of "the warre of all against all" tempered by a bit of nepotism, not observed human behavior. Apes fill the bill, but not humans.

Scientists have advanced two kinds of explanations for the high level of human cooperation. The **mismatch hypothesis** holds that the psychological machinery that supports human cooperation evolved in small hunter-gatherer societies with high genetic relatedness (**Figure 16.19**). Although high relatedness does not lead to very much cooperation in other primates, some special ecological situation may have favored it in early hominin populations. For example, a shift to hunting and the production of highly dependent infants may have favored male parental investment, food sharing, and cooperative hunting. In this kind of social environment, natural selection may have favored a psychology that made people more cooperative. Prosocial emotions, such as shame and guilt, may have been favored by selection because it motivated people to follow cooperative social norms. Because groups were small and made up of relatives, selection may have favored cooperation and psychological mechanisms that promote cooperation. Our evolved psychology misfires in contemporary societies in which most people live in groups with much lower degrees of relatedness.

The mismatch hypothesis has several weaknesses. First, surveys by Kim Hill and his colleagues indicate that the members of contemporary hunter-gatherer bands are not very closely related. People often move from one band to another so that the social world of modern hunter-gatherers typically encompasses about 500 people who all speak the same language. So the mismatch hypothesis is plausible only if ancestral hunter-gatherers lived in small, closed groups like other primates, not in the kinds of groups that characterize modern foragers. Second, the mismatch hypothesis cannot easily explain the scale of cooperation observed in contemporary societies. People, even people in small-scale societies, cooperate in large groups with people they do not know. The simplest version of the mismatch hypothesis suggests that people should be acutely sensitive to cues of kinship and reciprocity. They should be motivated to cooperate with relatives and people they know and be suspicious of strangers.

The **cultural group selection** hypothesis holds that extensive human cooperation is a side effect of rapid cultural adaptation. Systems of rewards and punishments can stabilize a variety of moral norms, including noncooperative ones, on different scales. As long as the cost of being punished exceeds the cost of following the norm, obeying the norm will be advantageous for

FIGURE 16.19

Hunter-gatherers often live in small, nomadic groups.

individuals. It doesn't matter what the norm requires. Mutually enforced sanctions could maintain cooperative or noncooperative norms: "You may steal your neighbor's cows to feed your family" or "You may not steal your neighbor's cows to feed your family." Similarly, punishment can maintain norms at different scales. "Do not steal a clan member's cattle, but the cattle of other clans are for brave men to steal" or "Do not steal the cattle of someone from your tribe, but the cattle of other tribes are for brave men to steal." These are both group-beneficial norms, but one benefits clans, whereas the other benefits tribes. The list of possible variations is nearly endless. As a result, different groups may tend to evolve toward different equilibria—one set of norms is enforced in one group, a different set in another group, a third set in a third group, and so on. This tendency will be opposed by migration and other kinds of social contact. Cultural adaptation is more rapid than genetic adaptation. Indeed, if we are correct, this is the reason we have culture—to allow different groups to accumulate different adaptations to diverse environments. As a result we expect that as culture became more important in the human lineage, behavioral differences between groups increased.

In Chapter 1 we saw that three conditions are necessary for adaptation by natural selection: First, there must be a struggle for existence so that not all individuals survive and reproduce. Second, there must be variation so that some types are more likely to survive and reproduce than others. And finally, variation must be heritable so that the offspring of survivors resemble their parents. We argued that selection typically occurs at the levels of individuals because these three conditions don't hold for groups. Groups may compete with one another and groups may vary in their ability to survive and grow, but the factors that lead to group-level variation in competitive ability are not transmitted from one generation to another, so there is no cumulative adaptation at the level of groups. The cultural group selection hypothesis emphasizes that once rapid cultural adaptation in human societies gave rise to stable (heritable), between-group differences, the stage was set for a variety of selective processes to generate adaptations at the group level.

Different human groups have different norms and values, and the cultural transmission of these traits can cause differences that can persist for long periods. The norms and values that predominate in a group may well affect the probability that the group survives, whether it is economically successful, whether it expands, and whether it is imitated by its neighbors (**Figure 16.20**). For example, suppose that groups with norms that promote military success are more likely to survive than

FIGURE 16.20

The Nuer (left) and Dinka (right) are two groups who live in South Sudan. Each of these groups includes several tribes that compete for grazing land. During the nineteenth century, the Nuer expanded at the expense of the Dinka because all the Nuer tribes shared norms about tribal membership and obligations that allowed them to organize much larger war parties than those organized by the Dinka.

groups lacking this sentiment. This creates a selective process that leads to the spread of such norms.

These two hypotheses are not mutually exclusive. Some of the psychological machinery necessary to create and enforce norms could easily have evolved in small groups of related individuals, and then these mechanisms made possible the norms that enforced more extensive and larger-scale cooperation. However, we argue that the mismatch hypothesis alone is not enough to understand the profound differences that have evolved between humans and other primates or to understand how humans have achieved such high levels of cooperation.

Is Human Evolution Over?

Students in our courses often wonder whether human evolution is over, and it seems a sensible question to consider as we come to the end of the story of human evolution. As we have seen, modern humans are the product of millions of years of evolutionary change. But so are cockroaches, peacocks, and orchids. All the organisms that we see around us, including people, are the products of evolution, but they are not finished products. They are simply works in progress, and this applies to us as well.

In one sense, however, human evolution is over. Because cultural change is much faster than genetic change, most of the changes in human societies since the origin of agriculture, almost 10 ka, and perhaps even before this point, have been the result of cultural, not genetic, evolution. Most of the evolution of human behavior and human societies is not driven by natural selection and the other processes of organic evolution; rather, it is driven by learning and other psychological mechanisms that shape cultural evolution. However, this fact does not mean that evolutionary theory or human evolutionary history is irrelevant to understanding contemporary human behavior. Natural selection has shaped the physiologic mechanisms and psychological machinery governing learning and other mechanisms of cultural change, and understanding human evolution can yield important insights into human nature and the behavior of modern peoples.

CHAPTER REVIEW

Key Terms

culture (p. 428)
social facilitation (p. 430)

observational learning
 (p. 430)
emulation (p. 430)

mismatch hypothesis
 (p. 441)

cultural group selection
 (p. 441)

Study Questions

1. The verb *ape* means "to copy or imitate." Is its meaning consistent with what we now know about the learning processes of other primates?

2. Why is cumulative cultural change likely to require emulation or observational learning?

3. Primatologists have documented many examples of behaviors that vary across populations, and some of those researchers have concluded that this variation is a form of culture. Explain why this is or is not a reasonable conclusion.

4. Some things that we do seem to be maladaptive (think about skydiving, drug abuse, and collecting classic cars). Some people would argue that these behaviors provide evidence that natural selection has no important impact on modern humans. Is this a reasonable argument? Why or why not?

5. Famous people, such as successful athletes and movie stars, are hired to sell all kinds of products—from underwear to cars—that have nothing to do with their professional accomplishments. How does the theory of cultural evolution help to explain why this might be a successful advertising strategy?

6. How do the dictator and ultimatum games provide evidence of prosocial sentiments? How crucial are anonymity and cross-cultural results for the conclusion that people are prosocial?

7. How does the pattern of cooperation in human groups differ from the pattern of cooperation we described in Chapter 7?

8. Thinking about warfare as a form of cooperation seems odd. Explain why evolutionary anthropologists think of war as a form of cooperation and why the existence of large-scale warfare poses a puzzle for evolutionary anthropologists.

9. What is the evidence that people cooperate in large, weakly related groups? Why is this phenomenon a puzzle from an evolutionary perspective?

10. In Chapter 7, we said that genetic group selection is usually not an important force in nature because migration reduces the amount of genetic variation between groups. Explain why movement between groups does not create the same problems for cultural group selection.

Further Reading

Boyd, R. 2018. *A Different Kind of Animal: How Culture Transformed Human Evolution*. Princeton: Princeton University Press.

Cronk, L., and B. Leech. 2013. *Meeting at Grand Central: Understanding the Social and Evolutionary Roots of Cooperation*. Princeton, N.J.: Princeton University Press.

Henrich, J. 2015. *The Secret of Our Success: How Culture Is Driving Human Evolution, Domesticating Our Species, and Making Us Smarter*. Princeton: Princeton University Press.

Hill, K. 2002. Altruistic cooperation during foraging by the Ache, and the evolved human predisposition to cooperate. *Human Nature*, 13, 105–128.

Hill, K., M. Barton, and M. Hurtado. 2009. "The Emergence of Human Uniqueness: Behavioral Characters Underlying Behavioral Modernity." *Evolutionary Anthropology* 18: 187–200.

Mesoudi, A. 2010. *Cultural Evolution*. Chicago: University of Chicago Press.

Richerson, P. J., and R. Boyd. 2005. *Not by Genes Alone: How Culture Transformed Human Evolution*. Chicago: University of Chicago Press.

 Review this chapter with personalized, interactive questions through INQUIZITIVE.

 This chapter also features a Guided Learning Exploration on the evolution of culture and cooperation.

EPILOGUE

THERE IS GRANDEUR IN THIS VIEW OF LIFE . . .

Here we end our account of how humans evolved. As we promised in the Prologue, the story has not been a simple one. We began, in Part One, by explaining how evolution works: how evolutionary processes create the exquisite complexity of organic design and how these processes give rise to the stunning diversity of life. Next we used these ideas in Part Two to understand the ecology and behavior of non-human primates: why they live in groups, why the behavior of males and females differs, why animals compete and cooperate, and why primates are so smart compared with other kinds of animals. Then, in Part Three, we combined our understanding of how evolution works and our knowledge of the behavior of other primates with information gleaned from the fossil record to reconstruct the history of the human lineage. We traced each step in the transformation from a shrewlike insectivore living at the time of dinosaurs; to a monkeylike creature inhabiting the Oligocene swamps of northern Africa; to an apelike creature living in the canopy of the Miocene forests; to the small-brained, bipedal hominins who ranged over Pliocene woodlands and savannas; to the large-brained and technically more skilled early members of the genus *Homo*, who migrated to most of Africa and Eurasia; and, finally, to creatures much like ourselves who created spectacular art, constructed simple structures, and hunted large and dangerous game just 100 ka. Finally, in Part Four, we turned to look at ourselves—to assess the magnitude and significance of genetic variation in the human species, and to try to explain how and why humans have become such a successful and unusual species.

FIGURE E.1

Charles Darwin died in 1882 and was buried in Westminster Abbey beneath the monument to Isaac Newton.

Evolutionary analyses of human behavior are not always well received. In Darwin's day, many people were deeply troubled by the implications of this theory. One Victorian matron, informed that Darwin believed humans to be descended from apes, is reported to have said, "Let us hope that it is not true, and if it is true, that it does not become widely known." Darwin's theory profoundly changed the way we see ourselves. Before Darwin, most people believed that humans were fundamentally different from other animals. Human uniqueness and human superiority were unquestioned. But we now know that all aspects of the human phenotype are products of organic evolution—the same processes that create the diversity of life around us. Nonetheless, many people still feel that we diminish ourselves by explaining human behavior in the same terms that we use to explain the behavior of chimpanzees or soapberry bugs or finches.

In contrast, we think the story of human evolution is breathtaking in its grandeur. With a few simple processes, we can explain how we arose, why we are the way we are, and how we relate to the rest of the universe. It is an amazing story. But perhaps Darwin himself (**Figure E.1**) put it best in the final pass of *On the Origin of Species*:

It is interesting to contemplate an entangled bank, clothed with many plants of many kinds, with birds singing on the bushes, with various insects flitting about, and with worms crawling through the damp earth, and to reflect that these elaborately constructed forms, so different from each other, and dependent on each other in so complex a manner, have all been produced by laws acting around us. These laws, taken in the largest sense, being Growth with Reproduction; Inheritance which is almost implied by reproduction; Variability from the indirect and direct action of the external conditions of life, and from use and disuse; a Ratio of Increase so high as to lead to a Struggle for Life, and as a consequence, Natural Selection, entailing Divergence of Character and the Extinction of less-improved forms. Thus, from the war of nature, from famine and death, the most exalted object which we are capable of conceiving, namely, the production of the higher animals, directly follows. There is grandeur in this view of life, with it several powers having been originally breathed into a few forms or only one; and that, whilst this planet has gone cycling on according to the fixed law of gravity, from so simple a beginning endless forms most beautiful and most wonderful have been, and are being evolved. [From C. Darwin, 1859, 1964, *On the Origin of Species*, facs. of 1st ed. (Cambridge, Mass.: Harvard University Press), p. 490.]

APPENDIX

The Skeletal Anatomy of Primates

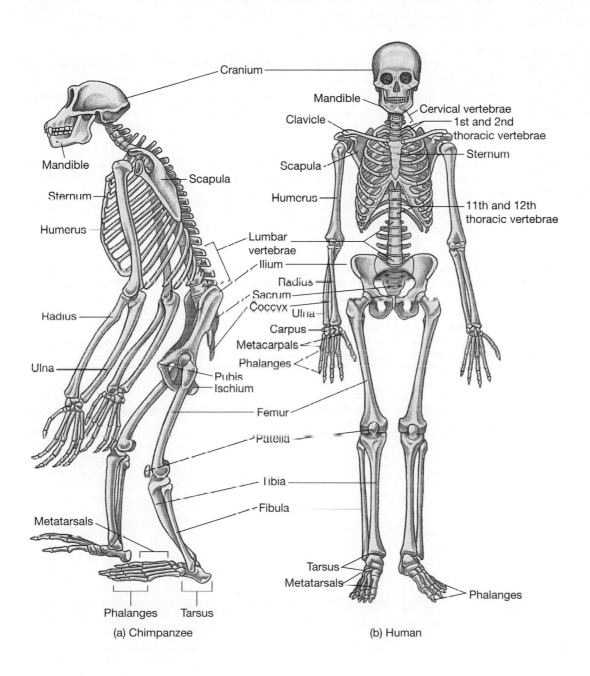

Cranium

Mandible

Sternum

Humerus

Radius

Ulna

Metatarsals

Phalanges — Tarsus

(a) Chimpanzee

Scapula

Lumbar vertebrae
Ilium
Radius
Sacrum
Coccyx — Ulna
Carpus
Metacarpals
Phalanges
Pubis
Ischium

Femur

Patella

Tibia

Fibula

Mandible
Clavicle

Scapula

Humerus

Cervical vertebrae
1st and 2nd thoracic vertebrae
Sternum

11th and 12th thoracic vertebrae

Tarsus
Metatarsals
Phalanges

(b) Human

GLOSSARY

abductor A muscle whose contraction moves a limb away from the midline of the body. The abductors that connect the pelvis to the femur act to keep the body upright during bipedal walking. (Ch. 10)

Acheulean industry A Mode 2 tool industry found at sites dated at 1.75 Ma to 0.3 Ma and associated with *Homo ergaster* and some archaic *Homo sapiens*. Named after the French village of Saint-Acheul, where it was first discovered, the Acheulean industry is dominated by teardrop-shaped hand axes and blunt cleavers. (Ch. 11)

activator A protein that increases transcription of a regulated gene. Compare *repressor*. (Ch. 2)

adaptation A feature of an organism created by the process of natural selection. (Ch. 1)

adaptive radiation The process in which a single lineage diversifies into several species, each characterized by distinctive adaptations. The diversification of the mammals at the beginning of the Cenozoic era is an example of an adaptive radiation. (Ch. 4)

adenine One of the four bases of the DNA molecule. The complementary base of adenine is thymine. (Ch. 2)

admixture The presence of DNA in an individual from a distantly related population or species, as a result of interbreeding between populations or species who have been reproductively isolated and genetically differentiated. (Ch. 14)

affiliative Friendly. (Ch. 7)

affinal kinship (Ch. 15) Kin ties through marriage, which we refer to as "in laws." (Ch. 15)

alkaloids Secondary compounds produced and kept in plant tissues to make the plant distasteful or even poisonous to herbivores. (Ch. 5)

allele One of two or more alternative forms of a gene. For example, the *A* and *S* alleles are two forms of the gene controlling the amino acid sequence of one of the subunits of hemoglobin. (Ch. 2)

alliance An interaction in which two or more animals jointly initiate aggression against or respond to aggression from, one or more other animals. Also called *coalition*. (Ch. 7)

allopatric speciation Speciation that occurs when two or more populations of a single species are geographically isolated from each other and then diverge to form two or more new species. Compare *parapatric speciation* and *sympatric speciation*. (Ch. 4)

altruism (altruistic, adj.) Behavior that reduces the fitness of the individual performing the behavior (the actor) but increases the fitness of the individual affected by the behavior (the recipient). (Ch. 7)

amino acids Molecules that are linked in a chain to form proteins. There are 20 amino acids, all of which share the same molecular backbone but have a different side chain. (Ch. 2)

analogy (analogous, adj.) Similarity between traits that is due to convergent evolution, not common descent. For example, the fact that humans and kangaroos are both bipedal is an analogy. Compare *homology*. (Ch. 4)

ancestral trait A trait that appears earlier in the evolution of a lineage or clade. Ancestral traits are contrasted with *derived traits*, which appear later in the evolution of a lineage or clade. For example, the presence of a tail is ancestral in the primate lineage, and the absence of a tail is derived. Systematists must avoid using ancestral similarities when constructing phylogenies. (Ch. 4)

angiosperms The flowering plants. The radiation of the angiosperms during the Cretaceous period may have played an important role in the evolution of the primates. (Ch. 9)

anticodon The sequence of bases on a transfer RNA molecule that binds complementarily to a particular mRNA *codon*. For example, for the mRNA codon AUC the corresponding anticodon is UAG because A binds to U, U binds to A, and C binds to G. (Ch. 2)

apatite crystal A crystalline material found in tooth enamel. (Ch. 9)

arboreal Active predominantly in trees. Compare *terrestrial*. (Ch. 4)

argon–argon dating A sophisticated variant of the potassium–argon dating method that allows very small samples to be dated accurately. (Ch. 9)

Aurignacian An early Upper Paleolithic stone tool industry found in Europe at sites that date to after 45 ka. (Ch. 13)

bachelor male A male that has not been able to establish residence in a bisexual group. Bachelor males may live alone or reside in all-male groups. (Ch. 6)

balanced polymorphism A steady state in which two or more alleles coexist in a population. This state occurs when heterozygotes have a higher fitness than any homozygote. (Ch. 14)

basal metabolic rate The rate of energy use required to maintain life when an animal is at rest. (Ch. 5)

base One of four molecules—adenine, guanine, cytosine, and thymine—that are bound to the DNA backbone. Different sequences of bases encode the information necessary for protein synthesis. (Ch. 2)

biface A flat stone tool made by working both sides of a core until there is an edge along the entire circumference. See also *hand ax*. (Ch. 11)

bilaterally symmetrical Describing an animal whose morphology on one side of the midline is a mirror image of the morphology on the other side. (Ch. 5)

binocular vision Vision in which both eyes can focus together on a distant object to produce three-dimensional images. See also *stereoscopic vision*. (Ch. 5)

biochemical pathway Any of the chains of chemical reactions by which organisms regulate their structure and chemistry. (Ch. 2)

biological species concept The concept that species is defined as a group of organisms that interbreed in nature and are reproductively isolated. Adherents of the biological species concept believe that gene flow tends to maintain similarities among members of the same species, and lack of gene flow is necessary to maintain differences between closely related species. Compare *ecological species concept*. (Ch. 4)

bipedal Describing locomotion in which the animal walks upright on two (hind) legs. Compare *quadrupedal*. (Ch. 9)

blade A stone tool made from a flake that is at least twice as long as it is wide. Blades dominate the tool traditions of the Upper Paleolithic. (Ch. 13)

blending inheritance A model of inheritance, widely accepted during the nineteenth century, in which the hereditary material of the mother and father was thought to combine irreversibly in the offspring. (Ch. 1)

camera-type eye An eye in which light passes through a transparent opening and is then focused by a lens on photosensitive tissue. Camera-type eyes are found in vertebrates, mollusks, and some arthropods. (Ch. 3)

canalized Describing traits that are very insensitive to environmental conditions during development, resulting in similar phenotypes in a variety of environments. Compare *plastic*. (Ch. 3)

canine The sharp, pointed tooth that lies between the incisors and the premolars in primates. (Ch. 5)

carbohydrates Certain organic molecules with the formula $C_nH_{2n}O_n$, including common sugars and starches. (Ch. 5)

carbon-14 dating A dating method based on an unstable isotope of carbon with an atomic weight of 14. Carbon-14 is produced in the atmosphere by cosmic radiation and is taken up by living organisms. After organisms die, the carbon-14 present in their bodies decays to a stable isotope (nitrogen-14) at a constant rate. By measuring the ratio of carbon-14 to the stable isotope of carbon (carbon-12) in organic remains, scientists can estimate the length of time that has passed since the organism died. The carbon-14 method is useful for dating specimens that are younger than about 40,000 years. Also called *radiocarbon dating*. (Ch. 9)

character A trait or attribute of the phenotype of an organism. (Ch. 1)

character displacement The result of competition between two species that causes the members of different species to become morphologically or behaviorally more different from each other. (Ch. 4)

Châtelperronian An Upper Paleolithic tool industry found in France and Spain that dates from 36 ka to 32 ka and is associated with Neanderthal fossil remains. (Ch. 13)

chromosome A linear body in the cell nucleus that carries genes and appears during cell division. Staining cells with dyes reveals that different chromosomes are marked by different banding patterns. (Ch. 2)

cladistic taxonomy A system for classifying organisms in which patterns of descent are the only criteria used. Compare *evolutionary taxonomy*. (Ch. 4)

cleaver A biface stone tool with a broad, flat edge. Cleavers are common at Acheulean sites. (Ch. 11)

coalition See *alliance*. (Ch. 7)

codon A sequence of three DNA bases on a DNA molecule that constitutes one "word" in the message used to create a specific protein. There are 64 codons. Compare *anticodon*. (Ch. 2)

coefficient of relatedness (*r*) An index measuring the degree of genetic closeness between two individuals. The index ranges from 0 (for no relation) to 1 (which occurs only between an individual and itself or between identical twins). For example, the coefficient of relatedness between an individual and its parents or its siblings is 0.5. (Ch. 7)

collected food A type of food resource, such as a leaf or fruit, that can be gathered and eaten directly. (Ch. 11)

combinatorial control The control of gene expression in which more than one regulatory protein is used and expression is allowed only in a specific combination of conditions. (Ch. 2)

common variation Refers to genetic loci where the allele with the lowest frequency in a population has a frequency of at least 1%. SNP chips only assay common variation, while whole-genome sequencing assays all loci, even those whose most rare allele is less frequent than 1%. (Ch. 14)

comparative method A method for establishing the function of a phenotypic trait by comparing species. (Ch. 4)

compound eye An eye in which the image is formed by many discrete photoreceptors. Compound eyes are found in insects and other arthropods. (Ch. 3)

conspecifics Members of the same species. (Ch. 5)

continental drift The movement over the surface of the globe of the immense plates of relatively light material that make up the continents. (Ch. 9)

continuous variation Phenotypic variation in which there is a continuum of types. Height in humans is an example of continuous variation. Compare *discontinuous variation*. (Ch. 1)

convergence The evolution of similar adaptations in unrelated species. The evolution of camera-type eyes in both vertebrates and mollusks is an example of convergence. See also *analogy*. (Ch. 1)

cooperative breeders A mating system in which there is one breeding female and all group members help care for the offspring. (Ch. 5)

core A piece of stone from which smaller flakes are removed. Cores and/or flakes may themselves be useful tools. (Ch. 11)

correlated response An evolutionary change in one character caused by selection on a second, correlated character. For example, selection favoring only long legs will also increase arm length if arm length and leg length are positively correlated. (Ch. 3)

correlation structure Refers to a pattern of genetic or phenotypic variation wherein individuals that have a particular variant at a genetic locus or in phenotypic trait are likely to have other particular variants at other genetic loci or in other phenotypic traits. In the presence of high correlation structure examined across many genetic loci or phenotypic traits, individuals can be accurately assigned to their group (for example, ancestral population, sex) even if the groups do not differ much on any single genetic locus or phenotypic trait. (Ch. 14)

cortex The original, unmodified surface of a stone used to make stone tools. (Ch. 11)

cross In genetics, a mating between chosen parents. (Ch. 2)

crossing-over The exchange of genetic material between homologous chromosomes during meiosis. Crossing-over causes recombination of genes carried on the same chromosome. (Ch. 2)

crural index The ratio of the length of the shin bone (tibia) to the length of the thigh bone (femur). (Ch. 12)

cultural group selection A process in which competition between culturally different groups leads to the spread of cultural practices prevalent in the successful groups. (Ch. 16)

culture Information stored in human brains that is acquired by imitation, teaching, or some other form of social learning and that can affect behavior or some other aspect of the individual's phenotype. (Ch. 16)

cytosine One of the four bases of the DNA molecule. The complementary base of cytosine is guanine. (Ch. 2)

dental calculus Mineralized plaque on teeth; plaque develops on teeth when food containing carbohydrates is eaten. (Ch. 12)

dental formula The number of incisors, canines, premolars, and molars on one side of the upper and lower jaws. (Ch. 5)

deoxyribonucleic acid See *DNA*. (Ch. 2)

derived trait A trait that appears later in the evolution of a lineage or clade. Derived traits are contrasted with *ancestral traits*, which appear earlier in the evolution of a lineage or clade. For example, the absence of a tail is derived in the hominin lineage, and the presence of a tail is ancestral. Systematists seek to use derived similarities when constructing phylogenies. (Ch. 4)

development All the processes by which the single-celled zygote is transformed into a multicellular adult. (Ch. 3)

diastema (diastemata, pl.) A gap between adjacent teeth. (Ch. 10)

diploid Referring to cells containing pairs of homologous chromosomes, in which one chromosome of each pair is inherited from each parent. Also referring to organisms whose somatic (body) cells are diploid; all primates are diploid. Compare *haploid*. (Ch. 2)

discontinuous variation Phenotypic variation in which there is a discrete number of phenotypes with no intermediate types. Pea color in Mendel's experiments is an example of discontinuous variation. Compare *continuous variation*. (Ch. 1)

diurnal Active only during the day. Compare *nocturnal*. (Ch. 5)

dizygotic twins Twins that result from the fertilization of two separate eggs by two separate sperm. Dizygotic twins are no more closely related than other full siblings. Compare *monozygotic twins*. (Ch. 14)

DNA Deoxyribonucleic acid, the molecule that carries hereditary information in almost all living organisms. DNA consists of two very long sugar–phosphate backbones (called "strands") to which the bases adenine, cytosine, guanine, and thymine are bound. Hydrogen bonds between the bases bind the two strands. (Ch. 2)

dominance The ability of one individual to intimidate or defeat another individual in a pairwise (dyadic) encounter. In some cases, dominance is assessed from the outcome of aggressive encounters; in other cases, dominance is assessed from the outcome of competitive encounters. (Ch. 6)

dominance matrix A square table constructed to keep track of dominance interactions among a group of individuals. Usually winners are listed down the left side and losers are listed across the top, and the number of times each individual defeats another is entered in the cells of the matrix. Individuals are ordered in the matrix so as to minimize the number of entries below the diagonal. This ordering is then used to construct the dominance hierarchy. (Ch. 6)

dominant Describing an allele that results in the same phenotype whether in the homozygous or the heterozygous state. Compare *recessive*. (Ch. 2)

ecological species concept The concept that natural selection plays an important role in maintaining the differences between species and that the absence of interbreeding between two populations is not a necessary condition for defining them as separate species. Compare *biological species concept*. (Ch. 4)

electron-spin-resonance dating A technique used to date fossil teeth by measuring the density of electrons trapped in apatite crystals in teeth. (Ch. 9)

emulation A form of social learning in which naive individuals acquire information about the end state of behavior but do not acquire information about the process required to generate the end state. (Ch. 16)

endocranial volume The volume inside the braincase. (Ch. 10)

environmental covariation The effect on phenotypes that occurs when the environments of parents and offspring are similar. Because environmental covariation causes the phenotypes of parents and offspring to be similar, it can falsely increase estimates of heritability. (Ch. 14)

environmental variation Phenotypic differences between individuals that exist because those individuals developed in different environments. Compare *genetic variation*. (Ch. 3)

enzyme A protein that serves as a catalyst, increasing the rate at which particular chemical reactions occur at a given temperature. Enzymes can control the chemical composition of cells by causing some chemical reactions to occur much faster than others. (Ch. 2)

equilibrium A steady state in which the composition of the population does not change. (Ch. 1)

estrus A period during the reproductive cycle of most mammals (and most primates) when the female is receptive to mating and can conceive. (Ch. 6)

eukaryotes Organisms whose cells have cellular organelles, cell nuclei, and chromosomes. All plants and animals are eukaryotes. Compare *prokaryotes*. (Ch. 2)

evolutionary taxonomy A system for classifying organisms that uses both patterns of descent and patterns of overall similarity. Compare *cladistic taxonomy*. (Ch. 4)

exon A segment of the DNA in eukaryotes that is translated into protein. Compare *intron*. (Ch. 2)

extracted food Food that is embedded in a matrix, encased in a hard shell, or otherwise difficult to extract. Extracted foods require complicated, carefully coordinated techniques to process. (Ch. 8, 11)

F_0, F_1, and F_2 generations A system for keeping track of generations in breeding experiments. The initial generation is called the F_0 generation, the offspring of the F_0 generation constitute the F_1 generation, and the offspring of the F_1 generation constitute the F_2 generation. (Ch. 2)

falciparum malaria A severe form of malaria. The sickle-cell allele for hemoglobin is common in West Africa because it confers resistance to falciparum malaria in the heterozygous state. (Ch. 14)

family A taxonomic level above genus but below order. A family may contain several genera, and an order may contain several families. Humans belong to the family Hominidae, and the other great apes belong to the family Pongidae. (Ch. 4)

fecundity The biological capacity to reproduce. In humans, fecundity may be greater than fertility (the actual number of children produced) when people limit family size. (Ch. 1)

femur The thigh bone. (Ch. 10)

fixation A state that occurs when all the individuals in a population are homozygous for the same allele at a particular locus. (Ch. 3)

flake A small chip of stone knocked from a larger stone core. (Ch. 11)

folivore (folivorous, adj.) An animal whose diet consists mostly of leaves. (Ch. 5)

foramen magnum The large hole in the bottom of the cranium through which the spinal cord passes. (Ch. 10)

fossil A trace of life more than 10,000 years old preserved in rock. Fossils can be mineralized bones, plant parts, impressions of soft body parts, or tracks. (Ch. 9)

founder effect A form of genetic drift that occurs when a small population colonizes a new habitat and then greatly increases in number. Random genetic changes due to the small size of the initial population are amplified by later population growth. (Ch. 14)

frugivore (frugivorous, adj.) An animal whose diet consists mostly of fruit. (Ch. 5)

gametes In animals, eggs and sperm. (Ch. 2)

gene A segment of the chromosome that produces a recognizable effect on phenotype and segregates as a unit during gamete formation. (Ch. 2)

gene flow The movement of genes from one population to another or from one part of a population to another as the result of interbreeding. (Ch. 4)

gene frequency The fraction of the genes at a genetic locus that are a particular allele (also called allele frequency). For example, a population that contains 250 *AA* individuals, 200 *AS* individuals, and 50 *SS* individuals has 700 copies of the *A* allele and 300 copies of the *S* allele; therefore, the frequency of the *S* allele is 0.3. (Ch. 3)

gene tree A phylogenetic tree tracing the pattern of descent for a particular gene. (Ch. 13)

genetic distance A measure of the overall genetic similarity of individuals or species. The best estimates of genetic distance use many genes. (Ch. 4)

genetic drift Random change in gene frequencies due to sampling variation that occurs in any finite population. Genetic drift is more rapid in small populations than in large populations. (Ch. 3)

genetic variation Phenotypic differences between individuals that result from the fact that those individuals have inherited different genes from their parents. Compare *environmental variation*. (Ch. 14)

genome All the genetic information carried by an organism. (Ch. 2)

genomewide association studies Studies that look for statistical associations between phenotypic traits (for example, stature) and a great many genetic markers located throughout the genome. An association between a particular marker and a phenotypic trait indicates that a gene near that marker affects the trait. (Ch. 14)

genotype The combination of alleles that characterizes an individual at some set of genetic loci. For example, in populations with only the *A* and *S* alleles at the hemoglobin locus, that locus has only three possible genotypes: *AA*, *AS*, and *SS*. (*SA* is the same as *AS*.) Compare *phenotype*. (Ch. 2)

genotypic frequency The fraction of individuals in a population that have a particular genotype. (Ch. 3)

genus (genera, pl.) A taxonomic category below family and above species. There may be several species in a genus and several genera in a family. (Ch. 4)

ghost lineage A population or species from whom we do not have direct ancient DNA sequence in unadmixed form, but whose existence can be statistically inferred on the basis of patterns of genetic variation in populations or species with which they interbred. (Ch. 12)

Gondwanaland The more southerly of the two supercontinents that existed from about 120 Ma to 100 Ma. Gondwanaland included the continental plates that now make up Africa, South America, Antarctica, Australia, New Guinea, Madagascar, and the Indian subcontinent. (Ch. 9)

grooming The process of picking through hair to remove dirt, dead skin, ectoparasites, and other material. Grooming is a common form of affiliative behavior among primates. (Ch. 7)

guanine One of the four bases of the DNA molecule. The complementary base of guanine is cytosine. (Ch. 2)

gum A sticky carbohydrate that some trees produce in response to physical damage. Gum is an important food for many primates. (Ch. 5)

gummivore (gummivorous, adj.) An animal whose diet consists mostly of gum. (Ch. 5)

gymnosperms A group of plants that reproduce without flowering. Modern gymnosperms include pines, redwoods, and firs. (Ch. 9)

haft To attach a spear point, ax head, or similar implement to a handle. Hafting greatly increases the force that can be applied to the tool. (Ch. 12)

Hamilton's rule A rule predicting that altruistic behavior among relatives will be favored by natural selection if $rb > c$, where r is the *coefficient of relatedness* between actor and recipient, b is the sum of the benefits of performing the behavior on the fitness of the recipient(s), and c is the cost, in decreased fitness of the donor, of performing the behavior. See also *kin selection*. (Ch. 7)

hand ax The most common type of biface stone tool found in Acheulean sites. It is flat and teardrop shaped with a sharp point at the narrow end. (Ch. 11)

haplogroup The mtDNA genotypes that descend from a node in the phylogenetic tree. (Ch. 13)

haploid A cell with only one copy of each chromosome. Gametes are haploid, as are the cells of some asexual organisms. Compare *diploid*. (Ch. 2)

haplorrhine Any member of the group containing tarsiers and anthropoid primates. The system that classifies primates into haplorrhines and strepsirrhines is a cladistic alternative to the evolutionary systematic taxonomy, in which primates are divided into prosimians and anthropoids, and tarsiers are grouped with prosimians. Compare *strepsirrhine*. (Ch. 5)

haplotype A particular set of alleles at some number of genetic loci that are transmitted together on the same chromosome. (Ch. 13, 14)

Hardy–Weinberg equilibrium The unchanging frequency of genotypes that results from sexual reproduction and occurs in the absence of other evolutionary forces such as natural selection, mutation, or genetic drift. (Ch. 3)

hemoglobin A protein in blood that carries oxygen, including two α (alpha) and two β (beta) subunits. (Ch. 2)

heritability The fraction of the phenotypic variation in the population that is the result of genetic variation. (Ch. 14)

heterozygous Referring to a diploid organism whose cells carry copies of two different alleles for a particular genetic locus. Organisms that are heterozygous are called heterozygotes. Compare *homozygous*. (Ch. 2)

hind-limb dominated A form of locomotion that depends mainly on the hind legs for power and propulsion. (Ch. 5)

home base A temporary camp that members of a group return to each day. At the home base, food is shared, processed, cooked, and eaten; subsistence tools are manufactured and repaired; and social life is conducted. (Ch. 11)

hominin Any member of the tribe Hominini, including all species of *Australopithecus* and *Homo*. (Ch. 10)

hominoid Any member of the superfamily Hominoidea, which includes humans, all the living apes, and many extinct apelike and human-like species from the Miocene, Pliocene, and Pleistocene epochs. (Ch. 4)

Homo heidelbergensis Middle Pleistocene hominins from Africa and western Eurasia. These hominins had large brains and very robust skulls and postcrania. (Ch. 12)

homologous chromosomes Sets of chromosomes have the same genetic loci, but often these loci contain different alleles. Human cells contain 23 pairs of homologous chromosomes. One member of each pair comes from the mother and the other from the father. (Ch. 2)

homology (homologous, adj.) Similarity between traits that is due to common ancestry, not convergence. For example, the reason that gorillas and baboons are both quadrupedal is that they are both descended from a quadrupedal ancestor. Compare *analogy*. (Ch. 4)

homozygous Referring to a diploid organism whose chromosomes carry two copies of the same allele at a single genetic locus. Organisms that are homozygous are called homozygotes. Compare *heterozygous*. (Ch. 2)

humerus The bone in the upper part of the forelimb (arm). (Ch. 10)

hunted food Live animal prey captured by human foragers or nonhuman primates. (Ch. 11)

hybrid zone A geographic region where two or more populations of the same species or two species overlap and interbreed. Hybrid zones usually occur at the habitat margins of the respective populations. (Ch. 4)

ilium (ilia, pl.) One of the three bones in the pelvis. (Ch. 10)

inbred mating Mating between closely related individuals. Also called *inbreeding*. Compare *outbred mating*. (Ch. 15)

incisors The front teeth in mammals. In anthropoid primates, incisors are used for cutting, and there are two on each side of the upper and lower jaws. (Ch. 5)

independent assortment The principle, discovered by Mendel, that each of the genes at a single locus on a pair of homologous chromosomes is equally likely to be transmitted when gametes (eggs and sperm) are formed. This happens because during meiosis, the probability that a particular chromosome will enter a gamete is 0.5 and is independent of whether other nonhomologous chromosomes enter the same gamete. Thus, knowing that an individual received a particular chromosome from its mother (and thus a particular allele) tells nothing about the probability that it received other, nonhomologous chromosomes from its mother. (Ch. 2)

infraorder The taxonomic level between order and superfamily. An order may contain several infraorders, and an infraorder may contain several superfamilies. (Ch. 5)

Initial Upper Paleolithic The earliest Upper Paleolithic industry in Europe found at sites that date to between 50 ka and 45 ka. (Ch. 13)

insectivore (insectivorous, adj.) An animal whose diet consists mostly of insects. (Ch. 5)

insulin A protein that is created by the pancreas and is involved in the regulation of blood sugar. (Ch. 14)

interbirth interval The period of time between the birth of one infant and the birth of the next infant. (Ch. 6)

intersexual selection A form of sexual selection in which females choose with whom they mate. The result is that traits making males more attractive to females are selected for. Compare *intrasexual selection*. (Ch. 6)

intrasexual selection A form of sexual selection in which males compete with other males for access to females. The result is that traits making males more successful in such competition are selected for, such as large body size or large canines. Compare *intersexual selection*. (Ch. 6)

introgression The transfer of genetic information from one species to another as a result of hybridization between them (Ch. 13)

intron A segment of the DNA in eukaryotes that is not translated into protein. Compare *exon*. (Ch. 2)

isotope A chemical element with the same atomic number as another element but having a different atomic weight. Unstable isotopes spontaneously change into more stable isotopes. (Ch. 9)

kibbutz (kibbutzim, pl.) An agricultural settlement in Israel, usually organized according to collectivist principles. (Ch. 15)

kin selection A theory stating that altruistic acts will be favored by selection if the product of the benefit to the recipient and the degree of relatedness (*r*) between the actor and recipient exceed the cost to the actor. See also *Hamilton's rule*. (Ch. 7)

knapping The process of manufacturing stone tools. (Ch. 11)

knuckle walking A form of quadrupedal locomotion in which, in the forelimbs, weight is supported by the knuckles rather than by the palm or outstretched fingers. Chimpanzees and gorillas are knuckle walkers. (Ch. 4)

lactase persistence Retention of the capacity to synthesize the enzyme lactase, which is necessary to digest the main carbohydrates in fresh milk after weaning. (Ch. 14)

lactase-phlorizin hydrolase (LPH) An enzyme produced in the small intestine that breaks down lactose in milk. The term is frequently shortened to lactase. (Ch. 14)

lactation (lactate, v.) Production of milk by the mammary glands in females; also the period during which milk is produced for nursing offspring. Lactation is a characteristic feature of mammals. (Ch. 3)

lactose A sugar present in mammalian milk. Most mammals—including most humans—lose the ability to digest lactose as adults. (Ch. 14)

Laurasia The more northerly of the two supercontinents that existed from roughly 150 Ma to 120 Ma. Laurasia included what are now North America, Greenland, Europe, and parts of Asia. (Ch. 9)

LCT The structural gene that codes for lactase-phlorizin hydrolase (lactase). (Ch. 14)

Levallois technique A three-step toolmaking method used by Neanderthals. The knapper first makes a core having a precisely shaped convex surface, then makes a striking platform at one end of the core, and finally knocks a flake off the striking platform. (Ch. 12)

linkage disequilibrium Occurs between loci when within an individual, the presence of an allele at one locus can be predicted by the presence of another allele at a different locus. (Ch. 14)

linked Referring to genes located on the same chromosome. The closer together two loci are, the more likely they are to be linked. Compare *unlinked*. (Ch. 2)

locus (loci, pl.) The position that a particular gene occupies on a chromosome. (Ch. 2)

long noncoding RNA (lncRNA) RNA molecules longer than 200 nucleotides. LncRNA has many functions, including gene regulation. (Ch. 2)

lordosis The S-shaped curvature typical of the human spine. (Ch. 10)

macroevolution Evolution of new species, families, and higher taxa. Compare *microevolution*. (Ch. 4)

maladaptive Detrimental to fitness. (Ch. 3)

mandible The lower jaw. Compare *maxilla*. (Ch. 5)

marsupial A mammal that gives birth to live young that continue their development in a pouch equipped with mammary glands. Marsupials include kangaroos and opossums. (Ch. 1)

mate guarding A behavior in which the male defends his mate after copulation to prevent other males from mating with her. (Ch. 3)

mating effort Refers to all of the activities that are related to conception, including the effort required to find mates and gain access to them. This may involve courtship displays, competition with rivals, or establishing a territory. (Ch. 6)

mating system The form of courtship, mating, and parenting behavior that characterizes a particular species or population. An example is polygyny. (Ch. 5)

matrilineage Individuals related through the maternal line. (Ch. 7)

maxilla The upper jaw. Compare *mandible*. (Ch. 5)

meiosis The process of cell division in which haploid gametes (eggs and sperm) are created. Compare *mitosis*. (Ch. 2)

messenger RNA (mRNA) A form of RNA that carries specifications for protein synthesis from DNA to the ribosomes. (Ch. 2)

microevolution Evolution of populations within a species. Compare *macroevolution*. (Ch. 4)

microlith A very small stone flake. Typical of African Later Stone Age (LSA) industries, microliths were probably hafted onto wood handles to make spears and axes. (Ch. 13)

microRNA (miRNA) Short segments of RNA that are involved in the translation of mRNA into protein and gene expression. Some are involved in regulating development and cell differentiation in complex organisms. (Ch. 2)

microsatellite loci Regions within DNA sequences in which short sequences of DNA are repeated multiple times—GTGTGT or ACTACTACT. They are also known as short tandem repeats. (Ch. 13)

Middle Stone Age (MSA) The stone tool industries of sub-Saharan Africa and southern and eastern Asia that existed 250 ka to 40 ka. The MSA is the counterpart of the Middle Paleolithic (Mousterian) in Europe. The MSA industries varied, but flake tools were manufactured in all of them. (Ch. 13)

mineralization (mineralized, adj.) The process by which organic material in the bones of dead animals is replaced by minerals from the surrounding rock, creating fossils. (Ch. 9)

minor marriage A form of marriage, formerly widespread in China, in which children were betrothed in infancy and then raised together in the household of the prospective groom. (Ch. 15)

mismatch hypothesis The idea that human minds are adapted to life in small-scale foraging societies and that this causes maladaptive behavior in complex, urban societies. (Ch. 16)

mitochondrial DNA (mtDNA) DNA in the mitochondria that is particularly useful for evolutionary analyses for two reasons: (1) Mitochondria are inherited only from the mother and, thus, there is no recombination, and (2) mtDNA accumulates mutations at relatively high rates, thus serving as a more accurate molecular clock for changes in the past few million years. (Ch. 12)

mitosis The process of division of somatic (normal body) cells through which new diploid cells are created. Compare *meiosis*. (Ch. 2)

Mode 1 A category of simple stone tools made by removing flakes from cores without any systematic shaping of the core. Both the flakes and the cores were probably used as tools themselves. Tools in the Oldowan industry are Mode 1 tools. (Ch. 11)

Mode 2 A category of stone tools in which cores are shaped into symmetric bifaces by the removal of flakes. The Acheulean industry is typified by Mode 2 tools. (Ch. 11)

Mode 3 A category of stone tools made by striking large symmetric flakes from carefully prepared stone cores by using the Levallois technique. The Mousterian industry in Europe and the Middle Stone Age industries in Africa are typified by Mode 3 tools. (Ch. 12)

Mode 4 A category of stone tools in which blades are common. Mode 4 tools are found in some Middle Stone Age industries in Africa, and they predominate in the Upper Paleolithic industries of Europe. (Ch. 13)

Mode 5 A category of stone tools in which microliths are common. The African Later Stone Age (LSA) industries are typified by Mode 5 tools. (Ch. 13)

modern synthesis An explanation for the evolution of continuously varying traits that combines the theory and empirical evidence of both Mendelian genetics and Darwinism. (Ch. 3)

molars The broad, square back teeth that are generally adapted for crushing and grinding in primates. Anthropoid primates have three molars on each side of the upper and lower jaws. (Ch. 5)

molecular clock The hypothesis that genetic change occurs at a constant rate and thus can be used to measure the time elapsed since two species shared a common ancestor. The molecular clock is based on observed regularities in the rate of genetic change along different phylogenetic lines. (Ch. 4)

monozygotic twins Twins that result from the fertilization of one egg by a single sperm. Early in development the fertilized egg splits to create two zygotes. Compare *dizygotic twins*. (Ch. 14)

morphology The form and structure of an organism; also a field of study that focuses on the form and structure of organisms. (Ch. 1)

most recent common ancestor (MRCA) The most immediate ancestor of individuals belonging to two different species or lineages. (Ch. 13)

Mousterian industry A stone tool industry characterized by points, side scrapers, and denticulates (tools with small toothlike notches on the working edge) but an absence of hand axes. The Mousterian is generally associated with Neanderthals in Europe. (Ch. 12)

mRNA See *messenger RNA*.

MSA See *Middle Stone Age*.

mtDNA See *mitochondrial DNA*.

multiparous A female who has had more than one pregnancy. (Ch. 6)

mutation A spontaneous change in the chemical structure of DNA. (Ch. 3)

natural selection The process that produces adaptation. Natural selection is based on three postulates: (1) The availability of resources is limited; (2) organisms vary in the ability to survive and reproduce; and (3) traits that influence survival and reproduction are transmitted from parents to offspring. When these three postulates hold, natural selection produces adaptation. (Ch. 1)

Neanderthal A form of archaic *Homo sapiens* found in western Eurasia from about 127 ka to about 30 ka. Neanderthals had large brains and elongated skulls with very large faces. They were also characterized by very robust bodies. (Ch. 12)

negative selection Selection against novel mutants that preserves the existing genotype. (Ch. 14)

negatively correlated Describing a statistical relationship between two variables in which larger values of one variable tend to co-occur with smaller values of the other variable. For example, the size and number of seeds produced by an individual plant are negatively correlated in some plant populations. Compare *positively correlated*. (Ch. 3)

neocortex Part of the cerebral cortex; generally thought to be most closely associated with problem solving and behavioral flexibility. In mammals, the neocortex covers virtually the entire surface of the forebrain. (Ch. 8)

neocortex ratio The size of the neocortex in relation to the rest of the brain. (Ch. 8)

neoteny The retention of juvenile traits into later stages of life. (Ch. 14)

niche The way of life, or "trade," of a particular species—what foods it eats and how the food is acquired. (Ch. 4)

NIDD See *non-insulin-dependent diabetes*.

nocturnal Active only during the night. Compare *diurnal*. (Ch. 5)

noncoding RNA (ncRNA) Molecules of RNA that do not code for proteins, including transfer RNA, ribosomal RNA, and microRNAs. (Ch. 2)

non-insulin-dependent diabetes (NIDD) A form of diabetes in which cells of the body do not respond properly to levels of insulin in the blood. NIDD has a genetic basis. (Ch. 14)

nonsynonymous substitution Substitution of one nucleotide for another in a DNA sequence that changes the amino acid coded for. (Ch. 14)

nucleus (nuclei, pl.) The distinct part of the cell that contains the chromosomes. Eukaryotes (fungi, protozoans, plants, and animals) all have nucleated cells; prokaryotes (bacteria) do not. (Ch. 2)

observational learning A form of learning in which animals observe the behavior of other individuals and thereby learn to perform a new behavior. Compare *social facilitation*. (Ch. 16)

occipital torus A horizontal ridge at the back of the skull in *Homo ergaster*, *Homo erectus*, and archaic *Homo sapiens*. (Ch. 11)

Oceania A region of the South Pacific that includes Polynesia, Melanesia, and Micronesia. (Ch. 14)

Oldowan tool industry A set of simple stone tools made by removing flakes from cores without any systematic shaping of the core. Both the flakes and the cores were probably used as tools. This industry is found in Africa at sites that date from about 2.6 Ma. (Ch. 11)

olfaction (olfactory, adj.) The sense of smell. (Ch. 5)

opposable Most primates, including humans, have an opposable thumb, which means that they touch all of their other fingers on the same hand with the thumb. Most primates, but not humans, also have an opposable big toe and can bend their big toe to touch the other toes on the same foot. (Ch. 5)

organelle A portion of the cell that is enclosed in a membrane and has a specific function; examples are mitochondria and the nucleus. (Ch. 2)

out-group A taxonomic group that is related to a group of interest and can be used to determine which traits are ancestral and which are derived. (Ch. 4)

outbred mating Mating between unrelated individuals. Compare *inbred mating*. (Ch. 15)

pair bonding A mating system in which a male and female form an exclusive mating relationship. Most primates that live in pairs mate mainly with one another but may sometimes mate with outsiders. Thus, most pair-living primates are not strictly monogamous. (Ch. 5)

paleontologist A scientist who studies fossilized remains of plant and animal species. (Ch. 9)

Pangaea The massive single continent that contained all of Earth's dry land until about 120 Ma. (Ch. 9)

parapatric speciation A two-step process of speciation in which (1) selection causes the differentiation of geographically separate, partially isolated populations of a species and (2) later the populations become reproductively isolated as a result of reinforcement. Compare *allopatric speciation* and *sympatric speciation*. (Ch. 4)

parent–offspring conflict Conflict that arises between parents and their offspring over how much the parents will invest in the offspring. These conflicts stem from the opposing genetic interests of parents and offspring. (Ch. 7)

parenting effort Includes all of the activities that are related to offspring care after conception occurs, such as sitting on the nest after eggs are laid or nursing infants after birth. (Ch. 6)

pastoralists People who make a living herding livestock. (Ch. 14)

personalized (or precision) genetic medicine Involves using an individual patient's genome to guide diagnosis and treatment in health care. (Ch. 14)

phenotype The observable characteristics of organisms. Individuals with the same phenotype may have different genotypes. Compare *genotype*. (Ch. 2)

phenotypic matching A mechanism for kin recognition in which animals assess similarities between themselves and others. (Ch. 7)

phylogeny The evolutionary relationships among a group of species, usually diagrammed as a "family tree." (Ch. 4)

pick A triangle-shaped biface stone tool found in Acheulean sites. (Ch. 11)

placental mammal A mammal that gives birth to live young that developed in the uterus and were nourished by blood delivered to a placenta. (Ch. 1)

plastic Describing traits that are very sensitive to environmental conditions during development, resulting in different phenotypes in different environments. Compare *canalized*. (Ch. 3)

pleiotropic effects Phenotypic effects created by genes that influence multiple characters. (Ch. 3)

plesiadapiform Any member of a group of primatelike mammals that lived during the Paleocene (65 Ma to 55 Ma). Although many paleontologists do not consider them to have been primates, the plesiadapiforms probably were similar to the earliest primates that lived around the same time. (Ch. 9)

polyandry A mating system in which there is one breeding female and more than one breeding male. Polyandry is generally rare in mammals, but characterizes some species of marmosets and tamarins. Compare *polygyny*. (Ch. 5)

polygenic risk score An estimate of the likelihood of an individual developing some particular phenotypic outcome, such as a particular disease, height, or IQ score, on the basis of his or her genotype. (Ch. 14)

polygyny A mating system in which a single male mates with many females. Polygyny is the most common mating system among primate species. Compare *polyandry*. (Ch. 5)

poor portability Refers to the lack of correspondence between results of genomewide association studies conducted in one population and another population. The more distantly related the populations, the worse the portability. (Ch. 14)

population genetics The branch of biology dealing with the processes that change the genetic composition of populations through time. (Ch. 3)

porphyria variegata A genetic disease caused by a dominant gene in which carriers of the gene develop a severe reaction to certain anesthetics. (Ch. 14)

positively correlated Describing a statistical relationship between two variables in which larger values of one variable tend to co-occur with larger values of the other variable. For example, in human populations the height and weight of individuals are positively correlated. Compare *negatively correlated*. (Ch. 3)

positively selected Describing selection that favors new genotypes and thus leads to genetic change. (Ch. 14)

postcranium (postcrania, pl.; postcranial, adj.) The skeleton excluding the skull. (Ch. 9)

potassium–argon dating A radiometric method of dating the age of a rock or mineral by measuring the rate at which potassium-40, an unstable isotope of potassium, is transformed into argon. This method can be used to date volcanic rocks that are at least 500,000 years old. (Ch. 9)

prehensile Describing the ability of hands, feet, or tails to grasp objects, such as food items or branches. (Ch. 5)

premolars The teeth that lie between the canines and molars. (Ch. 5)

pressure flaking A method for finishing stone tools. The toolmaker presses the edge of the tool with a sharp item, such as a piece of bone or antler, to remove small flakes. (Ch. 13)

primary structure The sequence of amino acids that make up a protein. (Ch. 2)

primiparous Refers to a female who has given birth for the first time. (Ch. 6)

proconsulid Any member of a group of early Miocene hominoids that includes the genus *Proconsul*. (Ch. 9)

prokaryotes Organisms that lack a cell nucleus or separate chromosomes. Bacteria are prokaryotes. Compare *eukaryotes*. (Ch. 2)

protein A large molecule that consists of a long chain of amino acids. Many proteins are enzyme catalysts; others perform structural functions. (Ch. 2)

protein-coding genes Genes that encode instructions for making proteins. (Ch. 2)

Punnett square A diagram that uses gene (or allele) frequencies to calculate the genotypic frequencies for the next generation. (Ch. 2)

quadrupedal Describing locomotion in which the animal moves on all four limbs. Compare *bipedal*. (Ch. 4)

radioactive decay Spontaneous change from one isotope of an element to another isotope of the same element or to an entirely different element. Radioactive decay occurs at a constant rate that can be measured precisely in the laboratory. (Ch. 9)

radiocarbon dating See *carbon-14 dating*. (Ch. 9)

radiometric method Any dating method that takes advantage of the fact that isotopes of certain elements change spontaneously from one isotope to another at a constant rate. (Ch. 9)

rain shadow An area of reduced rainfall found on the lee (downwind) side of large mountains and mountain ranges. (Ch. 9)

recessive Describing an allele that is expressed in the phenotype only when it is in the homozygous state. Compare *dominant*. (Ch. 2)

reciprocal altruism A theory that altruism can evolve if pairs of individuals take turns giving and receiving altruism during many encounters. (Ch. 7)

recombination The creation of new genotypes as a result of the random segregation of chromosomes and of crossing over. (Ch. 2)

redirected aggression A behavior in which the recipient of aggression threatens or attacks a previously uninvolved party. For instance, if A attacks B and B then attacks C, B's attacks are an example of redirected aggression. (Ch. 8)

regulatory gene A DNA sequence that regulates the expression of a structural gene, often by binding to an activator or repressor. (Ch. 2)

reinforcement The process in which selection acts against the likelihood of hybrids occurring between members of two phenotypically distinctive populations, leading to the evolution of mechanisms that prevent interbreeding. (Ch. 4)

repressor A protein that decreases transcription of a regulated gene. Compare *activator*. (Ch. 2)

reproductive isolation A relationship in which no gene flow occurs between two populations. (Ch. 4)

ribonucleic acid See *RNA*. (Ch. 2)

ribosome A small organelle composed of protein and nucleic acid that temporarily holds together the messenger RNA and transfer RNAs during protein synthesis. (Ch. 2)

RNA Ribonucleic acid, a long molecule that plays several important roles in protein synthesis. RNA differs from DNA in that it has a slightly different chemical backbone and it contains the base uracil instead of thymine. (Ch. 2)

rock shelter A site sheltered by an overhang of rock. (Ch. 12)

sagittal crest A sharp fin of bone that runs along the midline of the skull and increases the area available for the attachment of chewing muscles. (Ch. 10)

sampling variation The variation in the composition of small samples drawn from a large population. (Ch. 3)

scapula (scapulae, pl.) Shoulder blade. (Ch. 12)

secondary compounds Toxic (poisonous) chemical compounds produced by plants and concentrated in plant tissues to prevent animals from eating the plant. (Ch. 5)

selection–mutation balance An equilibrium that occurs when the rate at which selection removes a deleterious gene is balanced by the rate at which mutation introduces that gene. The frequency of genes at selection–mutation balance is typically quite low. (Ch. 14)

selective sweep A process in which one allele increases in a population as a result of positive selection. (Ch. 14)

sex ratio The number of individuals of one sex in relation to the number of the opposite sex. By convention, sex ratios are generally expressed as the number of males to the number of females. (Ch. 3)

sexual dimorphism Differences in body size or morphology between sexually mature males and females. (Ch. 5)

sexual selection A form of natural selection that results from differential mating success in one sex. In mammals, sexual selection usually occurs in males and may be due to male–male competition. (Ch. 6)

sexual selection infanticide hypothesis A hypothesis postulating that infanticide has been favored by sexual selection because males who kill unweaned infants can enhance their own reproductive prospects if they (1) kill infants whose deaths hasten their mothers' resumption of cycling, (2) do not kill their own infants, and (3) can mate with the mothers of the infants that they kill. (Ch. 6)

sickle-cell anemia A severe form of anemia that afflicts people who are homozygous for the sickle-cell gene. (Ch. 2)

single-nucleotide polymorphism (SNP, pronounced "snip") Occurs when members of a population differ at a particular nucleotide position in the genome. (Ch. 14)

SLI See *specific language impairment*.

SNP See *single-nucleotide polymorphism*.

social facilitation The situation that occurs when the performance of a behavior by older individuals increases the probability that younger individuals will acquire that behavior on their own. Social facilitation does not mean that young individuals copy the behavior of older individuals. For example, the feeding behavior of older individuals may bring younger individuals in contact with the foods that adults are eating and, therefore, increase the chance that younger individuals acquire a preference for those foods. Compare *observational learning*. (Ch. 16)

social intelligence hypothesis The hypothesis that the relatively sophisticated cognitive abilities of higher primates are the outcome of selective pressures that favored intelligence as a means to gain advantages in social groups. (Ch. 8)

social organization The size, age–sex composition, and degree of cohesiveness of primate social groups. (Ch. 5)

solitary A term used for animals that do not live in social groups and do not form regular associations with conspecifics. (Ch. 5)

species (sing. and pl.) A group of organisms classified together at the lowest level of the taxonomic hierarchy. Biologists disagree about how to define a species. See *biological species concept* and *ecological species concept*. (Ch. 1)

specific language impairment (SLI) A family of language disorders in which the affected person experiences difficulty using language but is of otherwise normal intelligence. Evidence suggests that at least some cases of SLI are hereditary. (Ch. 14)

spliceosomes Organelles that splice the mRNA in eukaryotes after the introns have been snipped out. (Ch. 2)

stabilizing selection Selection pressures that favor average phenotypes. Stabilizing selection reduces the amount of variation in the population but does not alter the mean value of the trait. (Ch. 1)

stasis A state or period of stability during which little or no evolutionary change in a lineage occurs. (Ch. 1)

stereoscopic vision Vision in which three-dimensional images are produced because each eye sends a signal of the visual image to both hemispheres in the brain. Stereoscopic vision requires binocular vision. (Ch. 5)

strategy A complex of behaviors deployed in a specific functional context, such as mating, parenting, or foraging. (Ch. 6)

stratum (strata, pl.) A geologic layer. (Ch. 9)

strepsirrhine Any member of the group containing lemurs and lorises. The system classifying primates into haplorrhines and strepsirrhines is a cladistic alternative to the evolutionary systematic taxonomy, in which primates are divided into prosimians and anthropoids, and tarsiers are grouped with prosimians. Compare *haplorrhine*. (Ch. 5)

subnasal prognathism The condition in which the part of the face below the nose is pushed out. (Ch. 10)

superfamily The taxonomic level that lies between infraorder and family. An infraorder may contain several superfamilies, and a superfamily may contain several families. For example, humans are a member of the superfamily Hominoidea, which contains the families Hominidae and Pongidae. (Ch. 4)

sutures Wavy joints between bones that mesh together and are separated by fibrous tissue. (Ch. 12)

sympatric speciation A hypothesis that speciation can result from selective pressures favoring different phenotypes within a population without positing geographic isolation as a factor. Compare *allopatric speciation* and *parapatric speciation*. (Ch. 4)

synonymous substitution Substitution of one nucleotide for another in a DNA sequence that does not change the amino acid coded for. (Ch. 14)

systematics A branch of biology that is concerned with the procedures for constructing phylogenies. Compare *taxonomy*. (Ch. 4)

tapetum (tapeta, pl.) A layer behind the retina in some organisms that reflects light. (Ch. 9)

taphonomy The study of the processes that affect the state of the remains of an organism from the time the organism dies until it is fossilized. (Ch. 11)

taurodont root A single broad tooth root in molars, resulting from the fusion of three roots. Taurodont roots were characteristic of Neanderthals. (Ch. 12)

taxonomy A branch of biology that is concerned with the use of phylogenies for naming and classifying organisms. Compare *systematics*. (Ch. 4)

temporalis muscle A large muscle involved in chewing. The temporalis muscles attach to the side of the cranium and to the mandible. (Ch. 10)

terrestrial Active predominantly on the ground. Compare *arboreal*. (Ch. 4)

territory A fixed area occupied by animals that defend the boundaries against intrusion by other individuals or groups of the same species. (Ch. 5)

tertiary structure The three-dimensional folded shape of a protein. (Ch. 2)

testes (testis, sing.) The male organs responsible for producing sperm. (Ch. 4)

theory of mind The capacity to be aware of the thoughts, knowledge, or perceptions of other individuals. A theory of mind may be a prerequisite for deception, imitation, teaching, and empathy. Researchers generally think that humans, and possibly chimpanzees, are the only primates to possess a theory of mind. (Ch. 8)

thermoluminescence dating A technique used to date crystalline materials by measuring the density of trapped electrons in the crystal lattice. (Ch. 9)

third-party relationships Relationships among other individuals. For example, monkeys and apes are believed to understand something about the nature of kinship relationships among other group members. (Ch. 8)

thymine One of the four bases of the DNA molecule. The complementary base of thymine is adenine. (Ch. 2)

torque A twisting force that generates rotary motion. (Ch. 10)

toxin A chemical compound that is poisonous or toxic. (Ch. 5)

trait A characteristic of an organism. (Ch. 1)

transfer RNA (tRNA) A form of RNA that facilitates protein synthesis by first binding to amino acids in the cytoplasm and then binding to the appropriate site on the mRNA molecule. There is at least one distinct form of tRNA for each amino acid. (Ch. 2)

transitive Describing a property of triadic (three-way) relationships in which the relationships between the first and second elements and the second and third elements automatically determine the relationship between the first and third elements. For example, if A is greater than B and B is greater than C, then A is greater than C. In many primate species, dominance relationships are transitive. (Ch. 6)

transposable elements Segments of DNA that move from one location to another within the genome of a single individual. (Ch. 14)

tRNA See *transfer RNA*.

Uluzzian A stone tool industry found in Italy and the Balkans at sites that date to after 45 ka. The Uluzzian industry has characteristics of both Middle and Upper Paleolithic industries. (Ch. 13)

unlinked Referring to genes on different chromosomes. Compare *linked*. (Ch. 2)

Upper Paleolithic The period from about 45 ka to about 10 ka in Europe, North Africa, and parts of Asia. The tool kits from this period are dominated by blades. (Ch. 13)

uracil One of the four bases of the RNA molecule. Uracil corresponds to the base thymine in DNA; as with thymine, its complementary base is adenine. (Ch. 2)

uranium–lead dating A method of dating zirconium crystals in igneous rocks that is based on the ratio of uranium to lead. This method can be used to date stalactites, stalagmites, and flow stone formed by precipitation in limestone caves and has been particularly useful for dating hominin remains found in caves in South Africa. (Ch. 9)

variant The particular form of a trait. For example, blue eyes, brown eyes, and gray eyes are variants of the trait eye color. (Ch. 2)

variation among groups Differences in the average phenotype or genotype between groups. (Ch. 14)

variation within groups Differences in phenotype or genotype between individuals in a group. (Ch. 14)

viviparity Giving birth to live young. (Ch. 5)

zygomatic arch A cheekbone. (Ch. 9)

zygote The cell formed by the union of an egg and a sperm. (Ch. 2)

CREDITS

TITLE PAGE
Photos: **Page iii**: by toonman/Getty Images.

ABOUT THE AUTHORS
Photos: **Page v** top: Brock Wagstaff; bottom: Jessica Gunson.

CONTENTS
Photos: **Page vii** top to bottom: Luciano Andres Richino/EyeEm/Getty Images, blickwinkel R. Koenig/Alamy Stock Photo, Thomas Shahan/Science Source; **p. viii** top: Dietmar Nill/Nature Picture Library/Alamy Stock Photo, bottom: Manoj Shah/Getty Images; **p. ix** top to bottom: Anup Shah/Getty Images, Anup Shah/Nature Picture Library/Alamy Stock Photo, Fiona Rogers/Nature Picture Library; **p. x** top to bottom: Dorling Kindersley: Andrew Kerr, Anup Shah/Animals Animals/age fotostock, Peter Bostrom; **p. xi** top to bottom: Javier Trueba/Madrid Scientific Films/Science Source, Javier Trueba/MSF/Science Source, Mira/Alamy Stock Photo; **p. xii** top: Lee Jae-WonN/Reuters/Newscom, bottom: Library of Congress, Prints & Photographs Division, Edward S. Curtis Collection.

PROLOGUE
Photos: **Page xx** top to bottom: Charles Darwin (w/c on paper)English School(19th century)/Down House Downe Kent UK/© Historic England/Bridgeman Images, Heritage Image Partnership Ltd/Alamy Stock Photo, Portrait of Isaac Newton (1642–1727) c.1726 (oil on canvas) Seeman Enoch (1694–1744)(after)/National Portrait Gallery London UK/Bridgeman Images; **p. xx** top: Image Source/Corbis, bottom: Robert Boyd; **p. xxii** top: Cyril Ruoso/Minden Pictures/Newscom, **p. xxii** bottom: The Natural History Museum/Alamy Stock Photo.

CHAPTER 1
Photos: **Pages 2–3**: Luciano Andres Richino/EyeEm/Getty Images; **p. 5**: bcampbell65/Shutterstock; **p. 7** left: Jess Kraft/Shutterstock, right: Pictorial Press Ltd/Alamy Stock Photo; **p. 8** top: Michael Stubblefield/Alamy Stock Photo, left and right: Peter Grant; **p. 12**: Anup Shah/Corbis Documentary/Getty Images; **p. 17** left: Paul Kay/Getty Images, right: Westend61 GmbH/Alamy Stock Photo; **p. 18**: Tui De Roy/Minden Pictures/Newscom; **p. 19** The Natural History Museum/Alamy Stock Photo; **p. 20**: Bill Radke/USFWS; **p. 22** left to right: CreativeNature_nl/iStock/Getty Images Plus, Kevin Langergraber, Juniors Bildarchiv/age fotostock.

CHAPTER 2
Photos: **Pages 24–25**: blickwinkel R. Koenig/Alamy Stock Photo; **p. 26** top: Gregor Mendel (oil on canvas)Austrian School (19th century) Private Collection/De Agostini Picture Library/The Bridgeman Art Library, bottom: De Agostini Picture Library/Getty Images; **p. 28** top: ISM/Pr. Philippe Vago/medicalimages.com, bottom: Dr. Jeremy Burgess/Science Source.

CHAPTER 3
Photos: **Pages 52–53**: Thomas Shahan/Science Source; **p. 54**: G. Büttner/OKAPIA/Science Source; **p. 60** top to bottom: A. Barrington Brown/Science Photo Library/Science Source, AP Photo/Jacob Harris, AP Photo/Claudio Luffoli; **p. 65**: Courtesy of Scott Carroll; **p. 71**: John Warburton-Lee Photography/Alamy Stock Photo; **p. 77**: Robert Boyd.

CHAPTER 4
Photos: **Pages 80–81**: Dietmar Nill/Nature Picture Library/Alamy Stock Photo; **p. 83** left © Gallo Images/Corbis via Getty Images, right: Ibrahim Süha Derbent/Getty Images; **p. 85** left to right: Miguel Castro/Science Source, FLPA/Alamy Stock Photo, Steve Gettle/Minden Pictures/Newscom; **p. 86**: Stuart Wilson/Science Source; **p. 90**: Joan Silk; **p. 91** clockwise from top left: Mark Jones/Roving Tortoise Photos/Getty Images, Tui De Roy/Minden Pictures/Newscom, Jose B. Ruiz/Nature Picture Library/Alamy Stock Photo, Painting © H. Douglas Pratt; **p. 92** top: Richard Owen, bottom: ©Ben Cropp/AUSCAPE All rights reserved; **p. 93** left to right: T. Whittaker/Premaphotos Wildlife, Andrey Gudkov/Shutterstock; Terry Whittaker/Science Source, Sergey Uryadnikov/Shutterstock; **p. 94** top: Courtesy of Robert Boyd & Joan Silk, bottom: Sylvain Cordier/BIOSPHOTO/Alamy Stock Photo; **p. 99**: Tom McHugh/Science Source; **p. 101**: John Giannicchi/Science Source.

CHAPTER 5
Photos: **Pages 107–108**: Manoj Shah/Getty Images; **p. 112** clockwise from top left: ardea.com/Liz Bomford, ftamas/Shutterstock, sphoom/Shutterstock, Herbert Kehrer/imageBROKER/Alamy, Jan Lindblad/Science Source; **p. 113** clockwise from left: Suzi Eszterhas/Minden Pictures/Newscom, Karl & Kay Ammann/Bruce Coleman Inc./Photoshot, ehtesham/Shutterstock; **p. 114** clockwise from top left: Stephen Dalton/Science Source; Anup Shah/Nature Picture Library/Alamy Stock Photo, Sergey Uryadnikov/Alamy Stock Photo, 4FR/Getty Images, Dallin Willden/EyeEm/Alamy Stock Photo, bottom: Photograph courtesy of Carola Borries; **p. 115**: Natalia Pryanishnikova/Alamy Stock Photo; **p. 120** top to bottom: Millard H. Sharp/Science Source, Jacana/Science Source, Robert Boyd, Solvin Zankl/Nature Picture Library/Alamy Stock Photo; **p. 121** clockwise from top left: Luiz Claudio Marigo/Nature Picture Library; **p. 121** top right: Rob Boyd and Joan Silk, center right: tristan tan/Shutterstock, center left: Elena Korchenko/Alamy Stock Photo, bottom: Nick Gordon/naturepl.com; **p. 122** left: Tui De Roy/Minden Pictures/Newscom, right: Photograph courtesy of Carola Borries; **p. 123** left to

right: Courtesy of Kathy West, David Wall/DanitaDelimont.com/Newscom, Marina Cords; **p. 124** top: Anna Yu/Getty Images, bottom: Kerstin Layer/age footstock, left:Sergey Podlesnov/Shutterstock, right: Nora Carol Photography/Getty Images; **p. 125** left: Laura Romin & Larry Dalton/Alamy Stock Photo, right: Andy Rouse/Nature Picture Library; **p. 126** left to right: Anup Shah/Getty Images, Kristin Mosher/Danita Delimont/Newscom, Frans Lanting/MINT Images/Science Source, bottom: Courtesy of Joan Silk; **p. 128**: Courtesy of Joan Silk; **p. 130** clockwise from top left: Martin Mecnarowski/Shutterstock, Ger Bosma/Getty Images, Michel Gunther/Science Source, Photograph courtesy of Carola Borries, Fabio Colombini Medeiros/Animals Animals, Courtesy of Joan Silk; **p. 131** left to right: Westend61/Getty Images, Manoj Shah/Getty Images, iPics/Shutterstock; **p. 132** top: Megan Whittaker/Alamy Stock Photo, bottom: Satellite image by DigitalGlobe; **p. 134** clockwise from top left: Robert Boyd, Robert Boyd, Robert Boyd, David Tipling Photo Library/Alamy Stock Photo, Robert Boyd, bottom left to right: Robert Boyd; **p. 140**: Tim Davenport/Wildlife Conservation Society.

Drawn Art: Figure 5.27: From Crofoot, M. C., I.C. Gilby, M.C. Wikelski, & R.W. Kays. "Interaction Location Outweighs the Competitive Advantage of Numerical Superiority in Cebus capucinus Intergroup Contests." *Proceedings of the National Academy of Sciences*, 105(2), 577–581. Copyright (2008) National Academy of Sciences, U.S.A; **Figure 5.32:** Reprinted with permission from AAAS. From Estrada, A.et al. "Impending Extinction Crisis of the World's Primates: Why Primates Matter." *Science Advances* 3(1). ©2017 The Authors, some rights reserved; exclusive licensee American Association for the Advancement of Science. Distributed under a Creative Commons Attribution NonCommercial License 4.0 (CC BY-NC) http://creativecommons.org/licenses/by-nc/4.0/; **Figure 5.34:** Republished with permission of Cell Press and Elsevier Science & Technology Journals. From Voigt et al." Global Demand for Natural Resources Eliminated More Than 100,000 Bornean Orangutans." *Current Biology: CB*, 28(5), 761–769. © 2018; permission conveyed through Copyright Clearance Center, Inc.

CHAPTER 6

Photos: Pages 142–143: Anup Shah/Getty Images; **p. 144:** Courtesy of Joan Silk; **p. 146** top: GP232/Getty Images, bottom: Courtesy of Kathy West; **p. 147:** Mark Conlin/Getty Images; **p. 148** top: Courtesy of Kathy West, bottom: Yasuo Tomishige/The Asahi Shimbun via Getty Images; **p. 152:** The Dian Fossey Gorilla Fund International; **p. 154:** Jürgen & Christine Sohns/imageBROKER/Alamy Stock Photo; **p. 155:** Joan Silk; **p. 156:** top to bottom: bluelake/Shutterstock, Matt Gibson/Shutterstock, Jose Francisco Arias Fernandez/EyeEm/Getty Images; **p. 157:** top: Robert Boyd/Joan Silk, bottom: Courtesy of Robert Boyd; **p. 158:** Joan Silk; **p. 159:** Martin Harvey/Photolibrary/Getty Images; **p. 160:** Toni Angermayer/Science Source; **p. 161:** Sylvain Cordier/hemis. fr/Alamy Stock Photo; **p. 162:** The Dian Fossey Gorilla Fund International; **p. 163:** Joan Silk; **p. 164:** Mark Moffett/Minden Pictures/Newscom; **p. 165:** Joan Silk; **p. 167:** Joan Silk.

Drawn art: Figure 6.5: Reprinted by permission of the University of Chicago Press. Figure 10.3 from John C. Mitani, et al. *The Evolution of Primate Societies.* © 2012 by The University of Chicago; permission conveyed through Copyright Clearance Center, Inc.; **Figure 6.9a&b:** Republished with permission of the American Association of Physical Anthropologists and John Wiley & Sons. From Robbins, A. M., Robbins, M. M., Gerald-Steklis, N., & Steklis, H. D." Age-Related Patterns of Reproductive Success Among Female Mountain Gorillas." *American Journal of Physical Anthropology: The Official Publication of the American Association of Physical Anthropologists*, 131(4), 511–521. © 2006; permission conveyed through Copyright Clearance Center, Inc.; **Figure 6.10:** Figure 3 from S. C. Alberts, et. al (2013) "Reproductive aging patterns in primates reveal that humans are distinct." *Proceedings of the National Academy of Sciences* 110(33), 13440–1344. Reprinted with permission; **Figure 6.12a:** Republished with permission of Cell Press and Elsevier Science & Technology Journals. From Walker, Kara K., Christopher S. Walker, Jane Goodalle, and Anne E. Pusey. "Maturation is Prolonged and Variable in Female Chimpanzees." *Journal of Human Evolution* Vol 114, 131–140. © 2018; permission conveyed through Copyright Clearance Center, Inc.; **Figure 6.12b:** Republished with permission of British Ecological Society and John Wiley & Sons. Jones, J. H., Wilson, M. L., Murray, C., & Pusey, A. "Phenotypic Quality Influences Fertility in Gombe Chimpanzees". *Journal of Animal Ecology*, 79(6) 1262–1269. © 2011; permission conveyed through Copyright Clearance Center, Inc.; **Figure 6.17b:** Reprinted by permission of the Royal Society. Figure from Archie, Elizabeth A. et al. "Social Affiliation Matters: Both Same-Sex and Opposite-Sex Relationships Predict Survival in Wild Female Baboon." *Proceedings of the Royal Society of London. Series B: Biological Sciences* 281:1793. © 2014; permission conveyed through Copyright Clearance Center, Inc.; **Figure 6.19:** Reprinted by permission of John Wiley & Sons. Figure from L. Barrett & S. P. Henzi. "Are baboon infants Sir Philip Sydney's offspring?" *Ethology* 106(7), 645–658, © 2000; permission conveyed through Copyright Clearance Center, Inc.; **Figure 6.24:** Republished with permission of The International Society of Behavioral Ecology Oxford University Press-Journals. From Dubuq, C, A. Ruiz-Lambides, A. Widdig. "Variance in Male Lifetime Reproductive Success and Estimation of the Degree of Polygyny in a Primates." *Behavioral Ecology*, 25(4), 878–889. © 2014; permission conveyed through Copyright Clearance Center, Inc.; **Figure 6.29:** Republished with permission of Springer Nature BV. From Piper, B.A.H, J.M. Dietz, & B.E. Raboy." Multi-Male Groups Positively Linked to Infant Survival and Growth in a Cooperatively Breeding Primate." *Behavioral Ecology and Sociobiology*, 71(12). © 2017; permission conveyed through Copyright Clearance Center, Inc.; **Figure 6.39a:** Reprinted by permission of the Royal Society. Figure from M. Heistermann, et al. "Loss of oestrus concealed ovulation and paternity confusion in free-ranging Hanuman langurs." *Proceedings of the Royal Society of London. Series B: Biological Sciences* 268:1484, 2445–2451. © 2001; permission conveyed through Copyright Clearance Center, Inc.; **Figure 6.39b:** Reprinted by permission of the Royal Society. Figure from M. Heistermann, et al. "Loss of oestrus concealed ovulation and paternity confusion in free-ranging Hanuman langurs." *Proceedings of the Royal Society of London. Series B: Biological Sciences* 268:1484, 2445–2451. © 2001; permission conveyed through Copyright Clearance Center, Inc.; **Figure 6.40:** Republished with permission of AAAS. From Roberts, E. K. et al. "A Bruce Effect in Wild Geladas."

Science 335(6073), 1222–1225. © 2012; permission conveyed through Copyright Clearance Center, Inc.

CHAPTER 7

Photos: Pages 170–171: Anup Shah/Nature Picture Library/Alamy Stock Photo; **p. 172:** K. G. Preston-Mafham/Premaphotos Wildlife; **p. 173** left: AP Photo/Rajesh Nirgude, right: Joan Silk; **p. 179** left: Kathy West, right: Joan Silk, top and bottom: Joan Silk; **p. 180:** image courtesy of Dr. Dana Pfefferle. Figure 2 in Pfefferle, Dana & Kazem, Anahita & Brockhausen, Ralf & Ruiz-Lambides, Angelina & Widdig, Anja. (2014). Monkeys Spontaneously Discriminate Their Unfamiliar Paternal Kin under Natural Conditions Using Facial Cues. Current Biology. 24. 10.1016/j.cub.2014.06.058.; **p. 181** clockwise from top left: Adrian Hepworth/Alamy Stock Photo, Marina Cords, Robert Ross/Getty Images, Joan Silk; **p. 182:** Joan Silk; **p. 184:** Joan Silk; **p. 185:** Terry Whitaker/Frank Lake Picture Agency/Corbis via Getty Images; **p. 187:** Joan Silk.

Drawn art: Figure 7.11b: Figure from A. J. Kazem & A. Widdig (2013) "Visual phenotype matching: cues to paternity are present in rhesus macaque faces." *PLOS One*, 8(2), e55846.; **Figure 7.18a:** Reprinted by permission of the Royal Society. Figure from Lukas, D., & Clutton-Brock, T.. "Cooperative Breeding and Monogamy in Mammalian Societies." *Proceedings of the Royal Society of London. Series B: Biological Sciences* 279: 1736, 2151–2156. © 2012; permission conveyed through Copyright Clearance Center, Inc.

CHAPTER 8

Photos: Pages 190–191: Fiona Rogers/Nature Picture Library; **p. 192:** Bettmann/Corbis via Getty Images; **p. 193** top: W. Perry Conway/Corbis via Getty Images, bottom: Joan Silk; **p. 194:** Gallo Images/Corbis via Getty Images; **p. 196:** Michael Nichols/Nat Geo Creative; **p. 197** clockwise from top right: Wolfgang Kohler The Mentality of Apes. Routledge & Kegan Paul Ltd. London 1927. Reproduced by permission of Taylor & Francis Books UK., woddle1000/iStock/Getty Images Plus, Frans Lanting/Mint Images/age footstock; **p. 199** clockwise from top left: Mary Beth Angelo/Science Source, Michael Nichols/Nat Geo Creative; Robert Boyd & Joan Silk; **p. 200:** Courtesy of Joan Silk; **p. 201** left: Courtesy of Joan Silk, right: Ross/Tom Stack Associates/Alamy Stock Photo; **p. 202:** Courtesy E. Menzel; **p. 204:** E. Hermann J. Call M.V. Hernandez-Lloreda B. Hare and M. Tomacello "Humans Have Evolved Specialized Skills of Social Cognition: The Cultural Intelligence Hypothesis." 2007 Science 317:1360–1366.

Drawn art: Figure 8.5: Reprinted by permission of the University of Chicago Press. Figure 10.3 and 10.11 from Van Schaik, CP and K. Isler. "Life-History Evolution in Primates" in *The Evolution of Primate Societies* by John C. Mitani, et al. © 2012 by The University of Chicago; permission conveyed through Copyright Clearance Center, Inc.; **Figure 8.6:** Republished with permission of Cell Press and Elsevier Science & Technology Journals. From Jones, J.H." Primates and the Evolution of Long, Slow Life Histories." *Current Biology: CB*, 21(18), R708-R717. © 2011; permission conveyed through Copyright Clearance Center, Inc.; **Figure 8.18:** Republished with permission of Cell Press and Elsevier Science &

Technology Journals. From Flombaum, Jonathan I. and Laurie R. Santos. Rhesus Monkeys Attribute Perceptions to Others." *Current Biology: CB*, 15(5), 447–452. © 2005; permission conveyed through Copyright Clearance Center, Inc.; **Figure 8.19:** Republished with permission of AAAS. from Hermann, E., J. Call, M. V. Hernandez-Lloreda, B. Hare, and M. Tomasello. "Humans Have Evolved Specialized Skills of Social Cognition," *Science* 317 (5843), 1360–1366. © 2007; permission conveyed through Copyright Clearance Center, Inc.

CHAPTER 9

Photos: Pages 208–209: Dorling Kindersley: Andrew Kerr; **p. 212:** kjekol/iStock/Getty Images Plus; **p. 218:** Doug M. Boyer/Duke University Department of Evolutionary Anthropology; **p. 222:** Shaun Curry/AFP/Getty Images; **p. 230** top: Christophe Ratier/NHPA.UK, bottom: Photograph courtesy Laura MacLatchy. MacLatchy L. "The Oldest Ape" 2004. Evolutionary Anthropology 13: 90–103; **p. 232:** Photo courtesy Salvador Moyà-Solà. Salvador Moyà-Solà Meike Köhler David M. Alba Isaac Casanovas-Vilar Jordi Galindo "Pierolapithecus catalaunicus a New Middle Miocene Great Ape from Spain" Science 19 Nov 2004: Vol. 306 Issue 5700 pp. 1339–1344 DOI: 10.1126/science.1103094.

Drawn art: Figure 9.1a & b: Figure from *Mammal Evolution: An Illustrated Guide* by R. J. G. Savage, pp. 38–39, 1986. Copyright © 1986 by Facts On File, Inc., an imprint of Infobase Publishing. Reprinted with permission of the publisher; **Figure 9.2:** Reprinted by permission of Princeton University Press. Figure 5.3 from Robert D. Martin. *Primate Origins and Evolution: A Phylogenetic Reconstruction.* © 1990 R. D. Martin. permission conveyed through Copyright Clearance Center, Inc.; **Figure 9.7:** Artwork of Carpolestes simpsoni by Doug M. Boyer, from "Paleontology: Primate Origins Nailed" by Eric J. Sargis, *Science* 298, Nov. 22, 2002, p. 1564. Reprinted with permission; **Figure 9.8:** Figure from *The Cambridge Encyclopedia of Human Evolution*, edited by Steve Jones, Robert Martin, and David Pilbeam, p. 200. Copyright © Cambridge University Press 1992. Reprinted with the permission of Cambridge University Press. Adapted from a figure published in *Primate Adaptation and Evolution* by J. G. Fleagle, (Academic Press, 1988), p 4. Copyright © 1988 Elsevier Ltd. Reprinted by permission; **Figure 9.10a & b:** Figure 3 from Kenneth D. Rose, "The Earliest Primates," *Evolutionary Anthropology*, Vol. 3, Issue 5 (1994): 159–173. Copyright © 1994 Wiley-Liss, Inc., A Wiley Company. Reprinted with permission of Wiley-Liss, Inc., a subsidiary of John Wiley & Sons, Inc.; **Figure 9.11a & b:** This figure (11.9 a & b) was published in *Primate Adaptation and Evolution*, J. G. Fleagle, (Academic Press, 1988). Copyright © 1988 Elsevier Ltd. Reprinted by permission; **Figure 9.23:** This figure (13.7) was published in *Primate Adaptation and Evolution*, J. G. Fleagle, (Academic Press, 1988). Copyright © 1988 Elsevier Ltd. Reprinted by permission; **Figure 9.24:** Reproduced with permission of Blackwell Publishing Ltd. and John Wiley & Sons. Figure 8.21 from Roger Lewin and Robert Foley, *Principles of Human Evolution, 2nd Edition.* © 2004 by Blackwell Science Ltd, a Blackwell Publishing company; permission conveyed through Copyright Clearance Center, Inc.; **Figure 9.25:** Reprinted by permission of Princeton University Press. Figure 2.23 from Robert D. Martin. *Primate Origins* and Evolution: A Phylogenetic Reconstruction. © 1990 R. D. Martin; permission conveyed through Copyright Clearance Center, Inc.

CHAPTER 10

Photos: Pages 236–237: Anup Shah/Animals Animals/age footstock; **p. 239** top: Sabena Jane Blackbird/Alamy Stock Photo, **p. 239** bottom: Brigitte Senuta, Martin Pickford, Dominique Gommery, Pierre Meind, Kiptalam Cheboie, Yves Coppens. "First hominid from the Miocene (Lukeino Formation, Kenya)" Comptes Rendus de l'Académie des Sciences - Series IIA - Earth and Planetary Science 332 (2001) 137–144 2001 Académie des sciences/Éditions scientifiques et médicales Elsevier SAS. Tous droits réservés S1251–8050(01)01529-4/FL A; **p. 241**: David Brill; **p. 242**: left: Image courtesy Robert Eckhardt. Galik K Senut B Pickford M Gommery D Treil J Kuperavage AJ Eckhardt RB. "External and Internal Morphology of the BAR 1002'00 Orrorin tugenensis Femur" 2004 Science 305 1450–1453; **p. 242** center and right: C. Owen Lovejoy "The natural history of human gait and posture: Part 2. Hip and thigh" Gait & Posture Volume 21 Issue 1 January 2005 Pages 113–124; **p. 243**: Redrawn from Y. Haile-Selassie G. Suwa and T. D. White "Late Miocene Teeth from Middle Awash Ethiopia and Early Hominid Dental Evolution" 2004 Science 05 Mar 2004: Vol. 30 Issue 5663 pp. 1503–1505; **p. 243**: David Brill; **p. 244** top: Tim D. White/Provided by David Brill, bottom: ©Tim D. White & Gen Suwa/Provided by David Brill; **p. 246**: C. Owen Lovejoy Gen Suwa and colleagues/Provided by David Brill; **p. 249**: © National Museums of Kenya; **p. 250** top: Eric Lafforgue/Art in All of Us/Corbis via Getty Images, bottom: John Reader/Science Source; **p. 251** top: courtesy Professor Zeray AlemSeged/DRP, bottom: Yohannes Haile-Selassie Luis Gibert Stephanie M. Melillo Timothy M. Ryan Mulugeta Alene Alan Deino Naomi E. Levin Gary Scott & Beverly Z. Saylor. "New species from Ethiopia further expands Middle Pliocene hominin diversity" Nature 521 483–488 (28 May 2015) doi:10.1038/nature14448; **p. 252**: Courtesy of Lee R. Berger and The University of The Witwatersrand'; **p. 253**: David Brill; p. 254 right: David Brill; **p. 255**: John Reader/Science Source; **p. 256**: Martin Harvey/Getty Images; **p. 258**: © National Museums of Kenya; **p. 259**: Album/Prisma/Newscom; **p. 261**: © National Museums of Kenya/Courtesy of Meave Leakey/photo by Robert Campbell; **p. 262**: David Brill.

Drawn art: Figure 10.3: Figure from "The evolution of human bipedality: ecology and functional morphology," by K. D. Hunt. Reprinted by permission of the author; **Figure 10.22**: Reprinted by permission of the University of Chicago Press. Figure 4.2 by Kathryn Cruz-Uribe, from *The Human Career: Human Biological and Cultural Origins, Second Edition* by Richard G. Klein. © 1989, 1999 by The University of Chicago; permission conveyed through Copyright Clearance Center, Inc.; **Figure 10.23**: Reprinted by permission of the University of Chicago Press. Figure 4.24 from Richard G. Klein, *The Human Career: Human Biological and Cultural Origins, Second Edition* © 1989, 1999 by The University of Chicago. permission conveyed through Copyright Clearance Center, Inc.; **Figure 10.34**: Republished with permission of AAAS. From Ungar, Peter S and Matt Sponheimer. "The Diets of Early Hominins," *Science* 334 (6053) © 2011; permission conveyed through Copyright Clearance Center, Inc.

CHAPTER 11

Photos: Pages 264–265: Peter Bostrom; **p. 267** top: Fred Spoor Philipp Gunz Simon Neubauer Stefanie Stelzer Nadia Scott Amandus Kwekason & M. Christopher Dean. Reconstructed Homo habilis type OH 7 suggests deep-rooted species diversity in early Homo. Nature 519 83–86 (05 March 2015) doi:10.1038/nature14224, bottom: Javier Trueba/MSF/Science Photo Library/Science Source; **p. 270**: imageBROKER/Alamy Stock Photo; **p. 271** top: David Lordkipanidze, Georgian State Museum, Georgian Academy of Sciences/Leo Gabunia, Abesalom Vekua, David Lordkipanidze, Carl C. Swisher, Reid Ferring, Antje Justus, Medea Nioradze, Merab Tvalchrelidze, Susan C. Antón, Gerhard Bosinski, Olaf Jöris, Marie-A.-de Lumley, Givi Majsuradze, Aleksander Mouskhelishvili. "Earliest Pleistocene Hominid Cranial Remains from Dmanisi, Republic of Georgia: Taxonomy, Geological Setting, and Age" Science 12 May 2000: Vol. 288, Issue 5468, pp. 1019–1025 DOI: 10.1126/science.288.5468.1019, left: Guram Bumbiashvili Georgian National Museum/David Lordkipanidze Marcia S. Ponce de León Ann Margvelashvili Yoel Rak G. Philip Rightmire Abesalom Vekua Christoph P. E. Zollikofer. "A Complete Skull from Dmanisi Georgia and the Evolutionary Biology of Early Homo" Science 18 October 2013: Vol. 342 no. 6156 pp. 326–331 DOI: 10.1126/science.1238484, right: Jay Matternes; **p. 272** top: National Anthropological Archive Negative no. 01019100/Smithsonian Museum of Natural History, bottom: The Natural History Museum/Alamy Stock Photo; **p. 273**: Alan Walker; **p. 274**: West Turkana Archeological Project-Mission Préhistorique au Kenya; **p. 279**: R. Potts, Smithsonian Institution; **p. 280**: Kim Hill and Magdalena Hurtado; **p. 283**: Anup Shah/Nature Picture Library; **p. 287** : Francisco d'Errico and Lucinda Backwell; **p. 289**: National Museum of Natural History Smithsonian Institution Washington D.C. (Rdg. 1–2) Courtesy of Kathy D. Schick and Nicholas Toth CRAFT Research Center Indiana University; **p. 290** top: Biosphoto/Superstock, bottom: Manuel Dominguez-Rodrigo; **p. 291**: © National Museums of Kenya. Photograph courtesy of Alan Walker; **p. 292** left: Biosphoto/Superstock, right: James Tyrrell/Gallo Images/Getty Images; **p. 293** top: Robert Boyd & Joan Silk, left: Joan Silk, right: Robert Boyd; **p. 294**: Robert Boyd; **p. 296**: Wendy Stone/Corbis via Getty Images.

Drawn art: Figure 11.6: Reprinted by permission of the University of Chicago Press Figure 5.10, © 1999 by Kathryn Cruz-Uribe from *The Human Career: Human Biological and Cultural Origins, Second Edition* by Richard G. Klein. © 1989, 1999 by The University of Chicago; permission conveyed through Copyright Clearance Center, Inc.; **Figure 11.7a-d**: Reprinted by permission of the University of Chicago Press. Figures 4.2 (© 1999 by Kathryn Cruz-Uribe), 4.23, 5.10 (© 1999 by Kathryn Cruz-Uribe), 7.2 from *The Human Career: Human Biological and Cultural Origins, Second Edition* by Richard G. Klein. © 1989, 1999 by The University of Chicago; permission conveyed through Copyright Clearance Center, Inc.; **Figure 11.45**: Reprinted by permission of the University of Chicago Press. Figure 4.41 from *The Human Career: Human Biological and Cultural Origins, Second Edition* by Richard G. Klein. © 1989, 1999 by The University of Chicago; permission conveyed through Copyright Clearance Center, Inc.; **Figure 11.46**: Reprinted by permission of the University of Chicago Press. Figure from Richard G. Klein, *The Human Career: Human Biological and Cultural Origins, Second Edition.* © 1989, 1999 by The University of Chicago; permission conveyed through Copyright Clearance Center, Inc.

CHAPTER 12

Photos: Pages 300–301: Javier Trueba/Madrid Scientific Films/Science Source; **p. 309** top: Javier Trueba/MSF/Science Photo Library/ Science Source, bottom: William Jungers, Stony Brook University; **p. 310:** Brumm A. Aziz F. van den Bergh G.D. Morwood M.J. Moore M.W. Kurniawan I. Hobbs D.R. Fullagar R. "Early stone technology on Flores and its implications for Homo floresiensis" 2006. Nature 441:624–628; **p. 311:** Javier Trueba/MSF/Science Photo Library/Science Source; **p. 312:** © Kennis& Kennis/Natural History Museum London; **p. 314** left: Damian Kuzdak/Getty Images, right: Aleksander Bolbot/Alamy Stock Photo, bottom: Len Rue Jr./ Science Source; **p. 315:** ToonBeeld/Frans de Vries. citation: Niekus MJLT, Kozowyk PRB, Langejans GHJ, et al. Middle Paleolithic complex technology and a Neandertal tar-backed tool from the Dutch North Sea. Proceedings of the National Academy of Sciences of the United States of America. 2019 Oct;116(44): 2; **p. 316** top: Photo from Joao Zilhao; **p. 317** top: Erik Trinkaus; bottom: Courtesy of Joan Silk; **p. 319:** left and right: James Steele Margaret Clegg and Sandra Martelli "Comparative Morphology of the Hominin and African Ape Hyoid Bone a Possible Marker of the Evolution of Speech" Human Biology 2013 85 (5) 639–672, center: David Brill; **p. 322:** Eric Delson; **p. 323:** Homo naledi a new species of the genus Homo from the Dinaledi Chamber South Africa Berger et al eLife 2015;4.e09560 https://elifesciences.org/content/4/e09560/article info CC by 4.0 Attribution https://creativecommons.org/licenses/by/4.0/, https://elifesciences.org/content/4/e09560/article-info; **p. 324:** Natural History Museum London/Science Photo Library/ Science Source.

Drawn art: Figure 12.21: Republished with permission of Academic Press and Elsevier Science & Technology Journals. From Wales et. al. "Modeling Neanderthal Clothing Using Ethnographic Analogue." *Journal of Human Evolution,* 63(6), 781–95. © 2012; permission conveyed through Copyright Clearance Center, Inc.; **Figure 12.3a & b:** Reprinted by permission of the University of Chicago Press. Figures 5.22, 5.26, © 1999 by Kathryn Cruz-Uribe, from *The Human Career: Human Biological and Cultural Origins, Second Edition* by Richard G. Klein. © 1989, 1999 by The University of Chicago; permission conveyed through Copyright Clearance Center, Inc.; **Figure 12.5:** Reprinted by permission of the University of Chicago Press. Figure 6.24 from Richard G. Klein, *The Human Career: Human Biological and Cultural Origins, Second Edition.* © 1989, 1999 by The University of Chicago; permission conveyed through Copyright Clearance Center, Inc.; **Figure 12.6:** Reprinted by permission of the University of Chicago Press. Figure 5.32, © 1999 by Kathryn Cruz-Uribe, from *The Human Career: Human Biological and Cultural Origins, Second Edition* by Richard G. Klein. © 1989, 1999 by The University of Chicago; permission conveyed through Copyright Clearance Center, Inc.; **Figure 12.13:** Reprinted by permission of the University of Chicago Press. Figure 6.48, © 1999 by Kathryn Cruz-Uribe, from *The Human Career: Human Biological and Cultural*

Origins, Second Edition by Richard G. Klein. © 1989, 1999 by The University of Chicago; permission conveyed through Copyright Clearance Center, Inc.

CHAPTER 13

Pages 328–329: Javier Trueba/MSF/Science Source; **p. 332** left and right: Housed in National Museum of Ethiopia Addis Ababa. Photo © 2001 David L. Brill/Brill Atlanta; **p. 334:** top to bottom: Image courtesy of Prof Christopher Henshilwood/University of Bergen Norway, Curtis Marean/Institute of Human Origins ASU, after Larsson & Sjöström Antiquity Volume 085 Issue 330 December 2011/©Antiquity Publications Ltd.; **p. 335:** The Granger Collection NYC: **p. 336** top: Figure 3 from Bouzouggar A. et al. 2007 "82000-year-old shell beads from North Africa and implications for the origins of modern human behavior." PNAS 104:9964–9969, bottom: Pierre-Jean Texier Diepkloof Project (MAE) CNRS UMR 5199-PACEA/Redrawn from Figure 1 in Texier et al 2009. A Howieson's Poort tradition of engraving ostrich eggshell containers dated to 60000 years ago at Diekloof Rock Shelter South Africa. PNAS 107 6181; **p. 337** top: Image Francesco d'Errico/Courtesy of Professor Christopher Henshilwood, bottom: Liu W Martinon-Torres M Cai Y-j Xing S Tong H-w Pei S-w Sier MJ Wu X-h Edwards RL Cheng H Li Y-y Yang X-x de Castro JMB Wu X-j. 2015. The earliest unequivocally modern humans in southern China. Nature 526 696–699 (29 October 2015) doi:10.1038/nature15696; **p. 349:** fig 4 from Mellars P. et al Genetic and archaeological perspectives on the initial modern human colonization of southern Asia. PNAS v110 no. 26 10699–10704 2013 and Mellars P. Going east: New genetic and archaeological perspectives on the modern human colonization of Eurasia.2006 Science 313(5788):796–800; **p. 350:** From V. V. Pitulko et al. 2004 "The Yana RHS Site: Humans in the Arctic before the Last Glacial Maximum" Science 303:52–56; **p. 354:** © Jean-Michel Labat/AUSCAPE All rights reserved; **p. 358:** Jeff Pachoud/AFP/Getty Images; **p. 359:** Album/Alamy Stock Photo.

Drawn art: Figure 13.1: Reprinted by permission of the University of Chicago Press. Figure 7.2 from by Richard G. Klein, *The Human Career: Human Biological and Cultural Origins, Second Edition.* © 1989, 1999 by The University of Chicago; permission conveyed through Copyright Clearance Center, Inc.; **Figure 13.6:** Drawing: "An atlatl is a tool that lengthens the arm," from The Testimony of Hands: Atlatls (http://hands.unm.edu/atlatls.html). Reprinted by permission of James Dixon; **Figure 13.19:** Reprinted with permission from AAAS. From Mellars, P. "Going East: New Genetic and Archaeological Perspectives on the Modern Human Colonization of Eurasia." *Science* 313(5788), 796–800, ©2006; permission conveyed through Copyright Clearance Center, Inc.; **Figure 13.26:** Reprinted by permission of Princeton University Press. Figure 34.8 from Soffer, Olga, "The Middle to Upper Patheolithic Transition on the Russian Plain," from Mellars, Paul; *The Human Revolution.* © 1989 Edinburgh University Press; permission conveyed through Copyright Clearance Center, Inc.; **Figure 13.27:** Reprinted by permission of the University of Chicago Press. Figure 7.22 from Richard G. Klein, *The Human Career: Human Biological and Cultural Origins, Second Edition.* © 1989, 1999 by The University of Chicago; permission conveyed through Copyright Clearance Center, Inc.;

Figure 13.28: Reprinted by permission of Princeton University Press. Figure 13.4 from Paul Mellars. *The Neanderthal Legacy: An Archaeological Perspective from Western Europe.* © 1996 Princeton University Press; permission conveyed through Copyright Clearance Center, Inc.

CHAPTER 14

Photos: Pages 362–363: Mira/Alamy; **p. 369**: AP Photo/Gregory Bull; **p. 370** top: Kenjiro Matsuo/AFLO SPORT/Alamy Live News, bottom: Orange Pics BV/Alamy Stock Photo; **p. 373** left: Eye of Science/Science Source, right: Andrew Syred/Getty Images; **p. 377**: Hermes Images/AGF Srl/Alamy Stock Photo; **p. 380**: H. Mark Weidman Photography/Alamy Stock Photo; **p. 385**: Jan Halaska/Science Source; **p. 397** left: Robert Shantz/Alamy Stock Photo, right: Mark Weidman Photography/Alamy Stock Photo.

Drawn art: Figure 14.1: Figure 3A from "Extension of Cortical Synaptic Development Distinguishes Humans from Chimpanzees and Macaques," by Xiling Liu, *Genome Research*, April 2012, 22(4). Copyright © 2012 Cold Spring Harbor Laboratory Press; **Figure 14.8:** Reprinted by permission of the Royal Society. Figure from Pascale Gerbault, et al. "Evolution of lactase persistence: an example of human niche construction." *Philosophical Transactions of the Royal Society (B)* 366, 863–877. (c) 2011; permission conveyed through Copyright Clearance Center, Inc.; **Figure 14.13:** Republished with permission of Cell Press and Elsevier Science &Technology Journals. From F. Prugnolle et al." Geography Predicts Neutral Genetic Diversity of Human Populations." Current Biology: CB, 15(5), R159-R160. © 2005; permission conveyed through Copyright Clearance Center, Inc.; **Figure 14.20:** Figure 2 from "What Controls Variation in Human Skin Color?" by Gregory S Barsh. PLOS Biology 1(3) e91.

CHAPTER 15

Photos: Pages 402–403: Lee Jae-Won/Reuters/Newscom; p. **405** top: Andrey Gudkov/Shutterstock, bottom: Herb Roe/www.chromesun.com; **p. 406**: Tom McHugh/Science Source; **p. 407**: Wertz, A. E. (2019). How plants shape the mind. Trends in cognitive sciences, 23(7), 528–531; **p. 410**: Anup Shah/Nature Picture Library; **p. 413**: ART Collection/Alamy Stock Photo.

Drawn art: Figure 15.5: Republished with permission of Elsevier Science & Technology Journals. From Barrett, H. C., & Broesch, J." Prepared Social Learning About Dangerous Animals in Children." *Evolution and Human Behavior,* 33(5), 499–508. © 2012; permission conveyed through Copyright Clearance Center, Inc.; **Figure 15.13:** Reprinted with permission from AAAS. From Hill, K.R., R.S. Walker, M. Božicevic, J. Eder, T. Headland, B. Hewlett, A.M. Hurtado, F. Marlowe, P.W. Wiessner. "Co-Residence Patterns in Hunter-Gatherer Societies Show Unique Human Social Structure." *Science* 331(6022), 1286–9, ©2011; permission conveyed through Copyright Clearance Center, Inc.; **Figure 15.14**: Reprinted with permission from AAAS. From Levinson, S. C. "Kinship and Human Thought." *Science* 336(6084), 988–989, ©2012; permission conveyed through Copyright Clearance Center, Inc.; **Figure 15.15**: Republished with permission of Cell Press and Elsevier Science & Technology Journals. From M. Dyble et al." Networks of Food Sharing Reveal the Functional Significance of Multilevel Sociality in Two Hunter-Gatherer Groups." *Current Biology: CB*, 26(15), 2017–21. © 2016; permission conveyed through Copyright Clearance Center, Inc.; **Figure 15.16:** Republished with permission of Elsevier Science & Technology Journals. From M. Dyble et al." A Friend in Need is a Friend Indeed: Need-Based Sharing, Rather than Cooperative Assortment, Predicts Experimental Resource Transfers Among Agta Hunter-Gatherers." *Evolution and Human Behavior,* 40(1), 82–89. © 2019; permission conveyed through Copyright Clearance Center, Inc.

CHAPTER 16

Photos: Pages 424–425: Library of Congress, Prints & Photographs Division, Edward S. Curtis Collection; **p. 427**: B&C Alexander/Arcticphoto; **p. 428**: Classic Image/Alamy; **p. 429**: © Peter Hudson; **p. 430** left: De Agostini Picture Library/Bridgeman Images, right: Pictures from History/Bridgeman Images; **p. 431** top: Anup Shah/Nature Picture Library; **p. 431**: bottom: Whiten A. "The second inheritance system of chimpanzees and humans" Nature 437 52–55 (1 September 2005); **p. 432**: Robert Boyd; **p. 434**: Kennan Harvey Photography Durango CO; **p. 435** top: Portrait of Thomas Hobbes (Westport 1588-Hardwick Hall 1679) English philosopher 1669–1670 by John Michael Wright (1617–1694) oil on canvas 66x54 cm Wright John Michael (1617–94)/National Portrait Gallery London UK/De Agostini Picture Library/Bridgeman Images, bottom: Kim Hill and Magdalena Hurtado; **p. 436**: Raymond Mendez/Animals Animals/Earth Scenes; **p. 437** top: Charles Deering McCormick Library of Special Collections Northwestern University Library, bottom: Sarah Mathew; **p. 441**: Alan Compton/Jon Arnold Images Ltd/Alamy Stock Photo; **p. 442** left: Copyright Pitt Rivers Museum University of Oxford. Accession Number: 1998.355.395.2, right: Copyright Pitt Rivers Museum University of Oxford. Accession Number: 2005.51.87.1.

EPILOGUE

Photos: Page 446: Portrait of Charles Darwin (1809–1882) 1883 (oil on canvas) Collier John (1850–1934)(after)/National Portrait Gallery London UK/Bridgeman Images

INDEX

Note: Page numbers in *italics* refer to figures and tables.

hemoglobin A, 370, 373, 374, *374,* 375
hemoglobin S, 372–73, *373,* 374, *374,* 375, *376*
Henrich, Joseph, 439
Henry, Amanda, 314
Henshilwood, Christopher, 336
Henslow, John Stevens, 6
heredity
    cell division and role of chromosomes in, 27–37, *28*
    chromosomes and Mendel's experimental results in,
        29–32, *30–33*
    linkage and recombination in, 33–37, *34–36*
    Mendelian genetics in, *26,* 26–27, *27*
    mitosis and meiosis in, *28,* 28–29, *29*
heritability, 383, 385
    missing, 386
Herrmann, Esther, 204
Herto (Ethiopia), 332, *332*
heterozygotes, 30–32, *30–33*
Hexian (China), 308
He Xian (China), 308
hidden variation, 64–65, *65*
Hill, Kim, 279, 418–19, 435, 440, 441
Hill, Reuben, 414
hindbrain, *198*
hind-limb dominated locomotion, *113*
Hirschhorn, Neil, 389
Hiwi people, *281,* 283–84
Hobbes, Thomas, 435, *435*
Holocene epoch, *210*
home bases of early toolmakers, 295–98, *296, 297*
Hominidae, *93,* 123, 124–26, *124–26*
hominins
    Ardipithecus as, 242–46, *243–46*
    Australopithecus as, 248, 249–57, *249–57*
    bipedal locomotion in, 238, *240,* 240–41, *241,* 246–48, *247*
    diversification of, *248,* 248–61
    earliest, 238, 239–46
    Homo as, 248–49
    Kenyanthropus as, 248, 260–61, *261*
    modern humans *vs.,* 238–39
    Orrorin tugenensis as, *239,* 239–42, *241, 242*
    Paranthropus as, 248, 257–60, *258–60*
    phylogenies of, 261–-262, *262*
    Sahelanthropus tchadensis as, 239, *239*
    sites of, 249, *249*
Hominoidea
    emergence of, 227–34, *230–32*
    taxonomy of, 122, 123–26, *124–26*
hominoids
    emergence of, 227–34, *230–32*
    family tree of, 92–94, *93*
Homo, early, 248–49, 266–68, *266–68*
Homo erectus, 268–73
    arrival in eastern Asia of, 272, *272*
    brain of, 270
    classification of, 326, *326*
    cooking by, 294–95, *295*
    dentition of, 270
    endocranial volume of, *268*
    evolutionary transition to modern humans from, 304
    extension from Africa into Eurasia of, *270,* 270–72, *271*

    first appearance in fossil records of, 248, *249,* 268–70, *269*
    life history of, 272
    meat eating by, 290–91, *291*
    in Middle Pleistocene, 307–8
    postcranial skeleton of, 272–73, *273*
    skull of, *269,* 269–70
Homo ergaster, 326, *326*
Homo floresiensis, 308, 308–9, *309*
Homo habilis, 248, 268, *268,* 270
Homo heidelbergensis, 304–7
    brain and skull of, 304, *305*
    classification of, 325, *326*
    diet of, 307
    in eastern Asia, 307–8, *308*
    hunting by, 305–6
    hyoid bone of, 319
    location of, 304–5, *306*
    and Neanderthals, 310–11, *311*
homologous characters, 99
    in primates, 110–11
homologous chromosomes, 28, 29, *29*
homology, reasoning by, 110
Homo luzonensis, 309–10
Homo naledi, 323, *323*
Homo neanderthalensis, 325–26, *326*
Homo rudolfensis, 248, 268, *268,* 270
Homo sapiens. see also modern humans
    in Africa before 60 ka, *330–32,* 330–37, *334–37*
    archaeological evidence of, 348–51, *349, 350*
    behavior of, 333–37, *334–37*
    classification of, 326, *326*
    dating of earliest, 331–33, *332*
    decorative arts by, 335–36, *336, 337*
    in European Upper Paleolithic, 351–59, *352–59*
    features of, *330,* 330–31, *331*
    fossil evidence of, 337–38, *338*
    genetic evidence of, 338–48, *340–45*
    locations of, *306*
    move out of Africa by, 337–51
    oldest fossils classified as, 324, *324*
    one *vs.* two expansions of, 351
    skull of, *269, 330,* 331
    social networks of, 335
    toolmaking by, 333–35, *334, 335*
homozygotes, 30–32, *30–33*
honey, 256
Howieson's Port (South Africa), 336
howler monkeys, *112,* 120
Hoyle, Fred, 13
Hrdy, Sarah Blaffer, 164–65, 167–68
human(s), modern. *see* modern humans
human genome, sequencing of, 365, 378–80, *379, 380*
Human Genome Diversity Project, 344, *382*
human variation
    and balanced polymorphism, 372–73, *373,* 374, *374,* 375, *376*
    in complex phenotypic traits, *383,* 383–90
    dimensions of, 368–70
    and genetic drift, *380,* 380–81
    genetic *vs.* environmental, 368–69
    and genomewide association studies, 385–86, 389–90
    among groups, 369–70, 375–81, 386–90

within groups, 369, 372–75, 383–86
heritability and, 383, 385
in lactase persistence, 375–78, *377*
and migration and population growth, 381, *382*
in non-insulin-dependent diabetes, 373–75
and race concept, *390*, 390–99, *393, 397*
selection-mutation balance in, 372
and selective sweep, 378–80, *379, 380*
in specific language impairment, 370–72, *371*
in stature, 367, 386, *386*
in traits influenced by single genes, 370–82
twin studies of, 383–85, *384*
humerus, 256
Hunt, Kevin, 247
Hunter, Rick, 323
hunter-gatherers
and European genetic admixture, 394–95
mismatch hypothesis of cooperation in, 441, *441*
hunting
by Central Inuit, 427, *427*
by chimpanzees *vs.* humans, 280, *280*
defined, 280
evidence for, 286–91, *287–91*
and food sharing and division of labor, *281*, 281–86, *283, 284*
and home bases of early toolmakers, 295–98, *296, 297*
by *Homo heidelbergensis*, 305–6
by Neanderthals, 313, *314*
vs. scavenging, 292–95, *292–95*, 313, *314*
skills needed for, *280*, 280–81, *281*
in Upper Paleolithic, 355, *355*
Huntington's disease, 397
Hupa tribe, *437*
Hurtado, A. Magdalena, 279
huts, 356
Hutterites, genetic drift in, 380–81
Huxley, Thomas Henry, 110, 311
hybrid zone, 89
hydrogen bonds, 38, *39*
Hylobatidae, *93*, 123, 124, *124*
hyoid bone and spoken language, 319, *319*

identical twins, 384, *384*, 399
Ileret (Kenya), 267, 268, 271, 272
ilium(ia), 240
imitation, 430–32, *432*
inbred matings, 409
inbreeding avoidance, 409–12, *410, 411*
incest
avoidance of, 409–12, *410, 411*
third-party attitudes toward, 411
incisors
of *Ardipithecus ramidus*, 244, *244*
of early haplorrhines, 224, *224*
of Neanderthals, 312
of primates, *113*, 118, *119*
independent assortment, 27, 29, 33–36, *34–36*
independent segregation, 27, 29, 33–36, *34–36*
individual selection, 11–12, *12*
indri, brain of, *198*
infanticide, 123, 164–68, *166–68*
infidelity, 416–17, *417*

infraorder, 117
inheritance
blending, 21, 63, *63*
cell division and role of chromosomes in, 27–37, *28*
chromosomes and Mendel's experimental results in, 29–32, *30–33*
linkage and recombination in, 33–37, *34–36*
Mendelian genetics in, *26*, 26–27, *27*
mitosis and meiosis in, *28*, 28–29, *29*
Initial Upper Paleolithic (IUP), 352, *352*
injuries in Upper Paleolithic, 357
in-laws, 418
insect(s), eusocial, 436
insectivores, 119, *119*, 129
insulin, 374
insulin-dependent diabetes, 374–75
interbirth intervals, 148, *149*
interbreeding
of modern humans with Neanderthals and Denisovans, 345–48
and race, 394–96, *397*
intermediate steps favored by selection, 14–17, *15–17*
intersexual selection, 156, *156*
intrasexual selection, 156–59, *157–59*
introgression, 346
introns, 45, 46, *46*, 47
Inuit communities, 396, *427*, 427–28, *428*
invasion of new habitats, 18–19
investment in offspring, 140–48, *140–48*
iris, 4, *4*
iron in primate diet, 127
Isaac, Glyn, 296
isotopes, 214
Israel, modern humans in, 332–33
IUP (Initial Upper Paleolithic), 352, *352*

Jadera haematoloma, 65–67, *65–68*, 405
Java, 308
jaws of early *Homo*, 267–68, *268*
jealousy, 416–17, *417*
Jebel Irhoud (Morocco), 324, *324*, 332
Jenkin, Fleeming, 21, 64
Jerimalai cave site (Timor), 349
Jersey (island), 18–19
Jinniushan site (China), 308, *308*
Johanson, Donald, 250
Jolly, Alison, 115
Ju/'hoansi bands, 419
Jurassic period, *210*
juvenile period, prolonged, 285

Kabwe cranium, 304, *305*
Kanapoi (Kenya), 249–50
kangaroos, *256*, 257
Kaplan, Hillard, 279, 280, 284
Kapsalis, Ellen, 181
Kapthurin Formation (Kenya), 334, 336
kapunji monkeys, 140, *140*
Katanda (Democratic Republic of the Congo), 334–35, *335*
Kebara (Israel), 333
Keeley, Lawrence, 278

polygyny, 137, 160–62, *161, 162,* 418

polyhedron, *275*

polymorphism, balanced, 373, 374, *374*

Pongidae, *93*

Pongo abelii, 124

Pongo pygmaeus, 124

Pongo tapanuliensis, 124

poor portability, 390

Pope, Teresa, 184–85

population bottleneck, 321

population genetics, 54–60

 defined, 54

 gene frequency in, 55–56

 genetic composition of population in, 54–55, *55*

 genotypic frequencies in, 54–55, *55,* 57–59, *57–59*

 and Mendel's experiments, 54, *54*

 natural selection in, 59–60

 random mating and sexual reproduction in, 55–59, *57–59*

population growth and genetic variation, 381, *382*

population sizes of modern humans, 342–43, *343,* 356–57

population splits in modern humans, 342

Population Y, 395–96

porphyria variegata, 381

portability, poor, 390

positively correlated characters, 68

positive selection, 366–68

postcranial bones

 of early *Homo,* 268

 of Eocene primates, 221

 of *Homo erectus, 272,* 272–73, *273*

 of modern humans, 331

 of proconsulids, 230

potassium-argon dating, 214

pottos, 117

Potts, Richard, 278, 297

Precambrian era, *210*

precision genetic medicine, 397–99

predation by primates, 133–35, *134, 135*

Předmostí (Czech Republic), *331,* 356

pregnancy

 duration of, 147, *147*

 energy costs of, 148

prehensile hands, 112–14, *113*

premolars

 of early haplorrhines, 224, *224*

 of primates, *113,* 118, *119*

pressure flaking, 334, *334*

Preutz, Jill, 273

primary structure of proteins, 42, *43,* 43–44

primate(s)

 activity patterns in, 130–31, *131*

 biogeography of, 115–16, *116*

 as closest relatives, 110–11

 conservation of, 138–40, *138–40*

 dentition of, *113,* 115, *118,* 118–19, *119*

 digestive systems of, 119, *119*

 distribution of, *115,* 115–16

 diversity of, 111, 117–26, *120–26*

 ecology of, 126–35

 evolution of early, 218–22, *218–22*

 features that define, 112–15, *112–15, 118,* 118–19, *119*

 folivorous, 119, *119*

 food distribution in, 126–30, *127–30*

 frugivorous, 119, *119*

 gummivorous, 118–19, *119*

 haplorrhine, 115, 120–26, *120–26*

 insectivorous, 119, *119*

 predation in, 133–35, *134, 135*

 ranging behavior in, 131–33, *132, 133*

 reasons to study, 110–11

 sociality of, *134,* 135–38

 strepsirrhine, 115, 117, *120*

 taxonomy of, *116,* 116–17

primate behavior, value of studying, 205

primiparous mothers, 148–49, *149*

Proconsul, 227–30, *230*

proconsulids, 227–30, *230*

prognathism, subnasal, 252

prokaryotes, 44

proofreading, 39

propliopithecids, 223, *223*

Propliopithecus chirobates, 223

prosocial sentiments, regulation of cooperation by, 438–41

protein(s)

 activator, 47, *48*

 defined, 42

 primary structure of, 42, *43,* 43–44

 in primate diet, 127, 128, *128*

 repressor, 47, *48*

 roles of, 41–42

 sequence of amino acids in, *42,* 42–43, *43*

 tertiary structure of, 42, *43*

protein-coding genes, 40–47, *41–43, 45, 46*

 in humans vs. chimpanzees, 366–68, *367*

Prugnolle, Franck, 343–44

punishment for violation of culturally transmitted norms, 440–41

Punnett square, 32, *33*

Pusey, Anne, 156

pygmy chimpanzees, 125–26, *126*

Qafzeh Caves (Israel), *331,* 332–33

quadrupeds, 94

Quaternary period, *210*

race, 390–99

 and genetic admixture, 394–96, *397*

 gradual genetic variation and, *393,* 393–94

 historical background of, 391

 and measuring amount of variation within and between populations, 391–92

 patterns of genetic variation and, 392–93

 and personalized (precision) genetic medicine, 397–99

 questioning scientific validity of, 396–99

 and skin color, 390, *390*

 widely held views of, *390,* 390–91

radioactive decay, 214

radiocarbon dating, 214

radiometric methods to estimate age of fossils, 213–17

rafting hypothesis, 225, *226*